Manual of Applied Field Hydrogeology

Willis D. Weight, Ph.D., P.E.

Montana Tech of the University of Montana
Butte, Montana

John L. Sonderegger, Ph.D.

Montana Tech of the University of Montana
Butte, Montana
Montana State University
Bozeman, Montana

McGraw-Hill

New York San Francisco Washington, D.C. Aukland Bogotá
Caracas Lisbon London Madrid Mexico City Milan
Montreal New Delhi San Juan Singapore
Sydney Tokyo Toronto

Library of Congress Cataloging-in-Publication Data

Weight, Willis D.
Manual of applied field hydrogeology/Willis D. Weight, John L. Sonderegger.
p. cm.
Includes bibliographic references (p.)
ISBN 0-07-069639-X
1. Hydrogeology. I. Sonderegger, John L. II. Title.

GB1003.2.W535 2000
551.49—dc21

McGraw-Hill 

*A Division of The **McGraw-Hill** Companies*

1 2 3 4 5 6 7 8 9 0 DOC/DOC 0 6 5 4 3 2 1 0

ISBN 0-07-069639-X

The sponsoring editor for this book was Larry S. Hager, the editing supervisor was David E. Fogarty, and the production supervisor was Sherri Souffrance.

Printed and bound by R. R. Donnelley & Sons Company.

This book was printed on recycled, acid-free paper containing a minimum of 50% recycled, de-inked fiber.

McGraw-Hill books are available at special quantity discounts to use as premiums and sales promotions, or for use in corporate training programs. For more information, please write to the Director of Special Sales, Professional Publishing, McGraw-Hill, Two Penn Plaza, New York, NY 10121-2298. Or contact your local bookstore.

Manual of Applied Field Hydrogeology

To our parents, our wives Stephanie and Brenda, our children, our teachers, and our students, without whose influence this work could not have been possible

Contents

Preface

This book resulted from an enquiry by Bob Esposito of the McGraw-Hill Professional Book Group about our long-running Hydrogeology Field Camp at Montana Tech of the University of Montana, which was started 1985 by Dr. Marek Zaluski. We have been directing it since 1989. Bob asked whether we would consider taking the course notes and field tasks and putting them into book form. A detailed outline and proposal was submitted, which was professionally reviewed. A distinguished reviewer was very enthusiastic about a book of this scope being produced. During his review he expressed the view that "every hydrogeologist with less than 10 years of experience should own this book." It was also his opinion that there was a burning need to have a reference that inexperienced field hydrogeologists could refer to that would explain things in real world terms. This has been our perspective in putting together the chapters. We have also added discussion of the fundamental principles of hydrogeology to provide a more complete scope. The *Manual of Applied Field Hydrogeology* is intended to be a text for a course in field hydrogeology with sufficient coverage of the basics in hydrogeology to be used as an upper level undergraduate introductory hydrogeology course or lower level graduate course with a strong field perspective. The word applied is important, because of the many practical examples presented.

Readers are encouraged to study the examples throughout the book. They stand out in a different font and represent helpful hints and examples from many years of experience. They also contain little anecdotes and solutions to problems that can save hours of mistakes or provide an experienced perspective. Calculations in the examples are intended to illustrate proper applications of the principles being discussed. To help illustrate field examples, the authors have taken many photos and created line drawings in the hope of making the reading more understandable and interesting. Many of the geologic settings that occur in nature are found in Montana, so most of the examples come from there, although international examples are included. Examples come previous field camps, consulting, and work experience.

It is useful for the practicing hydrogeologist to be able to read up on a topic in field hydrogeology without having to wade through hundreds of pages. Students and other entry-level professionals have needed a reference that can help them overcome the panic of the first few times of performing a task such as logging a drill hole, supervising the installation of a monitoring well, or analyzing slug-test data. We feel that if this book will help someone save time in the field or reduce someone's panic in performing a field task, then it will have been worth the effort.

It has been our experience that the only way to understand how to apply hydrogeologic principles correctly is to have a field perspective. Persons that use hydrogeologic data are responsible for their content, including inherent errors and mistakes that have occurred in the field. Without knowing what difficulties there are in collecting field data and what may go wrong, an office person may ignorantly use poor data in a design problem. It is also our experience that there are many people performing "bad science" because the fundamentals are not well understood. When one confuses the basic principles and concepts associated with hydrogeology, all sorts of strange interpretations result. This generally leads to trouble. If the fundamentals of hydrogeology and field hydrogeology are well understood, better interpretations and field decisions in hydrogeologic studies will result.

When people go fishing, they may try all sorts of bait, tackle, and fishing techniques; however, without some basic instruction on proper methodologies, either the lines get tangled, the fish always take the bait, or the person fishing gets frustrated. Even people who are considered to be fishing experts have bad days, however they always seem to catch fish most of the time. The *Manual of Applied Field Hydrogeology* takes a "teaching how to fish" approach.

It is always a dilemma to decide what to include and what to leave out in writing a book like this. It is our hope that the content is useful. We look forward to the ideas and feedback that will come from our readers and thank you in advance for your thoughts. Any errors found is this work are ours.

Acknowledgments

We would like to thank the many individuals who helped review the chapters and provide some of the photos used in this book. Peter Norbeck and John Metesh of the Montana Bureau of Mines and Geology in Butte, Montana, provided several helpful comments on some of the chapters. Dr. Mark Sholes and Dr. Hugh Dresser from the Department of Geological Engineering at Montana Tech of the University of Montana provided helpful suggestions on Chapter 2. Dr. Dresser provided all of the stereo pairs found in the book. Dr. Chris Gammons, who took John Sonderegger's place in helping direct the Montana Tech Hydrogeology Field Camp since 1997, provided many helpful comments for several of the chapters, and his help has been invaluable. Dr. William "Bill" Woessner was instrumental in providing significant help in Chapter 6 and in providing some of the photos. Dr. Woessner teaches a week of our hydrogeology field camp. Dr. Peter Huntoon provided some of the photos for Chapter 2 as well as helpful encouragement and examples. Dan O'Keefe, of O'Keefe Drilling in Butte, Montana, was very helpful in reviewing the drilling section and provided photos for Chapter 8. Kevin Mellot was very helpful in scanning the many slides used in the figures. We wish to thank our many students over the years for being the subjects of photos and providing good questions and discussion and the cooperating specialists, especially Herb and Dave Potts, Dan O'Keefe, Fred Schmidt, and Hayden Fergeson. Bob Bergantino of the Montana Bureau of Mines and Geology in Butte, Montana, designed to logo we have used for our hydrogeology field camp. Finally, Kay Eccleston was instrumental in transforming all of our chapters into desktop publishing form, providing the format for all the text, and typesetting the final pages for the whole book.

We would also like to thank Dr. Marek Zaluski for his contribution of Chapter 13 and some of the photos for this chapter. Dr. Curtis Link contributed all of the content for Chapter 4, including all the line drawings. We thank Dr. Marvin Speece for reviewing Chapter 4.

WILLIS D. WEIGHT, *Ramsay, Montana*

JOHN L. SONDEREGGER, *Bozeman, Montana*

Chapter 1

Field Hydrogeology

Water is a natural resource unique to the planet Earth. Water is life to us and all living things. After discounting the volumes represented by oceans and polar ice, groundwater is the next most significant source. It is approximately 50 to 70 times more plentiful than surface water (Fetter 1994). Understanding the character, occurrence, and movement of groundwater in the subsurface and its interaction with surface water is the study of hydrogeology. Field hydrogeology encompasses the methods performed in the field to understand groundwater systems and their connection to surface water sources and sinks.

A hydrogeologist must have a background in all aspects of the hydrologic cycle. They are concerned with precipitation, evaporation, surface water, and groundwater. Those who call themselves hydrogeologists may also have some area of specialization, such as the vadose zone, computer mapping, well hydraulics, public water supply, underground storage tanks, source-water protection areas, and surface-water groundwater interaction, actually each of the chapters named in this book and beyond.

The fun and challenge of hydrogeology is that each geologic setting, each hole in the ground, each project is different. Hydrogeologic principles are applied to solve problems that always have a degree of uncertainty. The reason is that no one can know exactly what is occurring in the subsurface. Hence, the challenge and fun of it. Those who are fainthearted, do not want to get their hands dirty, or cannot live with some amount of uncertainty are not cut out to be field hydrogeologists. The "buck" stops with the hydrogeologist or geologist. It always seems to be their fault if the design does not go right. Properly designed field work using correct principles is one key to being a successful field hydrogeologist. Another important aspect is being able to make simple common-sense adjustments in the field to allow the collection of usable data. The more level-headed and adaptable one is, the more smoothly and cost-effectively field operations can be run.

Hydrogeology is a fairly broad topic. Entry-level professionals and even well-seasoned, practicing hydrogeologists may not have attempted one or more of the topics described in this manual. New field tasks can be stressful and having to read a large reference book on one subject can be cumbersome. The objective of this book is to provide a brief presentation on the general topics associated with field hydrogeology. Field methods, tasks, pitfalls, and examples using ideal and nonideal behavior will be presented, in hopes of reducing stress and panic the first few times performing a new task.

1.1 Hydrologic Cycle

The hydrologic cycle is an open system powered by solar radiation. Water from the oceans evaporates into the atmosphere, is carried to land as precipitation, and eventually returns back to the oceans. Solar radiation is more intense nearer the equator, where rising air condenses and falls back onto the world's rain forests. The movement of moisture into the atmosphere and back onto the land surface is an endless cycle. Approximately five-sixths of the water that evaporates upward comes from our oceans; however, only three-fourths of the water that falls from the sky, in the form of precipitation, falls back into the oceans (Tarbuck and Lutgens 1993). This means one-fourth of all water that falls to Earth falls on land. Some of this water is stored in ice caps and glaciers, some runs off from the earth's surface and collects in lakes and various drainage networks, some replenishes the soil moisture, and some seeps into the ground. This is important in supplying the land masses with fresh water. Once the water reaches the land, it takes a variety of pathways back to the oceans, thus completing the hydrologic cycle (Figure 1.1). Other than ocean water (97.2%) and frozen water (2.1%), groundwater (0.6%) accounts for a significant volume of the Earth's water (Fetter 1994).

Having a general understanding of the hydrologic cycle is important for perspective, for keeping the big picture in mind. The occurrence and behavior of groundwater in the field can be tied back to this big picture. For example, global climatic conditions may contribute to why there are more or less wet years or help explain why dry years occur, which later affect water availability in storage. Are decreasing trends in hydrographs in wells tied to water use or drought conditions that may have depleted storage?

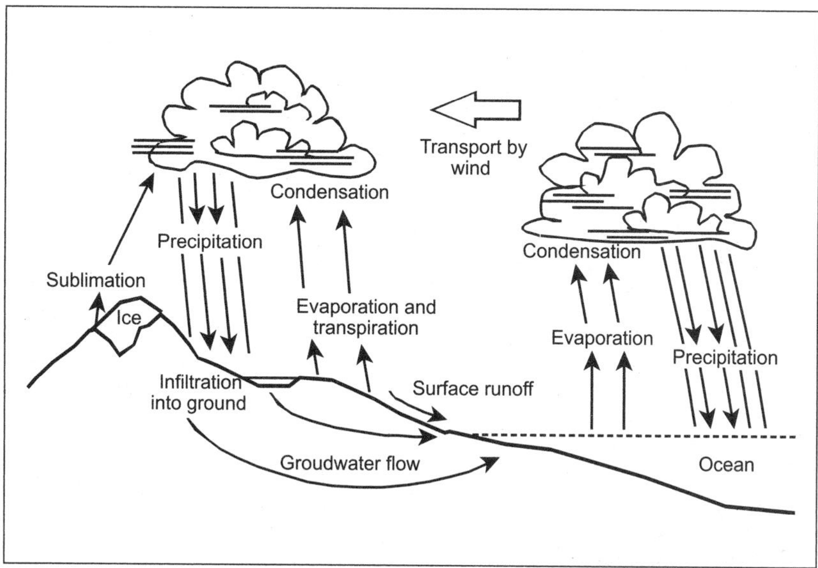

Figure 1.1 The hydrologic cycle.

Example 1.1

The sea temperatures in the equatorial region are being measured in real time on a continual basis (Hayes et al. 1991). There is a significant relationship between the atmosphere and the ocean temperatures that affect the weather around the globe. In normal years, the sea surface temperature is about 8 °C warmer in the west, with cool temperatures off the South America coast from cold water upwelling from the deep (NOAA 2000). The upwelling brings nutrient-rich waters important for fisheries and other marine ecosystems. Cool waters are normally within 50 m of the surface. The trade winds blow toward the west across the tropical Pacific, resulting in the surface sea elevation being 0.5 m higher in Indonesia than in Ecuador (Philander 1990). During the year of an El Niño, the warm water off the coast of South America deepens to approximately 150 m, effectively cutting off the flow of nutrients to near-surface fisheries. The trade winds relax and the rainfall follows the warm water eastward, resulting in flooding in Peru and the Southern United States and drought in Indonesia and Australia (NOAA 2000). A strong El Niño year occurred during 1997 to 1998.

Many hydrogeologists may be aware that global conditions are changing, but fail to apply this to the local drainage at hand. It is easy to become focused on only local phenomenon, such as within a given watershed. Sometimes one can get too close to a subject to be able to have the proper perspective to understand it.

There is the story of the four blind men who came in contact with an elephant. Each described what they thought an elephant looks like. One felt the trunk and exclaimed that the elephant must be like a vacuum at a local car wash. Another felt the tail and said that an elephant was like a rope. Yet, another felt the leg and said an elephant must be like a tree. Finally, one felt the ears, so big and broad, and thought the elephant must look like an umbrella. In their own way each was right, but presented only a part of the picture. Understanding the big picture can be helpful in explaining local phenomena.

1.2 Water-Budget Analysis

Most groundwater studies that a typical consulting firm may be involved with take place within a given watershed area (Figure 1.2). The hydrologic cycle is conceptually helpful, but a more quantitative approach is to perform a **water-budget analysis**, which will account for all of the inputs and outputs to the system. It is a conservation of mass approach. This can be expressed simply as:

$$\text{INPUT} - \text{OUTPUT} = \pm\Delta\text{STORAGE} \quad [1.1]$$

Figure 1.2 Aerial view of a dendritic patterned watershed.

The term ΔSTORAGE (change in storage) refers to any difference between inflow and outflow. An analogy with financial accounting can be used to illustrate. In a bank checking account, there are inputs (deposits) and outputs (writing checks or debits). Each month, if there are more inputs placed into the account than outputs, there is a net increase in savings. If more checks are written than there is money, one runs the risk of getting arrested. In a water-budget scenario, if more water is leaving the system than is entering, mining or dewatering of groundwater will take place. Dewatering may possibly cause permanent changes to an aquifer, such as a decrease in porosity, or compaction resulting in surface subsidence (USGS 1999). Areas where this has been significant include the San Fernando Valley, California; Phoenix, Arizona; and Houston, Texas. Several of the major components of inflow and outflow are listed in Table 1.1.

Table 1.1 Major Inflow and Outflow Components

Inflow	Outflow
Precipitation	Evapotranspiration
Surface water	Surface water
Groundwater flux	Groundwater flux
Imported water	Exported water
Injection wells*	Consumptive use
Infiltration from irrigation	Extraction wells

* Important for imported water.

It is often difficult to separate transpiration from plants and evaporation from a water surface, therefore, they are combined together into a term called **evapotranspiration** or **ET**. In any area with a significant amount of vegetation close to the water table, there may be diurnal effects in water levels. Plants and trees act like little pumps, which are active during daylight hours. During the day, ET is intense and nearby water levels drop and then later recover during the night. Diurnal changes from plants can also cause changes in water quality in streams (Chapter 6). Accounting for all the components within water-budget analyses are difficult to put closure on, although they should be attempted. Simplifying assumptions can sometimes be helpful in getting a general idea of water storage and availability. For example, it can be assumed that over a long period of time (e.g., more than one year) that changes in storage are negligible. This approach was taken by Toth (1962) to form a conceptual model for groundwater flow, by assuming the gradient of the water table was uniform over a one-year period, although the surface may fluctuate up and down. This model is also used when performing back-of-the-envelope calculations for water availability.

Example 1.2

The Sand Creek drainage basin is located 7 miles (11.4 km) west of Butte, Montana (Figure 1.3). The basin covers approximately 30 mi^2 (7,770 ha). In 1992, the land was zoned as heavy industrial. In 1995 there were two existing factories with significant consumptive use. The author's phone rang one afternoon, and the local city manager calling from a meeting on a speaker phone wanted to know how much additional water was available for development. The question was posed as to whether anyone was willing to pay for drilling a test hole so that a pumping test could be performed. After the laughter from the group subsided, they were informed that information to provide a quantitative answer was limited but that a number would be provided as a rough guess until better information could be obtained and that an answer would be forthcoming in a few minutes. Fortunately, there were some water-level data from which a potentiometric surface could be constructed (Chapters 3 and 5). From the potentiometric surface and a topo-

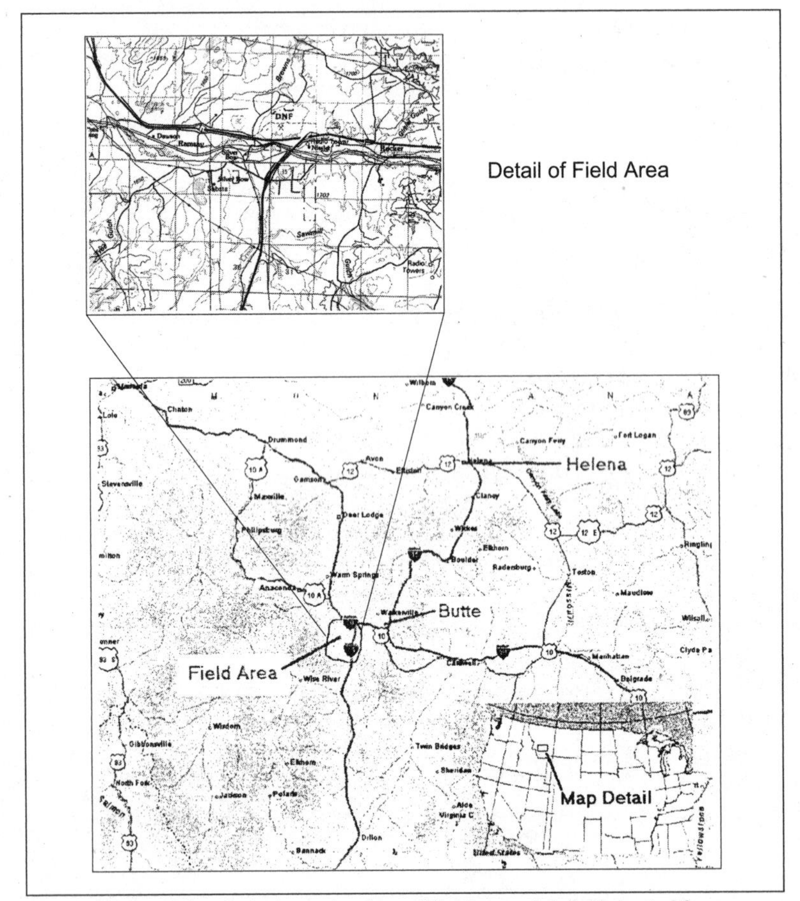

Figure 1.3 Sand Creek drainage basin and field area detail. [Adapted from Borduin (1999).]

graphic map, a hydraulic gradient and an aquifer width were estimated. A probable range of values was estimated for the hydraulic conductivity (Chapter 3), and a guess was given for aquifer depth. Darcy's Law was used to estimate the volume of water moving through a cross-sectional area within the watershed per unit of time (Chapter 5). This quantity was compared with the water already being used by the existing industrial sources. It was reasoned that if the existing consumptive use was a significant portion of the Darcian flow volume (greater than 20%), it wouldn't look like much additional development could be tolerated, particularly if the estimated contribution from precipitation did not look all that great. The local city manager was called back and provided with a preliminary rough guess of volume ranges. The caveat was that the answer provided was an extremely rough estimate, but did have some scientific basis. It was also mentioned that the estimate could be greatly strengthened by drilling test wells and performing additional studies.

Performing water-budget analyses is more difficult if there is significant consumptive use or if water is being exported. Sometimes there is a change in storage from groundwater occupying saturated media that ends up in a surface-water body. For example, in the Butte, Montana, area, short-term changes in storage can generally be attributed to groundwater flowing into a large open pit known as the Berkeley Pit (Burgher 1992). Water that was occupying a porosity from less than 2% in granitic materials and greater than 25% in alluvial materials was being converted into 100% porosity in a pit lake.

Example 1.3

Many investigations have been conducted in the Butte, Montana, area as a result of mining, smelting, and associated cleanup activities. It was desirable that a water-budget analysis be conducted in the Upper Silver Bow Creek Drainage to better manage the water resources available in the area. Field stations were established at two elevations, 5,410 ft and 6,760 ft (1,650 m and 2,060 m) to evaluate whether precipitation and evaporation rates varied according to elevation. Within the 123 mi^2 (31,857 ha) area, there were no historical pan evaporation data (Burgher 1992) (Figure 1.4). The period of study was from August 1990 to August 1991.

Part of the water balance required accounting for two sources of imported water from outside the area. One source was water from the Big Hole River, imported over the Continental Divide to the Butte public water supply system. Another source was water from Silver Lake, a mountain lake west of Anaconda, Montana, connected via a 30-mile (49-km) pipeline to mining operations northeast of Butte.

The water-budget equation used was:

$$P + Q_{imp} - E_T - Q_{so} - Q_{uo} - Q_{exp} - \Delta S \pm n = 0 \qquad [1.2]$$

Where:

P = precipitation from rain or snow

Q_{imp} = imported water from the Big Hole River and Fish Creek (in parenthesis below)

E_T = evapotranspiration

Q_{so} = surface outflow at the western edge of the valley

Q_{uo} = estimated groundwater outflow

Q_{exp} = exported water through mining activities

S = change in storage in the system groundwater to surface water (Berkeley Pit)

N = error term, net loss or gain

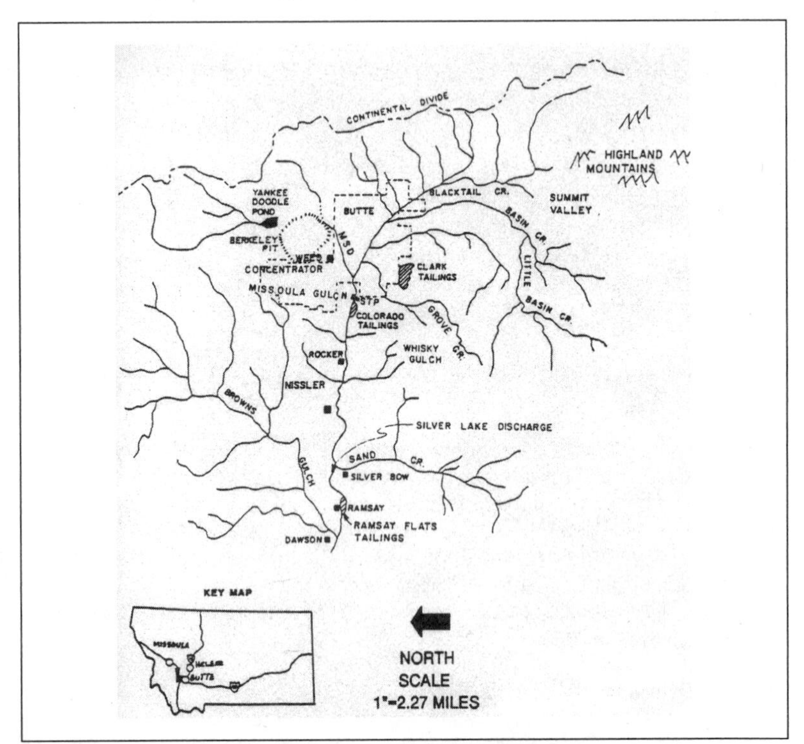

Figure 1.4 Silver Bow Creek drainage. [Adapted from Burgher (1992).]

Precipitation was higher at the upper site (13.35 in., 339 mm) compared to the lower site (10.5 in., 267 mm), while evaporation values were similar (23.79 in., 604 mm and 23.46 in., 596 mm). All values in Equation 1.2 were calculated in units of millions of gallons per day, where the error term is used to balance the equation. The results are shown below:

$$112.73 + (9.18 + 5.20) - 113.34 - 12.25 - 0.15 - 1.40 - 5.32 + 5.35 = 0$$

One question that could be asked is, "is more water within the region being used than is coming in?" Some areas have such an abundance of water that much development can still take place with little effect, while other areas are already consuming more water than their system can stand. An inventory of water use and demand needs to be taken into account if proper groundwater management is to take place.

Example 1.4

The Edwards aquifer of central Texas is an extensive karstified system (Chapter 2) in Cretaceous carbonate rocks (Sharp and Banner 1997). Historical water-balance analysis shows that this aquifer receives approximately 80% of its recharge via losing streams (Chapter 6) that flow over the unconfined portions of the aquifer (Chapter 3). The amount of recharge has varied significantly over time and seems to be connected to amount of stream flow (Figure 1.5). The average recharge between 1938 and 1992 is 682,800 acre-ft/year (26.6 m^3/sec, reaching a maximum of 2,486,000 acre-ft/year (97 m^3/sec) and a minimum of 43,700 acre-ft/year (1.7 m^3/sec) during 1956 (Sharp and Banner 1997). Other sources of recharge include leakage from water mains and sewage lines in urban areas and cross-formational flow where the aquifer thins, especially to the north (Sharp and Banner 1997).

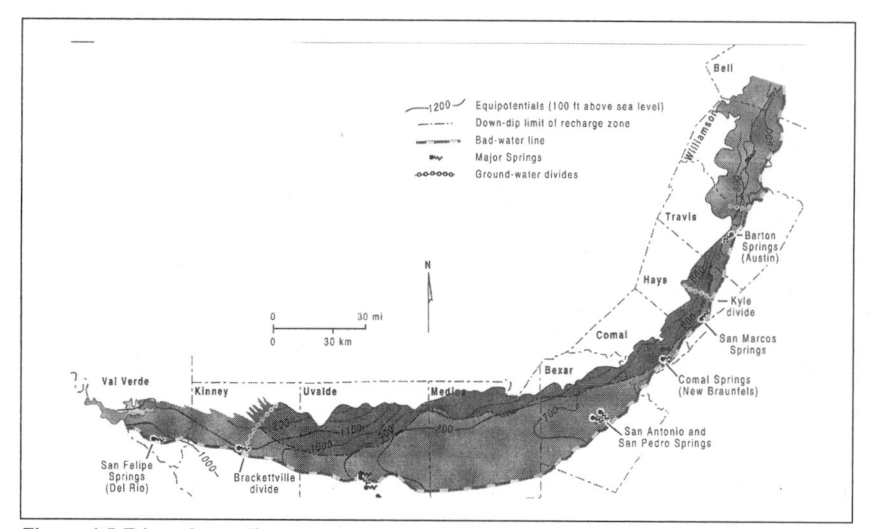

Figure 1.5 Edwards aquifer showing hydraulic boundaries, major springs, and equipotential lines. [From Sharp and Banner (1997). Reprinted with permission from GSA Today, 1997.]

Figure 1.5 also reflects discharge by springs and wells (the lower dashed and solid line respectively). Spring discharge follows a subdued pattern of recharge, while pumping discharge indicates an increasing trend over time. Peaks in the trend in pumping rate are inversely proportional to minima in recharge (Sharp and Banner 1997). Individual well yields are incredible. A single well drilled in San Antonio is reported to have a natural flow rate of 16,800 gpm (1.06 m^3/sec, Livingston 1942) and another well drilled in 1991 is likely the highest yielding flowing well in the world at 25,000 gpm (1.58 m^3/sec, Swanson 1991).

Figure 1.5 also indicates that other than a few years during the mid-1950s there has always been enough water to meet demands. With the current growth rates in the corridor between San Antonio and Austin Texas, there is a concern about water demands being able to keep up. For example, in 1996 the underdeveloped land north of Austin was being subdivided at a rate of 1 acre every 3 hours (Sharp and Banner 1997)!

Complicating matters are the additional roles the Edwards Aquifer plays in supplying water for recreation areas, fresh water critical for nurseries in estuaries for shrimp, redfish, and other marine animals, and spring water for threatened and endangered species that dwell in them (Sharp and Banner 1997). The Edwards Aquifer is a good example of how water-balance studies can assist in addressing the significant decisions that are continually badgered by special and political interests.

The previous discussion and examples point out that water-balance studies are complicated and difficult to understand. For this reason, many water-balance studies are being evaluated numerically (Anderson and Woessner 1992). A numerical groundwater-flow model organizes all available field information into a single system. If the model is calibrated and matched with historical field data (Broedehoeft and Konikow 1993), one can evaluate a variety of "what if" scenarios for management purposes. For example, what would happen if recharge rates decreased to a particular level when production rates are increased? The areal effects of different management scenarios can be observed in the output (Erickson 1995). A simpler approach can be taken by evaluating flow nets (Chapter 5).

1.3 Sources of Information on Hydrogeology

Information on hydrogeology can be gathered from direct and indirect sources. Direct sources would include specific reports where field data have been collected to evaluate the hydrogeology of an area (Todd 1983). Information may include water level (Chapter 5), geology (Chapter 2), pumping test (Chapters 9 and 10), and groundwater flow direction (Chapter 5). Indirect information may include data sources that are used to project into areas with no information. For example, a few well logs placed on a geologic map and correlated with specific units can be used to project target depths for drilling scenarios.

The following list suggests some ideas on where to start locating sources of hydrogeologic information. It is by no means an exhaustive list and no specific order is intended:

- Oil field logs or geophysical logs.
- Published or unpublished geologic maps.
- U.S. Geological Survey (www.usgs.gov)—topographical maps and other published information, such as water supply information, including surface water stations and flows.
- U.S. Environmental Protection Agency (U.S. EPA) (www.epa.gov).
- National Oceanographic and Atmospheric Association (NOAA)—precipitation data information (www.NOAA.gov).
- Topographic maps—helpful in locating wells and evaluating topography and making geologic inferences.
- Structure-contour maps that project the top or bottom of a formation and their respective elevations.
- State surveys and agencies—published information within a given state, well-log data bases, and other production information, often organized by county or section, township, and range.
- Libraries—current and older published information. Sometimes the older published information is very insightful; additionally, many libraries have search engines for geologic and engineering references, such as GEOREF and COMPEDEX.
- Landowners—valuable historical perspectives and current observations.
- Masters and Ph.D. theses—These can be of varying quality, so glean what you can with a grain of salt. Some are very well done while others should not have been published.
- Summary reports prepared by state agencies on groundwater, surface water, or water use and demand.
- Consulting firms and other experienced hydrogeologists.
- Index maps listing geologic information, well logs, and water rights information by county or watershed.
- U.S. Geological Survey Water Supply Papers and Open File Reports.
- County planning offices—information in geographical information system (GIS) formats, well ownership listings, or other useful information.
- Some authors or editors attempt to evaluate all references associated with groundwater and then group the references according to topics (for example, van der Leeden 1991).
- Internet search engines.

In most states, a well log is required to be filed each time a well is drilled. Included is the depth, lithologic description, perforation or screened interval, static water level (SWL), and brief pumping or bailing test information. This can be used to evaluate the depth to water, determine the lithologies, and get a general idea of well yield. Sometimes this information is not reported, as some drillers are hesitant to report unsuccessful wells. If no water is found, sometimes the well logs do not get filed, when, in reality, no water reported in a well log is good information. Experienced hydrogeologists learn how to combine a variety of geologic and well-log database sources into a conceptual model, from which field decisions can be made.

1.4 Site Location for Hydrogeologic Investigations

As simple as it sounds, the first task is to know where the site is. As a hydrogeologist, you may be investigating a "spill," evaluating a property for a client who is considering buying the property, locating a production well, or participating in a construction dewatering project. It is imperative that you know where the site is, so you can assess what existing information there might be. If this is a preexisting site, there will be some information available; however, if you are helping to "site in" a well for a homeowner or a client for commercial purposes on an undeveloped property, you must know where the property is (see well drilling in Section 1.8). Part of knowing where a site is includes both its geography and its geology.

Many water-well drillers are successful at finding water for their clients without the help of a hydrogeologist. Either the geologic setting is simple and groundwater is generally available at a particular depth, or they have experience drilling in a particular area. If they don't feel comfortable with the drilling location, they will always ask the client where they want the hole drilled so that they can't be held liable for problems when they go wrong. The problem is that most home owners don't have much of an idea about the geology of their site and choose a location based on convenience to their project rather than using field or geologic information. The phone call to the hydrogeologist generally comes after a "dry" hole or one with a disappointingly low yield has been drilled. The phone call may come from the driller or the client. Before carefully looking into the situation, that is, answering questions about what happened and what the drilling was like, you must know where the location of the site is. Unfortunately, sometimes you are given the wrong site-location information and you end up doing an initial geologic investigation in the wrong place. A personal example will help illustrate this concept.

Example 1.5

A phone call came after drilling a 340-ft duster (dry hole). The drill site was chosen near an old, existing homestead cabin. The client (landowner) figured that since the homestead had water, and it was an ideal location for his ranching operation, it would probably make a good site. The driller asked for the location and drilling commenced. After going 340 ft, with no water in sight, the driller decided to call and get some recommendations. The property consisted of more than 600 acres near a small town in western Montana. The section, township, and range were provided. After an initial investigation in the library, geologic and topographic maps were located. The geologic information was superimposed onto the topographic map, and a couple of cross sections were constructed. The target zone would be a coarser-grained member of a lower Cretaceous sandstone. A meeting time and place were arranged. Within minutes of driving down the road from the meeting site, it became evident that the driller was heading to a location different from what was described. Instead of slightly undulating Cretaceous sedimentary rocks, the outcrops were basalt, rhyolite, and Paleozoic carbonates that had been tilted at a high angle. The dry hole had been drilled into a basalt unit, down geologic dip (Chapter 2). It becomes difficult to recommend anything when the structure and geologic setting are uncertain.

The purpose for well drilling was for stock watering. An initial design was recommended that would not require electrical power. A local drainage area could be excavated with a backhoe and cased with 24-in. galvanized culvert material. A 3-in. PVC pipe could be plumbed into the "culvert well" at depth and run to a stock tank (Figure 1.6). It was a gravity-feed design and would help keep cows at the far end of his property in the summer.

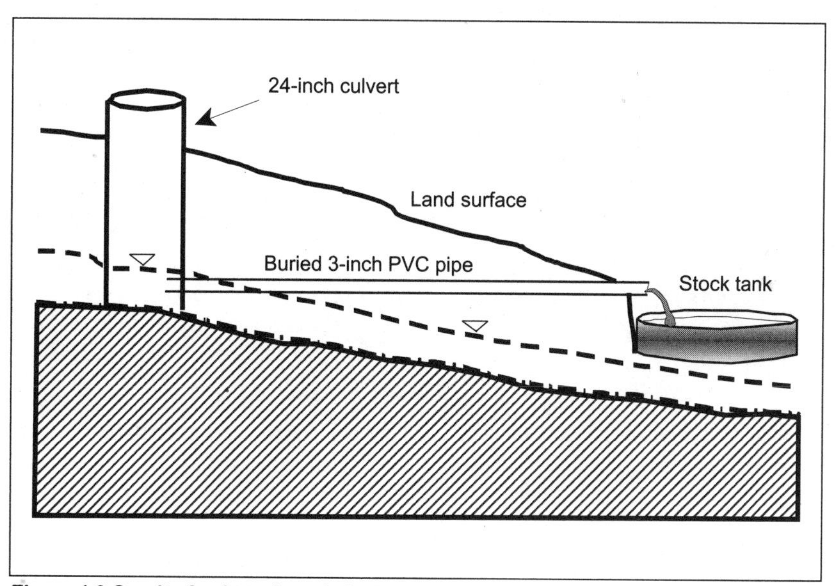

Figure 1.6 Gravity-feed stock tank design from Example 1.5 (not drawn to scale).

After this initial meeting, the rancher informed the author that they tried the backhoe method and found that the area had good aquifer materials, but no water. Once the proper location was known, we went back in the office and evaluated the geology. We determined that this drainage was situated on a large fault and that any infiltrating water collected would be lost quickly into the Paleozoic limestone formations, and would be found only by drilling well below the surface.

In Example 1.5, an understanding of the geology was essential to locating a target for water development. Even if the geologic setting is known, modifications due to structural features may need to be made. Discussion of the geology of hydrogeology is further discussed in Chapter 2.

1.5 Taking Field Notes

A field notebook is your memory of events, locations, and figures. Without decent field notes, it will be difficult to reconstruct what happened in the field, let alone remember the important details that are necessary for billing out time and completing reports back at the office. Notebooks should be accessible, sturdy and weatherproof, or at least you should write with a waterproof pen. Some people prefer to use a pencil (with something hard like a 6H lead), while others prefer pens. What is the problem with using a pencil? Pencils can smear as the pages chafe back and forth in a field bag. After a few times in the field, the information becomes hard to read. A pen can usually be clearly visible for years. After all, one should never erase in a field book. If your field notebook is needed in court, erase marks will be questioned and the book later thrown out as being tampered with. Don't use whiteout or some other coverup to make your field book look prettier for the same reason. After all, it is a field book. Mistakes will be made, and when they occur, neatly draw a line through the mistake and continue on. Your notebook should not be a disorganized mess, because it will be needed to help you remember what happened later on. Some people have a "working" field book and a second field book that the day's work is translated into for neatness. It is a good idea to record or add information each evening or during slack time, or details will be missed. You take notes so you don't have to remember everything.

Inside the cover, put your name, address, and phone number, so you can be reached in the event you lose your field book. Some prefer to staple a business card inside the front cover. The back of the field book should have some blank pages, reserved to write the names, addresses, and phone numbers of contacts or distributors of equipment, supplies, or parts frequently needed. Again, business cards can be stapled directly inside these back pages, for ready access. For some people, it is also helpful to put conversion tables in the field book so that they are accessible for calculations. Your field book is your memory.

Although there are a host of topics that can be recorded in a field book, some examples of the more commonly encountered tasks are presented below.

Daily Information

There is some basic information that should appear in a field book each time one ventures out. The location, purpose, and objective of the day's work should be noted. For example, soil sampling along Skunk Creek for ACME remediation study. The day's date and time departed can be very important. It is a good idea to record the times people arrive and leave a site as a daily itinerary. This could be significant in a law suit if claims are made about who was where, when. It is a good idea to record the people in the party who are present and their phone numbers and names (including the last names). Saying that Jim or Susan was there may not be very helpful a few years down the road. The basic who?, when?, where?, why?, and how? approach to field notes is a good way to remember what should be recorded.

Comments on the weather and how it changes during the day can also be significant in field interpretation. For example, a rainstorm moving in will change the air pressure. In confined aquifers, this may help explain why the water levels increased or slowed down during a pumping test (discussed in Chapter 9). Rainstorms may be the source of why the stage in a stream changed. If the weather is very cold, this may contribute to errors committed in the field from stiff fingers, trying to get the job done quickly, or trying to work with gloves on. Conversely, hot weather can also affect the quality of field data collected. Equipment is also temperature-sensitive, particularly water-quality equipment. This may contribute to data values drifting out of calibration. Of course, this can occur as one bounces down a bumpy road!

It is a good idea to record detailed observations and descriptions during slack time while your mind is still fresh. Field sketches and diagrams are useful. Many field books have grid lines so that well logs can be drawn vertically to scale. If your drawing or sketch is not to scale, say so in the field book. You may have a field book dedicated to a particular project or get a new one each year. It is advisable to organize or record information in a systematic way. For example, some people like to number the pages sequentially so that a table of contents can be made at the front of the field book. This is helpful if there are a variety of jobs over a broad area. Some of us just prefer to mark the beginning and end of a particular day's effort by noting page 1 of 4, 2 of 4, for example. Whatever system you use, strive to be consistent, complete, and neat.

Lithologic Logs

Most companies or agencies have their own forms for filling out lithologic information during drilling (Figure 1.7). This is desirable, but if for some reason you forgot your forms or none are provided, you may have to record lithologic log information in your field book. If you forgot your field book too, then, shame on you, you forgot your memory. It may be, for example, that you are making field descriptions of surface geology during a site characterization or recording lithologic information from monitoring wells. (How to log a drillhole is discussed in Chapter 8 and a more detailed discussion of geologic information is presented in Chapter 2.)

Lithologic logs are generally recorded during the drilling process. The type of drilling affects the time that you have to record information. If you are logging and bagging core, collecting chips, or doing a variety of other tasks, then it can get tricky getting everything done while still recording a meaningful lithologic log.

Example 1.6

I can recall my first attempt at recording lithologic logs for drill holes in overburden on a coal property near Hanna, Wyoming. It was a forward rotary drilling rig, and drillers could drill with air for the first 200 ft or so in relatively soft layered sediments. The mast was almost free-falling at a rate of approximately 20 ft per minute. The drillers laughed as I frantically tried to record information. It was bewildering as the various lithologies changed in texture and color before my eyes.

Besides learning to work with your driller, it is helpful to have a shorthand set of descriptions for lithologies and textures (this is discussed further in Chapter 8). Many companies may have their own system, and these can later be incorporated into a software package for visualization. Rock or soil descriptions should be indicated by the primary lithology first, followed by a series of descriptors. Information recorded could include the following:

- Lithology name.
- Grain size and degree of sorting (is this the actual crushed bedrock or soil particles?).
- Color (can be affected by drilling fluids, such as mud).
- Mineralogy (HCl fizz test, or mineral grains observed in the cuttings).
- Probable formation name, etc. For example, sandstone: very-fine to fine-grained, tan, with dark cherty rock fragments, the basal Member of the Eagle sandstone.

Project Hillcrest School Well No. HC-8 1/2

Location 3000 Continental Dr Butte

Engineer W. WEIGHT Installed By D. CROWLEY Date 14 SEP 96

Method of Installation H.S.A. B61-HDX

LOG OF BORING AND WELL

BORING

Depth In ft.	Description	H. Nu Reading
0	Asphalt Surface Gravel	
	Silty sand drk gray	
5	Silt, dark carbonaceous micaceous	
	GRAVEL BRN 3/4" MINUS wood chips	
10	Silt, brn clayey, silty	
	Clayey Gravel	
	Med Sand with fine gravel 3/8" minus	
15	Gravel 1/2"-3/4"	
	Sand med coarse @ fine gravel 3/8" minus	
	water @ 19'	
20		
25	med Sand - Silty matrix	

OBSERVATION WELL INFORMATION

Type of WELL dewatering

Ground Elev. ______ Top of Riser Elev. ______

Vented Cap

S.W.L. ______

I.D. of Riser Pipe ______

Type of Pipe ______

Type of Backfill Around Riser ______

L_1 = -0.3'

L_2 = 1.1

L_3 = 16.4

L_4 = 14.0

L_5 = 17.8

L_6 = 10

L_7 = 0.7

L_8 = 10

L_9 = 40

Top of Seal Elev. ______

Type of Seal Material ______

Top of Filter Elev. ______

Type of Filter Material 6-9 mesh

Size of Openings 40 Slot

Diameter of Screened Tip ______

Bottom of Well Elev. ______

Bottom of Boring Elev. ______

Diameter of Boring 10 3/4"

Remarks ______

L-1 Flush MT/Well Protector
L-2 Surface to Bentonite Seal
L-3 Bentonite Seal
L-4 Filter Pack
L-5 Surface to Top of Screen
L-6 Screen
L-7 Sump
L8 Rat Hole
L-9 Total Depth

Prepared by W. Weight

Figure1. 7 Form for well logs and well completion diagrams.

You may notice things, such as whether there were coarsening-upward or coarsening-downward sequences, unconformities, or abrupt coloration changes, as would occur if the rocks changed from a terrestrial aerobic origin (red, yellow, tan) to those of an anaerobic origin (grey, dark or greenish) (Figure 1.8). The relative ability for rocks to react with dilute HCl can be used to help distinguish one formation from another or which member you may be in. Record dates and how samples were taken. For example, were they washed? (The fines would wash through a sieve, while the chips or coarse fraction would remain behind.) Were they grab samples, split-spoon, or core samples?

Figure 1.8 Cross-sectional view of changing lithologies, showing a small-scale (several meters wide) thrust fault.

Another useful thing to note along with the lithologic log is the drilling rate. There may be a variety of similar rock types, but some may be well cemented and drill slower or faster than others. This may affect the ability for vertical groundwater communication between units, or it may be helpful for a blasting engineer, who needs to know about a very hard sandstone or igneous unit that will require extra blasting agents. Drilling rates can be roughly compared to penetrometer tests. Some drilling may be smooth or "chatterier". For example, a hard sandstone may cause the drill string to chatter because the bit bounces somewhat as it chews up the formation, but the drilling will proceed slowly. Contrasted to this, a coal bed will chatter when it is soft and is chewed up quickly. Notice how long it takes between the first "chatter" sound and the appearance of black inky water from the coal bed at the land surface. Use your senses to notice changes. Additional comments about logging during drilling are made in Chapter 8.

Well Drilling

While performing well drilling, you not only need to record the lithologic information, but also the well completion information. Depending on the complexity and depth of the well, there may be other key people involved: other geologists, tool pushers, mud loggers, or engineers. Once again, write down the names, addresses, and phone numbers of the driller and these key people, or better yet get their cards and staple them into your book. Hopefully, you have performed some background geologic work and have an idea about what formation you are drilling and the targeted depth. What is the purpose of the well (monitoring, production, stock well, etc.)? Were there other wells drilled in the area? How deep did they drill for water or product contamination? What were the production zones? What will the conditions likely be (hard, slow drilling, heaving sands, etc.)?

Record a detailed location of the well, using coordinates if possible, township descriptions, or other identifiers, such as latitude and longitude from a global positioning system (GPS) (Figure 1.9). It should be noted that GPS systems vary dramatically in accuracy. Hand held units will get you in the ball park, but to accurately locate (within inches or centimeters) a well, it is necessary to have a base station and a rover unit with expensive software to correct for the global changes that occur during the day. Will the elevations be surveyed or estimated from a topographic map?

Write down the make and model of the rig and any drilling fluids that were used and when. Did you start with air or begin drilling with mud? It is a good idea to write down a summary of the work completed by your driller before you arrive. Big projects may need to be drilled continuously in shift work format until the job is done.

Example 1.7

In western Wyoming in the early 1980s, some deep (1,200 ft, 370 m) monitoring wells were being placed in the structural dip of a mining property. Each well took several days, so drilling and well completion took place in 12-hour shifts by the geologists and drillers on a continuous basis until the wells were completed. A new drill crew and geologist would arrive every 12 hours. It was critical that communication took place between the ones leaving and the ones arriving for continuity.

It is also important for billing purposes to write down incidents of slack time (standby), equipment breakdowns, or runs for water. The work sub-

Figure 1.9 A global positioning system used in the field. The rover unit is tied to a base station located within 6 kilometers.

mitted should match the field notes taken or inquiries might be made. Each morning or during slack time, measure and record the static water levels. This is particularly important first thing in the morning after the rig sat all night. The size and type of bit is also important for knowing the hole diameter. Any unusual or problematic situations or conditions should be recorded and discussed with your driller.

Well Completion

Once the hole is drilled, the well is completed from the bottom up. Your field book needs to contain sufficient information to construct well completion diagrams back at the office or in the motel room at night. It is helpful to review the information on forms required by the local government, so that all details are covered. For example, was surface casing used? If so, what was the diameter? What about screen type or slot size? What are the diameters of the borehole and pump liner? The mechanics of well completion will be discussed in Chapter 8; however, the following items should probably be included in your field book:

- Total depth drilled (TDD) and hole diameter (important for well hydraulics calculations).
- Indication of type and length of bottom cap (was it left open hole?).

- Is there a sump or tailpipe (a blank section of casing above the bottom cap)?
- Interval of screened or perforated section, slot size and type or size of perforations, outer diameter (OD) and inner diameter (ID) of screen, and material type. Is the screen telescoped with a packer, threaded or welded on?
- Number of sections of casing above the screen, length of each interval, and height of stick up. Twenty-foot (6.1-m) pipe can vary in length. Fractions of a foot can add up when going several hundred feet into the ground.
- Was the well naturally developed or was packing material used? If packed, what volume was used, height above the screen (number of 100-lb, 45.4-kg, or 50-lb, 22.7-kg, bags)? Interval of packing material? This is helpful for planning future wells to be drilled.
- Was grouting material used? If so, over what interval?
- What kind of surface seal was used (concrete, neat cement, cuttings)?
- What kind of security system is there for the well cap?
- Was the well completed at multiple depths, and if so, what are the individual screened intervals and respective grouting intervals?
- Was the well developed? What was the static water level before development?
- What method was used (pumping, bailing, etc.) and how long? How many purge volumes or for what duration did development take place?

Pumping Tests

There are many details that need to be remembered during a pumping test, let alone many pieces of equipment to gather together. In addition to your field book, for example, it is imperative that you have well logs for all wells that will be included in the pumping test, with well-completion details. Your field book will play a critical role in remembering the details. One of the first items of business is to measure the static-water levels of all wells to be used during the test. Any changes will be important for interpretation. The weather conditions may also affect the water-level responses. Is there a storm moving in? This can result in a low-pressure setting where water levels may rise. In confined systems, the author has observed changes measuring .5 to 1 ft (0.15 to 0.3 m) depending on the atmospheric pressure drop and confining conditions. Is there a river or stream nearby? Did the stage remain constant or raise or drop? During Montana Tech's field camps, there have been observed stage drops of 2 ft during a 24-hour pumping test. How does this affect the results?

In October 1998, while attending a MODFLOW 98 conference in Boulder, Colorado, several groundwater modelers from around the world were informally interviewed and asked questions about what they thought should be included in a field book. One resounding remark stood out. "Please tell your readers that they should always make backup 'hand' measurements in their field books on changing water levels in wells, during a pumping test! Don't just rely on the fancy equipment." Another comment was, plot the data in the field so that one knows whether the test has gone on long enough (see Chapter 9). The data would be obtained from measurements recorded in field books or downloaded from software. Along this line, there should be a notation of the pumping rate during the test. Were there changes in discharge rate that took place during the test? Remember that theoretically the pumping rate is supposed to be constant.

Troubles in the field need to be recorded. Sometimes individuals don't want to remember something goofy or embarrassing that happened in the field, but it may be the key to making a proper interpretation. Why didn't anyone get the first 5 minutes of the recovery phase of the pumping test? The reasons may range from being asleep and didn't notice the silence of the generator as the fuel ran out, or perhaps off taking care of a personal matter, or couldn't get the darn data logger to step the test properly. Perhaps a friendly Rottweiler powered through the equipment area and managed to knock out one of the connecting cables (Figure 1.10). I'm sure a documentary on the things that can go wrong during a pumping test would make interesting reading. The following bulleted items will serve as a reminder of the most important general items to record but is by no means an exhaustive list. Additional discussion can be found in Chapter 9.

- Personnel on hand at the site, either assisting or present for observation.
- Static-water levels of all of the wells, prior to emplacement of any equipment.
- Sketch map showing the orientation and distances of observation wells to the pumping well, also include possible boundary effect features, such as steams, bedrock outcrops, other pumping wells, etc., that may affect the results.
- Weather conditions at the beginning, during, and end of the test.
- List of all equipment used. Type of pump, riser pipe, discharge line, flow meters, or devices used to measure flow (e.g., bucket and stop watch?) Again a sketch or photograph in the field is helpful.

Figure 1.10 Rottweiler harassing the author during a pumping test.

- Exact time the pump was turned on and off for the recovery phase. (It is helpful to synchronize watches for this purpose.)
- Manual water-level measurements of observation wells. Include the well ID, point of reference (top of casing), method of measurement (e-tape or steel tape), and a careful systematic recording of times and water levels in a column format.
- Record of the pumping rate. How was this measured and how often during the test?
- Recovery phase of the test. Again, manual measurements, conditions, and times.
- Any other problems or observations that may prove helpful in the interpretation of the results.

Water-Quality Measurements

Any type of field measurement can have errors. Errors can be made by the equipment being out of calibration or bumped around in the field, or the operator may just get tired during the day. Maybe the weather conditions rendered the collecting of data fairly intolerable. Honest observations in a field book can help someone remember the conditions under which the data were collected and help someone else reviewing the data or their interpretation. The field book can explain why the numbers look strange. For example, many parameters are temperature-sensitive. Along with a pH measurement, there should be a knowledge of what the temperature is to set the calibration knob correctly (Figure 1.11). Many specific conductance meters have a temperature corrections to 25° C internal within the device. Failing to take into account the temperature can significantly affect the results recorded. Chain of command and other details of the inherent problems that can occur in the field will be discussed in greater detail in Chapter 7. The following items should be considered for field notes:

- What is the sample type (e.g., surface water, domestic well, monitoring well)?
- What were the weather conditions throughout the day?
- What were the date and exact time when the sample was take? (Many samples are required to reach a lab within 24 hours.)
- Who were the personnel present (was the operator alone or was there someone else helping)? Who was present at the respective sampling locations? Did the well owner come out? It may be important to be able to tell who was there.
- What were the methods of calibration and correction for drift?
- What were the units? Were the data temperature-corrected? Remember, many parameters are temperature-dependent and temperature-sensitive!

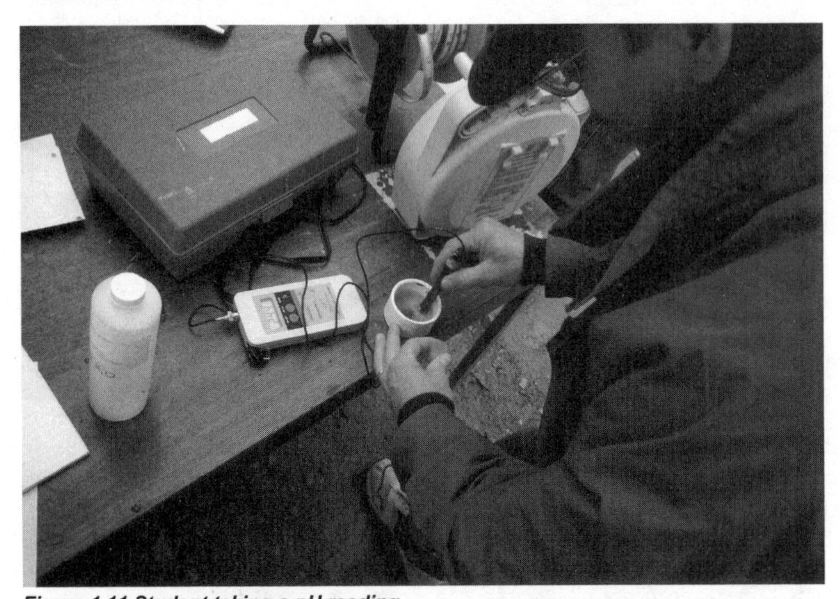

Figure 1.11 Student taking a pH reading.

- If it is a well sample, what volume of water was evacuated before the samples were collected? What were the purge times and respective volumes? How was the sample taken? Was it via a bailer, a discharge hose from a pump, or a tap? Which parameters were monitored during the purging (pH, conductivity, temperature)? Where was the discharge water diverted to?
- How long did it take to reach a stable reading? Was the reading taken prematurely before stabilization took place? This can vary significantly between pieces of equipment. Was a stir rod going during the reading or was the container just swished around every once in a while?
- What were the methods of end-point detection? Was a pH meter or color reagents used.
- Were the samples filtered or acidified? How?
- Was the laboratory protocol followed? Were the chain of command papers filled out?
- Units! Units! Units!

1.6 Groundwater Use

It is instructive to evaluate the requirements of water for different purposes. How much water is needed for domestic purposes? In the author's home, we have a water softener that has a built-in digital readout flow meter. When the shower is on, one can read the rate flowing in the pipe (4 to 6 gpm). It also provides an average total use per day. In our case, in the first quarter of 2000 the daily amount was 398 gal. This is for 7 people and includes consumption in the form of showers, hand washing, toilet flushing, drinks of water, and water used for cooking and laundry. Or approximately 57 gallons per day (gpd) per person. This does not account for water used in the yard from the hose. This seems to fit in well with other estimates of water in the range of 50 to 80 gpd (189 to 302 L) per person per day (Fetter 1994; Driscoll 1986).

In Montana, domestic well users can have a well drilled on their private property that yields up to 35 gpm (132 L/min) and does not exceed a volume of 10 acre-ft/year (12,300 m^3/yr) without a special permit. Yields and volumes in other states will vary, but specific values are established through the legislative process. Wells used for irrigation or public water supply exceeding 100 gpm (378 L/min) require a water-use permit. As part of the technical evaluation the well location (point of diversion), purpose, impacts to other existing users, demonstration of the water being there, and the well design (Chapter 8) are described by a hydrogeologist for evaluation by a state agency. As part of the process a technical review is made of the application and a notice is placed in the local newspaper to advise other users of this taking place. Any complaints must be filed and addressed. Once the water-use permit is granted it is viewed as a water right for groundwater. In the western United States, all water rights are determined using the following philosophy "first in right, first in time." A maximum rate and a maximum volume are applied for, conditional on the water use. The rate is given in gallons per minute (gpm) and the volume is given in acre-feet over a specified time.

Irrigation waters and water rights in the eastern united States are distributed following a Riparian doctrine. In this case, all water users in a given drainage basin share the allocatable water. If flow rates decrease by 10%, the users proportionally decrease water use by 10%. This works in areas where the precipitation is higher and streams are sustained by a fairly continuous source of base flow from groundwater discharge.

The use of groundwater for public water supply along with coordinating where to locate potential contaminating sources (sanitary landfills and industrial parks) falls under the area of groundwater resource management.

1.7 Groundwater Planning

As the demand for groundwater supply increases, there will be a need for improved management practices. This is being addressed in part by evaluating groundwater resources from the bigger picture. Basins and watersheds are being delineated and the information stored as layers within databases in a geographic information system (GIS). Each information layer may contain point information, such as wells and points of diversion, or linear features such as roads, fence lines, or property boundaries. Polygon shapes that delineate geology or other features can also be stored. Tables of information that may include depth to water, production rates, or lithologic information can be stored and retrieved. Sources of information for these databases may come from drillers, landowners, or state agencies as groundwater characterization studies or aquifer vulnerability studies are performed.

A significant source of information within each state is being generated from source-water protection (SWP) area studies defined in the 1996 Federal Safe Drink Water Act Amendments. By 2000 most states had to have a source-water protection plan for all public water-supply wells. This was designed to be a practical and cost-effective approach to protecting public water supplies from contamination. Each source-water plan provides a wealth of information to each state's database and will allow a more comprehensive management plan.

Source-Water Protection Studies

Recharge areas to aquifers or areas that contribute water to surface waters used for public drinking water are delineated on a map. Geologic and hydrologic conditions are evaluated in the delineation study to provide physical meaning to the SWP areas. Assessments are made for contaminant sources within the area. Included may be businesses, surface or subsurface activities, or land uses within SWP areas where chemicals or other regulated contaminants are generated, used, transported, or stored that may impact the public water supply.

Each state develops its own delineation and assessment process, so that information is reported consistently and a minimum of technical requirements have been met. As a hydrogeologist, you may be hired to perform the technical assessment of the geologic and hydrologic information within an SWP area report. An example of the type of information needed is presented in the Montana Department of Environmental Quality Circulars PWS 5 and PWS 6 (Montana DEQ 1999). The following discussion is included for comparison with other State programs and to see what hydrogeologic topics are involved.

PWS 5 is a circular that helps an evaluator determine whether a public water supply has the potential to be directly influenced by surface water. In the circular, surface water is defined as any water that is open to the atmosphere and subject to surface runoff. Included as surface water are perennial streams, ponds, lakes, ditches, some wetlands, intermittent streams, and natural or artificial impoundments that receive water from runoff. Influence to groundwater supply can occur through infiltration of water through the stream bed or the bottom of an impoundment (Chapter 6).

A scoring system of points is used during the preliminary assessment to determine whether the water supply can be classified as groundwater or whether there may be some influence from surface water. For preexisting wells, surface water impacts are scored points by historical pathogenic or microbial organisms that have been detected in a well. Additionally, turbidity, distances from a surface-water source, depths of perforations below surface, and depth to the static water level also contribute to points accumulated. In this case, like the game of golf, the fewer the points, the less likely a surface-water source impacts a given water-supply well.

If the number of points exceeds a limit, then further analysis is required. Springs or infiltration galleries immediately fall under the requirement of performing additional water-quality monitoring. A well may also be required to undergo intensive monitoring for two months following the completion of construction to determine its suitability as a public water supply source. This would include weekly sampling for bacterial content and field parameters of temperature, turbidity, specific conductivity, and pH. The field parameters are also performed on the nearby surface-water source. A hydrograph of water quality parameters versus time is plotted to compare similarities between surface water and groundwater. It is reasoned that groundwater parameters will not change much, while there may be significant variations in the chemistry of the surface water. Surface-water sources may be used for public water supply if the source water passes the biological and microbial tests.

If a microbial particulate analysis is required for a well, the client is required to conduct two to four analysis over a 12 to 18 month period, according to method EPA 910/9-92-029. The possibility of connection with surface water is indicated by the presence of "insects, algae, or other large diameter pathogens." A risk factor is also specified by the following bio-indicators: Giardia, coccidia, diatoms, algae, insect larvae, rotifera, and plant debris. The sampling method is performed using the following steps:

- Connect the sampling devise as close to the source as possible.
- Assemble the sampling apparatus and other equipment *without* a filter in the housing to check whether the correct direction of flow is occurring.
- Flush the equipment using water from the source to be filtered, for a minimum of 3 minutes. Check all connections for leaks. An in-line flow restrictor is desirable to reduce flow to 1 gpm (3.8 L/min).
- Filtering should occur at a flow rate of 1 gpm (3.8 L/min). During the flushing stage, the flow can be checked using a calibrated bucket and stop watch.
- Shut off flow to the sampler. Put on gloves or wash hands and install the filter in the housing. Make sure a rubber washer or o-ring is in place between the filter housing and the base.
- Turn on the water slowly with the unit in the upright position. Invert the unit to make sure all air within the housing has been expelled. When the housing is full of water, return the unit to the upright position and turn on flow to the desired rate.
- Filtering should be conducted at a pressure of 10 psi. Adjustment of the pressure regulator may be necessary.
- Allow the sampler to run until 1,000 to 1,500 gal (3.78 to 5.67 m^3) have been filtered, mark the time when water was turned on and off.

- Disconnect the filter housing and pour the water from the housing into a ziplock plastic bag. Carefully remove the housing filter and place it in the bag with the water. Seal the bag, trying to evacuate all the air and place it in a second bag and make sure neither bag leaks.

- Pack this into a cooler, making sure the filter does not freeze, as frozen filter fibers cannot be analyzed. Send the filter and data sheet to an acceptable laboratory within 48 hours.

If a public water supply source meets the criteria as an appropriate source, a SWP area must be delineated, as defined in the 1996 Federal Safe Drinking Water Act Amendments. In this process, recharge sources that contribute water to aquifers or surface water used for drinking water are delineated on a base map. Included is a narrative that describes the land uses within the delineated area along with a description of the characteristics of the community and nature of the water supply. Methods and sources of information used to delineate the SWP and potential contaminant sources should also be described. An example of the type of information is presented from PWS 6 (1999).

- **Introduction**—describes the purpose and benefit of the SWP plan.

- **Background**—includes a discussion of the community, the geographic setting, description of the water source, the number of residents to be served, well completion details, pumping cycles, the water quality, and natural conditions that may influence water quality at the public water system.

- **Delineation of water sources**—presents the hydrogeologic conditions (aquifer properties and boundaries), source-water sensitivity to contamination (high, moderate of low), conceptual model based upon the hydrogeologic conditions, method of delineation, and model input parameters (Table 1.2).

- **Inventory of potential contaminant sources**—requires an inventory sheet for the control zone for each well. All land uses need to be listed. These may be classified as residential or commercial (sewered, unsewered, mixed), industrial, railroad or highway right-of-way, and agriculture (dry-land, irrigated, pasture) or forest. A list of contamination sources are listed in Table 1.3.

Once the SWP area has been defined, the property owners are advised. If property owners do not like the outcome of the investigation, they may sue under a Takings proviso. Takings issues are not something just for attorneys, but something that should also be familiar to a hydrogeologist. Your greatest security in a courtroom situation is to be able to demonstrate that best possible practices have been used in the evaluation. Part of these practices include field studies and field data from which interpretations are made. As a field hydrogeologist, keep in mind the big picture and realize that your work may be revisited in court.

Table 1.2 Model Input Parameters for a Source-Water Protection Area

Input Parameter	Value(s) Used	Units	How Derived	Remarks
Elevation at well				
Static water level				
Transmissivity				
Thickness				
Hydraulic conductivity				
Hydraulic gradient				
Flow direction				
Effective porosity				
Pumping rate				
100-day total				
1-Year total				
3- Year total				

1.8 Summary

Field hydrogeology is an interesting subject. Each time you go into the field, each time you drill a new well or go into a new area, the geology and hydrogeologic conditions change. This is the fun and challenge of it. Keep in mind the big picture of the hydrologic cycle while also paying attention to the detailed items, such as diurnal water-quality changes in surface streams or storm fronts coming in during a pumping test, and write these observations in your field book. By synthesizing the data, a conceptual model of a given area will emerge. The conceptual model will take shape as additional field data is collected.

Prior to heading for the field, check your sources of hydrogeologic information and make sure you have your field book or you won't be able to remember the details of your daily experiences. Work hard, keep your wits about you, and be safe, and soon your personal experience database will turn you into a valuable team member.

Table 1.3 Common Sources of Groundwater Contamination (U.S. EPA 1990)

Agricultural	**Commercial**
Animal burial areas	Airports
Animal feedlots	Auto repair shops
Chemical applications (pesticides,	Beauty parlors
fungicides, fertilizers, etc.)	Boat yards
Chemical storage facilities	Car washes
Irrigation systems	Cemeteries
Manure spreading and pits	Construction areas
Industrial	Dry cleaning establishments
Asphalt plants	Educational institutions (labs, storage)
Chemical manufacturing, warehouses,	Gasoline stations
and distribution	Golf courses (chemical applications)
Electrical and electronic products and	Jewelry and metal plating
manufacturing	Laundromats
Electroplates and metal fabrication	Medical institutions
Foundries	Mortuaries
Machine and metalworking shops	Paint shops
Manufacturing and distribution of	Photography establishments, printers
cleaning supplies	Railroad tracks and railyards
Mining and mine drainage	Research laboratories
Paper mills	Road de-icing activities (road salt)
Petroleum products and distribution	Scrap and junkyards
Pipelines (oil, gas, other)	Storage tanks (above and below ground)
Septic lagoons and sludge	**Residential**
Storage tanks	Fuel storage systems
Timber facilities	Furniture, wood strippers, refinishers
Toxic and hazardous spills	Household hazardous products
Transformers and power systems	Lawns, chemical applications
Wells (operating and abandoned)	Septic systems, cesspools
Wood preserving facilities	Water softeners
Naturally Occurring	Sewer lines
Groundwater surface-water interaction	Swimming pools (chlorine)
Iron and magnesium	**Waste Management**
Natural leaching (uranium, radon gas)	Fire training facilities
Saltwater intrusion	Hazardous waste management units
Brackish water circulation	Municipal waste incinerators
	Landfills and transfer stations
	Wastewater and sewer lines
	Recycling reduction facilities

References

Anderson, M.P., and Woessner, W.W., 1992. *Applied Groundwater Modeling—Simulation of Flow and Transport.* Academic Press, San Diego, CA, 381 pp.

Borduin, M.W., 1999. *Geology and Hydrogeology of the Sand Creek Drainage Basin, Southwest of Butte, Montana.* Master's Thesis, Montana Tech of the University of Montana, Butte, MT, 103 pp.

Bredehoeft, J., and Konikow, L., 1993. Ground-Water Models: Validate or Invalidate. *Ground Water,* Vol. 31, No. 2, pp. 178–179.

Burgher, K., 1992. *Water Budget Analysis of the Upper Silver Bow Creek Drainage, Butte, Montana.* Master's Thesis, Montana Tech of the University of Montana, Butte, MT, 133 pp.

Driscoll, F.G., 1986. *Groundwater and Wells.* Johnson Screens, St Paul, Minnesota, 1108 pp.

Erickson, E.J., 1995. *Water-Resource Evaluation and Groundwater-Flow Model from Sypes Canyon, Gallatin County, Montana.* Master's Thesis, Montana Tech of the University of Montana, Butte, MT, 69 pp.

Fetter, C. W., 1994. *Applied Hydrogeology, 3rd Edition.* Macmillan College Publishing Company, New York, 691 pp.

Hayes, S.P., Mangum, L.J., Picaut, J., Sumi, A. and Takeuchi, K., 1991. TOGA-TAO: A Moored Array for Real-Time Measurements in the Tropical Pacific Ocean. *Bulletin of the American Meteorological Society,* Vol. 72, pp. 339–347.

van der Leeden, F., 1991. *Geraghty & Miller's Groundwater Bibliography 5th Edition.* Water Information Center, Plainview, NY.

Livingston, P., 1942. A Few Interesting Facts Regarding Natural Flow from Artesian Well 4, Owned by the San Antonio Public Service Company, San Antonio, Texas, *U.S. Geological Survey Open-File Report,* 7 pp.

Montana DEQ, 1999. *Groundwater Under the Direct Influence of Surface Water.* Montana Department of Environmental Quality Circular PWS 5, 1999 Edition, 30 pp.

Montana DEQ, 1999. *Source Water Protection Delineation,* Montana Department of Environmental Quality Circular PWS 6, 1999 Edition, 21 pp.

NOAA, 2000. *The El Nino Story.* http://www.pmel.noaa.gov, 4 pp.

Philander, S.G.H., 1990. *El Nino, La Nina and the Southern Oscillation.* Academic Press, San Diego, CA, 289 pp.

Sharp, J.M., Jr., and Banner, J.L., 1997. The Edwards Aquifer: A Resource in Conflict. *GSA Today,* Vol. 7, No. 8, Geological Society of America, pp. 2–9.

Swanson, G.L., 1991. Super Well is Deep in the Heart of Texas. *Water Well Journal,* Vol. 45, No. 7, pp. 56–58.

Tarbuck, E.J., and Lutgens F.K., 1993. *The Earth, An Introduction to Physical Geology.* Macmillan Publishing Company, New York, NY, 654 pp.

Todd, D.K., 1983. *Groundwater Resources of the United States.* Premier Press, Berkeley, CA, 749 pp.

Toth, J.A., 1962. A Theory of Ground-Water Motion in Small Drainage Basins in Central Alberta, Canada. *Journal of Geophysical Research,* Vol. 67, pp. 4375–4381.

U.S. EPA, 1990. *Guide to Groundwater Supply Contingency Planning for Local and State Governments.* EPA-440/6-90-003. U.S. EPA Office of Groundwater Protection, Washington, DC, 83 pp.

USGS 1999. *Land Subsidence in the United States.* Galloway, D., Jones, D.R., and Ingebritsen, S.E., (eds.), U.S. Geological Survey, Circular 1182, 177 pp.

Chapter 2

The Geology of Hydrogeology

Before going into the field or performing field work, the hydrogeologist needs to have a general understanding and a knowledge of subsurface information. Engineers who work with geologic information may have forgotten geologic terms or can't remember what they mean. This chapter is a reference on geologic topics and how they are specifically applied to hydrogeology.

Why is it important to understand the geologic setting? Perhaps the following questions will help. Is the area structurally complex? Are the rocks metamorphic, igneous, or sedimentary? What geologic mapping has been published or conducted in the area? Is there faulting or structural lineaments in the area? Are the formations flat lying or tilted? What surface-water sources in the area pass over formational outcrops? What are the topographic conditions like? Is the area flat lying, rolling hills, or mountainous? What is the climate like? Is it arid or humid? (This will make a big difference on recharge to the system.) What is the location and accessibility to the property? A few hours spent gathering this information will help in developing a conceptual model and in preventing mistakes one would make during field interpretations. Not understanding the general geology or field conditions before going into the field can be disastrous.

If possible, the hydrogeologist should put together a geologic model or conceptual geologic model. The geologic model forms the basis for a conceptual groundwater-flow model, and includes which units may dominate flow and which units or features may inhibit groundwater flow. The geologic stratigraphic units are broken into hydrostratigraphic units, or in other words, aquifers and confining units (discussed in Chapter 3). Metamorphic and igneous rocks may have zones within them that are productive or have significant localized fracture zones. Locations of water production are also areas of concern for potential groundwater contamination. What is the vulnerability of the water supply to contamination? Conceptual geologic models aid in predicting the direction of groundwater flow, its occurrence, and its interaction with surface water.

2.1 Geologic Properties of Igneous Rocks

Igneous rocks are an important source of water in some regions. For the most part, **extrusive** igneous rocks (those that have erupted and formed on the land surface) have a greater capacity for water transmission and storage than do **intrusive** igneous rocks (those that formed and cooled beneath the earth's surface). Extrusive and intrusive igneous rocks can generally be distinguished by their texture and mineral composition. Extrusive igneous rocks will generally be finer grained and have fewer distinguishable minerals than intrusive igneous rocks. Both will vary in appearance according to mineral content, with quartz- and feldspar-rich rocks being lighter in color than rocks with a more **mafic** mineral content. Mafic minerals are richer in iron and magnesium. For example, rhyolites and granites will tend to be light in overall appearance and have visible quartz and potassium feldspar, although rhyolite will be fine-grained from rapid cooling, and the granite will be coarse grained. A classification scheme for the various igneous rocks is summarized in Figure 2.1. Because extrusive igneous rocks generally have more important water-bearing zones, these will be discussed first.

Extrusive Rocks

Extrusive rocks are also known as volcanic rocks. They have erupted to the surface either through volcanoes or fissures. The major volcanic mountain ranges around the world form near the margins of plate boundaries. The Cascade Range in Washington and Oregon and the Andes Mountains of South America are forming where crustal plates are colliding. The mineralogically more mafic and denser oceanic crust is subducting (passing) under the lighter continental crust. At depths of 60 mi (100 km), temperatures and pressures are high enough to cause partial melting and mixing of crustal materials, forming molten rock under the surface known as **magma.** Magma is more buoyant than the surrounding solid rock and seeks a pathway to the surface where it emerges as **lava**. As magma rises, volatile gases within the fluid become less constricted and rapidly expand, resulting in an explosive eruption at the surface (Figure 2.2). Sometimes the volatile gases fracture the overlying rocks pneumatically into a **breccia pipe** (Figure 2.3). Fractures in the breccia pipes may later become mineralized, an important concept for mineral exploration and mining development. Gas bubbles that expand in lava are preserved as the lava freezes to

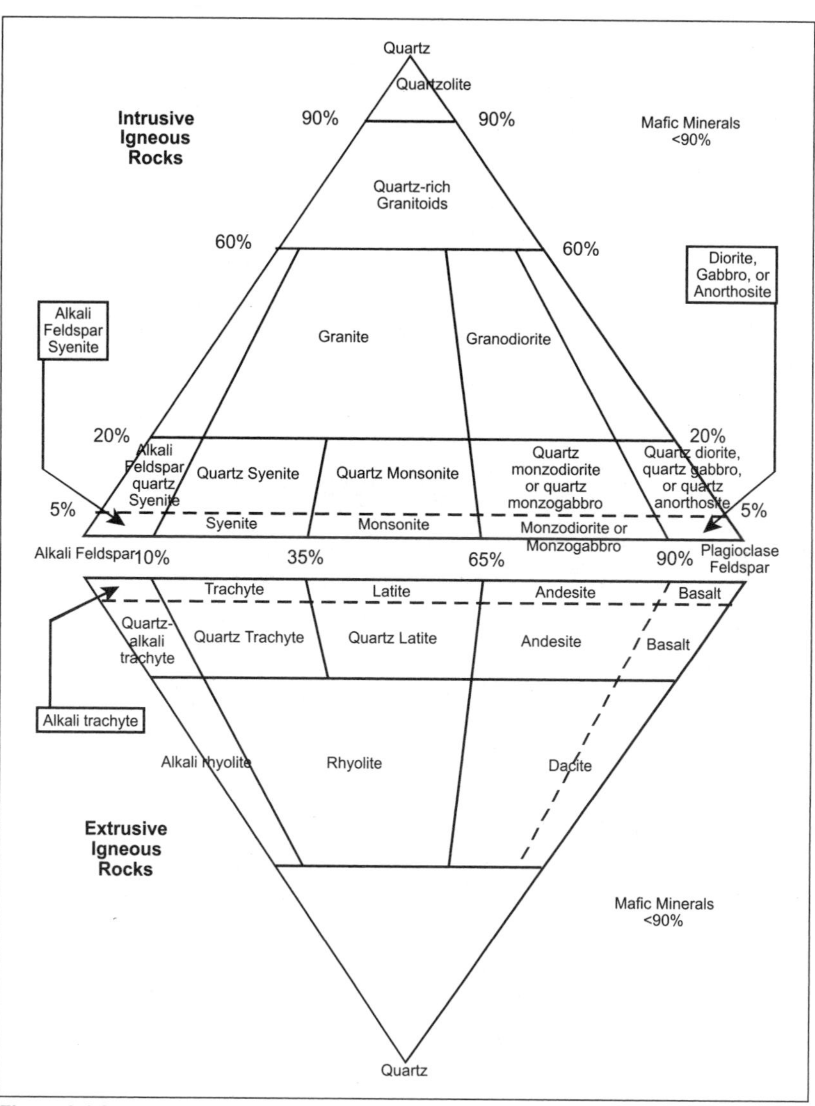

Figure 2.1 Summary classification scheme for the various igneous rocks. [Modified from Streckeisen (1976).]

form **vesicular** lavas (Figure 2.4). Gas bubbles increase the porosity but tend to be poorly connected, and therefore do not increase permeability. Some magmas literally expel fluids so quickly that they turn fluids into a foamy froth that cools. This is **pumice,** which may have a porosity of 80% and floats because of the poor interstitial connections.

Figure 2.2 Mount St. Helens eruption, May 18, 1990. (David Frank, USGS.)

Figure 2.3 Breccia pipe in southwestern Montana (note angular fragments).

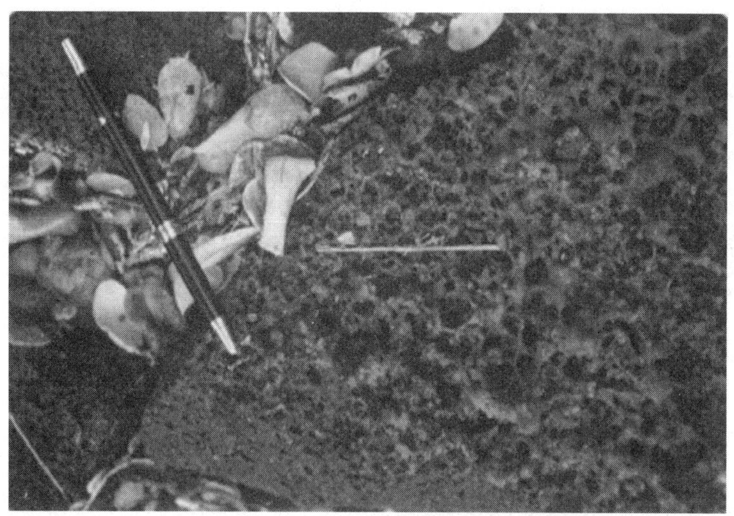

Figure 2.4 Vesicular lava from southeast Hawaii. (Photo courtesy of Hugh Dresser.)

Other locations where igneous rocks commonly form are places where the crust is extending or pulling apart, known as **rifting**. Igneous activity has occurred, for example, in the Basin and Range province of the United States and in eastern Africa as a result of these forces. Extensional areas allow easier routes for deep-sourced magmas to rise to the surface. Most of these magmas are of basaltic composition. Sometimes rifting is initiated in a region and then suddenly ceases from changes in plate-movement direction or heat-flow patterns in the asthenosphere.

The most common extrusive igneous rocks are basalt, andesite, and rhyolite, listed in order of increasing silica (SiO_2) content (Figure 2.1). Silica increases the viscosity of the igneous melt, much as corn starch or flour thickens gravy. The greater the silica content, the thicker the magma. Thicker magmas that are gas-rich explode more violently. Compositionally, rhyolite is similar to granite, the most common intrusive igneous rock. Generally, granites are more common because magmas of this composition seldom reach the land surface. Rhyolites generally form in areas of high heat flow, for example, in Yellowstone National Park (a continental hot spot) or in New Zealand. Rhyolite eruptions are characteristically violent explosions followed by viscous lava flows that don't move far from the vent (Figure 2.5). Interstitial openings within rhyolitic tuffs and from multiple layers may result in production wells in the tens of gallons per minute range (50 to several hundred m^3/day).

Figure 2.5 Stereo pair of rhyolite flow, north end of Mono Craters, California. (Photos courtesy of Hugh Dresser.)

Areas of high heat flow in geothermal settings may also cause the dissolution of minerals. As these waters rise to the surface, they cool and encounter cold-water recharge resulting in near-surface precipitation. This precipitation zone may create a confining layer that seals deeper aquifers from shallow aquifers or from interaction with surface water.

Example 2.1

Near Gardiner, Montana, at the north end of Yellowstone National Park, the Yellowstone River flows northwestward into the Corwin Springs known geothermal area (Figure 2.6). In 1986, a production well on the west side of the Yellowstone River was drilled to a depth of 460 ft (140 m) and aquifer-tested in September of the same year (Sorey 1991). After pumping at a production rate of 400 gpm (25 L/s) for 13 hours, La Duke Hot Springs on the east side of the Yellowstone River began to decrease in flow. This prompted a temporary moratorium on the drilling of production wells near Yellowstone Park (Custer et al. 1993). It is interesting that surface sealing from mineral precipitation resulted in a separation of the shallow alluvial system of the Yellowstone River from the deeper rhyolitic rocks. A similar surface-sealing phenomenon occurs in the Rotorua area of New Zealand (Allis and Lumb 1992).

Andesite

Andesitic rocks are intermediate in silica composition between rhyolite and basalt and are associated with the largest, most beautiful volcanoes in the world. Mount Fujiyama of Japan and Mount Rainier are examples (Figure 2.7). Andesitic rock typically forms where partial melting of oceanic and continental crust occur together. The Andes Mountains represent a type locality for this kind of rock. Andesites usually have a violent eruption, from explosively escaping volatile gases, followed by lava flows. During the explosive part, ash and other ejecta spews out onto the flanks of the volcano and surrounding area. The ensuing lava flows cover parts of the ejecta, providing a protective blanket. Subsequent eruptions result in a layering of

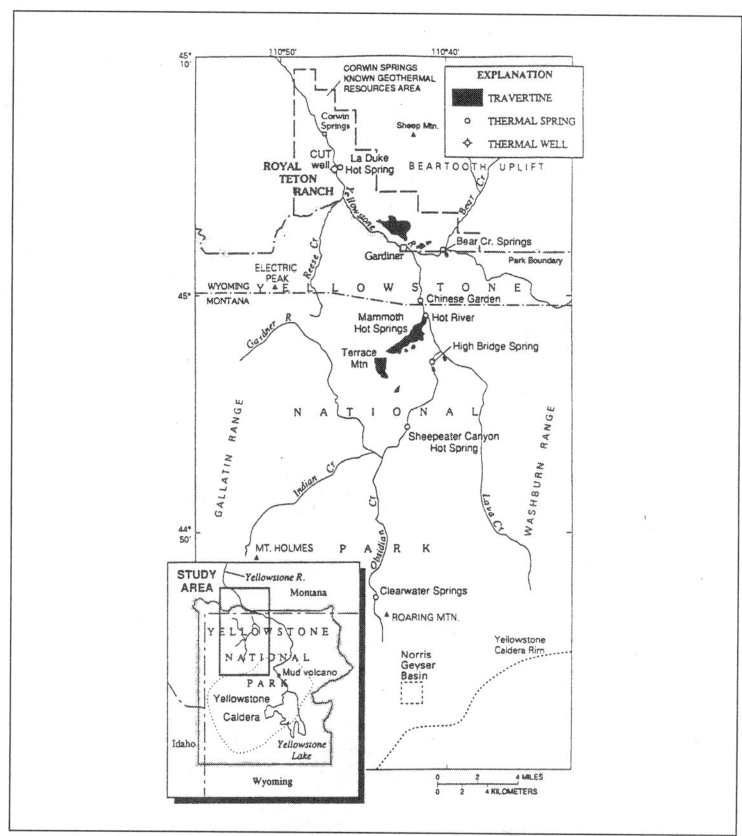

Figure 2.6 Corwin Springs known geothermal resource area near Yellowstone National Park from Example 2.1 (Sorey 1991).

Figure 2.7 Stereo pair of Mount Rainier, Washington. (Photos courtesy of Hugh Dresser.)

ejecta and lava that builds up into large steep-sided volcanoes. Given the layered nature of andesitic terranes, there is a potential for water production among the layers. Another source of water production can be associated with fractures from cooling. Liquid rock, as it cools, tends to form polygonal shaped columns, known as **columnar jointing** (Figure 2.8). These may provide significant permeability for groundwater flow. Columnar jointing is also commonly observed in basaltic lava flows.

Figure 2.8 Columnar jointing in a basalt flow.

Basalt

The most common extrusive igneous rock is basalt. It is the rock that constitutes the oceanic crust and is common within the Pacific Rim in a series of island arcs known as the ring of fire (Figure 2.9). Basalts are common because they form from partial melting of the mantle and constitute the oceanic crust. In areas such as the mid-oceanic ridges and areas where the continental crust is extended, basaltic magmas rise quickly up their vents and flow onto the surface as a dark gray to black lava. These magmas are very fluid because of their low silica content. Many basaltic lava flows, where they occur repetitively, are fairly thin, on the order of 15 to 40 ft thick (4.5 to 12.2 m). At the top and bottom of these flows are scoria zones of high porosity from burning vegetation and cooling processes. The middle part of basalt flows can be quite dense (Figure 2.10). Some regions can have thicknesses of multiple flows that are in excess of thousands of feet (several hundreds of meters). Many of these are in areas where fissure eruptions have occurred. Fissure eruptions like those of the Deccan Plateau in India and the Oregon–Washington area may have resulted from meteorite impact (Alt and Hyndman 1995). Saturated, thinly layered basalt lava flows may result in some of the most prolific aquifers.

Figure 2.9 Stereo pair of steaming Redoubt Volcano, Chismit Mountains, Alaska. (Photos courtesy of Hugh Dresser.)

Figure 2.10 Basalt with multiple flows and columnar joining.

Example 2.2

In the eastern Snake River Plain, Idaho, is a very prolific aquifer known as the Eastern Snake River Plain Aquifer. As an undergraduate student, the author worked for Jack Barraclough of the U.S. Geological Survey, Water Resources Division, at the Idaho National Engineering Laboratory (INEL), while participating in several field geologic and hydrologic studies. (It is now known as the Idaho National Engineering and Environmental Laboratory (INEEL)). One project involved logging cores from deep drill holes (greater than 1,000 ft, or 300 m). Basaltic lava flows ranged from 15 to 35 ft thick (4.5–10.7 m). Most were clearly marked by a basal vitrophere or obsidian zone with some **scoria** (a high vesicular or bubbly zone) on the order of 1 to 2 ft thick (.3 to .6 m), above which was a dense basalt with occasional gas bubbles frozen into position as they tried to rise to the surface (bubble trains). Each dense zone could be distinguished by its olivine content (a mineral characteristic of rock from the mantle), which ranged from approximately 2 to 12%. The top was distinguished by another scoria zone. These layers extended from the surface down to the bottoms of the core holes, separated only by a few thin sedimentary layers, representing erosional hiatuses. It was obvious that there were abundant sources of connected permeability to allow free flow of groundwater.

During the summer of 1979, a deep geothermal production well was attempted, known as INEL#1. Geologists thought they could identify the location of a caldera ring, from surface mapping. A **caldera** is a collapsed magma camber characterized by ring faults and a crater rim more than 1 mile (1.6 km) across. The concept was that at sufficient depth, fracturing and heat would allow the production of a significant geothermal production well. The proposed purpose of the well was for power generation and heating. The drilling required a large oil rig with a 17-ft platform and 90-ft (27.4-m) drill pipe. Surface casing was set with a 36-in (91-cm) diameter and eventually "telescoped" down to an open hole 12 $^1/_4$ inches (31.1 cm) in diameter at the total depth of 10,380 ft (3,164 m). The layered basalt flows are over 1,700 ft (518 m) thick at this location. These are underlain by dacitic and rhyodacitic rocks (Figure 2.1). Although the temperatures at depth were up to 325 °F (149 °C), there was poor permeability and production.

Wells less than 150 meters deep at the INEEL can produce in excess of 4,000 gpm (21,800 m^3/day). Some high-capacity production wells start out with a pumping level approximately 1 foot (0.3 m) below static conditions that then recover back to static conditions (Jack Barraclough, personal communication, INEEL, July 1978)! Not all basaltic rocks can be thought of as potentially prolific aquifers. They can vary greatly in their ability to yield water. For example, Driscoll (1986) points out that incomplete rifting about one billion years ago in the central United States produced a massive belt of basalt extending from Kansas to the Lake Superior region. This massive belt contains relatively few fractures or vesicular zones, resulting in poor water production (only a few gpm, 10 to 20 m^3/day). A wide range of hydrologic properties for basalts in Washington State has been reported by Freeze and Cherry (1979) in Table 2.1.

Intrusive Rocks

Intrusive igneous rocks are also known as plutonic rocks because they form large bloblike shapes underground that may result from partial melting of colliding plates, rifting, and melting above hot spots. They cool under the Earth's surface. Characteristically, these rocks form a tight network of interlocking grains or **phenocrysts** that compete for space during the cooling process (Figure 2.11). (Phenocrysts are also observed in extrusive igneous rocks, but have a matrix around them that is very fine-grained.) Because cooling is slow and volatile gases such as water vapor are present, mineral growth is enhanced. The three most common intrusive igneous rocks whose extrusive counterparts have already been discussed are granites, diorites, and gabbros. Gabbro is compositionally equivalent to basalt. Magma chambers that are larger than 60 mi^2 (100 km^2) are known as **batholiths**. Smaller bloblike bodies are known as stocks. Intrusive features that are discordant and fill vertical to angled fractures are known as **dikes**, and concordant intrusive bodies that are injected in between layers are known as **sills** (Figure 2.12). Each of these may affect groundwater flow directions.

Table 2.1 Range of Hydrologic Properties of Basalts in Washington State

	Hydraulic Conductivity (cm/sec)	Porosity (%)
Dense basalt	10^{-9}–10^{-7}	0.1–1
Vesicular basalt	10^{-7}–10^{-5}	5
Fractured basalt	10^{-7}–10^{-3}	10
Interlayered zones	10^{-6}–10^{-3}	20

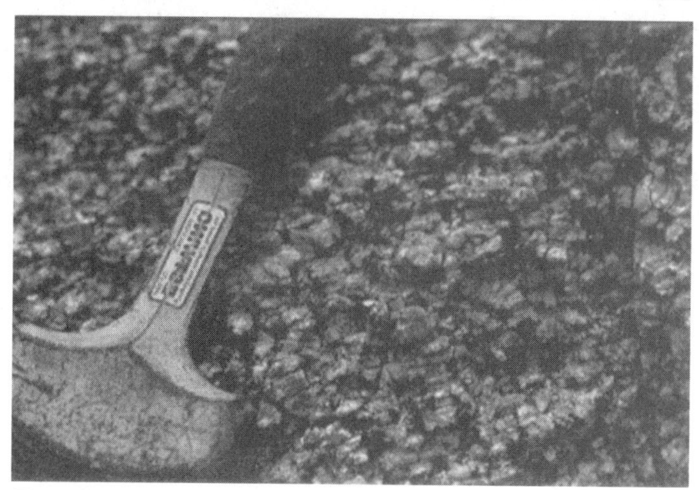

Figure 2.11 Interlocking crystals in weathering granite visible to the naked eye.

Figure 2.12 Columnar joining in basalt dike with trapped volatile gas vent, Grose Ventre Range, Wyoming.

The primary porosity of such rocks tends to be low, less than 1%, although granites with porosities greater than 1% are known (Fetter 1994). The ability of intrusive igneous rocks to produce water generally comes from secondary porosity, generated by fracturing and faulting. Large igneous bodies are subject to the stresses involved in mountain building. These forces may produce fracture patterns oriented obliquely to the principle directions of stress. Fracturing tends to occur in a characteristic crossing pattern that can be observed in the field (Figure 2.13). Sometimes fractures in intrusive igneous rocks are enhanced by **exfoliation,** a process of expansion from unloading as the weathering process strips off overlying materials. A list of porosity types and additional discussion of porosity is found in Chapter 3.

Minor drainages often develop in weaker fracture zones promoting recharge to granitic aquifers. Minor fractures can produce some water, however, larger sustained yields require more extensive fracture networks, such as fault zones. Large faults or fractures extend for distances of a mile (1.6 km) or more and are visible on aerial photographs. These longer fracture features are known as **lineaments** (Figure 2.14). Minor drainages that are controlled by these features tend to be abnormally straight, and thus can be recognized as potential target drilling areas. Another perspective of fracture zones being productive, is that they are also the most vulnerable to surface contamination. This is important for well-head or source-water protection issues.

Figure 2.13 Fracture patterns and weathering in granite, Laramie Range, Wyoming.

Figure 2.14 Stereo pair Nez Perce lineament in southwestern Montana. (Photos courtesy of Hugh Dresser.)

Example 2.3

In southwestern Montana a large granitic body known as the Boulder Batholith was once a magma chamber for a large volcanic system that has been stripped away by weathering, erosion, and uplift. Its associated mineralization is responsible for a significant amount of the colorful mining history of the old west and current mining activity for precious metals. It extends from Butte to Helena, Montana, and is nicely exposed along Interstate 90 near Homestake Pass (Figure 2.15). This igneous body is disrupted by numerous faults.

A homeowner from Pipestone, Montana, on the east side of Homestake Pass called seeking the opinion of a hydrogeologist on where to drill a domestic well. The local State agency that keeps records of wells drilled in Montana was contacted to obtain information regarding any existing wells, their drilling depths, and production rates. After plotting this information on a topographic map, it was observed that the most productive wells are aligned with the significant canyons that have exceedingly straight drainage patterns. The projection lineaments of these drainages extend westward into the Homestake Pass area. Production rates varied from 8 to 20 gpm. (43.6 to 109 m^3/day). Domestic wells that were located between drainages and away from major lineament patterns indicated production rates in the 2 to 3 gpm (10.9 to 16.4 m^3/day) range or less. Static water levels in wells were approximately 30 to 70 ft (9.1 to 21.3 m) below ground surface.

Figure 2.15 Fracturing in granitic rocks near Homestake Pass, southwestern Montana. (See also Figure 2.16.)

This homeowner's property is located between the major lineaments, so prospects for a higher productive well were not very good. The assumption was that away from major fracture zones the smaller patterns were somewhat random. The hope was to drill a deeper well (400 to 500 ft, 122 to 152 m) and intersect sufficient minor fractures to yield a couple of gpm (10 m^3/day). This approach yielded a well 450 ft (137 m) deep, producing 2 gpm (11 m^3/day). The static water was only 40 ft, so there was sufficient water from casing storage to yield significant quantities for a family.

Away from mountain fronts in southwestern Montana are intermontane valleys whose valley fill deposits are asymmetrical. Typically, eastern margins have greater depths to basement rocks than do western margins. In the Pipestone, Montana, area, granitic rocks are covered by a thin layer of sedimentary rocks, so the land surface features appears to be flat lying. The question comes up "how come they can find water and we can't find much?" The answer comes from understanding the subsurface geology and where the major productive fracture zones are.

Previously it was mentioned that intrusive igneous features such as dikes and sills can affect the flow patterns. Many magma chambers experience intermittent periods of activity, with multiple periods of intrusion. Older granitic bodies can be intruded by younger magmas of similar or differing composition. Some fractures may be filled by younger intrusives, inhibiting groundwater flow.

Example 2.4

Located at the south end of Butte, Montana, on the west side of Homestake Pass is a subdivision known as Terra Verde (Figure 2.16). Homeowners prefer to build at as high an elevation as possible for the best possible view of the nearby Highland mountains (Figure 2.17). This results in having to drill deeper wells with a greater uncertainty of success, particularly if the locations they wish to place their homes are not near any lineament features.

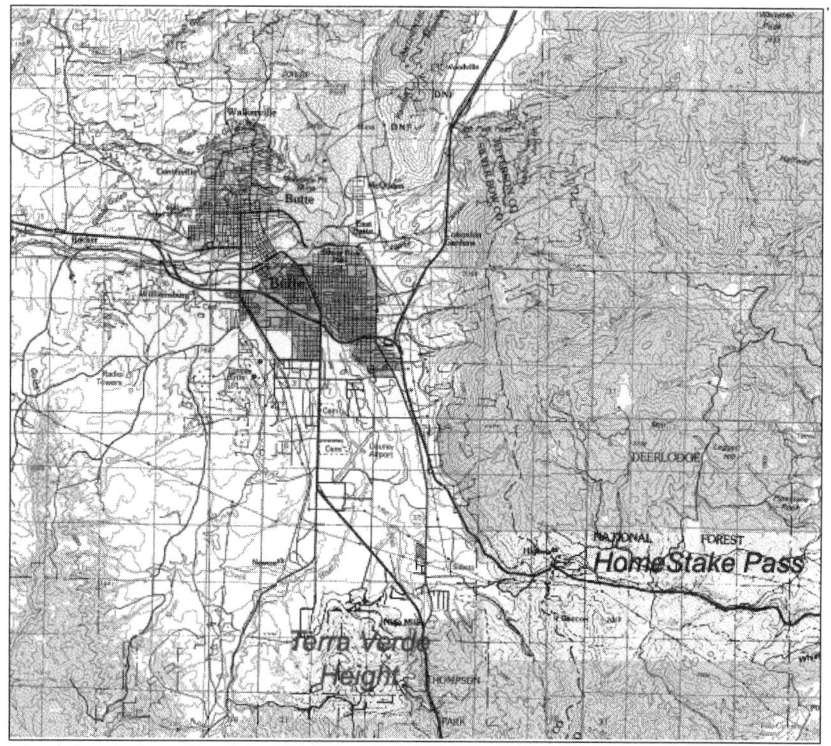

Figure 2.16 Location map of Terra Verde Height subdivision and Homestake Pass, south of Butte, Montana.

In walking around the property of a prospective home builder, the author noticed that there were a series of large pine trees that grew in a straight pattern for at least $^1/_4$ mile (400 m). It occurred to the author that this might represent a significant lineament even though it was not readily observable on an air photo. The author noticed (imagined?) a second such pattern that crossed the first on the prospective homeowner's property. The second observed or imagined lineament formed a subdued depression at the surface. Down slope there are several **aplite** (very fine-grained, light-colored) dikes. The thinking was that perhaps these dikes resulted in inhibiting groundwater flow allowing greater storage capacity in the "up-gradient" direction. The recommendation was to drill in the lineament crossing pattern area a few hundred feet (several tens of meters) uphill

Figure 2.17 Highland Mountains looking south, near Butte, Montana.

from the dike's outcropping. This formed the reasoning for a best educated guess, presuming that the orientations of the intersecting fractures were nearly vertical. This illustrates the importance of understanding the fracture orientations (Section 2.5) and intersections.

Another recommendation made to the homeowner was that if no significant water was found after drilling 200 ft (60 m), then the probability of finding additional water-bearing fractures was even less likely (lithostatic pressures would tend to squeeze fractures closed). The well was drilled to a depth of 198 ft. At 185 ft, a significant fracture zone was encountered yielding 15 gpm. Encountering the fracture zone can be considered as serendipity, but perhaps the aplite dikes did provide a mechanism for keeping water in the "uphill" area.

This success story prompted another neighbor to call for advice. Unfortunately, the second property owner had already started the expensive process of blasting and constructing a basement for his home, located approximately $^1/_4$ mi (400 m) downslope and to the east of his neighbor. This significantly limited the range of area available to recommend drilling locations, because in Montana it is necessary to dig a trench from the well head to the basement to keep water lines from freezing. After performing a field investigation, surface lineaments were not observed and the aplite dikes located on the first property were also not observed. There was not much to go on. A recommendation was provided with a low but unknown probability for success. The hole was drilled 270 ft, and although the formation was getting softer, the homeowner told the driller to stop. Instead, a water witch was hired to locate the next location, where a 5 gpm (27 m^3/day) well was drilled at a depth of 106 ft (32.2 m). This of course propagates wild ideas about the location and availability of water. More about water witching is found in Section 8.7.

2.2 Geologic Properties of Metamorphic Rocks

Metamorphic rocks are those that have "changed form" through changes in temperatures and pressures. Rocks that are sedimentary or igneous if subjected to sufficient temperatures and pressures will change form or recrystalize to form a new rock. It is important to realize that this process occurs without melting, although fluids may be present to aid in the recrystallization process known as **metasomatism**. The classification of metamorphic rocks is based upon the composition and texture. Particular mineral assemblages correlate with certain temperature and pressure conditions. Metamorphic rocks are named by describing their mineral content and observing whether they are foliated or not.

The geologic setting where temperatures and pressures are applied may be reflected in the texture of the rocks. Rocks that are involved in colliding plates tend to have a foliated texture (Figure 2.18), whereas rocks forming near hot magma chambers may be nonfoliated. Foliation occurs where platy minerals crystalize perpendicular to the stresses being applied.

Figure 2.18 Foliated rocks in Hoback Canyon, Wyoming.

Rocks that are deeply buried will metamorphose even though there may not be any relative tectonic movement. On the other hand, rocks relatively near to the earth's surface (2 mi, or 3 km) may indicate temperature and pressure conditions comparable to deep crustal rocks because they were involved in plate tectonic collisions. Foliation is an important property in controlling the direction of groundwater flow and in identifying potential production zones. Having an understanding of the regional geology is critical to understanding the potential for water development in a local area. Measuring the orientation of foliation structures is discussed in Section 2.5.

There are two general groups of metamorphic rocks, **regional** and **contact** metamorphic rocks. The most common are the regional metamorphic rocks. These are the ones formed from plate-tectonic movement and tend to occupy large areas.

Plate Tectonic Settings of Metamorphic Rocks

The lithosphere is broken into approximately 12 large plates (Figure 2.2). These all move relative to each other in a variety of ways. Some are divergent (moving apart), some are convergent (moving together), and some move past each other at transform boundaries. Relative plate movement causes a variety of stress conditions in crustal rocks. In convergent areas, such as where the Gorda plate is colliding with the North American plate, subduction of the Gorda plate under the North American plate results in partial melting at depth and the generation of a volcanic arc known as the Cascade Range (Figure 2.19). The volcanoes of the Cascade range erupt above areas of partial melting where active magma chambers are developed. The country rock next to these magma chambers is subject to high

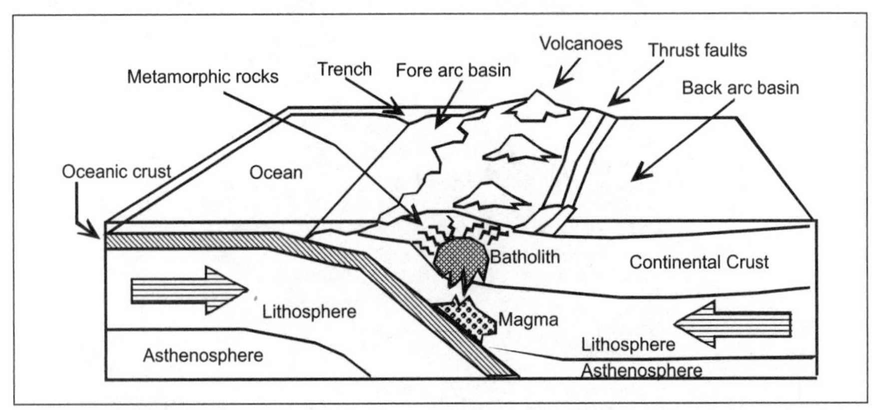

Figure 2.19 Schematic of a converging plate margin with oceanic and continental crust.

temperatures and hot fluids, forming zones of contact metamorphic rocks. The temperature and pressure conditions and available fluids determine the minerals that form, including ore minerals and explains why the country rock in some mining districts have such an altered and baked look. Magma being less dense than the solid country rock buoyantly seeks to rise to the surface. Fracturing of the overlying rocks may result in the injection of mineral bearing fluids that harden into dikes and sills. Modern mining districts are often connected to some igneous source that was active in the geologic past.

Rocks away from magma chambers that are involved in plate convergence are subject to tremendous stresses. The pressure conditions are usually measured in the thousands of atmospheres range, with units of **kilobars**. Temperatures are greater than 200 degrees celsius. A general classification scheme for metamorphic rocks is shown in Figure 2.20. Rocks are distinguished by minerals present, which form in a characteristic temperature and pressure environment and given a **facies** name. For example, lower temperature and pressure conditions are characteristic of the green schist facies. The name comes from the alteration of basalts to chlorite, with its typical greenish color.

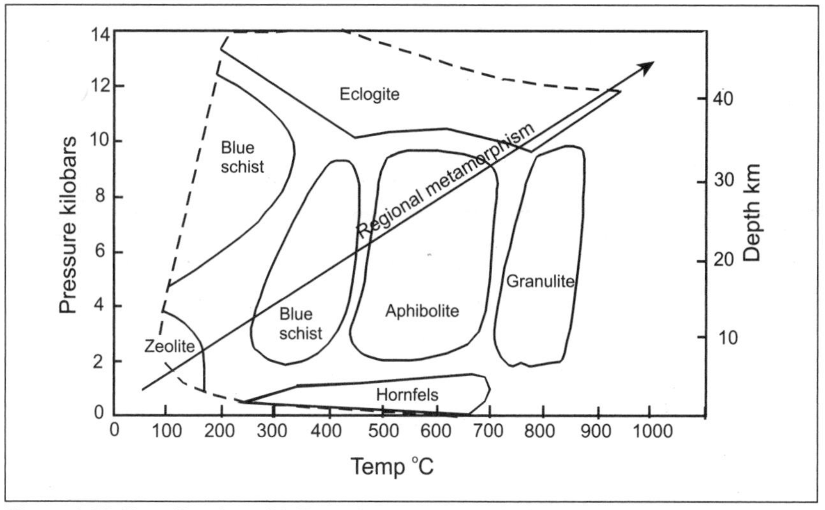

Figure 2.20 Classification of folicated metamorphic facies, showing path of regional metamorphism. Vertical axis is in kilobars of water pressure and depth in kilometers, and horizontal axis is in temperature °C. [Modified from Turner (1968).]

Regional metamorphic rocks usually have a foliated texture unless they are in close proximity to a magma chamber. The degree of metamorphism may affect a metamorphic rock's ability to transmit water. To illustrate the physical changes that occur in rock type with increases in temperature and pressure during metamorphism, an example using a sedimentary rock is provided.

Example 2.5

Suppose that a shale, the most common sedimentary rock, is subject to tectonic stresses. As the conditions of temperature and pressure increase, the clay minerals will recrystalize perpendicular to the applied stresses forming a slate (Figure 2.21). Slates are fine-grained and have poor primary water-yielding capacities. However, if fractured, slates can yield sufficient water for most domestic purposes (Chapter 1).

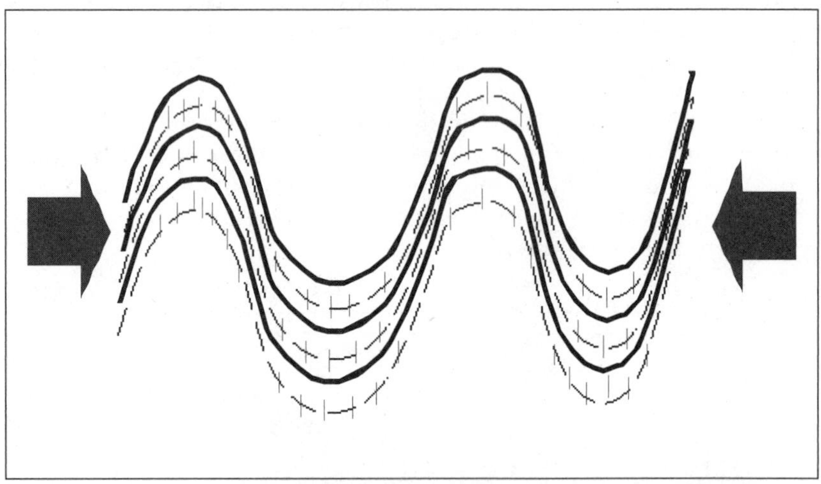

Figure 2.21 Schematic illustrating the reorientation of clay-mineral growth when a shale is subjected to horizontal stresses. (Heavy and dashed folded lines represent original bedding and clay orientation prior to applied stresses. Vertical markings represent foliation orientation of recrystalized minerals.)

If the process of increasing temperatures and pressures continues, the slate will first transform into a phyllite and then a schist. Essentially, the clay minerals of the original shale will grow into sheet silicates of the mica group, resulting in a fabric known as **foliation**. Foliated fabric results in an anisotropy for fluid flow. Flow parallel to foliation may be orders of magnitude greater than flow perpendicular to foliation. Continued increases in temperatures and pressures result in mineral separation into light and dark bands, forming a gneiss.

In the conditions described in Example 2.5 there is a tremendous competition for space during mineral growth, and the primary yielding capacity for metamorphic rocks tends to be very low. For example, Freeze and Cherry (1979) report primary porosities in the range of 10^{-9} to 10^{-11} cm/s for metasediments of the Marquette Mining district in Michigan. A characteristic feature of metamorphic rocks and intrusive igneous rocks is that the hydrologic properties of porosity, permeability, and well yield decrease with depth (Davis and Turk 1964) (Figure 2.22). This is true of most rocks. Generally, igneous and metamorphic rocks are not known as big water producers without secondary porosity and permeability being created by faulting and fracturing.

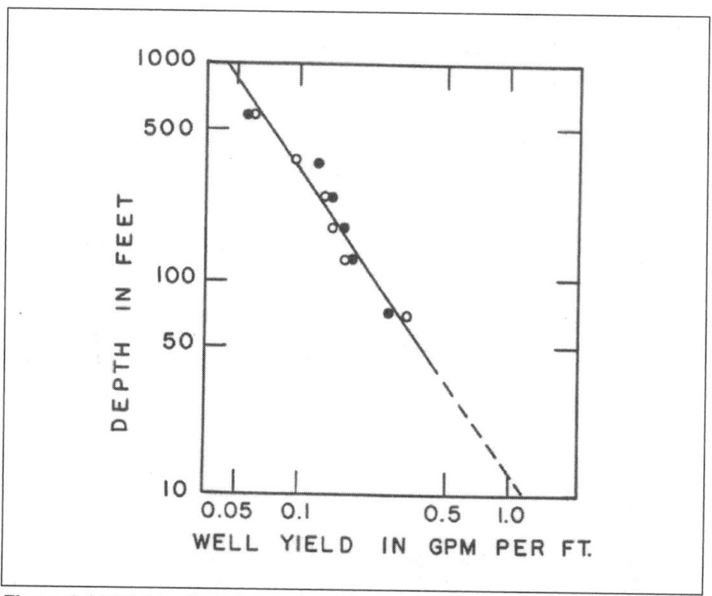

***Figure 2.22** Yields of wells in crystalline rock of eastern United States. Open circles represent mean yields of granitic rock based on a total record of 814 wells; black dots represent mean yields of schist based on a total record of 1,522 wells. [From Davis and Turk (1964). Reprinted with permission of Groundwater, 1964.}*

2.4 Geologic Properties of Sedimentary Rocks

The most common water-bearing materials that produce potable water are sedimentary rocks. These can be consolidated or unconsolidated. Sedimentary rocks, by nature, tend to have high primary porosity and, depending on the depositional environment and particle size, and they may have very high hydraulic conductivities. Sedimentary rocks are classified according to grain size and texture. Grains sizes are divided into gravel, sand, and mud according to a Wentworth-like classification scheme shown in Table 2.2 (Folk 1966). Mud includes all silt and clay-sized particles. The mud fraction is usually analyzed by a pipette or hydrometer method.

Table 2.2 Sediment Classification Based Upon Grain Size

Class	Other Names	Particle Size, mm	U.S. Sieve Size
Extremely coarse gravel	Boulders	> 256	Wire mesh
Very coarse gravel	Cobbles	64–256	Wire mesh
Coarse gravel	Pebbles	16–64	Wire mesh
Medium gravel	Pebbles	8–16	Wire mesh
Fine gravel	Pea gravel	4–8	Wire mesh
Very fine gravel	Granules	2–4	10–5
Very coarse sand		1–2	18–10
Coarse sand		0.5–1.0	35–18
Medium sand		0.25–0.5	60–35
Fine sand		0.125–0.25	60–120
Very fine sand		0.0625–0.125	230–120
Coarse to very fine silt		0.0039–0.0625	< 230
Clay		<0.0039	

Weathering

Sediment particles result from the weathering of igneous, metamorphic, and sedimentary rocks. The ease of weathering depends primarily on climatic conditions and rock type. Climates that are warm and moist produce the highest weathering rates. The composition of minerals in rocks is also a big factor. Rocks that crystalize at high temperatures and pressures tend to weather more quickly than minerals that form at lower temperatures and pressures. An example of this is known as Goldrich's (1938) weathering series and is illustrated in Figure 2.23.

Figure 2.23 is essentially the inverse of Bowen's reaction series (1928). Bowen performed a series of laboratory experiments to learn which silicate minerals form first from a molten state. The arrows between the calcium and sodium plagioclase indicate a continuous series because the ionic radii of these two are similar. Thus they can readily substitute for one another in the crystal lattice. Calcium plagioclase forms at a higher temperature and gives way to increasing sodium content with decreasing temperature. Potassium (K) feldspar is distinguished from plagioclase because potassium's ionic radius is much larger than sodium or calcium. This makes these two minerals immiscible in the molten state. The other side of the diagram represents a discontinuous series of minerals that form as temperatures

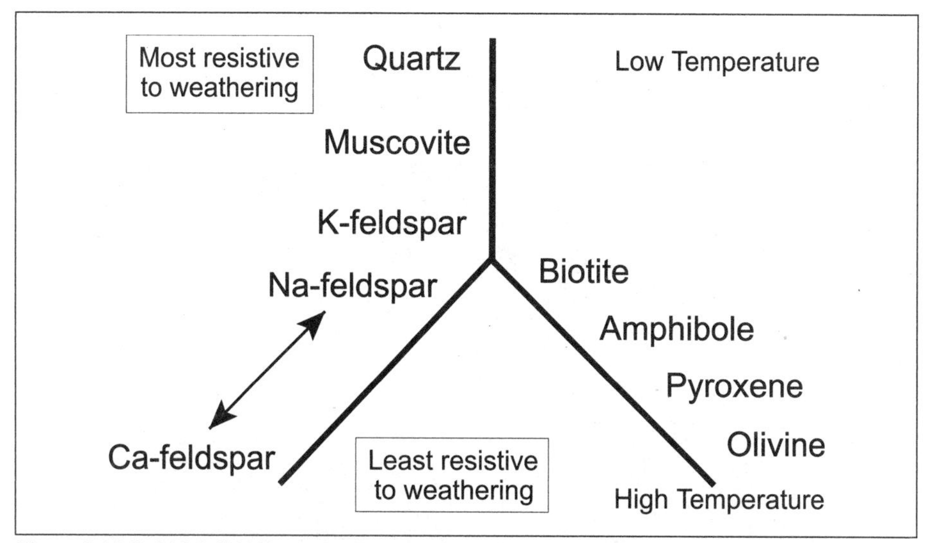

Figure 2.23 Goldrich (1938) weathering sequence. (Illustrates which minerals are more or less resistant to weathering. Quartz is the most resistant and olivine and calcium plagioclase are most easily weathered.)

change. At high temperatures near 1,400 to 1,500°C, olivine forms. As temperatures drop, the olivine is resorbed and pyroxene forms. The type of rocks that cool from a melt depend greatly on the original composition. Rocks that originate from the mantle form basaltic magmas, those that partially melt from a mixture of oceanic and continental crust are andesitic, and granitic rocks form from molten continental crust (near 600°C). Many metamorphic and sedimentary rocks are composed of the above minerals groups.

The most abundant minerals, the feldspars, weather to clay minerals, and this is why shale is the most common sedimentary rock. The other most common mineral in sedimentary rocks is quartz, a principal component in sandstone. Quartz, muscovite, and K-feldspar are non-ferromagnesian silicate minerals. The higher-temperature minerals of the discontinuous series (Figure 2.23) are ferromagnesian minerals because of their iron and magnesium content. They weather relatively quickly and color soils reddish, yellow, or brown.

Weathering has two general categories, mechanical and chemical. Mechanical weathering includes those processes that act to break down the larger rocks into smaller rocks without changes their physical properties. Another name could be disintegration. For example, a chunk of granite can be broken into smaller pieces by frost wedging, where water enters a crack and freezes. Because water in a solid state occupies more space than in the liquid state it will wedge or push the rock apart. This is why talus slopes form at the bottom of cliffs (Figure 2.24). When granitic rocks, formed at depth, are exposed to lower pressures at the surface they tend to spall off in onion-skin-like layers (exfoliation) (Figure 2.25).

Figure 2.24 Talus slope at the base of a cliff near the east entrance to Yellowstone Park, Wyoming.

Figure 2.25 Onion-skin-like layers of exfoliation in granite, Half Dome, Yosemite Park, California.

Chemical weathering includes processes that change the chemical lattice structure and composition of minerals to reach equilibrium with surface conditions. Another name could be decomposition. Hydrolysis and oxidation are examples. Chemical weathering through water and carbon dioxide act to convert feldspars into clays. Some of the mechanisms of weathering are listed in Table 2.3. Weathering breaks rocks into sediments, which then become available for transport. Mechanical weathering processes are important for enhancing the permeability of geologic materials and providing additional pathways of recharge into aquifer systems. Additionally, dissolution is important in adding dissolved minerals to groundwater and contributing to karstic conditions in carbonate rocks (Section 2.6).

Table 2.3 Examples of Mechanical and Chemical Weathering

Mechanical Weathering	Chemical Weathering
Frost wedging	Hydrolysis
Exfoliation	Oxidation
Heating and cooling	Dissolution
Plants and animals	

Transport of Sediment and Depositional Environments

Once sediments have been broken down and decomposed by the weathering process, they can be transported by wind, water, or ice into a depositional environment. Later, through **diagenesis**, unconsolidated sediments are compacted and cemented into sedimentary rock. Sedimentary rocks contain sedimentary structures that reveal the processes and mechanisms of transport. For example, ripples and trough cross-bedding indicate **fluvial** or stream processes and large cross-bedding up to 15 to 30 ft (4.6 to 9.1 m) high between bedding planes indicate **eolian** or desert wind conditions (Figure 2.26). Glacial deposits tend to be a mixture of large and small particles dumped together in no particular order (Figure 2.27). Having an understanding of depositional environments is necessary to produce a three-dimension picture of the sediment distribution. This can be a great aid when installing monitoring wells, drawing cross sections, or dividing up units into a hydrostratigraphy (Chapter 3). Is this sand channel likely to be continuous or will it quickly pinch out? This is also helpful in interpreting pumping test data and understanding groundwater flow.

Figure 2.26 Large cross-bedding typical of an eolian environment.

Figure 2.27 Glacial till showing poor sorting near Leadville, Colorado.

Sediments that have been transported by water and wind tend to be sorted and stratified. **Sorting** is a measure of the distribution of grain sizes. Sediments with a narrow range of grain size (all similar sized) are said to be well sorted. An example of a depositional environment where this occurs is a beach sand. Here, the wave action winnows out the smaller grain sizes, leaving coarser sediments of similar size. In contrast to a well-sorted sedimentary unit is a glacial till where grain sizes from boulders to clay-sized particles are mixed together. In this case, sediments are poorly sorted.

Tills, although they may have high porosity, tend to produce units that are poor conductors of water.

Engineering literature uses the term **grading** instead of sorting. A well-graded unit is one with a wide range of grain sizes, like the glacial till example above. The beach sand example would be considered poorly graded. The terms are nearly opposite, so one must be careful in the descriptions. A geologist may be more comfortable using sorting terms, and an engineer may be more comfortable using graded sedimentary terms. Drill logs used for constructing cross sections may have been described by either or both professionals, so it is important to pay attention.

It is recommended that a dictionary of geologic terms, such as that of the American Geologic Institute (Bates and Jackson1984), be kept in the office library. Table 2.4 presents some of the most common ones, along with generalized characteristics.

Stratigraphy

The hierarchy of names given to rocks is based upon the stratigraphic code. The breakdown of names can range from **group**, to **formation**, to **member**, to **bed**. Groups represent a collection of formations. For example, the Colorado Group represents several marine shales that have been grouped together.

Formations are names given to mappable rock units that have occurred in a similar depositional environment at a similar point in time. Generally, they consist of a certain lithologic type or combination of types (Bates and Jackson 1984). They are laterally extensive enough to be mapped and identified in the field. Formations are usually named from where they are best exposed and have been described in detail. For example, the Lahood Formation is a Precambrian marine fan deposit with turbidite sequences. This formation is best exposed near Lahood, Montana, where there are coarse boulders mixed with sandy turbidite beds. To the north, this formation grades into finer-grained, sandy turbidite beds.

Formations are divided into members if there are distinctive characteristics that can be mapped or identified over significantly large lateral distances. For example, the Madison Limestone is divided into the Mission Canyon and Lodgepole members. Each member is distinguished by fossils and paleokarstic features. They are best exposed in the Little Rocky mountains in northeastern Montana (Figure 2.28).

Table 2.4 Listing of Geologic Terms, Depositional Environments, and General Characteristics

Geologic Term	Environment	Comments
Alluvium	Stream, flood plains or alluvial fans	Coarser-grained channel sediments surrounded by finer-grained sediments away from the channels, including silts and clays. Changes in lithology are commonly abrupt and laterally discontinuous.
Colluvium	Topographic slopes	Loose, incoherent coarse to fine-grained deposits collecting on slopes by gravity.
Drift	Glacial	Geologic materials deposited by ice or melt water. Layered or stratified drift occurs from melt water streams.
Eolian	Desert or pertaining to wind	Well-sorted fine to medium sands, with large cross-bedding.
Fluvial	Streams, rivers, or stream action	Channels fine upward with fair to good sorting, a variety of cross-bedding is visible.
Karst	Limestone dissolution	A topography formed by the dissolution of limestone, dolomite, or gypsum creating caves, sinkholes, and underground drainage.
Lacustrine	Lake	Shales in thinly laminated beds.
	Beach	Well-sorted deposit that is longitudinally extensive along the ocean front but laterally limited in the landward direction. If prograding, these can form sheet sand units.
Paludal	Marsh or swamp	Both are organically rich. Marsh sediments produce fibrous peat, and swamps produce woody peat, which can eventually become coal.
Pelagic	Ocean	Sediments originating in ocean water.
Playa	Ephemeral lake	The lowest part of an undrained basin receives intermittent water. Characterized by clay, silt, sand, and soluble salts.
Turbidite	Continental Shelf, slope, or in a lake	Sediments well graded and laminated from moving down-slope in a body of water.

Figure 2.28 Mission Canyon member of Madison limestone exposed in the Little Rocky Mountains on the Fort Belnap Indian Reservation near Hayes, Montana.

Formation names often change at state boundaries, which can be confusing for regional studies. For example, the Madison Limestone is known as the Redwall Limestone in Arizona, where it weathers into a distinctive reddish wall exposed in the Grand Canyon.

Further subdividing of formations from members to beds is done if they are particularly distinctive. For example, within the Fort Union Formation in eastern Montana and Wyoming is the Upper Tolluck, Tongue River, and Lower Lebo Shale members. Within the Tongue River member are laterally extensive coal beds known as the Dietz and Monarch beds. These are important because they form regional aquifers (Figure 2.29). Having an understanding of the local stratigraphy and rock formations is helpful in setting up the geologic framework for a groundwater flow system.

Formation names are also given to distinct mappable units of igneous or metamorphic rocks. From the perspective of a hydrogeologist, formational units are grouped together based upon the hydraulic properties. Those with a similar enough hydraulic conductivity are combined together into one hydrostratigraphic unit (Chapter 3).

Figure 2.29 Monarch coal bed from the Tongue River member of the Fort Union formation near Sheridan, Wyoming.

2.5 Structural Geology

Geologic formations may be folded, faulted, or tilted. These represent the response of geological materials to stresses. We live on a dynamic earth. Plate tectonic movements result in applied stresses that can cause geologic formations to be pulled apart, folded, or rumpled like carpet (Figure 2.30). These expressions of deformation are known as **strain**. During these processes, the physical properties of formations can change, affecting porosity

Figure 2.30 Folded Cretaceous Kootenai formation in southwestern Montana.

and fluid conductivity. The ability to understand the physical orientation of rock relationships within an area is known as structural geology. The deformation and relative timing of tectonic stresses in large regional areas is the purvue of **tectonics**.

Disruptive changes in rock formations require well-developed observational skills. The surface expressions of geologic units are identified in the field and located and recorded on a map. Included are rock-type descriptions, observation of fossils that can be correlated with other locations, and the orientation relative to a flat plane. This is done by taking strikes and dips of the bedding planes in sedimentary rocks, foliation orientations in metamorphic rocks, or flow banding in igneous rocks. The locations and orientations of faults are also recorded. The structurally disturbed geological materials can control the direction of groundwater movement and provide areas of higher or lower well yield. Not understanding the orientation of geological formations may result in drilling in the wrong location.

Strike and Dip

A fundamental step in interpreting the orientation of formations in space is taking a strike and dip. The **strike** is the azimuth orientation of the intersection of a horizontal plane with any inclined plane or surface (Figure 2.31). This is usually taken with a Brunton compass. A bulls eye bubble indicates when the compass is being held level. The azimuth orientation is read from the compass (Figure 2.32). The strike is indicated with a line drawn on the map in the azimuth orientation. In order for the orientation of

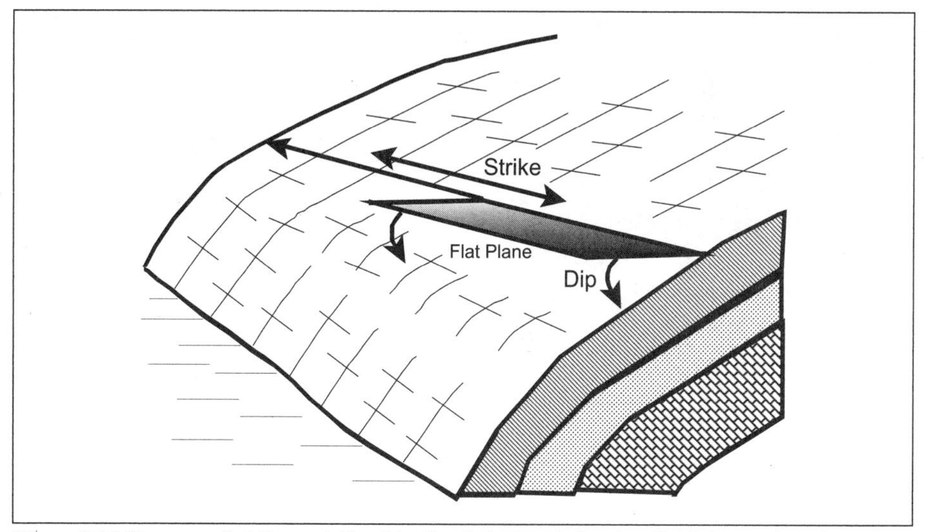

Figure 2.31 Diagram of strike and dip of inclined strata relative to a flat intersecting plane.

the compass to match the orientation of a topographic map in the field, one must first correct for the declination of the Earth's magnetic field. The declination is the horizontal angle between true north and magnetic north at a given location. This is always indicated at the lower left corner of a U.S. Geological Survey quadrangle topographic map.

Figure 2.32 Strike reading taken with a Brunton compass.

The **dip** is the angle from horizontal down to the inclined surface (Figure 2.33). The dip is read by holding the edge of the Brunton perpendicular to the strike line and moving a clinometer until its bubble is centered. Care must be taken to rotate the position of the Brunton back and forth slightly

Figure 2.33 Dip angle read from a clinometer on a Brunton compass.

to obtain a maximum inclination for dip. Anything less than the maximum would be an apparent dip and not a true dip.

A field geologist will take several strike and dip readings around an area. In addition to strikes and dips, markings are also made of contacts between formations and faults that are observed. These are usually plotted on an aerial photo or on a topographic map. The different formations are given a distinctive pattern or color code and assembled into a geologic map. Areas that are covered or where faults are inferred or projected are indicated with dashed lines.

The strike and dip of a formation can be estimated from a geologic map if the contacts are accurately drawn on a topographic map, and not in a straight line. This process is known as a **three-point problem**. Three-point problems can be solved with data that do not outcrop if they are from boreholes, mine shafts, or other subsurface information when the dip is uniform (Bennison 1990). The strike is determined by locating where the outcrop of a particular bed intersects the ground surface at the same elevation at two locations. A straight edge is used to draw a line connecting these two points. A perpendicular line is drawn to a third point where the outcrop elevation is known and the dip can be calculated or determined graphically. In Figure 2.34 the 2,200-ft contour is used to locate the strike line, and structure contours parallel to the 2,200-ft contour project the planar sur-

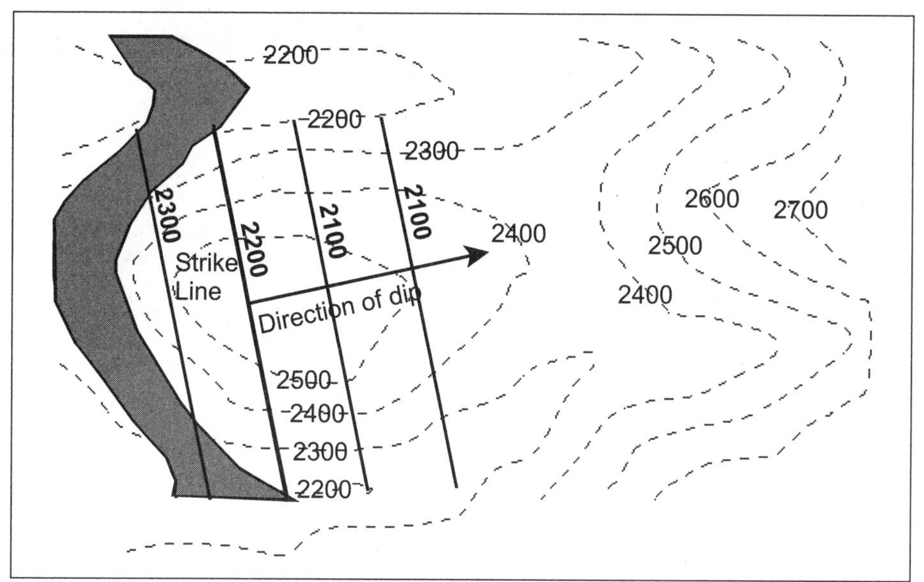

Figure 2.34 Illustration of a three-point problem used to determine the strike and dip from an outcrop and topographic map.

face of the bed. Spacing is determined from the 2,300-ft line (another outcrop of known elevation). Doing a three-point problem enables you to know how deep a target is when drilling, projecting inclined planes, or constructing cross sections.

Example 2.6

In 1997, summer students of the Montana Tech hydrogeology field camp were involved in logging a drill hole for the Meadow Village subdivision near the Big Sky Ski Resort south of Bozeman, Montana. The target zone was the basal Cretaceous Kootenai Formation at approximately 1,000 ft (300 m) below the land surface. It was hoped that sufficient permeabilities could produce a well in excess of 100 gpm (500 m^3/day). Since the geologic layers dip approximately coincident with the topographic slope, it was decided to see if the formations being drilled through would be exposed in the canyon walls near Ousel falls in the south fork of the Gallatin River (Figure 2.35). The canyon cuts down a stratigraphic section. At the level of the river, the Thermopolis Shale, the unit just above the Kootenai Formation was exposed. It helped to compare the sedimentary layers exposed at the surface with the drill cuttings. A three-point problem determined the strike and dip of the Thermopolis Shale from the elevation of two exposures in the canyon and the intersection of this same unit in the drillhole. Unfortunately, tight cementation at depth limited the productivity of this well.

Figure 2.35 Blackleaf formation overlying Thermopolis shale exposed in canyon wall near Big Sky ski resort in southwestern Montana.

Fold Geometry

One form of strain is manifested when rocks subjected to stresses respond by folding. When strata become folded there are some basic parts of the fold that define their geometry. A plane that bisects the structure is known as the axial plane. On either side of the axial plane are the limbs. If the fold axis is oriented vertically, then the fold is symmetrical. If the fold axis is rotated, the fold is asymmetrical. Another feature associated with folds is the orientation of the fold trace (Figure 2.36). The fold trace is aligned at the top of the fold. If the fold trace is inclined from horizontal, then it is said to be plunging. Strike measurements of the axial plane provide the overall orientation of the structure.

In Figure 2.36, two basic types of folds are illustrated. In diagram A the fold is a syncline, and in B and C anticlines are shown. In synclines the rocks in the middle are youngest, and the reverse is true of anticlines. Photographs of each are shown in Figures 2.37 and 2.38, respectively. The way to keep this straight is to try the exercise in Example 2.6.

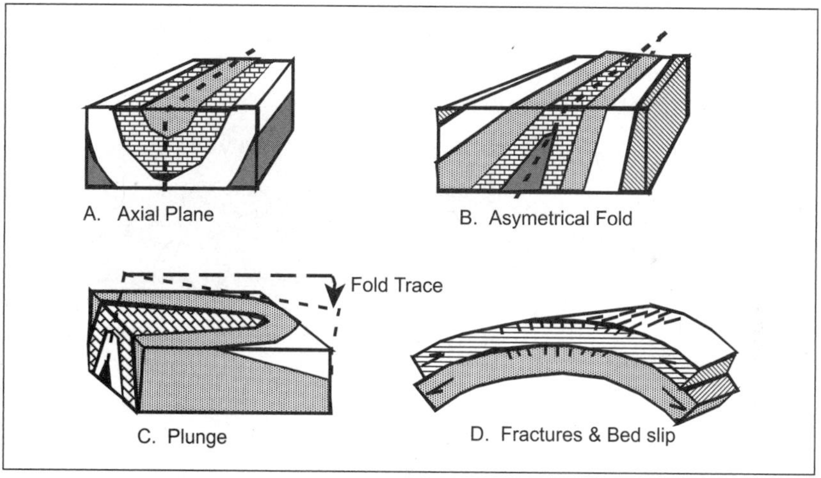

Figure 2.36 Fold geometry: A. axial plane, B. asymmetrical fold, C. plunging fold, and D. rotational fractures around axial plane and slip bed.

Example 2.7

Take a paperback book or manual. If we use the convention of saying page one is the oldest and larger numbers represent younger beds, page one will be on the bottom. If the book is pushed together and the fold is anticlinal, then page one will be to the inside and the larger numbers will be on the outside. This becomes more obvious if the book is cut along the crest of the fold (not recommended if you like this book). In folding the book into a syncline, page one will now be to the outside. Again, refer to Figure 2.36.

Figure 2.37 Student standing in the axis of a synclinal fold.

Figure 2.38 Stereo pair Big Sheep Mountain anticline, Bighorn Basin, Wyoming. (Photos courtesy of Hugh Dresser.)

Faulting

Another form of strain is a brittle response where rocks rupture from the applied stresses. Rocks that merely break into a pattern of fractures with no displacement are known as **joints** (Figure 2.39). When rock masses move relative to each other, faulting occurs. If a rock mass breaks into two blocks, then the plane separating the blocks is the fault plane. Typically, this plane is inclined. The block on the upside of the inclined fault plane is known as the **hanging wall**, and the lower block is known as the **footwall**. The term hanging wall originated from mining in mineralized areas where reverse faulting was common. The miner would stand on the footwall and hang his lantern on the hanging wall.

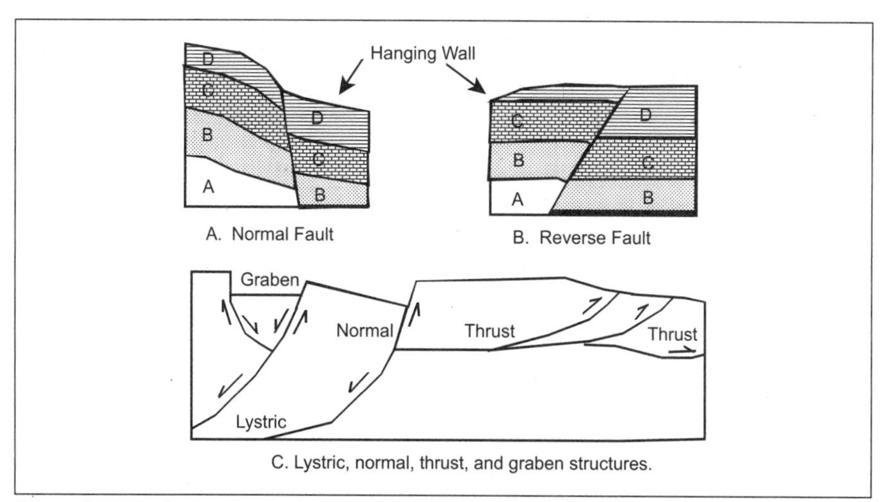

Figure 2.39 Fault structures: A. normal fault, layer A oldest, D youngest; B. reverse fault, layer A oldest; C. variety of structures in cross section, lystric (low-angle normal fault), normal fault, thrust fault (low-angle reverse fault), and a graben valley.

Imagine a stack of sedimentary layers from A to D, with layer A being the oldest. A fault divides the layers into two rock masses. If the hanging wall moves down relative to the footwall, the fault is said to be **normal** (Figure 2.39). In this scenario, younger rocks still overlie older rocks (a normal relationship). If the hanging wall has moved up relative to the footwall the fault is a **reverse** fault. Here, older rocks are overlying younger rocks (an abby-normal or reverse situation). Normal and reverse faults tend to be a relatively high angle (>45°) (Figure 2.40). In extensional areas, normal faulting is common with initial high-angle normal faults becoming more horizontal with depth. Where this occurs the fault is said to be **lystric** (Figure 2.39). Also, in extensional areas a series of blocks may move relative to each other to accommodate the additional space. When this occurs, a whole block may move downward bounded by two normal faults to form a **graben** valley (Figure 2.41). The adjoining uplifted blocks are known as **horsts**. A good example of where this has occurred is in the basin and range province in Idaho, Nevada, Utah, and Arizona (Figure 2.42).

In compressional regimes the rock masses tend to move in a more compact form. This is where reverse faulting is common. Low-angle reverse faults (commonly less than 15°) are known as thrust faults. Rock masses that move as thrust packages tend have fluid pressures involved to reduce the friction between rock masses (Hubbert and Rubey 1959). A reverse fault within the Koontenai formation is shown in Figure 2.43).

Figure 2.40 High-angle normal fault near the Ruby Mountains, Nevada.

Figure 2.41 Stereo pair graben valley structure near Divide, Montana. (Photos courtesy of Hugh Dresser.)

Faulting can occur within a thin zone or across a wide zone. This has important implications in forming boundary conditions for fluid flow. A wide shear zone may serve as a conduit to allow confined aquifer waters to move upward. If an aquifer unit is cut by a fault that brings a confining unit next to a permeable unit, the fault may represent a barrier plane to flow.

Figure 2.42 Horst mountains and graben valleys of the basin and range near Salt Lake City, Utah.

Figure 2.43 Reverse faulting in the Kootenai formation in southwestern Montana.

Other Observations in Structures

As a geologist maps a structure, such as an anticline, there are other smaller-scale features that can be observed that contribute to understanding the larger structure. For example, one can observe smaller-scale folds within the larger structure known as parasitic folds (Figure 2.44). The

Figure 2.44 Parasitic fold 1 meter across within an anticlinal fold several hundred meters wide.

strike of these will be similar to the larger structure, however, there is a systematic rotational component depending on which limb of the axis the minor structures are observed (Figure 2.36D). This same phenomenon is true of fractures. The outer edges of strata around a fold, whether the structure be an anticline or a syncline, are subject to extensional stresses. The strata inside a fold are subject to compressional forces. This can greatly affect the fluid flow properties along strike of a structure.

Example 2.8

Within the Paradox basin near Moab, Utah, along the Cane Creek anticline fracture permeabilites and associated enhanced flow conditions were described by Huntoon (Figure 2.45). A potash mine was developed 3,000 ft below the ground surface in a sylvite ore zone, accessible by a 2,790-ft deep shaft along the flank of the Cane Creek anticline. Mining was conducted by room and pillar method and was plagued by gaseous conditions that are common in the Pennsylvanian Paradox Formation and overlying Honaker Trail Formation (Huntoon 1986).

Geologic studies in the Paradox basin reveal that salts within the basin behave plastically. Burial of the salt beds by younger sediments provided continuous lithostatic loading, which when differentially applied resulted in flow structures forming during the accumulation period (Huntoon 1986). During the formation of the Cane Creek anticline, salt beds slowly migrated and bulged in the crest of anticlinal folds. While bulging occurred, the overlying rocks at the crest of fold experienced extensional fracturing and normal faulting (Figure 2.36D) very closely aligned with the strike of the Cane Creek Anticline (Figure 2.46). Beds closest to the mine area are under compression.

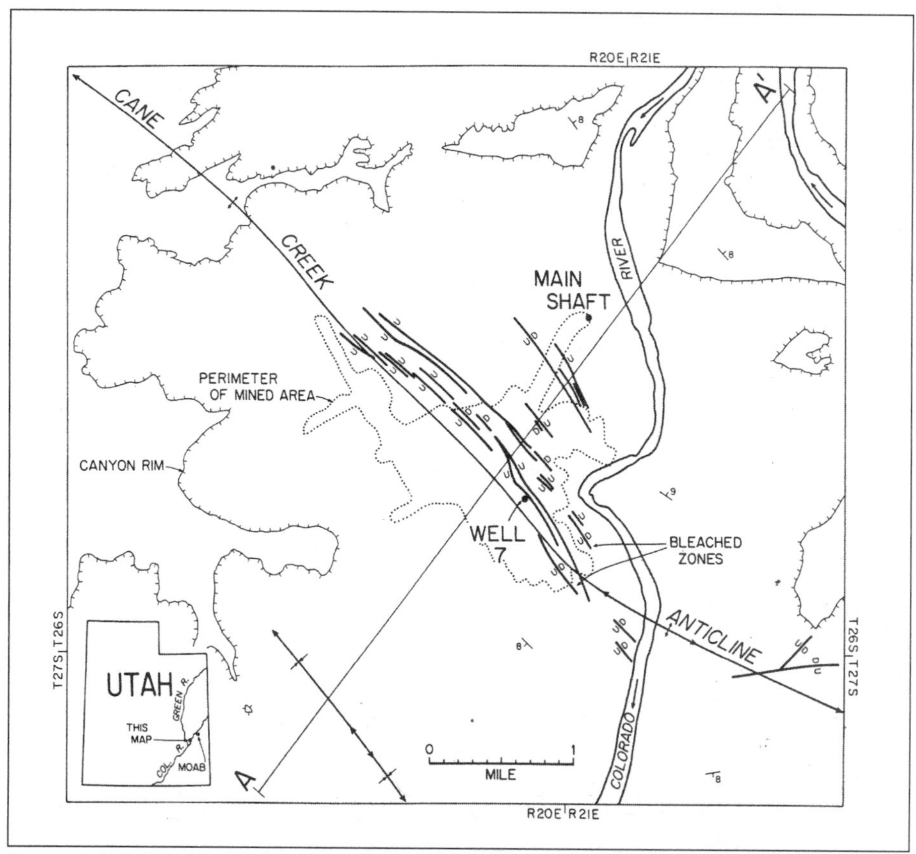

Figure 2.45 Extensional faults along the Cane Creek anticline near Moab, Utah. [From Huntoon (1986). Reprinted with permission of the National Groundwater Association (1986).]

It was decided that a solution mining method would be employed to remove the salt by flooding the 150 miles of mining cavities and pumping this to the surface (Huntoon 1986). The Texasgulf 7 well was drilled in the crest of the anticline. The loss of circulation during drilling in the upper extensional beds and subsequent breakthrough into the mine cavity and rapid draining of the drilling fluid provides an interesting story.

2.6 Karst Effects

Karst is a word that was derived from the word "kras" from vicinity of Trieste, Italy, and adjacent Slovenia, which means bare, stony ground (Huntoon 1995; Quinlan et al. 1996). It does not have a universally accepted definition, but dissolution is its primary process in developing a distinctive surface topography, a topography characterized by sinkholes,

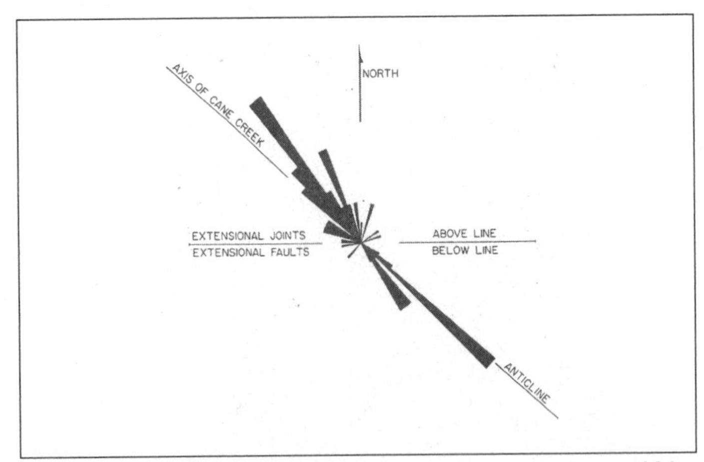

Figure 2.46 Strike relationship of 46 sets of extensional joints and 24 extensional faults near Moab, Utah. [From Huntoon (1986). Reprinted with permission of the National Groundwater Association (1986).]

caves, and underground drainage (Bates and Jackson 1984). Karstic conditions often develop from dissolution of the host rock, along fractures that become enlarged from millimeters to centimeters and even meters, resulting in a triple porosity system (Quinlan et al. 1996) (Figure 2.47). This is significantly different from a porous media system, which tends to have en-

Figure 2.47 Keyhole caves developed in the upper 300 ft of the Mississippian Redwall Limestone, Marble Canyon, Arizona, illustrating triple porosity in one location. (Photo courtesy of Peter Huntoon.)

hanced permeability from secondary porosity features such as jointing and faults.

In Chapter 5, in a discussion of groundwater flow, Darcy's Law describes how the volume of groundwater flow is proportional to the hydraulic conductivity and the hydraulic gradient. In porous media systems, the dominant controlling factor is the hydraulic conductivity. Hydraulic gradient is only of incidental concern because production zones in indurated nonsoluble rocks are concentrated within areas of the highest secondary porosity (Huntoon 1995). In Section 2.5, extensional stresses were described in fold structures resulting in enhanced permeability along the strike of such structures. Here, the extensional fracturing was responsible for creating the pathways that dominate fluid flow. In short, the largest fractures automatically have the greatest ability for fluid flow, thus resulting in the highest hydraulic conductivities.

It is a mistake to assume karstic flow systems developed in the same way. Many hydrogeologists or geologists may believe that cave systems and other karstic features developed where the fractures were largest. However, this is not necessarily true. In soluble host rock karst systems, the greatest fluid flow occurs in the areas of highest hydraulic gradients in concert with available carbon dioxide in the recharge area. Here, fractures and other conduits are enlarged by dissolution of the host rock that develop in conduit systems parallel to fluid flow. There is an organizational hierarchy of dissolution tubes within a complex network that is progressively more organized in the "down gradient" direction (Huntoon 1995). This development can ignore preexisting fracture systems developed by tectonic processes. Furthermore, karst systems change dynamically over time as a result of changes in stage levels and tectonic forces, which result in different flow paths being developed (Quinlan et al. 1996), that may even crosscut older systems (Huntoon 1995).

Hydrogeologic field work in karst systems is markedly different from the traditional approaches used in nonsoluble rock systems such as sedimentary rocks or igneous rock systems with significant secondary porosity. This is discussed next and in Chapter 13, where there is additional discussion on tracer tests. The purpose of presenting karst systems in this chapter is to contrast these with the geologic conditions characteristic of those described in the previous sections. We begin with some of the terms associated with karst studies.

Karst Aquifer

Formal definitions of nonkarst aquifers are presented in Chapter 3. A **karst aquifer** has been defined by Huntoon (1995), as follows:

A karst aquifer is an aquifer containing soluble rocks with a permeability structure dominated by interconnected conduits dissolved from the host rock which are organized to facilitate the circulation of fluid in the downgradient direction wherein the permeability structure evolved as a consequence of dissolution by the fluid.

The difference between a karst aquifer and karst topography is that the aquifer is saturated and karst topography may only be partially or nonsaturated. The field hydrogeologist should know that confined karst aquifers may not have any obvious surface topographic indications. The telling evidence of a confined karst aquifer is often found in large springs that have developed down-gradient (Huntoon 1997).

Epikarst

Epikarst is a term that describes a veneer surface topography characterized by wide, intensely dissolved openings in the upper part of carbonate bedrock (Mangin 1974–1975). The famous Kunning stone forest in the Yunnan Province in south China is a classic example (Figure 2.48). Epikarst aquifers can be characterized as shallow, thin, unconfined aquifers were lateral groundwater movement is more dominant than vertical, although functionally they move groundwater to down-gradient seeps and collector structures that feed deeper systems (Huntoon 1992). Alpine and subalpine epikarst tends to be unconfined and ephemerally saturated near the surface (Mills 1989).

Figure 2.48 Stone forest karst, Kunming, China. (Photo courtesy of Peter Huntoon.)

Paleokarst

Paleokarst is a name given to former circulation structures that have been destroyed by burial, infilling, collapse, compaction, brecciation, cementation, or structural fragmentation (Huntoon 1995). In this definition, the emphasis is in the inactivation of an organized permeability structure through destructive processes, although this term has also been used to reflect dewatering (James and Choquette 1988). Hence, when you hear someone refer to paleokarst, you should picture some collapsed feature that no longer circulates water (Figure 2.49).

Figure 2.49 Collapsed structure in limestone paleokarst.

Fieldwork in Karst Areas

The person required to characterize a karst aquifer has a difficult assignment. Traditional point-sampling methods such as drilling wells and performing pumping tests are ineffective in helping predict groundwater flow. Instead it becomes necessary to characterize the complex conduit system and aquifer architecture (Huntoon 1995). Surface mapping is difficult at best, given the topographic conditions, although field inspection of aperture widths is helpful in understanding the system. Aerial photographs are helpful in defining the epikarst areas, and discharge springs provide an indication of some of the low points in the aquifer system.

Epikarst zones in the Rocky Mountain region are commonly developed on surfaces of outcrops, under thin soils and vegetation, or are bare

(Huntoon 1997) (Figure 2.50). The deeper systems, often dominated by caves, are commonly linked by vertical features such as shafts or collector dissolution tubes. Outlet springs are generally controlled by local base levels in the adjoining basins or along the flanks of the mountain ranges. Spring discharges often vary two orders of magnitude between the spring and fall (Huntoon 1997). Flows through the aquifer system can vary many orders of magnitude. Flows in smaller apertures and through the rock matrix are assumed to follow Darcian conditions. However, flow in the larger apertures are dominantly turbulent (Huntoon 1995). Using Reynold's number as the onset for turbulent flow in fractures, Quinlan et al., (1996) describe how results from over 2,255 tracer tests in unconfined carbonate aquifers indicate velocities become turbulent in fractures in the few millimeters to few centimeters aperture range. (The field methodologies for this are described in Chapter 13.) This is consistent with the cubic law for fracture flow volume described by Dominico and Schwartz (1990) for hydraulic gradients greater than 0.001.

Figure 2.50 Classic epikarst developed on the Mississippian Madison Limestone, Tosi Basin, Gros Ventre Range, western Wyoming. (Photo courtesy of Peter Huntoon.)

Flow velocities from tracer tests generally indicate extremely short residence times through cave systems. Quinlan et al. (1996) report several studies from around the world that tracer flow velocities in the hundreds of meters per hour range are not uncommon. They cite examples from a chalk aquifer in France, the Edwards aquifer in Texas (Chapter 1), and other geologically young carbonate aquifers in the southeastern part of the United

States. A geometric mean between 0.02 and 0.05 m/sec is reported in Quinlan et al. (1996). This has serious implications for protection of recharge areas in these systems (Huntoon 1997). Should the recharge areas for these aquifers not be protected from surface contamination, they would quickly be polluted in the down gradient direction.

The network of conduit tubes within a karst aquifer network have multiple interconnections (Huntoon 1995). For this reason, there are numerous opportunities for water to shunt across to other areas in the flow system, particularly in unconfined karst aquifers. The direction of flow and multiple pathways appears to be stage dependent. Tracer studies reveal that although they are introduced at a single injection point, they can emerge at diverse exit points (Huntoon 1995). The scary thing is that at certain stage elevations, springs may be deemed safe because the water-quality data look clean. However, if the stage changes, then a once clean spring may become contaminated just because shunting within the flow system has changed.

Another application of field studies involves basin circulation and regional flow studies. Many carbonate systems outcrop along the upper elevations of mountains in higher precipitation areas. Persons performing fieldwork may erroneously believe that surface streams that enter karst systems account for the water that recharges a basin. In the Rocky Mountain region, cave systems have developed along the flanks of uplifted mountain ranges that reject recharge through a decrease in transmissive properties in the basinward direction (Huntoon 1985).

Example 2.9

Wyoming lies within an area characterized by crustal shortening. This results in elongated mountain uplifts and deep sediment-filled basins. The mountain uplifts are mountain blocks that have ridden up thrust-fault ramps from Laramide deformation (Huntoon 1985). One should notice that there is a homoclinal margin that parallels uplift and a fault-severed margin (Figure 2.51).

Rocks that outcrop along the homoclinal margins constitute an ideal setting for recharge because these rocks are hydraulically connected to rocks in the basins. In reality, transmissivities within the carbonate rocks decrease basinward, and much of the apparent recharge is rejected through surface springs that discharge at the toe of the mountain front (Huntoon 1985). Huntoon (1985) discusses over 10 cave or sinkhole systems within the Trapper Canyon and Medicine Lodge drainages that lie on the western slope of the Bighorn Mountains (Figure 2.52). Documented stream gauging and other fieldwork indicate that 25 to 35% of the water observed to enter these cave systems actually may be recharging the basin aquifers (Huntoon 1985).

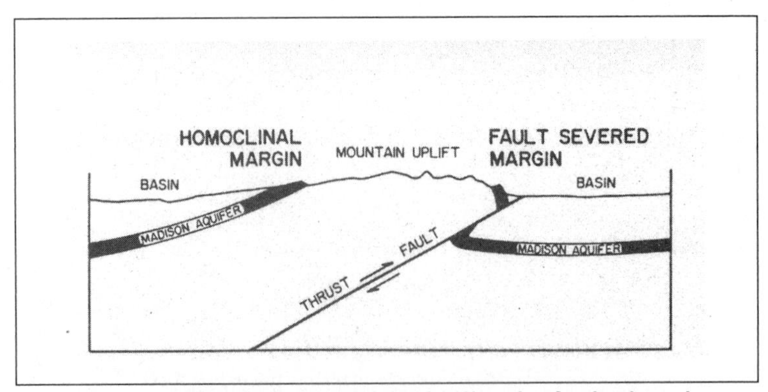

Figure 2.51 Generalized cross section in the Wyoming foreland province. [From Huntoon (1985). Reprinted with permission from the NGWA (1985).]

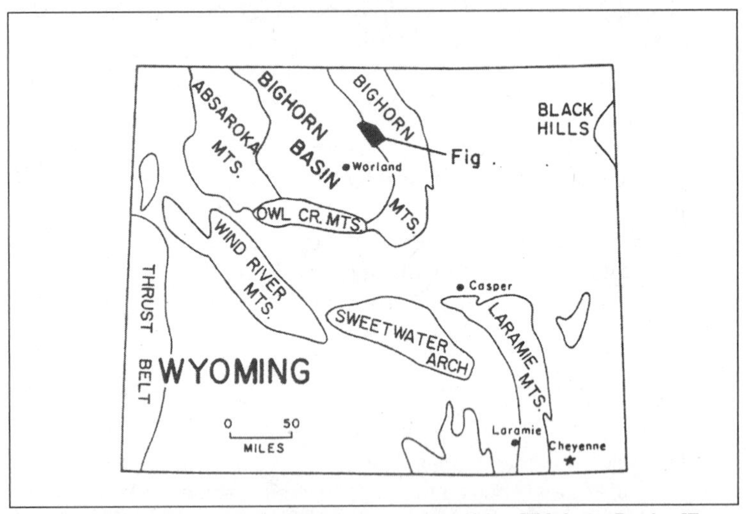

Figure 2.52 Location of recharge areas on east side of Bighorn Basin. [From Huntoon (1985). Reprinted with permission from the NGWA (1985).]

Apparently, the decreased transmissivities can be explained by the difference in hydraulic gradients found in the recharge areas compared to average gradients within the basins in the same aquifer. In the recharge areas, gradients averaged 400 ft/mi. (0.076 ft/ft) compared to an average of 40 ft/mi. (0.0076 ft/ft) in the basin (Huntoon 1985). The high gradients were responsible for developing a karst system with cave-sized conduits (Figure 2.53). The discharge springs at the toe of the drainages are coincident with the decreases in hydraulic gradient and subsequent decreases in transmissivities. A similar phenomenon was observed in the Madison aquifer in the Black Hills (Rahn and Gries 1973).

Figure 2.53 Stream in Thunder River cave, Grand Canyon, Arizona. (Photo courtesy of Peter Huntoon.)

2.7 Using Geologic Information

The geology of hydrogeology could easily be the topic of many chapters or a book. Geologic concepts and examples are scattered throughout this book, because geology is inseparable from groundwater systems. The geologic features form the framework and boundary conditions for groundwater-flow systems. The more case histories and field examples one studies and experiences, the larger the library of examples one can draw from to solve problems.

If you have avoided studying geology because of the inherent uncertainty in knowing what is occurring at depth, why study hydrogeology? It has the same types of uncertainties. The author's experience is that being able to know exactly what is occurring at depth is an uncomfortable area for many engineers, yet many will think nothing of using field data provided by geoscientists in a computer simulation model or design process. If you use geologic information, you have a responsibility to make sure that the constraints are also known. You are responsible for the information you use in a design or simulation. Take another course or two in geology or read up on it if it is a weakness in your life. It will make you a better professional.

References

Allis, R.G, and Lumb, J.T., 1992. The Rotarua Geothermal Field, New Zealand: Its Physical Setting, Hydrology, and Response to Exploitation. *Geothermics*, Vol. 21, No. ½, pp. 7–24.

Alt, D., and Hyndman, D.W. 1995. *Northwest Exposures, A Geologic Story of the Northwest.* Mountain Press, Missoula, MT, 443 pp.

Bates, R.L., and Jackson, J.A, (eds.), 1984. *Dictionary of Geological Terms 3rd Edition.* American Geological Institute, Anchor Press, Garden City, NY, 571 pp.

Bennison, G.M., 1990. *An Introduction to Geological Structures and Maps. Boutledge.* Chapman and Hall, New York, 69 pp.

Bowen, N.L., 1928. *The Evolution of the Igneous Rocks.* Princeton University Press, Princeton, NJ.

Custer, S.G., Michels, D.E., Sill, W., Sonderegger, J.L., Weight, W.D., and Woessner, W.W., 1994. *Recommended Boundary for a Controlled Groundwater Area in Montana Near Yellowstone Park.* Water Resources Division, National Park Service, Fort Collins, CO, 29 pp.

Davis, S.N. and Turk, L.J., 1964. Optimum Depth of Wells in Crystalline Rock. *Ground Water*, Vol. 2, No. 2, pp. 6–11.

Driscoll, F.G., 1986. *Groundwater and Wells.* Johnson Screens, St. Paul Minnesota, 1108 pp.

Domenico, T.A., and Schwartz, F.W., 1990. *Physical and Chemical Hydrogeology.* John Wiley & Sons, New York, 824 pp.

Fetter, C. W., 1994. *Applied Hydrogeology, 3rd Edition.* Macmillan College Publishing Company, New York, 691 pp.

Folk, R.L., 1966. A Review of Grain-size Parameters. *Sedimentology*, Vol. 6, Elsevier Publishing, Amsterdam, pp. 73–93.

Freeze, A., and Cherry, J., 1979. *Groundwater.* Prentice-Hall, Upper Saddle River, NJ, 604 pp.

Goldrich, S.S., 1938. A Study in Rock Weathering. *Journal of Geology*, Vol. 46, pp. 17–58.

Hubbert, M.K., and Rubey, W.W., 1959. Role of Fluid Pressure in Mechanics of Overthrust Faulting: I, Mechanics of Fluid-filled Porous Solids and its Application to Overthrust Faulting. *Geological Society American Bulletin*, Vol. 70, pp. 115–166.

Huntoon, W.P. 1985. Rejection of Recharge Water from Madison Aquifer Along Eastern Perimeter of Bighorn Basin, Wyoming. *Ground Water*, Vol 23, No. 3., pp. 345–353.

Huntoon, W.P., 1986. Incredible Tale of Texasgulf Well 7 and Fracture Permeability, Paradox Basin, Utah. *Ground Water*, Vol. 24, No. 5, pp. 644–653.

Huntoon, W.P., 1992. Hydrogeologic Characteristics and Deforestation of the Stone Forest Karst Aquifers of South China. *Ground Water*, Vol. 30, No. 2, pp. 162.

Huntoon, W.P., 1995. Is it Appropriate to Apply Porous Media Groundwater Circulation Models to Karstic Aquifers? *Groundwater Models for Resources Analysis and Management*, Aly I. El-Kadi (ed.), CRC Press, Boca Raton, FL, pp. 339–358.

Huntoon, W.P., 1997. The Case for Upland Recharge Area Protection in the Rocky Mountain Karsts of the Western United States. In *Karst Waters and Environmental Impacts*, Gunay, G., and Johnson, A.I., (eds.), A.A. Balkema, Rotterdam/Brookfield, 1997.

Mangin, A., 1974, 1975. Contributions of the Hydrodynamic Recharge of Karst Aquifers. *Annals of Speleology*, Vol. 29, No. 4, p. 495, 1974; Vol. 30, No. 1, p. 21, 1975.

James, N.P., and Choquette, P.W., (eds.), 1988. *Paleokarst*, Springer-Verlag, New York, 416 pp.

Mills, J.P., 1989. *Foreland Structure and Karstic Ground Water Circulation in the Eastern Gros Ventre Range, Wyoming*, Master's Thesis, University of Wyoming at Laramie, 101 pp.

Quinlan, J.F., Davies, G.J., Jones, S.W., and Huntoon, P.W., 1996. The Applicability of Numerical Models to Adequately Characterize Ground-Water Flow in Karstic and Other Triple-Porosity Aquifers, *Subsurface Fluid-Flow (Ground-Water and Vadose Zone) Modeling*, ASTM STP 1288, Joseph D. Ritchey and James O. Rumbaugh, (eds.), American Society for Testing and Materials, pp. 115–133.

Rahn, P.H., and Gries, J.P., 1973. Large Springs in the Black Hills, South Dakota and Wyoming, *South Dakota Geology Survey Report of Investigation 107*, 46 pp.

Sorey, M.L. (ed.), 1991. Effects of the Potential Geothermal Development in the Corwin Springs Known Geothermal Resources Area, Montana, on the Thermal Features of Yellowstone National Park, *U.S. Geological Survey Water-Resources Investigations Report 91-4052*, pp. A1–H12.

Streckeisen, A.L., 1976. To each Platonic Rock its Proper Name. *Earth Science Review*, Vol. 12, pp. 1–34.

Turner, F.J., 1968. *Metamorphy Petrology*. McGraw-Hill, New York, 366 pp.

Chapter 3

Aquifer Properties

It is important to be able to translate the geology (Chapter 2), when it becomes saturated, into hydrogeology. The physical properties of geologic materials control the storativity and ability of fluids to move through them. Rock units that do not allow fluids through them become barriers to fluid flow and in turn change the direction of groundwater movement (Chapter 5). Other features such as fault zones may serve as conduits to fluid flow or act as barriers. In this chapter the physical properties of saturated geologic materials are presented to provide a basic understanding of aquifers, confining layers, and boundary conditions as a basis for understanding groundwater flow presented in Chapter 5. Boundaries are often determined directly through drilling (Chapter 8), pumping tests (Chapters 9 and 10), or geophysical methods (Chapter 4).

3.1 From the Surface to the Water Table

When precipitation hits the land surface, some water enters the soil horizon. This process is known as **infiltration**. Water that accumulates on the surface faster than it can infiltrate becomes **runoff** (Chapter 1, Figure 1.1). The rate at which water infiltrates or runs off is a function of the physical properties of the surficial soils. Some of the important factors appear to be thickness, clay content, moisture content, and intrinsic permeability of the soils' materials (Baldwin 1997). (Additional discussion on intrinsic permeability is given in Section 3.2). Infiltrating water that encounters soils with higher clay content tends to clog the pores, causing precipitation to mound up and run off, unless they are exceedingly dry (Stephens 1996). Sandier soils promote infiltration and exhibit less vegetative growth, while soils with a higher clay content appear to promote plant growth. Glaciated areas provide an example of an environment where many soil types can be found.

Glacial sediments deposited via moving water become stratified or layered and tend to be well drained; examples include outwash deposits, kames, and eskers. Sediments transported by ice that accumulate along the sides and end of a glacier are poorly sorted and contain a higher content of clay and silt; examples include lateral and end moraines, which are poorly drained (Figure 3.1). In the field well-drained soils can be distinguished by lessor plant growth and a rougher appearance (Figure 3.2). Once infiltration occurs, any groundwater that descends below the rooting depth continues onto the regional water table.

Figure 3.1 Terminal moraine in southwestern Montana, indicating poorly sorted sediments.

Between the soil horizon and the regional water table is an area referred to as the vadose zone (Figure 3.3). The ability of the vadose zone to hold water depends upon the moisture content and grain size. Wells completed in the vadose zone will have no water in them, even though the geologic materials appear to be wet, while wells completed in saturated fine-grained soils will eventually contain groundwater. Chapter 12 is devoted to the vadose zone and its properties and field methodologies.

Another part of the vadose zone immediately above the regional water table is the capillary fringe. The capillary fringe is essentially saturated, but groundwater is being held against gravity under negative pressure (less than atmospheric). This same phenomenon is observed when one puts a paper towel into a pan of water. The water is attracted to the surfaces of the towel fibers being drawn up through very small pore tubes between the fibers. Similarly, in the capillary fringe groundwater seeks to wet the sur-

Figure 3.2 Gravel channel exposed in road cut near Sheridan, Wyoming. Coarser sands lack vegetation, while finer-grained upper and lateral floodplain deposits support more vegetative growth.

faces of geologic materials with an attraction greater than the force of gravity. The thickness of the capillary fringe is grain-size dependant. The finer grained the material, the thicker the capillary fringe because of the smaller pore throats, increased surface area, and surface tension.

When drilling wells or installing monitoring equipment, one must also be careful that the first water encountered is actually the regional water table and not a **perched aquifer**. Perched aquifers represent infiltrating groundwater that accumulates over confining layers of limited areal extent above the regional water table (Figure 3.3). Perched aquifers may be capable of sustaining enough water for a few residences, but generally not enough for many residences or long-term production. Several water levels in wells in the same area would help one determine whether a perched water table exists or not. Aquifers are defined and discussed in Section 3.4.

Example 3.1

A consulting company was evaluating the drilling depths for production wells for a proposed subdivision. Estimates were being made based upon existing wells in the area. The evaluator did not realize that there were wells completed in a local perched aquifer and regional unconfined aquifer. In the end, he averaged the well depths to estimate drilling costs. This of course resulted in bidding way too low and the consulting company losing money. The crux of the problem was in failing to understand the flow system and that the differences between well depths represented two separate aquifers.

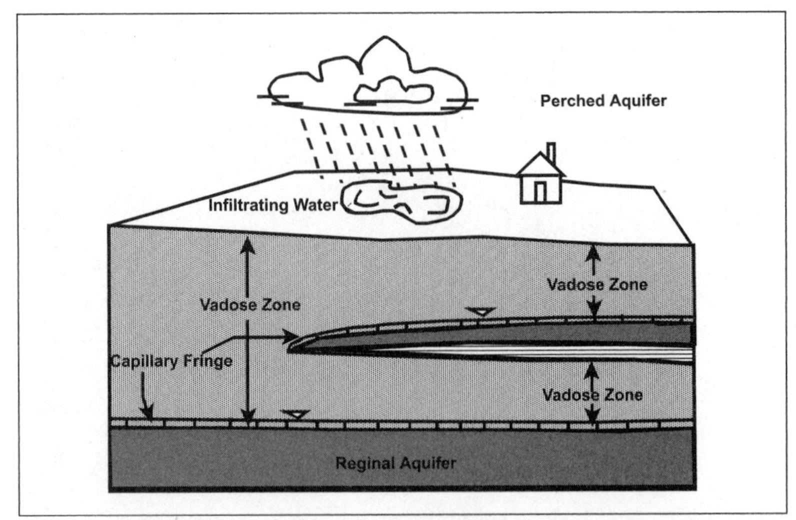

Figure 3.3 Schematic of the vadose zone, infiltrating water, and the capillary zone.

3.2 Porosity and Aquifer Storage

The volume of water that an aquifer can take in or release from for a given change in head in the system relates to storage. The amount of water an aquifer can hold in storage is often determined by its porosity. The porosity of earth materials is a function of size, shape, and arrangement or packing. The ability of water to move through an aquifer is described by its permeability or hydraulic conductivity (see Section 3.3).

Porosity

The **porosity** is represented as the nonsolid fraction of geologic materials. This is where fluids can be held. In the vadose zone, the porosity or open spaces are filled with air and water (Section 3.3). The total porosity of geologic materials expressed as a percent is represented by:

$$\eta\% = \frac{V_V}{V_T} \times 100\% \qquad [3.1]$$

where:

η = porosity

V_V = volume of the void

V_T = total volume

Porosity is often broken down into primary and secondary porosity. Primary porosity is the void space that occurred when the rock or geologic material formed. Secondary porosity refers to openings or void space created after the rock formed (Figure 3.4). Examples of these are given in Table 3.1.

Figure 3.4 Secondary porosity from faulting. A spring is emanating from the fault zone.

Table 3.1 Example of Primary and Secondary Porosity

Primary Porosity	Secondary Porosity
Vesicles	Faults
Intergranular pores	Fractures
Interlayer partings and unconformities	Solution channels
Between lava flows	Stylolites
Lava tubes	Enhanced pathway from plants and animals
Intercrystalline pores	

Generally, each aquifer or confining layer is modeled using a single overall porosity unless its lateral changes are known through a distribution of cores. In fractured rock, such as a fractured sandstone or granite, a "dual-porosity" model may be more appropriate. A dual-porosity model assigns a porosity to the fracture zone (secondary porosity) and to the geologic block materials (primary porosity)(see Chapter 10). Carbonate karstified systems may have a triple porosity where there are microscale and two levels of macroscale pore spaces (larger fractures and caves, see Chapter 2). The larger-scale fractures may lead to misleading interpretations of unlimited supply.

Example 3.2

When the other author worked for the Alabama Survey, he heard about a test on a Huntsville municipal well (pumped at approximately 2,000 gpm, 0.13 m^3/day) that ran for a week with only a couple of feet of drawdown. The test was continued to 10 days and ran out of water on the 8th or 9th day with a sustainable pumping rate of only a couple of hundred gpm or less (0.013 m^3/day).

Of the total porosity of geologic materials, there is a portion that will drain freely by gravity and an amount retained in the geologic materials. The volume of water that will drain by gravity for a unit drop in the water table from a unit volume of aquifer is referred to as the **specific yield** (S_y). The water that remains clinging to the surfaces of the solids is called the **specific retention** (S_r). Although they are strictly different things, the specific yield is often used as an estimate for the **effective porosity** (η_e), a term used to describe the porosity available for fluid flow. It should be noted that specific yields are estimated from vertical drainage tests, while effective porosities are generally used for horizontal flow calculations, such as to calculate groundwater velocities (Chapter 5). Although one may generally assume that the specific yield and the effective porosity are the same for a coarser-grained aquifer, in some soils there may be a high content of soil aggregates and water may become tied up in dead-end pore spaces. This results in the effective porosity and specific yield being significantly different (Cleary et al. 1992). The specific yield and specific retention make up the total porosity, expressed in the following relationship.

$$\eta = S_Y + S_R \qquad [3.2]$$

Example 3.3

Suppose you have a 5-gal bucket full of dry sandy material. You add just enough water until the sand becomes saturated at the 5-gal mark (Figure 3.5). Assume that it takes exactly 1 gal to fill all of the pores. From Equation 3.1, the percent porosity equals:

$$\eta = \frac{1 \text{ gal}}{5 \text{ gal}} \times 100\% = 20\% \qquad [3.3]$$

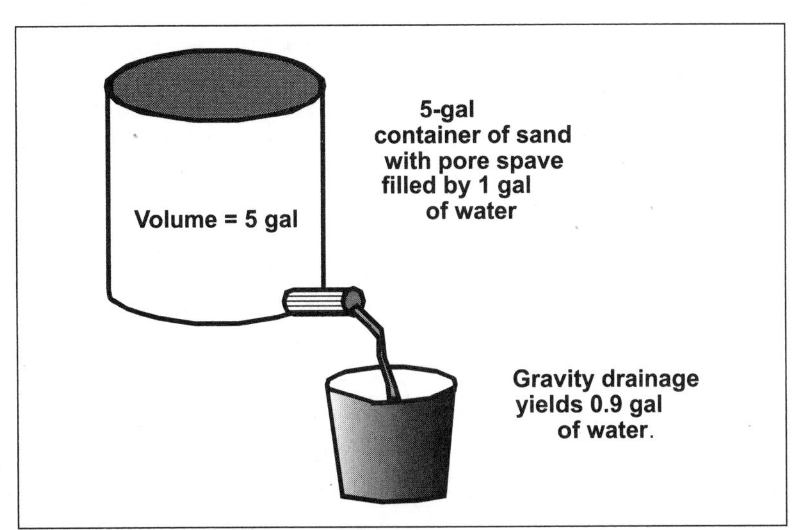

Figure 3.5 Illustration of the concept of specific yield. A 5-gal bucket of sand is filled to the top with 1 gal of water; 0.9 gal drains out by gravity, while 0.1 gal is retained in the pore spaces.

If a hole was drilled into the bottom of the bucket and the water allowed to drain into a container until it stopped, the specific yield and specific retention could be determined. Assume the amount of water that drained was 0.9 gal. By definition, the specific yield would be:

$$S_Y = \frac{0.9\,\text{gal}}{5\,\text{gal}} = 0.18 \quad [3.4]$$

From Equation 3.4 the effective porosity would be estimated to be 18%. The specific retention would be 0.02 or 2% of the total volume. In most sand and gravel aquifers, the specific retention is quite low (<5%). As the grain size decreases, the porosity actually increases (Table 3.2, Figure 3.6), but the specific retention increases as well. The smaller the grain size, the greater the surface area for water to cling to.

One notices from Table 3.2 that the porosity of sand and gravel together is less than that of either material separately. This is a function of the packing of grains and sorting. Sorting was described in Chapter 2 and is an expression of ranges of grain sizes. The packing of grains is a function of the size, shape, and arrangement of grains. If one takes spheres of equal size and arranges them, there are two end member packing arrangements that represent the most porous packing (cubic packing) and the least porous (rhombohedral packing) (Figure 3.7). Cubic packing is where grains are stacked vertically on top of one another with their edges touching. This yields a porosity of 47.6%. In rhombohedral packing the spheres are pushed together in their most compact form, yielding a porosity of 25.9%. When grains of differing sizes are mixed together, the smaller grains fit between the grains of the larger ones, thus reducing the overall porosity.

Table 3.2 Ranges of Porosities in Typical Earth Materials [Modified after Driscoll (1986); Freeze and Cherry (1979); Roscoe Moss (1990)]

Unconsolidated Materials	η%	Consolidated Rock	η%
Clay	40–70	Sandstone	5–35
Silt	35–50	Limestone/dolomite	<1–20
Sand	25–50	Shale	<1–10
Gravel	20–40	Crystalline rock (fractured)	<1–5
Sand and gravel	15–35	Vesicular basalt	5–50

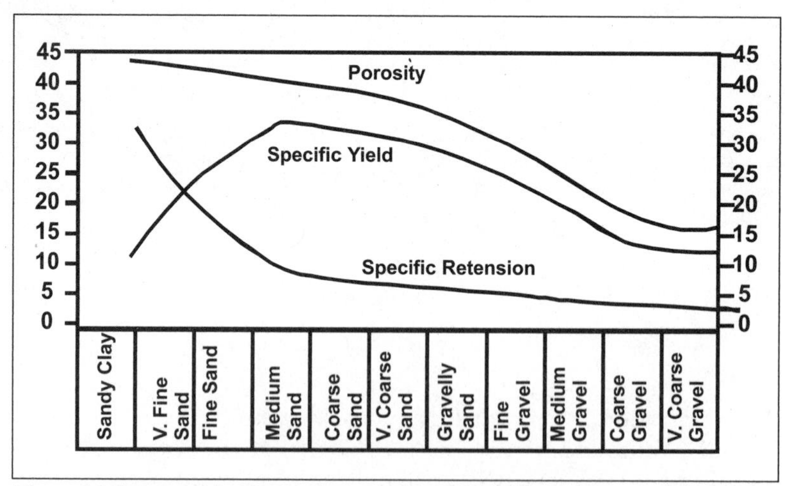

Figure 3.6 Relationship of porosity, specific yield, and specific retention with grain size. [Modified after Scott and Scalmanini (1978).]

In the saturated zone below the water table the porous openings are *completely* filled with water. The total porosity can also be calculated from the following relationship:

$$\eta\% = 100\left[1 - \frac{\rho_b}{\rho_p}\right] \quad [3.5]$$

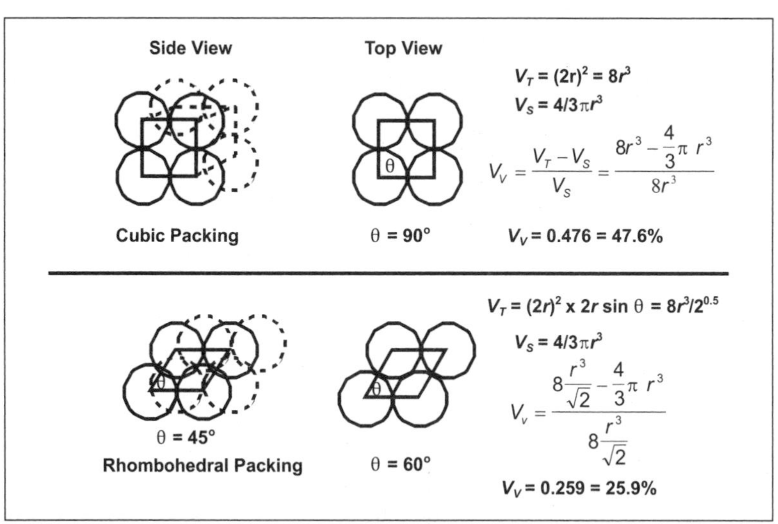

Figure 3.7 Diagram of cubic and rhombohedral packing and determination of porosity.

where:

η = porosity

ρ_b = dry bulk density (m/L^3)

ρ_p = particle density (m/L^3)

Equation 3.5 is determined by taking a core from the field. The volume of saturated material is measured and then weighed (the wet weight). The geologic materials are then placed in a container and cooked at 105°C until dry. The new dry weight compared to the volume represents the dry bulk density (ρ_b). The particle density (ρ_p) is the density of the solid materials. A common value for aquifer materials is that of quartz, 2.65 g/cm^3. The density of groundwater is close to 1.0 g/cm^3 depending on the water temperature (Appendix A). The ratio of the particle density to the density of water is known as the **specific gravity**.

Example 3.4

A core of sandy materials of approximately 80 cm^3 was taken from the field and weighed. The net soil material minus the tare weight was 166 g (wet weight). The soils were saturated and reweighed with a value of 172 g. After placing the sample in the oven at 105°C for several hours the dry sample weighed 148 g (dry weight). The water temperature in the field was 12°C, and the density is 0.999099 g/cm^3.

The dry bulk density:

$$\rho_b = 148 \text{ g}/80 \text{ cm}^3 = 1.85 \text{ g/cm}^3$$

If the particle density is assumed to be 2.65 g/cm^3, the porosity can be estimated using Equation 3.5 to be:

$$\eta = 100\left[1 - \frac{1.85 \text{ g/cm}^3}{2.65 \text{ g/cm}^3}\right] = 30.2\%$$

Another way the porosity can be estimated is by evaluating the volume occupied by the water in the core at saturation minus the dry weight.

$$\text{saturation} - \text{dry} = 172 \text{ g} - 148 \text{ g} = 24 \text{ g}$$

This occupies a volume of:

$$V = 24 \text{ g}/(0.999099 \text{ g/cm}^3) = 24.02 \text{ cm}^3$$

Using Equation 3.1:

$$\eta = 100 \frac{24.02 \text{ cm}^3}{80 \text{ cm}^3} = 30.0\%$$

The discrepancy in values can be attributed to an assumed particle density in the first case or because of volume errors. The above example was artificially created to illustrate the concept of porosity. In reality, the core must be tapped into a container in the laboratory. During the tapping process differential compaction results in field conditions being lost. It is always harder to place a field core into a container of equal volume.

In a study conducted at the U.S. Geological Survey laboratory reported by Morris and Johnson (1967), differing earth materials were tested and evaluated for the physical and hydrologic properties. Anderson and Woessner (1992) summarized the results of their findings for specific yield in Table 3.3. It is interesting how the means of unconsolidated materials follow the trend indicated in Figure 3.6. When the reported arithmetic means differ significantly from the midpoint of the range values, this indicates that the distribution is skewed. Notice also that the differences between the unconsolidated sedimentary materials, such as fine and medium sand, compared to their lithified counterparts, fine and medium sandstone, are different because of the volume occupied by cementing agents. Notice also that the specific yield is always less than the total porosity.

A more detailed example from Morris and Johnson (1967) illustrating the grain-size distributions of water-laid sandy materials and their physical and hydrologic properties are shown in Tables 3.4a and 3.4b. It is interesting to note the similarities of specific gravity regardless of the grain-size distribution and the range of grain-size distributions, dry bulk densities, and hydraulic conductivities. It is apparent from the grain-size distributions that there is a correlation between grain-size and hydraulic conductivity.

Table 3.3 Ranges of Values of Specific Yield [Adapted from Anderson and Woessner (1992)]

Material Class	Material	No. of Analysis	Range	Arithmetic Mean
Sedimentary	Clay	27	0.01–0.18	0.06
	Silt	299	0.01–0.39	0.20
	Sand (fine)	287	0.01–0.46	0.33
	Sand (Med)	297	0.16–0.46	0.32
	Sand (Coarse)	143	0.18–0.43	0.30
	Gravel (fine)	33	0.13.–-0.40	0.28
	Gravel (med)	13	0.17–0.44	0.24
	Gravel (coarse)	9	0.13–0.25	0.21
	Siltstone	13	0.01–0.33	0.12
	Sandstone (fine)	47	0.02–0.40	0.21
	Sandstone (med)	10	0.12–0.41	0.27
	Limestone	32	0–0.36	0.14
Wind Deposits	Loess	5	0.14–0.22	0.18
	Eolian Sand	14	0.32–0.47	0.38
Metamorphic	Schist	11	0.022–0.033	0.026
Igneous	Tuff	90	0.02–0.47	0.21

Table 3.4a Grain-Size Distribution of Differing Water-Laid Soils [From Morris and Johnson (1967)]

Location	Clay < 0.004 mm	Silt 0.004 0.063 mm	V. Fine Sand 0.063-0.125 mm	Fine Sand 0.125–0.25 mm	Med. Sand 0.25–0.5 mm	Coarse Sand 0.5–1.0 mm	V. Coarse Sand 1–2 mm	V. F. Gravel 2–4 mm	Fine Gravel 4-8 mm	Med. Gravel 8-16 mm
Brunswick, GA	10.4	19.6	57.2	7.2	3.4	1.6	0.6			
McCurtain Co., OK	2.8	8.8	19.2	64.6	4.4	0.2				
Gallaway Co., KY	4.0	0.2	1.1	19.4	73.4	1.8	0.1			
Arapahoe Co., CO	0.3	0.0	0.0	0.2	3.5	24.3	40.5	23.7	6.8	0.7

Table 3.4b Physical and Hydrologic Properties of Water-Laid Soils [From Morris and Johnson (1967)]

Location	Depth ft	Specific Gravity (g/cm³)	Dry Bulk Density (g/cm³)	Specific Retention (%)	Specific Yield (%)	Total Porosity (%)	Hydraulic Conduct. (ft/day)
Brunswick, GA	496–497	2.71	1.58	22.8	18.9	41.7	1.7
McCurtain, Co. OK	—	2.65	1.90	0.7	27.6	28.3	11
Gallaway, Co. KY	—	2.67	1.48	1.0	43.6	44.6	53
Arapahoe, Co. CO	33.5–34	2.61	1.67	2.9	33.1	36.0	802

Storativity

The amount of water an aquifer can release or take into storage is known as the **storage coefficient**. The numerical value assigned is to an aquifer is the **storativity**, a dimensionless value determined from pumping tests (Chapter 10). In a pumping test the storativity represents the storativity from the saturated thickness contributing to the well bore. Storativity is defined according to Equation 3.6.

$$S = S_Y + S_S \times b \qquad [3.6]$$

The first term has already been defined as the specific yield (S_y). The second term is the **specific storage** (S_s), dimensioned 1/L, multiplied by the saturated thickness (b). The specific yield is approximately equal to the total storativity value for most unconfined aquifers. The usual range of storativity values in unconfined aquifers is 0.03 to 0.3 (Fetter 1994). The value of S_s in unconfined aquifers is practically negligible, unless there are sections within the aquifer where the grain size is very small. In confined aquifers (Section 3.4), the water released from storage is a function of the compressibility of the aquifer materials and the compressibility of water (Equation 3.7).

$$S_S = \rho \times g(\alpha + \eta\beta) \quad [3.7]$$

where:

ρ = fluid density, m/L^3

g = gravitational force, L/t^2

α = compressibility of mineral skeleton, $(1/(M/Lt^2)$

β = compressibility of water, $(1/(M/Lt^2)$

Specific storage is also known as elastic storage. This is from the flexible nature of mineral skeletons when changes in stress are "felt," such as in a pumping or slug test (Chapter 10). Aquifer dewatering causes the mineral skeleton to be stressed sufficiently that compression of the mineral skeleton takes place, with or without gravity drainage. This phenomenon can also occur during a significant seismic event, such as an earthquake; however, this usually results in permanent structural changes to the mineral skeleton. Permanent structural changes in the mineral skeleton are not restorable. Compressional stresses on the mineral skeleton are also enhanced by overpumping. Earth materials within an aquifer most susceptible to permanent changes are the finer grained materials (Table 3.5).

Table 3.5 Range of Compressibility Values [Adapted from Freeze and Cherry (1979) and Roscoe Moss (1990)]

Material	Compressibility (m^2/N)	Compressibility (ft^2/lb)
Clay	10^{-6}–10^{-8}	10^{-5} –10^{-7}
Sand	10^{-7}–10^{-9}	10^{-6}–10^{-8}
Gravel	10^{-8}–10^{-10}	10^{-7} –10^{-9}
Fractured rock and sedimentary rocks	10^{-8}–10^{-10}	10^{-7} –10^{-9}
Igneous and metamorphic rock	10^{-9}–10^{-11}	10^{-8} –10^{-10}
Water	4.4×10^{-10}	2×10^{-9}

In confined aquifers, storativities range from 10^{-3} to 10^{-6} (0.001 to 0.000001). Because of the compressibility of water, storativity values less than 10^{-6} are not possible in porous media aquifers ($n > 0.03$). When evaluating a report, be aware that the calculated storativities from a software package may be much lower than possible in real life.

So now the question comes up, what about storativities between 0.001 and 0.03? These fall into the leaky confined or semiconfined range. The closer values are to 0.001, the more semiconfining the aquifer is and harder it is for leakage to occur. Contrasted to this, storativities nearer to 0.01 reflect a fairly leaky system (Chapter 10).

3.3 Movement of Fluids through Earth Materials

Now that we have considered the ability of earth materials to store water, another important term refers to the ability of fluids to move through them. The term most commonly used is **hydraulic conductivity**. It encompasses the ability of a material to conduct fluids under a *unit* hydraulic gradient considering the dynamic viscosity in units of length over time (L/t) (Fetter 1994). The **viscosity** is a term that indicates the resistence of fluid to flow. It has units of newton-seconds over meters squared (N-sec/m^2). Thick fluids like tar or cold molasses have high viscosity, where alcohol is an example of a low viscosity fluid. **Permeability** is another term commonly used to express the ability of fluids to pass through earth materials and oft times both hydraulic conductivity and permeability are used interchangeably in groundwater studies, although it should not be done. This needs an explanation.

The conductivity properties of fluids are related to the **specific weight** (γ) and the **dynamic viscosity** (μ) of the fluid. The specific weight represents the gravitational driving force of the fluid. The ability of fluids to move is also inversely proportional to the resistence of fluids to shearing (Fetter 1994). This is expressed in Equation 3.8.

$$K = k_i \frac{\gamma}{\mu} = k_i \frac{\rho g}{\mu} \qquad [3.8]$$

where:

k_i = intrinsic permeability, Darcy (9.87×10^{-9} cm^2)

γ = Specific weight = density × gravity

g = force due to gravity (L/t^2)

μ = dynamic viscosity (m/Lt)

The **intrinsic permeability** (**k_i**) represents the physical flow properties of the geologic materials. It is essentially a function of the pore size openings only (no fluids are included). The larger the pore opening, the larger the intrinsic permeability. This relationship can be in Tables 3.4a and 3.4b. The specific weight (γ) indicates how a fluid of a given density will move from gravity. The dynamic viscosity (μ) indicates that the less resistive the fluid, the more conductive the earth material is. Those that equate hydraulic conductivity to intrinsic permeability consider most fresh groundwater to have insignificant changes in specific weight and dynamic viscosity; therefore, the ability of groundwater to move is mainly proportional to the intrinsic permeability alone. This should not be done because the hydraulic conductivity will change by a factor of 3 just by changing the water temperature between 2°C and 30°C. This is also not true when contaminants interact with groundwater with different fluid properties, such as nonaqueous phase liquids (NAPLs) or with saline waters.

Hydraulic conductivity values for earth materials range over 12 orders of magnitude (Figure 3.8). The distribution of hydraulic conductivity is log-normally distributed, so averaged values should be geometric means or some other transformation. Obtaining precise or accurate values for hydraulic conductivity is unlikely, so when they are reported as such they should be viewed with a jaundiced eye. This leads to caution for those who

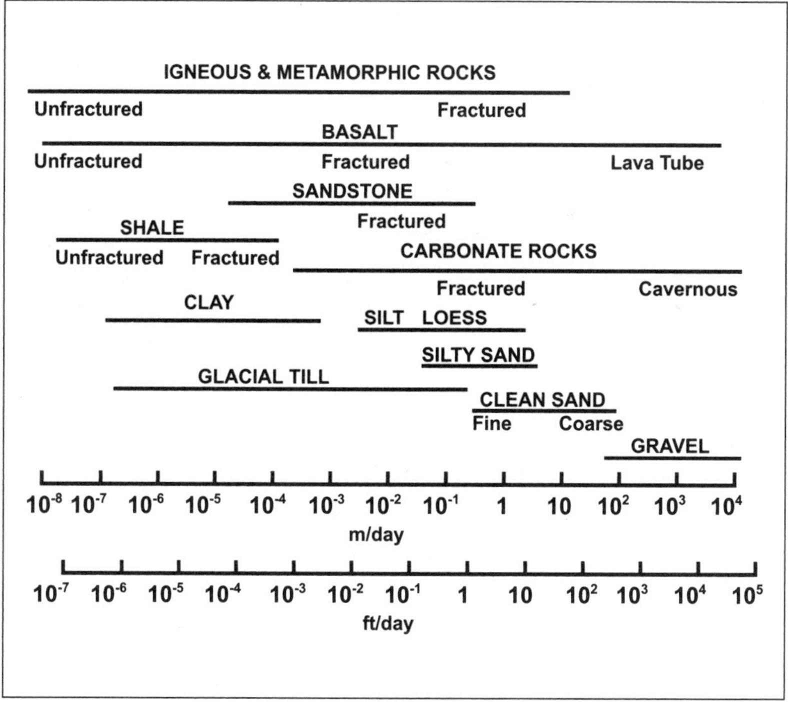

Figure 3.8 Ranges of hydraulic conductivity values for earth materials. [Adapted from Anderson and Woessner (1992).]

report hydraulic conductivity values. Just because your calculator gives a number to many decimal places, your responsibility as a hydrogeologist is to report only that level of precision that you feel is justified. Typically this would be with a maximum of two significant figures (Table 3.6).

Hydraulic conductivity is reported in a variety of different units in the literature and in software packages. A hydrogeologist needs to be able to convert back and forth from one unit system to another through unit conversions and by knowing a few simple conversion factors (Appendix B). It may be that you are out in the field with only your field book and a calculator and you have to make a decision based upon some numbers.

Table 3.6 Ranges of Hydraulic Conductivity Values for Various Earth Materials [Adapted from Domenico and Schwartz (1990)]

Material Type	Material	Hydraulic Conductivity cm/sec
Unconsolidated	Unweathered marine clay	8×10^{-11} – 2×10^{-7}
	Clay	1×10^{-11} – 4.7×10^{-7}
	Silt, loess	1×10^{-7} – 2×10^{-3}
	Fine sand	2×10^{-5} – 2×10^{-2}
	Medium sand	9×10^{-5} – 5×10^{-2}
	Coarse sand	9×10^{-5} – 6×10^{-1}
	Gravel	3×10^{-2} – 3
	Till	8×10^{-10} – 2×10^{-4}
Sedimentary rocks	Shale	1×10^{-11} – 2×10^{-7}
	Siltstone	1×10^{-9} – 1.4×10^{-6}
	Sandstone	3×10^{-8} – 6×10^{-4}
	Limestone, dolomite	1×10^{-7} – 6×10^{-4}
	Karst and reef limestone	1×10^{-4} – 2
	Anhydrite	4×10^{-11} – 2×10^{-6}
	Salt	1×10^{-10} – 1×10^{-8}
Crystalline rocks	Unfractured basalt	2×10^{-9} – 4.2×10^{-5}
	Fractured basalt	4×10^{-5} – 2
	Weathered granite	3.3×10^{-4} – 5.2×10^{-3}
	Weathered gabbro	5.5×10^{-5} – 3.8×10^{-4}
	Fractured igneous & metamorphic rocks	8×10^{-7} – 3×10^{-2}
	Unfractured igneous& metamorphic rocks	3×10^{-12} – 2×10^{-8}

Example 3.5

An estimate of hydraulic conductivity at 2 × 10^{-4} cm/sec was estimated for a sandy material. The number needed to be converted into ft/day to perform a calculation for average linear velocity during a tracer test (Chapter 13). The only resources you have are a writing utensil, a field book, and a calculator.

$$K = 2 \times 10^{-4} \frac{\text{cm}}{\text{sec}} \times \left(\frac{1\text{in.}}{2.54\,\text{cm}}\right) \times \left(\frac{1\,\text{ft}}{12\,\text{in.}}\right) \times \left(\frac{60\,\text{sec}}{1\text{ min}}\right) \times \left(\frac{1440\,\text{min}}{1\text{ day}}\right) = 0.6\text{ ft / day}$$

The average linear velocity was estimated to be:

$$V_{\text{ave}} = \frac{K}{\eta_e} \times \frac{\partial h}{\partial l} = \frac{0.57\text{ ft / day}}{0.26} \times \frac{3\text{ ft}}{145\text{ ft}} = 0.045\text{ ft / day}$$

The hydraulic conductivity can be estimated for sandy materials where the **effective grain size** (d_{10}) is between 0.1 mm and 3.0 mm (Hazen 1911), where d_{10} represents the smallest 10% of the sample. (It is important to pay attention to the limits over which this is applicable). The effective grain size is determined from a grain-size distribution plot (see Chapter 8 and Example 3.6). Grain size plots are helpful in determining the sorting. The sorting is estimated with the **uniformity coefficient** (C_u) expressed in Equation 3.9.

$$Cu = \frac{d_{60}}{d_{10}} \qquad \textbf{[3.9]}$$

Values less than 4 are well sorted, and values greater than 6 are considered to be poorly sorted (Fetter 1994). The Hazen equation (1911) relating hydraulic conductivity to effective grain size and a sorting coefficient is shown in Equation 3.10. The most common error made by users of this equation is to forget to convert the grain-size parameters from millimeters to centimeters.

$$K = C(d_{10})^2 \qquad \textbf{[3.10]}$$

Where:

K = hydraulic conductivity in (cm/sec)

d_{10} = effective grain size (cm)

C = sorting and grain-size coefficient in (1/cm/sec)

The coefficient C is assigned according to sorting *and* grain size (Table 3.7). The grain size is determined by evaluating the median grain size (d_{50}) from a grain-size distribution curve (Example 3.6). Values that are poorly sorted and finer grained receive smaller coefficient numbers. We recommend that the coefficients in Table 3.7 only be estimated to the nearest value of 10.

Table 3.7 Hazen Equation Coefficients in $(cm\text{-}sec)^{-1}$ Based on Sorting and Grain Size

Description	Coefficient
Poorly sorted to well-sorted very fine sand	40–80
Poorly sorted to moderately sorted fine sand	40–80
Moderately sorted to well-sorted medium sand	80–120
Poorly to moderately sorted coarse sand	80–120
Moderately sorted to well-sorted coarse sand	120–150

Shepard (1989) evaluated the data from published studies relating grain-size to hydraulic conductivity by plotting hydraulic conductivity (in ft/day) verses median grain size (d_{50}) on log-log paper. Various plots were made based upon sediments from different depositional environments (Chapter 2), each forming a straight-line plot. The slope of the plot was related to an exponent (Equation 3.11). The values of the exponent range between 2.0 and 1.5 for glass spheres of equal size to poorly sorted unconsolidated materials. Example C values and exponents are shown in Table 3.8.

$$K = C_F \times d_{50}^{\ i} \qquad [3.11]$$

Where:

K = hydraulic conductivity (ft/day)

C_F = shape factor, (based upon depositional environment) in units which convert mm^2 to ft/day

d_{50} = median grain size in mm

i = exponent (between 2.0 and 1.5) = slope on log-log plot

Table 3.8 Hydraulic Conductivity (K in ft/day) Related to Shape Factors (C), Based on Depositional Environment and Median Grain Size (mm) [Adapted from Shepard (1989)]

Deposit. Environ. (Shape Factor)	0.01 mm K ft/day	0.1 mm K ft/day	1.0 mm K ft/day	10.0 mm K ft/day	Exponent
Glass beads (40,000)	4.0	400	40,000	—	2
Eolian dunes (5,000)	1.0	70	5,000	—	1.85
Beach deposits (1,600)	0.5	28	1,600	—	1.75
Channel deposits (450)	0.23	10	450	20,100	1.65
Sedimentary rock (100)	0.1	3.2	100	3,160	1.5

Example 3.6

A grain-size distribution based upon water-laid deposits from McCurtain County, Oklahoma, and Arapahoe County, Colorado (from Morris and Johnson 1967), are plotted in Figure 3.9. The respective hydraulic conductivities are calculated based upon the Hazen (1911) method and the relationship developed by Shepard (1989). These are compared with the hydraulic conductivities determined in the laboratory by Morris and Johnson 1967).

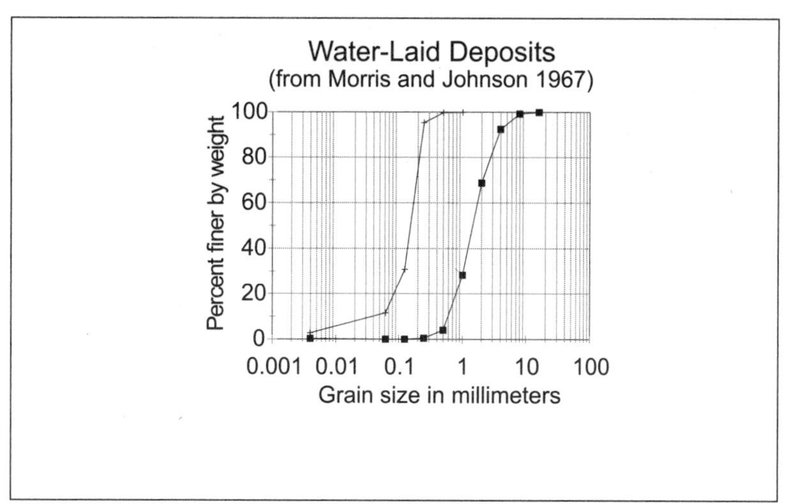

Figure 3.9 Plot of grain size versus percent finer by weight. Pluses represent McCurtain Co., Oklahoma, and boxes represent Arapaho Co., Colorado.

Equations 3.9 and 3.10, and Table 3.7 are used to evaluate the hydraulic conductivity using the Hazen (1911) method, and Equation 3.11 and Table 3.8 are used to evaluate the hydraulic conductivity using the Shepard (1989) relationship. Both are compared to the coefficient of permeability reported by Morris and Johnson (1967). The results are shown in Table 3.9 with an example calculation below.

Table 3.9 Summary of Calculations to Evaluate Hydraulic Conductivities in ft/day for Two Water-Laid Sand Deposits

Location	d_{10} mm	d_{50} mm	d_{60} mm	C_u	C	K_{hazen} ft/day	C_F	Exp	$K_{shepard}$ ft/day	K_{lab} ft/day
Oklahoma	0.05	0.17	0.18	3.6	90	6	1200	1.7	60	11
Colorado	0.6	1.5	1.7	2.8	150	1530	1200	1.75	2400	802

Although the sand from Oklahoma has an effective grain size less than 0.1 mm, a well-sorted sample may allow the calculations to be useable. The uniformity coefficient indicates very a well-sorted sand the with a d_{50} in the fine sand range (between 0.125 and 0.25 mm). Converting the median grain size into centimeters and using a Hazen (1911) *C* factor of 90 yields:

$$K_{\text{Hazen}} = 90(0.05\text{cm})^2 = 2.25 \times 10^{-3}\,\frac{\text{cm}}{\text{sec}}\left(\frac{86400}{1\text{ day}}\right)\left(\frac{1\text{ft}}{30.48\text{cm}}\right) = 6.4\text{ft}/\text{day}$$

For the Shepard (1989) method it is noticed that the sands are water-laid and well sorted, a shape factor C_F between beach and channel deposits was selected, with an exponent of 1.7 to yield the following hydraulic conductivity:

$$K_{\text{Shepard}} = 1200 \times (0.17\,\text{mm})^{1.7} = 59\text{ft}/\text{day}$$

All values were within one order of magnitude, but the Hazen (1911) method appears to have better agreement with the laboratory calculated value.

Example 3.6 illustrates how estimates for hydraulic conductivity vary based upon grain-size analysis. It is the author's experience that similar discrepancies also occur when performing pumping tests (Chapters 9 and 10) and slug tests. Conversions from one system to another are also illustrated in the example. Table 3.10 gives conversions for the more commonly used hydraulic conductivity values.

Table 3.10 Conversion Factors for Commonly Used Hydraulic Conductivity Values

Multiply	By	To Obtain
1 gal/day/ft^2	0.1337	ft/day
1 cm/sec	2835	ft/day
1 m/day	3.2808	ft/day
1 gal/day/ft^2	4.72×10^{-5}	cm/sec
1 ft/day	3.53×10^{-4}	cm/sec
1 m/day	1.16×10^{-3}	cm/sec
1 gal/day/ft^2	4.07×10^{-2}	m/day
1 cm/sec	864	m/day
1 ft/day	0.3048	m/day
1 cm/sec	21203	gal/day/ft^2
1 ft/day	7.481	gal/day/ft^2
1 m/day	24.54	gal/day/ft^2

Transmissivity

The hydraulic conductivity can be estimated based upon grain-size relationships or directly in the field by performing slug tests (Chapter 11). Hydraulic conductivities are also estimated from pumping tests (Chapters 9 and 10), by first obtaining a **transmissivity** (T) value. The transmissivity is directly related to the hydraulic conductivity by the following relationship:

$$T = K \times b \qquad [3.12]$$

where:

K = hydraulic conductivity

b = saturated thickness

During a pumping test the estimated saturated thickness contributing to the screened interval varies depending on the time of pumping, geology, and whether an aquifer is confined or unconfined. More discussion on this topic is found in Section 10.5. The common approach is to assume the **saturated thickness (*b*)** in an unconfined aquifer represents the thickness of saturated materials from bedrock to the water table and in a confined aquifer, b is simply the thickness of the aquifer between confining units. One must realize that this assumes there no low hydraulic conductivity barrier layers within the aquifer and/or that the entire thickness is screened. This could only be true in thin aquifers.

3.4 Aquifer Concepts

An **aquifer** is a formation, part of a formation or group of formations that contain *sufficient* saturated permeable material to yield *significant* quantities of water to wells or springs. Water within the zone of saturation is at a pressure greater than atmospheric pressure. The definition is useful because of its flexible application, depending on water use within an area. For example, in a dry semi-arid region, any aquifer that can produce 10 gpm. for stock watering purposes would be considered to be sufficient and significant (Figure 3.10). Contrasted to this, a center-pivot irrigation production must produce a minimum of 500 gpm for a 160-acre piece (quarter-section) of land to be considered sufficient and significant (Figure 3.11). In the latter case, a 10-gpm domestic well may not be considered very significant.

Figure 3.10 Well discharge for stock watering (windmill).

An aquifer can be made up of a single formation, for example, the Permian Quadrant Quartzite in southwestern Montana, which is tightly cemented at outcrop, but is fairly friable and porous at depth. Other aquifers that consist of groups of formations are probably better known as aquifer systems. Well-known examples are the Floridian Aquifer System (Miller 1986) and the High Plains Aquifer System (Gutentag et al. 1984), both of which consist of units of varying lithologies and ages. Some aquifers may only consist of one member within a formation or a single bed. Within the Fort Union Formation, exposed in eastern Montana and Wyoming, is the Tongue River member containing several thick sub-bituminous coal beds.

Figure 3.11 Pivot irrigation system in southwestern Montana.

The Dietz and Monarch coal beds extend regionally and are import sources of water for stock. Aquifer units are separated by confining units of varying physical properties.

Unconfined Aquifers

When water infiltrates into the ground, gravity pulls it downward until it encounters a confining layer. Because flow is inhibited, the zone of saturation literally piles up above this confining layer. The saturated thickness represents the height of the pile, and since the porosity of the overlying materials is connected to the land surface, it is called an **unconfined aquifer**. The top of this saturated material is called the **water table** and is located where the pressure is atmospheric (Figure 3.12). It also represents a physical moving boundary that fluctuates according the amount of recharge or discharge. Groundwater beneath the water table is under a pressure greater than atmospheric, because of the weight of water above a given point of reference. This concept will be explained in more detail in Chapter 5.

Confined and Artesian Aquifer

Geologic materials that outcrop at the surface at a relatively high elevation (recharge area) receive water by infiltration recharge (Figure 3.13). Materials of sufficient transmissivity that become saturated and are overlain and underlain by confining layers become aquifer units with trapped

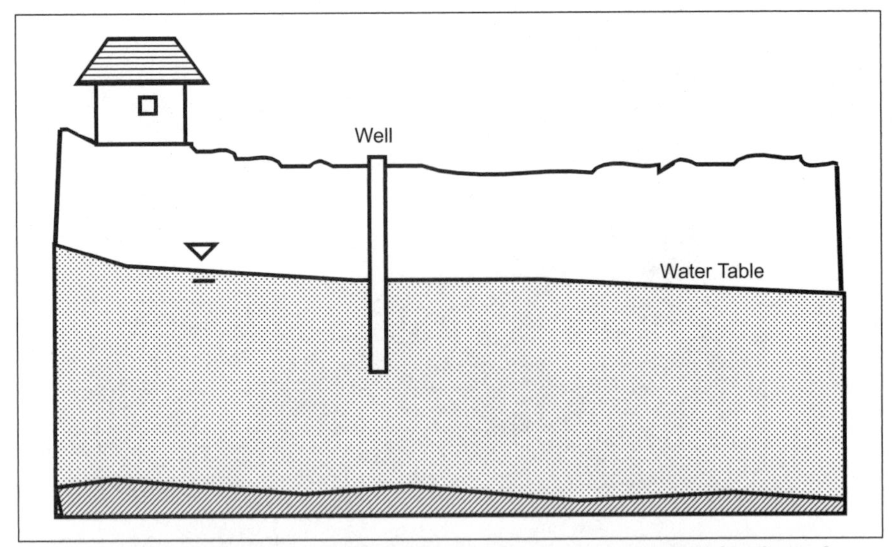

Figure 3.12 Unconfined or water table aquifer with pore spaces connected to the surface.

Figure 3.13 Tilted bedding of a recharge area in the Big Horn Mountains, Wyoming.

groundwater. The elevation of the zone of saturation propagates a pressure throughout the system from the weight of the water. When this occurs, there is sufficient pressure head within the system to lift the water within cased wells above the bottom of the upper confining layer (Figure 3.14). By

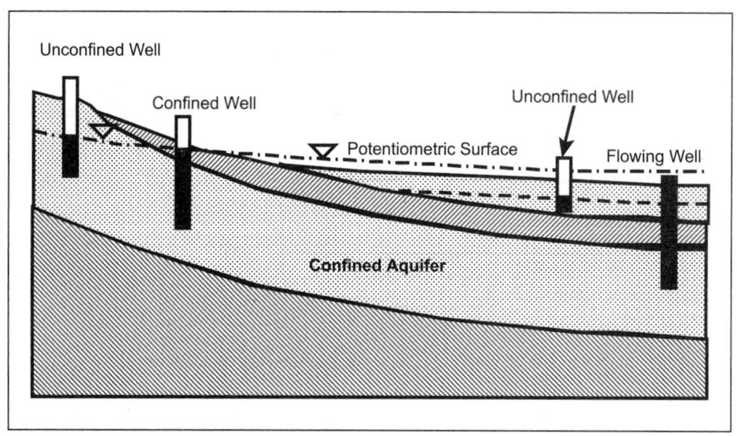

Figure 3.14 Confined aquifer and surficial unconfined aquifer setting. Three wells completed in the lower aquifer, two of which are unconfined. Overall head in the upper confined aquifer is less than the lower confined aquifer.

definition, this is a **confined** or **artesian** aquifer. Both terms refer to the same condition.

Generally, most confined aquifers receive recharge through vertically leaking geologic materials above or below them (Freeze and Cherry 1979). When water levels in cased wells rise to the land surface and become flowing wells, they are still considered to be from a confined aquifer (Figure 3.15). The height to which water levels rise forms a surface known as the **potentiometric surface** (Figure 3.14). In an unconfined aquifer the potentiometric surface is the water table. The slope of this surface is called the **hydraulic gradient**. The slope of the hydraulic gradient is proportional to the hydraulic conductivity. The lower the hydraulic conductivity, the greater the slope of the hydraulic gradient (Chapter 5). Each aquifer has its own potentiometric surface, from which groundwater flow and other interpretations are made (Chapter 5).

Confining Layers

If aquifers yield significant quantities of water, **confining layers**, although they store water, generally slow or inhibit fluid flow through them. Some prefer to use separate terminology for varying types of confining layers. For example, layers that pass water slowly may be referred to as **aquitards**, while those that are considered to be impermeable are called **aquicludes** (Fetter 1994; Freeze and Cherry 1979). Substances may have *very* low hydraulic conductivity, but nothing is really impermeable, including geotextiles, the most compacted clay, and the tightest bedrock. Hydraulic

Figure 3.15 Attempt to measure field parameters from a flowing well in Petroleum County, Montana.

conductivities that are six orders of magnitude less permeable than another substance may *just* appear to be impermeable.

For practical purposes, confining layers can be defined as those geologic materials that are more than two to three orders of magnitude less permeable than the aquifer above or below it. Those less than two orders of magnitude different are hydraulically connected enough to be combined. Using this as a basis, formational stratigraphy is regrouped and reevaluated in terms of hydrostratigraphy.

Hydrostratigraphy

When one considers the definitions of aquifers and confining layers in performing a site characterization or defining the geology of an area, it is also important to define the **hydrostratigraphy**. This involves the combining or separation of units with similar hydraulic conductivities into aquifers or confining layers. It may be, for example, that the Ordovician and Mississippian formations (group of formations) collectively form one large Madison Aquifer, while the Cretaceous Muddy Sandstone (part of a formation) forms another. These may be isolated from aquifers above or below by a part of a formation, a single formation or a group of formations or units. The hydrostratigraphy is used along with structural data to build a **conceptual model** of a flow system.

Example 3.7

On the Hualapai Plateau, Arizona, it was desired to drill wells for water development rather than depending upon temporary stock tanks behind earthen dams (Huntoon 1977). The earthen dams were only successful if flash floods occurred to trap relatively poor-quality water. The target zone was the lowermost Rampart Cave member of the Muav limestone, some 1,500 ft below the canyon rim. The Cambrian section was originally described by Nobel in 1914 and 1922 and then in detail by McKee and Resser in 1945. They honored the nomenclature designated by Nobel (1922) but assigned all shale units as part of the Bright Angel Shale and the limestone units as part of the Muav Limestone. This posed a problem because the depositional environment was that of the transgressive sea where the two formations interfinger and subunits extend for miles. The lowest three members of the Muav Limestone are interbedded with shale units that are unnamed (Figure 3.16). The hydrogeologist hired to define the target zones said that if they drilled past the limestone into upper part of the shale, they would reach maximum production (Huntoon 1977). The problem was that drilling took place through the Peach Springs member of the Muav Limestone and the driller terminated in the uppermost unnamed shale (Figure 3.16). The hydrogeologist failed to inform the driller that there were other limestone and shale units that needed to be penetrated before the target zone would be encountered. In most drill locations, they had stopped short of the target by 150 to 250 ft (Huntoon 1977).

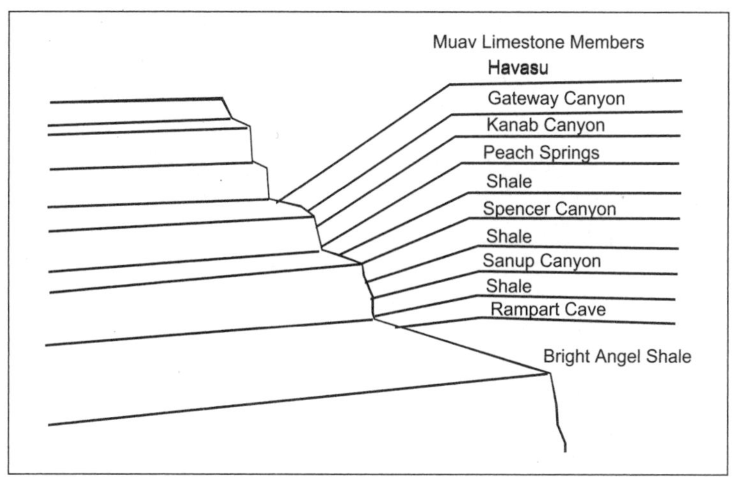

Figure 3.16 Detailed Cambrian section of the lower part of the Muav Limestone on the Hualapai Plateau, Arizona. [Adapted from Huntoon (1977).]

A given conceptual model of the physical behavior of groundwater movement depends somewhat on scale. There are regional groundwater studies and local groundwater studies. The hydrostratigraphy of a small site (e.g., 5 acres, 2.0 ha) would likely be defined by local drill-hole information.

3.5 Boundary Concepts

The hydrostratigraphy, structural changes, and adjacent earth materials all affect groundwater flow and the separation of saturated materials into aquifer units. Each aquifer will have its own potentiometric surface and a hydraulic gradient corresponding to its hydraulic conductivity. If there are multiple aquifers in a particular area, it is important to identify the potentiometric surface of each aquifer separately. Figure 3.17 shows the potentiometric surface of four aquifers: an unconfined bedrock aquifer, unconfined sand and gravel aquifer (dashed line), and two other confined aquifers (solid lines). Note how the potentiometric surface of the intermediate layer (well B) is lower than the others, but the potentiometric surface of the lowest layer (well C) is higher than that of wells A and D. The slope of the potentiometric surfaces are toward or away from the reader, except the bedrock aquifer, which slopes towards the alluvial system. Because groundwater always moves from areas of higher total head to areas of lower total head, the groundwater from the lowest system (well C) wants to move vertically through a confining layer towards the intermediate aquifer. Additionally, the groundwater from the surficial unconfined aquifer wants to leak downward towards the intermediate (well B). Groundwater will always move as long as there is a slope or head difference from one area to another, thus creating a hydraulic gradient. There may be a horizontal component to groundwater flow (within aquifers) and a vertical component (between aquifers and in recharge or discharge areas) of groundwater flow. Groundwater systems continue to move until equilibrium is reached.

In considering the boundary conditions of Figure 3.17, the granitic bedrock and the clay layers separate the three sand and gravel aquifers from each other. Given the low porosity of the granite, long-term yields from well D would be expected to be minimal compared to well C. It is apparent that the lowest layer is receiving recharge up river from a higher elevation than the other two sand and gravel aquifers. What would the water levels in wells be like if the bedrock were siltstone instead of granite? Think of the physical properties of siltstone compared to granite.

Another example shows how structural effects and changes in aquifer properties create aquifer boundaries (Figure 3.18) In the diagram there are two springs. Boulder Springs is a contact spring where a unit of higher hydraulic conductivity intersects the land surface. Why is Basin Springs there? What effect is the Basin Fault having on the system? Notice that the river is losing water. Are the slopes of the hydraulic gradient within the different units appropriate?

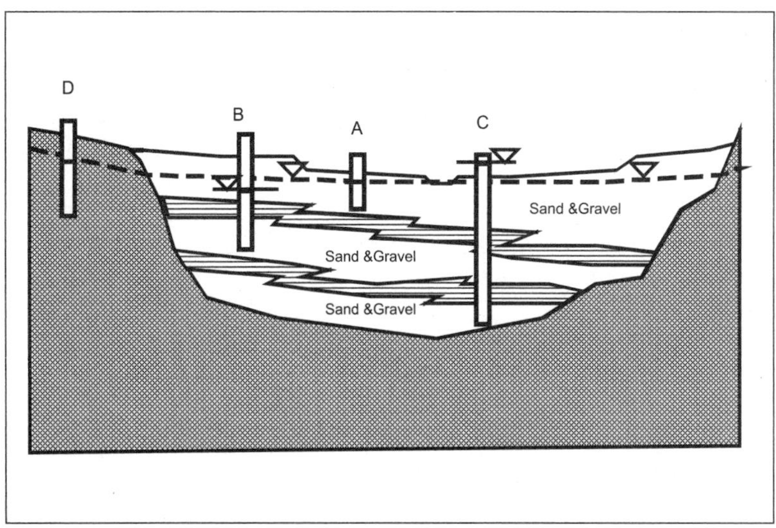

Figure 3.17 Schematic illustrating four aquifer units: well A, an unconfined aquifer connected to the river; well B, an intermediate confined aquifer; and well C, completed in a confined aquifer.

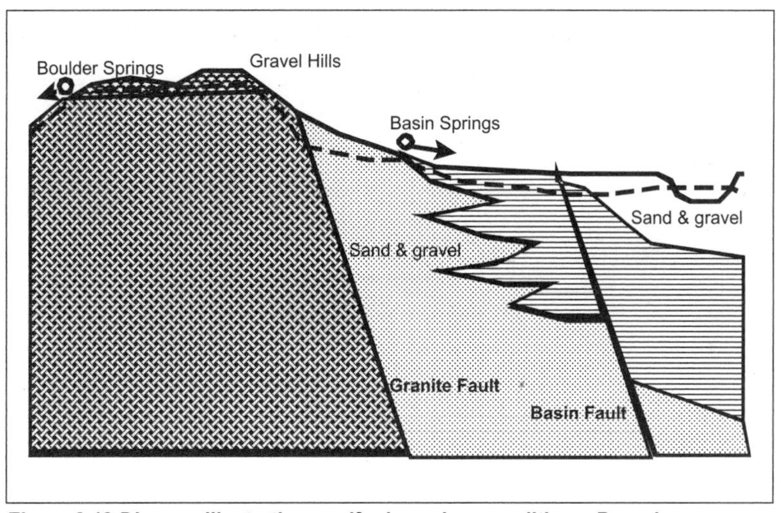

Figure 3.18 Diagram illustrating aquifer boundary conditions. Boundary conditions are caused by varying hydraulic conductivities of rock units and structural offsets.

Homogeneity and Isotropy

There are two terms that are often confused and/or ignored in hydrogeologic studies that are related to the physical and flow properties of an aquifer, **heterogeneity** and **anisotropy**. Heterogeneity refers to the lateral and vertical changes in the physical properties of an aquifer. For example, the grain-size packing, thickness, porosity, cemetation, and hydraulic conductivity at one location (site A) can be compared to the same physical properties at another location (site B) (Figure 3.19). If the variability of the physical properties relative to the volume of the system of interest is small, the aquifer is said to be **homogeneous**. A **heterogeneous** aquifer will have a different saturated thickness, porosity, or some other physical property from location to location. For example, if a sandstone becomes finer grained laterally (as in a facies change) or thins or thickens laterally, it is *not* homogeneous.

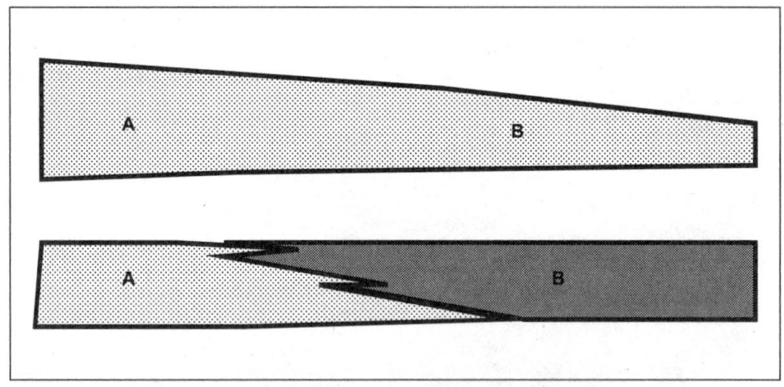

Figure 3.19 Schematic showing heterogeneity. In the upper diagram, the thickness changes from point A to point B. In the lower diagram, the grain-size and packing changes from point A to point B.

Anisotropy refers to differences in hydraulic properties depending on the direction of flow. If a porous media were made up of spheres, the permeability of the media would be the same in all directions. In most depositional environments the sediments are laid down in a stratified manner, so that the vertical hydraulic conductivity is *very* different from the horizontal hydraulic conductivity. This difference can be from percentages to an order of magnitude or more. In considering a Cartesian coordinate system, where *Z* represents the vertical dimension, even in an *X-Y* plane there is a preferential flow in one direction over the other. A summary of these concepts is shown in Figure 3.20. Faults, fractures, bedding planes, and other features are potentially important causes of anisotropy and heterogeneity.

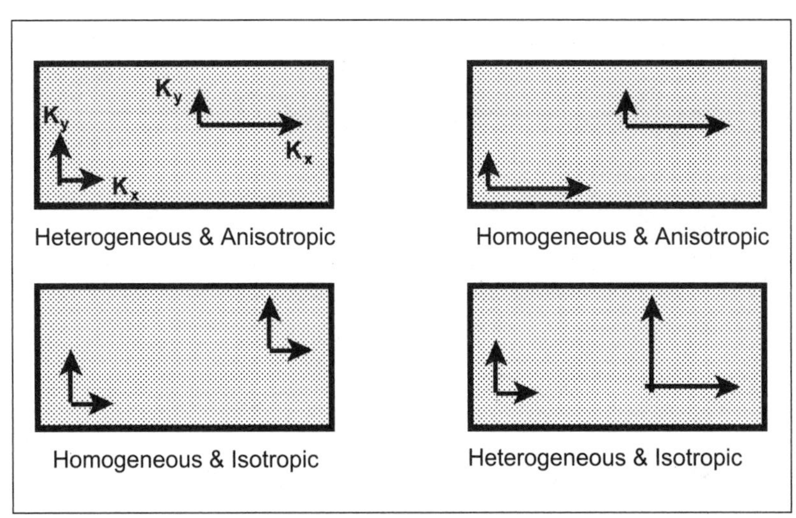

Figure 3.20 Combinations of heterogeneity and isotropy. [Adapted from Freeze and Cherry (1979).]

3.6 Springs

Springs are an important source of hydrogeologic information. They occur because the hydraulic head in the aquifer system intersects the land surface (Figure 3.21). By paying attention to their distribution, flow characteristics, and water qualities, much valuable information can be derived without drilling a single well. Springs can provide more information than wells.

Perhaps the most common springs are contact springs. In an area of layered sedimentary rock, springs will develop if recharge enters a higher hydraulic conductivity unit overlain by a lower hydraulic conductivity unit. In the field, these are observed as places where trees and other vegetation flourishes in an otherwise dry area. Look for game trails that lead into the vegetation, the animals know where water is. Springs from confined aquifers occur either in structurally complex areas where the recharge area is high, relative to the location of the rock outcrops lower in the system. They occur either from folded beds, depositional contacts, or fault planes. They represent places of discharge where minimum hydraulic head occurs locally within an aquifer system. A summary of the physical characteristics springs provide is listed in Table 3.11.

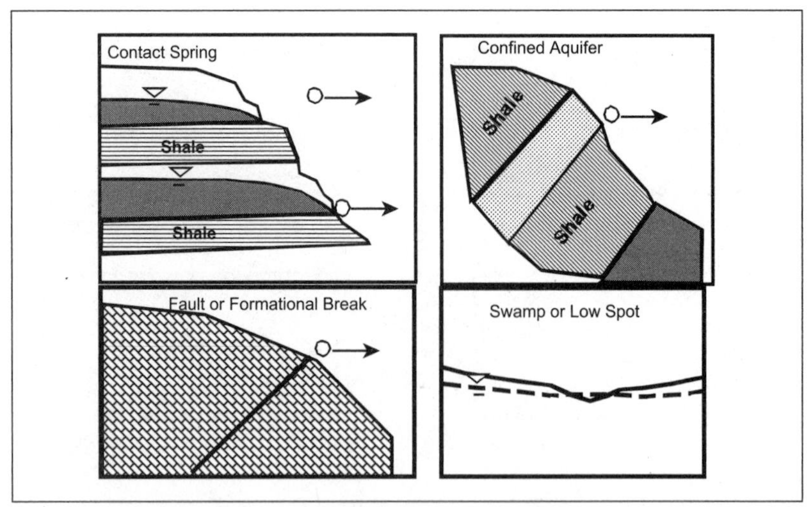

Figure 3.21 Cross-sectional views of a variety of springs: contact spring, confined aquifer is charge, fault, or depositional plane, and swamp or low spot.

Table 3.11 Physical Characteristics of Springs

Characteristic	Comments
Provides elevation of minimum head	As single points or a string of points on a topographic map
Hydrostratigraphy defined	Vegetation at discharge points separated by areas lacking in vegetation
A point of maximum hydraulic conductivity	
Discharge type, fractures, porous discharge	Type of aquifer permeability observed
Discharge rate	Volume and frequency may be steady or vary
Discharge time	Steady flow indicates a larger system than one that stops flowing after a short time
Type of aquifer	Confined or unconfined
Groundwater quality	General chemistry of aquifer
Place to age date	Provides a place of maximum residence time in the aquifer
Thermal data	Circulation depths or thermal influences
Periodicity	Is there a regular discharge interval?
Can be used to understand structural relationships	Discharge areas can be correlated with geologic structures

3.7 Summary

The subsurface geology when saturated is divided into hydrostratigraphic units consisting of aquifers and confining units. The physical properties of aquifers include their ability to store and transmit water while confining layers tend to inhibit flow and serve as barriers to flow. Porosity represents the volume of pore space that can be occupied by groundwater. Porosity can be estimated from cores taken in the field and then tested for their water content. The effective porosity is that part of porosity available for fluid flow. It is often estimated to be the value for specific yield.

Hydraulic conductivity is a measure of the transmissive qualities of saturated earth materials. It can be estimated based upon grain size or by stressing the aquifer through pumping tests. The rate at which water can move through an aquifer is reflected by the slope or hydraulic gradient of the potentiometric surface; high hydraulic conductivity materials have a flatter hydraulic gradient than low hydraulic conductivity units. The hydrostratigraphy is placed within the geologic framework to create a conceptual model of groundwater flow. Fault zones and bedrock may serve as barriers and control the distribution of hydraulic head. In deciphering the structure of an area and determining aquifer characteristics, springs in the area provide direct quantitative information about the aquifer and its chemistry.

References

Anderson, M.P. and Woessner, W.W., 1992. *Applied Groundwater Modeling: Simulation of Flow and Advective Transport.* Academic Press, San Diego, CA, 381 pp.

Baldwin, D.O., 1997. *Aquifer Vulnerability Assessment of the Big Sky Area, Montana.* Master's thesis, Montana Tech of the University of Montana, Butte, MT, 110 pp.

Cleary, R.W., Pinder, G.F., and Ungs, M.J., 1992. *IBM PC Applications in Ground Water Pollution & Hydrology,* Short course notes. National Ground Water Association, San Francisco, CA.

Domenico, P.A., and Schwartz, W.W., 1990. *Physical and Chemical Hydrogeology.* John Wiley & Sons, New York, 824 pp.

Driscoll, F., 1986. *Groundwater and Wells.* Johnson Filtration Systems, St. Paul, MN, 1108 pp.

Fetter, C. W., 1994. *Applied Hydrogeology, 3rd Edition.* Macmillan, New York, 691 pp.

Freeze, A., and Cherry, J., 1979. *Groundwater.* Prentice-Hall, Upper Saddle River, NJ, 604 pp.

Gutentag, E.D., Heimes, F.J., Krothe, N.C., Luckey, R.R, and Weeks, J.B, 1984. *Geohydrology of the High Plains Aquifer in Parts of Colorado, Kansas, Nebraska, New Mexico, Oklahoma, South Dakota, Texas, and Wyoming.* U.S. Geological Survey Professional Paper 1400-B.

Hazen, A., 1911. Discussion: Dams on Sand Foundations. *Transactions, American Society of Civil Engineers,* Vol. 73, 199 pp.

Huntoon, P.W., 1977. Cambrian Stratigraphic Nomenclature and Ground-water Prospecting Failures on the Hualapai Plateau, Arizona. *Ground Water,* Vol. 15, No. 6, pp. 426–433.

McKee, E.D., and Resser, C.E., 1945. *Cambrian History of the Grand Canyon Region.* Carnegie Institute, Washington Publication 563, 168 pp.

Miller, J.A. 1986. *Hydrogeological Framework of the Floridian Aquifer System in Florida and Parts of Georgia, Alabama, and South Carolina.* U.S. Geological Survey Professional Paper 1043-B.

Morris, D.A., and Johnson, A.I., 1967. *Summary of Hydrologic and Physical Properties of Rock and Soil Materials, as Analyzed by the Hydrologic Laboratory of the US Geological Survey 1948–1960.* U.S. Geological Water-Supply Paper 1839-D.

Nobel, L.F., 1914. *The Shinumo Quadrangle, Grand Canyon District, Arizona.* U.S. Geological Survey, Bulletin 549, 100 pp.

Nobel, L.F., 1922. *A Section of the Paleozoic Formations of the Grand Canyon at the Bass Trail.* U. S. Geological Survey Professional Paper 131, pp. 23–73.

Roscoe Moss Company, 1990. *Handbook of Ground Water Development.* John Wiley & Sons, New York, 493 pp.

Shepard, R.G. 1989. Correlations or Permeability and Grain Size. *Ground Water,* Vol. 27, No. 5, pp. 633–638.

Stephens, D.B., 1996. Vadose Zone Hydrology. Lewis Publishers, CRC Press, Boca Raton, FL, 347 pp.

Chapter 4

Basic Geophysics of the Shallow Subsurface

Curtis A. Link

This chapter provides a basic introduction to the principles and applications of geophysical techniques as they might be applied to problems related to hydrogeology. The goal of geophysics is to determine the properties and parameters of subsurface materials (usually rock or sometimes soils) from measurements made at the surface of the earth. Even though the scale of the problem may change, the physics of the basic techniques remains the same.

Many of the techniques and methods presented can be applied at various scales such as those encountered in oil exploration, in understanding shallow lithological parameters, or in determining properties at the mantle or core level.

Since this chapter covers a variety of techniques, it is impossible to explore each method in detail. Our goal is to provide readers with a basic understanding of the various geophysical techniques that are most commonly used such that they can determine which techniques are most applicable to particular problems.

The format of the chapter will be to introduce each technique; discuss data acquisition, processing, and interpretation principles; and provide one or more application examples from the literature.

4.1 Common Targets for Shallow Geophysical Investigation

This chapter is meant to provide basic information on the most common geophysical techniques. However, for those unfamiliar with the methods and techniques of geophysics, it might be useful at this point to list some of the applications suitable for shallow geophysical techniques (list from Greenhouse et al. 1998). Please note that specific methods are site-dependent and the quality of results is dependent on the target, geologic environment, surface features, and other factors that affect the contrast of the target to the surrounding medium.

- Archeology
- Bedrock topography
- Forensics
- Mine waste
- Stratigraphy
- Buried drums, tanks, etc.
- Unexploded ordnance
- Faults, fractures, voids
- Water table
- Landfills and pits
- Contaminants—organic, inorganic

4.2 Approaches to Shallow Subsurface Investigations

Of course, the best way to determine properties of the subsurface is to observe or measure the materials directly. This is generally very difficult and usually impossible. The next best approach is to sample the subsurface materials at selected locations and extrapolate or interpolate the properties between sample locations. This is a typical and effective process whose success is limited mostly by the number of sample locations and the spacing between them. Large sample location spacings lead to increasingly unreliable predictions. Another common problem is that the materials of interest are often hazardous and it is usually desirable and often critical to avoid contamination of a surrounding zone by the hazardous material. An obvious example is the location of a corroded steel drum containing toxic substances. Physically probing for the drum poses the risk of puncturing it and exacerbating the problem. Using geophysical magnetic surveying techniques, the drum can be located allowing a careful excavation of the drum and its remains (Reynolds 1997).

Geophysical techniques provide an alternative to the sample/ interpolate approach. Geophysical methods allow calculation of subsurface propertied without subsurface sampling. An effective approach to characterizing a site is to combine geophysical methods with observations from several sample locations to do a *guided* interpolation between the sample points, guided in the sense that the image of the properties between the sample points produced by the geophysical method does not have the level of detail obtained at the sample points themselves, but that the lower resolution geophysical image is derived from the actual intervening earth materials. This is in contrast to interpolating properties between sample points without any direct information from the intervening materials.

A further advantage of geophysical methods is that large areas can be surveyed quickly at a relatively low cost. Instead of being viewed as an additional cost, a geophysical survey should be considered as adding value by making the entire site investigation more cost-effective (Reynolds 1997).

4.3 Overview of Geophysical Techniques

A geophysical investigation often forms part of a larger study. As such, the geophysicist is responsible not only for the **acquisition** and **processing** of the geophysical data but also the **interpretation** of the final results for presentation to the client. Careful consideration and a close working relationship with the client is required if the geophysicist is to present results in a form consistent with the expected results. For example, in the case of a seismic refraction survey, the client probably is not interested in seismic velocity variations, but in the types of materials present and their properties (Reynolds 1997). The final form of the results should certainly be agreed upon before the investigation begins.

Methods

Geophysical techniques can be broadly classified into two groups. **Passive** methods, which detect variations in the natural earth fields such as gravitational or magnetic, and **active** methods, in which artificial signals are transmitted into the earth and subsequently recorded after passing through and being modified by earth materials (Reynolds 1997).

Limitations

Although geophysical methods are extremely useful in providing information about the subsurface, it is important to be aware of their limitations. A common limitation is the lack of sufficient contrast in earth properties (Burger 1992). This problem manifests itself in not being able to detect or distinguish zones or properties of interest in the subsurface. Sometimes it is possible to use a different method or combine methods to alleviate the problem.

Another difficulty arises due to the nature of the problems and the methodologies employed. Few, if any, of the problems encountered in geophysical investigations have unique solutions. This arises from the fact that most of the geophysical solutions developed result from a process called inverse modeling. **Inverse modeling** is the process of observing the effects (i.e., geophysical measurements) and then solving for the cause (e.g., seismic layer velocities). This is in contrast to **forward modeling**, which would be equivalent, for example, to knowing the seismic layer velocities and then calculating the travel times that would be measured (Burger 1992). This relationship is shown schematically in Figure 4.1. This inherent nonuniqueness makes it necessary to use all available information in the interpretation and sometimes the processing stage. For example, when performing a seismic refraction survey, it is desirable to have shallow well logs and some sense of the shallow geology to control and limit obviously incorrect solutions in the processing and interpretation steps.

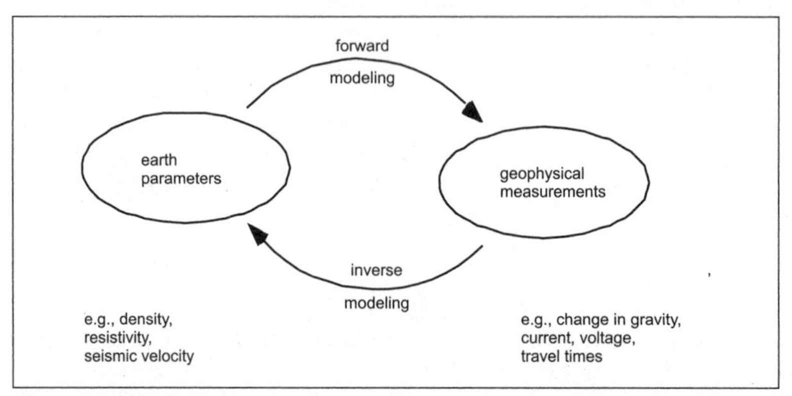

Figure 4.1 Schematic showing the complementary nature of forward and inverse modeling.

Still another limitation is **resolution**. Resolution, as defined by Sheriff (1991), is the ability to separate two features that are very close together or, alternatively, as the smallest change in input that will produce a detectable change in output. The desired resolution of relevant parameters may not be possible for specific geophysical methods or it may be cost prohibitive.

Finally, another limitation is **noise**. Noise can take a variety of forms. Sheriff (1991) defines noise as any unwanted signal or, as a disturbance that does not represent part of a signal from a specific source. Noise can arise in the instruments used, result from uncontrollable conditions, or be related to operator error. Noise can mask the expected contrasts of earth parameters or manifest itself as error or uncertainty in the inversion process.

4.4 Matching Geophysical Methods to Applications

Because of the different properties measured by the different geophysical methods, it is necessary to match the correct method for the desired application. Reynolds (1997) provides a useful listing the various geophysical methods and their main applications (Table 4.1). Further detail for each of the methods, including examples and case studies, will be shown in their respective chapters.

4.5 Geophysical Survey Planning

It cannot be overemphasized that the prime objective of the geophysical survey should be clear from the beginning; especially before planning of the survey begins. Constraints and methodologies of surveys vary as the objectives do, and it is impossible to establish guidelines that will work, for example, for all electrical resistivity surveys. It should also be emphasized that there may not be an ideal method to apply. Sometimes, even with the use of two or more methods, the best results that can be obtained may be only to eliminate possibilities, for example, determining the absence of the water table at a particular depth. In addition, clearly defining the objective before a project begins can simplify the final stages of the project, which often involve interpretation, by ensuring that the deliverables are agreeable to both parties.

There have been only a few published guidelines on geophysical survey planning and execution: British Standards Institute BS 5930 (1981), Hawkins (1986), the Geological Society Engineering Group Working Party Report on Engineering Geophysics (1988) and Reynolds (1997). Also, little attention has been paid to survey design, at least for smaller scale investigations. This is contrast to hydrocarbon exploration by the seismic industry, which devotes considerable resources to survey planning and design and even uses sophisticated survey design software packages to optimize recording parameters and strategies.

Table 4.1 Geophysical Methods and Main Applications [From Reynolds (1997)]

Geophysical Method	Dependent Physical Property	Applications (see key below)									
		1	2	3	4	5	6	7	8	9	10
Gravity	Density	P	P	s	s	s	s	!	!	s	!
Magnetic	Susceptibility	P	P	P	s	!	m	!	P	P	!
Seismic refraction	Elastic moduli; density	P	P	m	P	s	s	!	!	!	!
Seismic reflection	Elastic moduli; density	P	P	m	s	s	m	!	!	!	!
Resistivity (ER)	Resistivity	m	m	P	P	P	P	P	s	P	m
Spontaneous potential (SP)	Potential differences	!	!	P	m	P	m	m	m	!	!
Induced polarization (IP)	Resistivity; capacitance	m	m	P	m	s	m	m	m	m	m
Electromagnetic (EM)	Conductance; inductance	s	P	P	P	P	P	P	P	P	m
EM-VLF	Conductance; inductance	m	m	P	m	s	s	s	m	m	!
Ground penetrating radar (GPR)	Permitivity; conductivity	!	!	m	P	P	P	s	P	P	P
Magneto-telluric (MT)	Resistivity	s	P	P	m	m	!	!	!	!	!

P = primary method

s = secondary method

m = usable, but not necessarily the best approach, or undeveloped for this application

(!) = unsuitable

Applications:

1. Hydrocarbon exploration (coal, gas, oil)
2. Regional geological studies (100s of km^2)
3. Exploration/development of mineral deposits
4. Engineering site investigations
5. Hydrogeological investigations
6. Detection of subsurface cavities
7. Mapping leachate and contaminant plumes
8. Location definition of buried metallic objects
9. Archaeogeophysics
10. Forensic geophysics

It is instructive to include Reynolds's (1997) quote from Darracott and McCann (1986):

> Dissatisfied clients have frequently voiced their disappointment with geophysics as a site investigation method. However, close scrutiny of almost all such cases will show that the geophysical survey produced

poor results for one or a combination of the following reasons: inadequate and/or bad planning of the survey, incorrect choice or specification of technique, and insufficiently experienced personnel conducting the investigation.

It is the goal of this chapter to help increase the "satisfaction level" of individuals or organizations who have chosen to use geophysical methods as a sole or companion approach to a site investigation.

Cost is another factor associated with survey planning and design. Costs span the range from those for a simple one-method survey conducted by a geophysicist with local college student assistants requiring minimum travel to a multiple-method survey done in a remote area requiring a full field crew and various forms of logistic support, which might even include helicopters. Costs also vary with the amount of postacquisition processing or interpretation expected. Again, this can vary from a single product from a simple survey (e.g., a single refraction line interpretation) to multiple versions of numerous products that might require significant amounts of time spent on testing.

It is worthwhile to remember that the geophysics component of a site investigation is usually a relatively small fraction of the total cost and should be viewed with that perspective. Indeed, the judicious use of geophysics can save large amounts of money by allocating resources more effectively (Reynolds 1997). For example, a reconnaissance survey might be able to identify zones for more detailed work without the expense of performing a detailed investigation over the entire study area.

4.6 Seismics

This section on seismics will cover the basic principles and applications of **seismic refraction** and **reflection.**

Basic Principles

The physical basis of seismology can be summarized simply. A disturbance (e.g., a vibration like a sledge hammer blow) is created in a localized area, which then propagates as a vibration through the surrounding materials. The form of this propagating **wave** is spherical in a homogeneous medium. If the medium consists of different materials with different **elastic** properties, the shape of the wave front changes based on the nature of the materials. The propagating wave vibration is then recorded by an instrument called a **seismograph,** which is connected to sensing devices called

geophones. The recording is typically done at the earth's surface where the original **seismic source** disturbance is created.

A geophone consists of a coil and magnet device attached or coupled to the ground, usually by means of a metal spike. As the ground surface vibrates in response to a propagated wave disturbance, the case of the geophone, which usually contains the magnet, moves relative to the coil, which is suspended by top and bottom springs. This relative motion induces a voltage which is proportional to the relative velocity of the coil with respect to the magnet (Sheriff 1991; Evans 1997). The voltage is recorded and analyzed for (1) the voltage signal shape characteristics and (2) the times at which the characteristic parts of the signal occur. The timing information is used with the geometry (locations of the seismic sources and receivers) to determine the earth propagation velocities and structure. This velocity and structure information is then interpreted to represent a geological cross section.

A simple example illustrating the wave propagation process occurs when a pebble is thrown into a still pond. Everyone is familiar with the circular wave patterns spreading out from the point where the pebble entered the water. With a stop watch and measuring tape, one can in fact determine the **speed of propagation** of the surface water waves. Also, with careful observation, one can determine some of the common wave characteristics such as **wavelength**, **period**, and **frequency**.

An illustration of a simple waveform and its characteristics is shown in Figure 4.2.

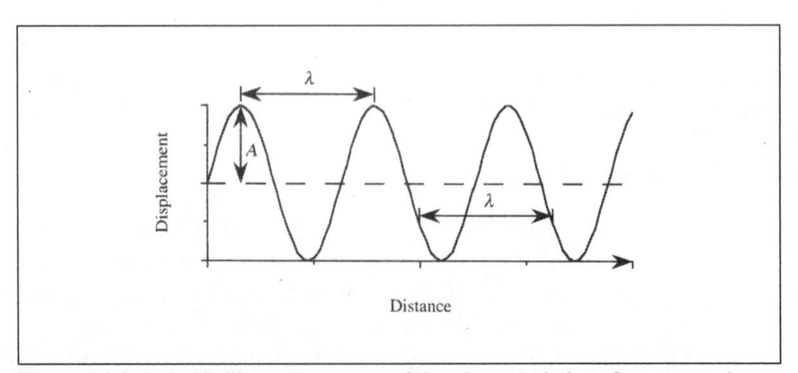

***Figure 4.2** Schematic illustrating some of the characteristics of a wave and common terminology: wavelength (λ), amplitude, crest or peak, and trough. [After Burger (1992).]*

For "real-world" propagating waves, amplitude and wavelength (related to frequency) vary depending on the propagation conditions and media. The relationship among wave propagation velocity, frequency, and wavelength is given by:

$$v = f\lambda \qquad [4.1]$$

where:

v = wave propagation velocity

f = frequency in cycles/second or Hz

λ = *wavelength*

Figure 4.3 shows a plot of 24 waveforms as recorded in a refraction survey. Each waveform is called a **trace**. The traces are numbered sequentially, horizontally across the top of the plot. The vertical dimension of each trace shows the voltage variation with time as recorded from individual geophones. The dark portion of the trace is called the **peak** and indicates positive voltage; the open **trough** is negative. Time increases downward in units of milliseconds (ms). Note the changing waveform for each trace and how the pattern changes from trace to trace. The trace variations are characteristic of the propagation medium. The change from zero voltage to a positive or negative voltage is called the **first break** (e.g., trace 1, where the

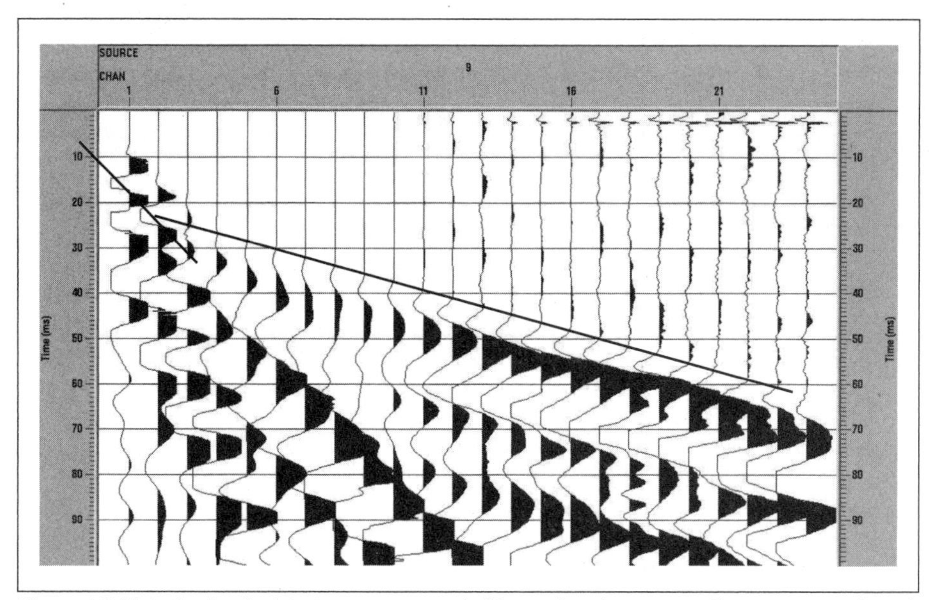

Figure 4.3 Plot of seismic data traces as recorded in a refraction survey. Traces are numbered 1 to 24, from left to right, and show signal variation with time from top to bottom (time in ms). Note the changing waveforms on each trace and the change in slope of first arrival times.

line intersects the trace) and indicates the **first arrival** of the propagated wave energy. The arrow points to the trace where there is a change in slope of the first arrival alignments. This indicates the occurrence of **refraction**. Analysis of the first arrival slopes (sloping lines) gives the velocities and thickness of the earth layers (see section on refraction).

Figure 4.4 depicts seismic wave propagation in a simple medium. Notice that what is labeled as the **ray** is perpendicular to the **wavefronts**. Many of the diagrams used to depict seismic waves show only rays to simplify the drawings. This is especially convenient when illustrating the concepts of reflection and refraction. However, it is important to keep in mind that even though the concept of the ray is useful, seismic waves propagate as spherical wavefronts, and the ray only approximates the wavefronts when the amount of curvature of the wavefronts is very large (i.e., tightly focused wavefronts). This is called the **high-frequency approximation**.

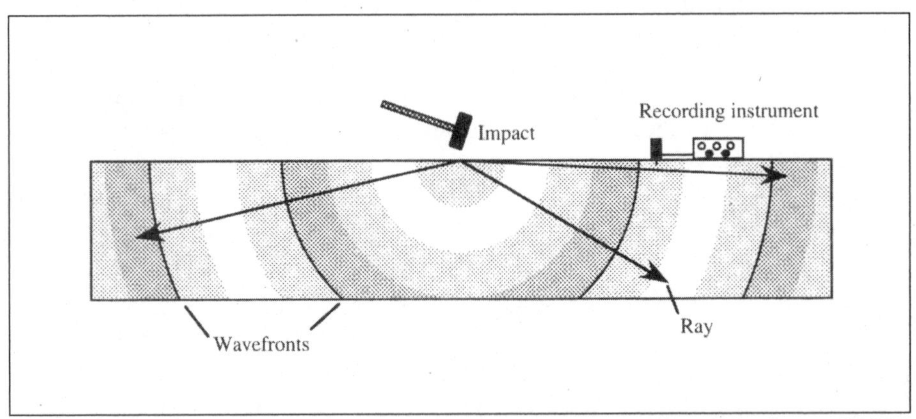

Figure 4.4 Illustration of wavefronts and a single ray in a simple medium. [After Burger (1992).]

Seismic waves can be categorized into two groups: body waves and surface waves. For many purposes, surface waves are considered noise.

Seismic body waves can propagate in two main modes: **compressional** (P or P-wave) and **shear** (S or S-wave) waves.. The modes are distinguished by the direction of particle movement associated with the particular mode. Figure 4.5 is an illustration of the particle motion associated with P-wave propagation. For P-waves, particle motion is in the direction of wave propagation; for S-waves, particle motion is perpendicular to the direction of propagation.

The most common types of geophones are sensitive to ground displacement in the vertical direction (see ground motion components in Figure

4.5). Depending on the angle of emergence of the seismic wave, the vertical component may only represent a small fraction of the energy associated with the propagating wave.

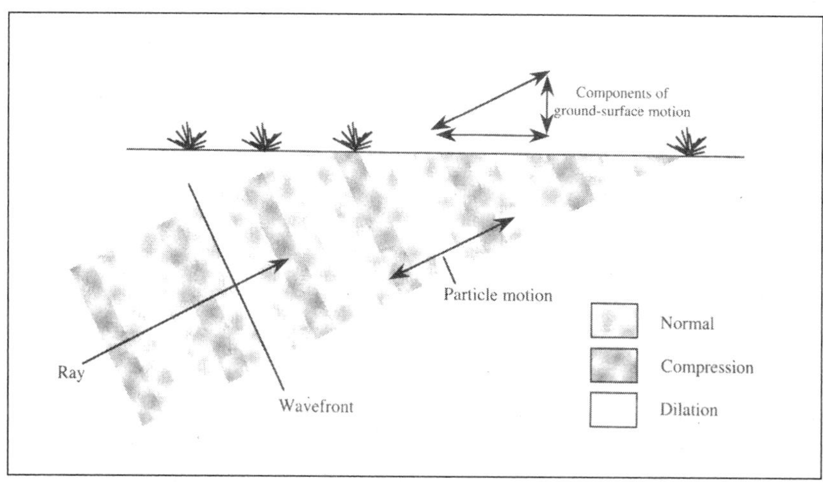

Figure 4.5 Schematic showing particle motion for compressional or P-wave. [After Burger (1992).]

Useful information about earth properties is often contained in S-wave recordings. Shear waves are often more sensitive to conditions such as anisotropy due to layering and fluid content in formations. To record S-waves, geophones that are sensitive to particle motion in the horizontal direction must be used. Horizontal direction is usually specified as **in-line** (in the direction of the line of geophones deployed) or **cross-line** (perpendicular to the line of deployed geophones). For recordings of S-waves, geophones are usually used that can record three components; the vertical as well as the two horizontal components. These are called **3-C** recordings.

Seismic surveys can generally be accomplished in three ways: **refraction** surveys, **reflection** surveys, and **tomography** surveys. Reflection and refraction surveys are described in the following sections.

Reflection principles

Even though refraction methods comprise the highest proportion of shallow seismic surveys, it is useful to look at the process of reflection first. The basic principle controlling reflection is shown in the Figure 4.6 and can be stated as, "the angle of incidence equals the angle of reflection." Figure 4.6 shows an incident seismic ray (recall that a ray is a geometric construct

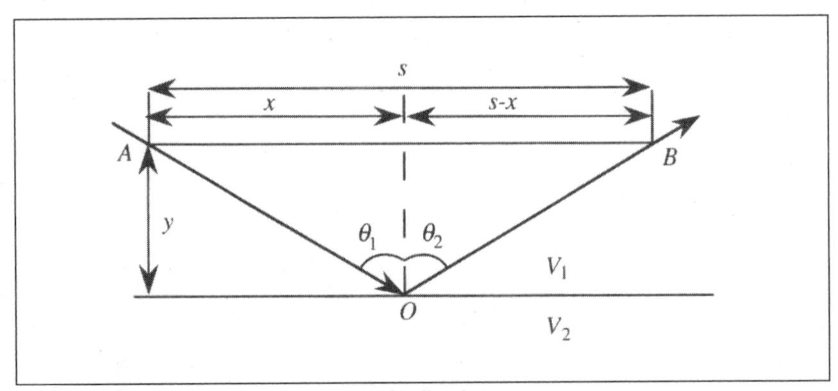

Figure 4.6 Illustration of the principle of reflection.

that is perpendicular to spherical wavefronts) on a plane interface separating two geologic media with different properties.

In this example, the properties that affect the seismic wave propagation are seismic velocity and bulk density. Because density doesn't vary as much as velocity in a normal sedimentary sequence, usually only seismic velocity is specified as the variable. Table 4.2 shows values for density (ρ) and compressional (υ_p), and shear (υ_s) seismic velocities for various rock types.

The convention for referring to the angles associated with propagation is to measure the angles from the normal to interface. Thus, the **Law of Reflection** can be stated as:

$$\theta_{incidence} = \theta_{reflection} \qquad [4.2]$$

It should also be noted that the amplitudes of reflected waves are lower than the amplitudes of the incident wave. This follows from the conservation of energy and looking at the redistribution of the incident wave energy.

Refraction principles

The law of refraction, or **Snell's Law**, describes what happens to waves as they are transmitted through an interface. Snell's Law can be stated as:

$$\frac{\sin(\theta_{incident})}{\upsilon_1} = \frac{\sin(\theta_{refracted})}{\upsilon_2} \qquad [4.3]$$

Table 4.2 Seismic Velocities and Densities for Common Rock Types [After Burger (1992)]

Rock Type	Density ρ (g/cm³)	v_p m/s	v_s m/s
Shale (AZ)	2.67	2124	1470
Siltstone (CO)	2.50	2319	1524
Limestone (PA)	2.71	3633	2319
Limestone (AZ)	2.44	2750	1718
Quartzite (MT)	2.66	4965	3274
Sandstone (WY)	2.28	2488	1702
Slate (MA)	2.67	4336	2860
Schist (MA)	2.70	4680	2921
Schist (CO)	2.70	5290	3239
Gneiss (MA)	2.64	3189	2053
Marble (MD)	2.87	5587	3136
Marble (VT)	2.71	3643	2355
Granite (MA)	2.66	3967	2722
Granite (MA)	2.65	3693	2469
Gabbro (PA)	3.05	5043	3203
Diabase (ME)	2.96	6569	3682
Basalt (OR)	2.74	5124	3070
Andesite (ID)	2.57	4776	2984
Tuff (OR)	1.45	996	659

Values selected from Press (1966).

This is illustrated in Figure 4.7, which shows Snell's Law for a single interface. It is important to note that the phenomenon occurs at every interface, plane or curved, where there is a difference in seismic velocities (and densities). Thus, for a multilayered case, refraction occurs at each interface and the ray can bend away from the interface normals ($v_1 < v_2$) or toward them ($v_1 > v_2$).

An example is shown in Figure 4.8 for seismic velocities ($v_1 < v_2$) and ($v_1 > v_2$)(assuming constant density). Note the direction of bending of the ray with respect to the normal: for ($v_1 < v_2$), the ray bends away from the normal; for ($v_1 > v_2$), the ray bends toward the normal.

A phenomenon that occurs in the process of refraction is called **critical refraction**. Critical refraction occurs when the refraction angle in Snell's Law is equal to 90°. A refraction angle of 90° means that the refracted wave travels along (parallel to) the medium interface. This phenomenon results in what are called **head waves**.

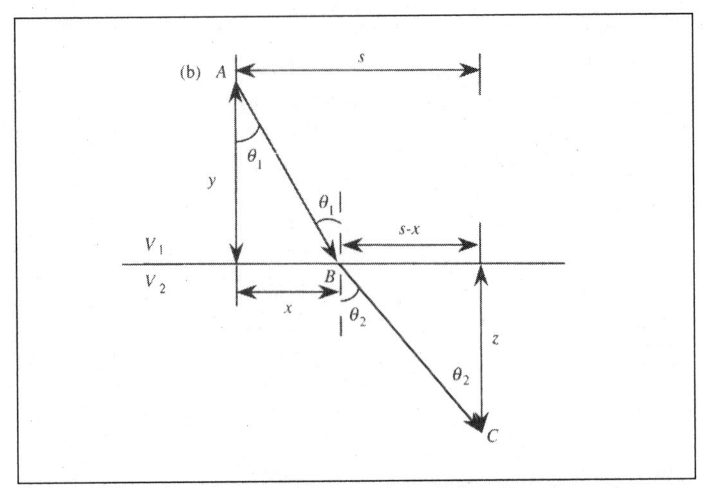

Figure 4.7 Snell's Law (refraction) for a single interface ($v_1 < v_2$).

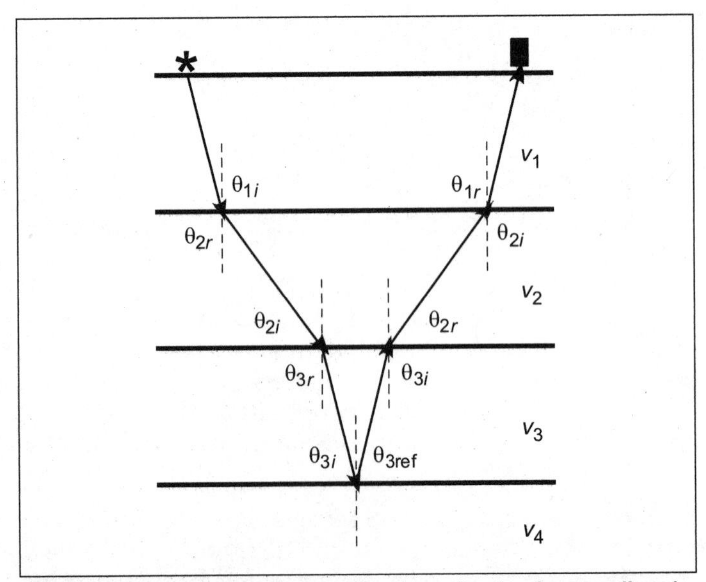

Figure 4.8 Example of Snell's Law showing refraction from medium 1 to medium 2, refraction from medium 2 to medium 3, reflection at the interface between medium 3 and medium 4, and then refraction back to the geophone at the surface.

Example 4.1

Critical angle calculation

For a 2-layered medium with v_1 = 1,000 m/s and v_2 = 1,500 m/s, what is the angle of incidence for a critically refracted ray?

Solution:

$$\frac{\sin\theta_1}{v_1} = \frac{\sin\theta_2}{v_2}$$

$$\sin\theta_1 = \frac{v_1}{v_2}\sin\theta$$

$$\sin\theta_1 = \frac{v_1}{v_2}\sin 90^0 = \frac{v_1}{v_2}$$

$$\theta_1 = \sin^{-1}\left({}^{v_1}\!/_{v_2}\right) = \sin^{-1}\left(1{,}000\,\tfrac{m}{s}\Big/1{,}500\,\tfrac{m}{s}\right)$$

$$\theta_1 = 41.8^\circ$$

Figure 4.9 shows critically refracted rays. Note how the critically refracted wave traveling along the interface between medium 1 and medium 2 gives rise to head waves in medium 1. Analysis of the recorded arrival times is based on calculations for three segments: time for energy to propagate from the seismic source to the interface at the seismic velocity v_1, time for the energy to propagate from the point of contact with the interface to the point of exit from the interface at v_2, and time for the energy to propagate to the surface at the speed v_1.

The simple geological model shown in Figure 4.9 would produce a refraction record similar to the actual one shown in Figure 4.3. The time/distance plot at the top of Figure 4.9 shows the first break times picked from the refraction record plotted against geophone offset (distance). This ***t-x*** plot is then used for analysis. The slopes of the lines drawn on the first breaks along with the **intercept time** solve the simple refraction problem:

- Velocity of layer 1
- Velocity of layer 2
- Thickness of layer 1

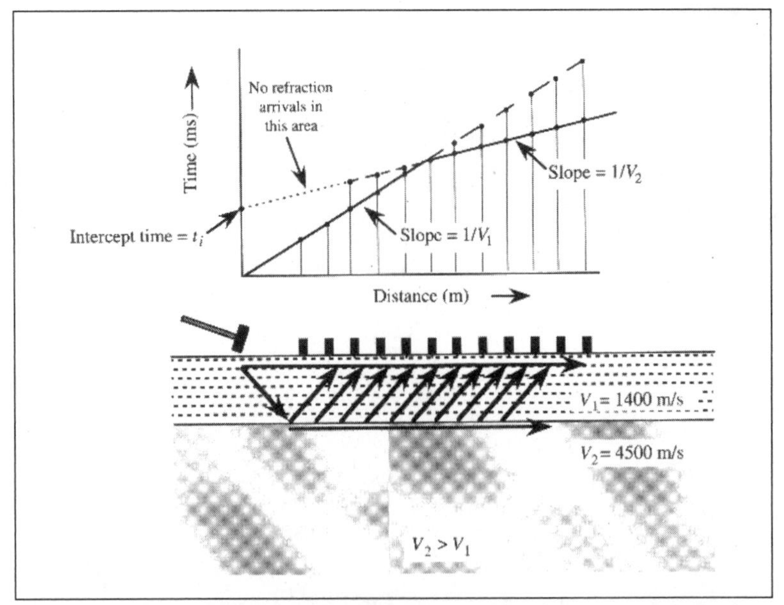

Figure 4.9 Example showing critical refraction and head wave propagation in a simple geological model.[From Burger (1992).]

The solution for this simple refraction problem is given by:

$$h_i = \frac{t_i}{2}\frac{v_2 v_1}{\left(v^2_2 - v^2_1\right)^{1/2}} \quad [4.4]$$

where:

t_i = intercept time picked from the *t-x* plot (Figure 4.9)

v_1 = inverse of the slope of line 1 (Figure 4.9)

v_2 = inverse of the slope of line 2

Refraction analysis increases in complexity as geological models become more realistic, starting from the single, horizontal subsurface just described to multiple horizontal interfaces, single dipping interfaces, and multiple dipping interfaces. In addition, heterogeneities in the subsurface make it difficult to fit straight lines to first break picks. For example, a "pod" of slower material will delay propagating energy such that the time increases for a particular geophone or geophones. Thus, those first breaks will not lie on a straight line.

Thus it can be seen that the process of picking the first breaks is actually an interpretation in itself. At this point, it is worth noting that the subsurface *does not* really look like the simple layers described. In fact, the subsurface is better described by properties that change continuously rather than discretely.

Refraction Surveying

Targets

The seismic refraction method is useful when the target is shallow and low-dipping and consists of materials with different elastic properties. Refraction surveying is often used to detect the water table or a shallow bedrock interface. It can be, and should be, if possible, used with other techniques such as electrical resistivity. This combination works especially well if the target is the water table. Mapping of the water table using this approach can be done relatively cheaply and quickly in areas with limited water well information.

Limitations

A significant limitation of the refraction method is that it requires that material velocities increase with depth. If a medium contains a layer with a lower velocity that the layer above it, the layer with the lower velocity is called a **hidden layer** and will not be detected by the refraction method. The reason this occurs can be easily seen by looking at the example for the critical angle calculation. The angle of incidence is undefined for $v_1 > v_2$ using the critical angle condition of $\theta_2 > 90°$. Thus, for a typical shallow geological scenario of interbedded sands, clays, and gravels, a low velocity layer would remain undetected using the refraction method.

Equipment

Refraction surveying can be accomplished with equipment as minimal as a sledgehammer, geophones and cabling, and a seismograph. It helps to have additional field hands to improve efficiency. In addition, it is necessary to have some means of analyzing the data. This could be done by hand, but it is convenient and more efficient to use software designed for that purpose.

Common seismographs available for purchase or rent typically have 24 channels, which means that 24 individual geophones or groups can be used. Unless you are doing refraction (or reflection) surveys on a regular basis, it is better to rent the equipment or to contract someone who can do the survey for you.

Seismic sources that can be used for shallow surveys can be as simple as the sledgehammer-plate combination (as shown in Figure 4.10 without the plate). For deeper targets or more penetration, explosive sources such as shotgun-type devices (e.g., Betsy Seisgun) or modified rifles are used.

Figure 4.10 Example of portable seismograph and sledgehammer source

Local universities and colleges are often good sources for rental equipment. One example of a company that rents equipment is K. D. Jones Instruments Corporation, Normangee, Texas.

Site selection and planning

For geophysical investigations, as well as others, it is always important to determine as much as possible about the local geology, both surface and subsurface, before undertaking fieldwork (Burger 1992). More effective surveys result from this a priori knowledge, and the information can be critical in processing and interpretation of the data. Geophysical solutions are implicitly nonunique and a priori information is often crucial in eliminating physically unrealizable solutions from the many possible.

Example 4.2

Figure 4.11 is an example of the results obtained from a refraction survey conducted by Lankston (1990) for a potential waste disposal site. With careful data acquisition and analysis, a geologic section can be formed showing excellent vertical resolution as well as horizontal resolution.

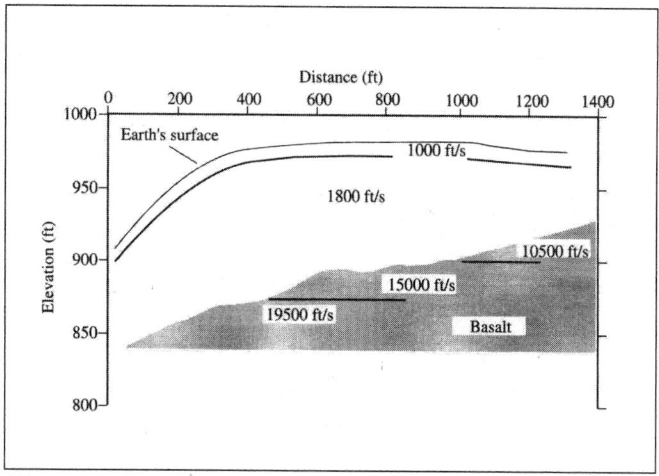

Figure 4.11 Example of a geologic section derived from a refraction study. The particular method used for refraction data analysis is able to determine lateral velocity changes as well as vertical. [From Burger (1992), after Lankston (1990).]

Reflection Surveying

Although refraction surveying is the predominant technique in shallow seismic investigations, the increasing capabilities of inexpensive computers and availability of processing software has made reflection surveying an attractive alternative in many cases.

Reflection surveying is fundamentally different from refraction surveying because it uses reflections from a geologic interface rather than recording the refracted head wave traveling along the interface. One of the major difficulties to deal with in reflection surveying is the lower amplitude of the recorded reflection compared to a refraction. This has led to innovations in processing reflection recordings, which include **stacking** as the most important noise-reducing technique. Stacking is the process of assembling source/receiver combinations from different short records that reflect from the same subsurface point. These combinations arise from different source/receiver locations; thus they have different travel path lengths (Figure 4.12). The reflection arrives at a different time for each pair. A correction is calculated to remove this time difference and the traces are then summed at each time increment. The effect is to enhance coherent events on the trace and suppress random and unwanted noise. This **zero-offset** trace is then plotted with other zero-offset traces at the x-location corresponding to the reflection point. This set of common depth point (CDP) traces is then plotted and called a **seismic section**. On the seismic section, adjacent reflection events line up, allowing an interpreter to analyze the section for geologic interfaces corresponding to the reflections (Figure 4.13).

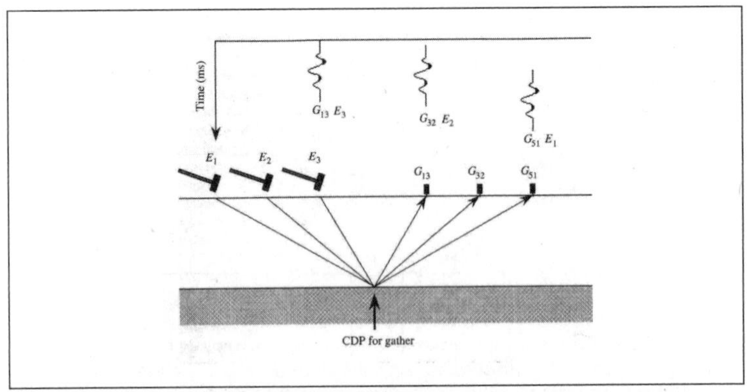

Figure 4.12 Schematic showing three source/receiver pairs that each record a reflection from a common point (CDP). The individual traces exhibit different arrival times which are corrected in subsequent processing. [From Burger (1992).]

Targets

The seismic reflection method is useful for mapping variable subsurface topography and structure and stratigraphy in the overburden (Greenhouse et al. 1998). The main benefit of the reflection method is that it is able to provide more detail and resolution in the interpreted results than the refraction method. However, this increased resolution comes at a price—the presence of favorable conditions that permit recording good reflection shot records. (A shot record is the set of traces resulting from a single source location and multiple geophone locations; see Figure 4.3.) It is usually necessary (or at least recommended) to perform tests to determine if reflections can be recorded in an area of interest. If no evidence of reflections occurs on a shot record, a refraction survey or other approach would be advisable.

Optimum conditions for the reflection method in shallow environmental and geotechnical studies occur when the overburden is fine-grained and saturated, conditions not conducive to ground-penetrating radar (Greenhouse et al. 1998). The reflection method can be particularly valuable in hazard assessment. Detailed information can be obtained from the shallow

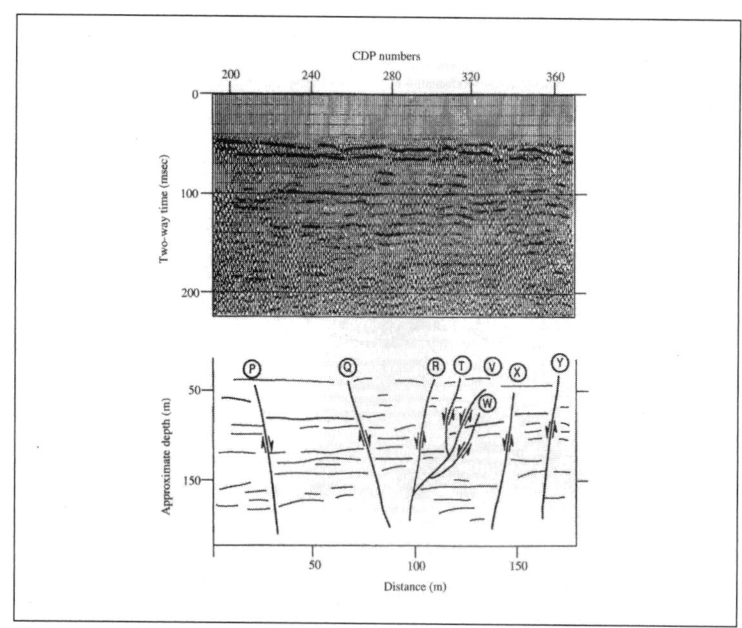

Figure 4.13 Example of a stacked seismic section and geologic interpretation for a shallow seismic reflection survey from southwestern Oklahoma. Stacked reflection traces in top panel; geologic interpretation in bottom panel. [From Burger (1992) and Miller et al. (1990).]

subsurface in zones of active faults or in areas of potential collapse due to sinkholes or mining activity (Burger 1992).

Limitations

Reflection surveying is mainly limited by lithologic conditions, in other words, whether reflections can be recorded or not. In addition, more resources are required to analyze the data records. Reflection surveying typically records more data than the refraction method. This requires increased computer storage capability.

Processing reflection data using the CDP method generally requires software to perform the proper trace sorting and corrections. A simpler approach, the **optimum offset** method (Hunter et al. 1984), can be applied without additional software and is simple to implement and quite efficient (Figure 4.14).

Equipment

Equipment for shallow reflection surveys is similar to that required for refraction surveys. A sledgehammer might be an effective source in some ar-

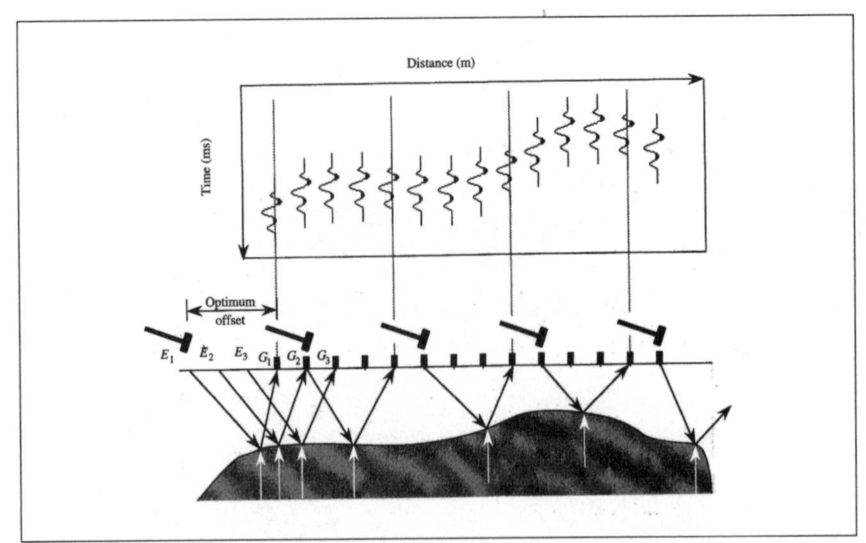

***Figure 4.14** An example of common-offset profiling. Position of light-colored arrows indicates reflection points relative to the position of the recording geophone. [From Burger (1992).]*

eas but often reflection surveying requires more seismic source energy. A shotgun or rifle type source can be used effectively in those cases. Other types of sources include weight dropping, elastic wave generators (using a large elastic band to input seismic energy to the ground), and small-scale vibratory sources.

Most of the processing concerns choosing appropriate parameters. Very sophisticated and complex packages are used by the oil industry. More modest packages useful for geotechnical applications include WINSEIS (Kansas Geological Survey), EAVESDROPPER, PROSEIS, and SPW. Seismic UNIX, or SU, is a free processing software package available from the Colorado School of Mines Center for Wave Phenomena.

Following are two examples showing the application of the seismic reflection method.

Example 4.3

Figure 4.15 shows a seismic section resulting from a survey done by the shallow seismic group at the University of Kansas/Kansas Geological Survey. This group is well known for their innovations and work in the area of shallow seismics. The section shows evidence of dissolution structures in the discontinuity of stratigraphic boundaries.

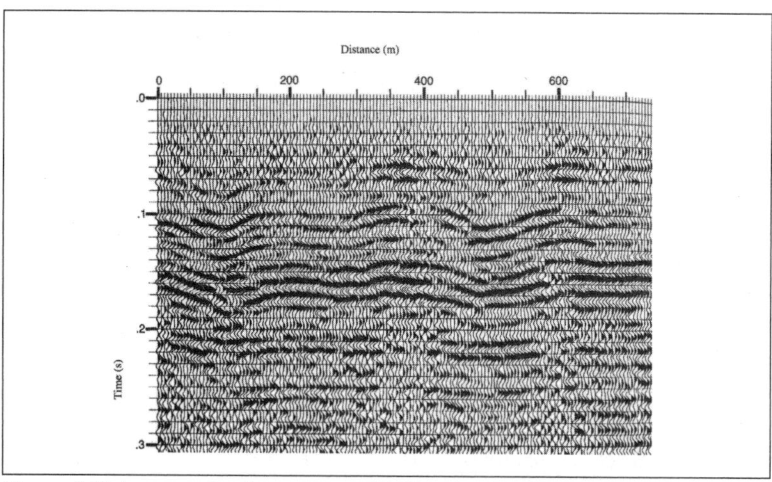

Figure 4.15 An example of a processed CDP seismic reflection record showing stratigraphic changes caused by sinkholes formed by salt dissolution. [From Burger (1992). Figure courtesy of R. D. Miller and D. W. Steeples, Kansas Geological Survey.]

Example 4.4

Figure 4.16 illustrates the detection of faults cutting coal seams using the seismic reflection method. The subvertical lines on the section are interpreted fault locations and do not result from the seismic data processing. Other useful examples and guidelines can be found in Ward Volumes I–III (1990).

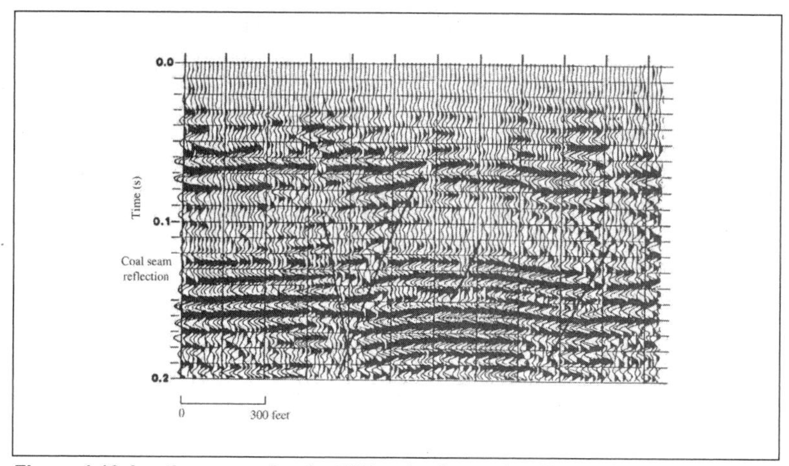

Figure 4.16 Another example of a CDP seismic section illustrating locations of faults cutting coal seams. [From Burger (1992), after Gochioco and Cotton (1989).]

4.7 Electrical

Basic Principles and Units

The methods introduced in this section are based on measurements and interpretations resulting from earth materials with different electric properties. The quantities measured are either **conductivity** (ability of a material to conduct current under an applied electric force, i.e., voltage) or **resistivity** (property of a material that is a measure of resistance to current flow).

Resistivity is the reciprocal of conductivity and its units are ohm·m, or Ω·m. The units of conductivity are mhos/m, or more commonly given as millimhos/m. For example, a commonly used unit for groundwater conductivity is the microSiemen per centimeter. A conductivity of 1 mS/m is equivalent to 10 μS/cm (Greenhouse et al. 1998).

The commonly used term **resistance** depends not only on the material but also its dimensions and is measured in the familiar unit ohm. The fundamental property of the material is its resistivity, usually denoted by the variable ρ. **Conductance** is the inverse of resistance, and its unit is the Siemen, which is the reciprocal of the ohm.

The most fundamental issue in undertaking a geophysical survey is the detectability of the target. As discussed earlier, detection requires sufficient contrast in physical properties and suitable factors of scale, shape, and depth of the target (Greenhouse et al. 1998). The other factor that goes hand-in-hand with detectability is resolution. Detecting a target and resolving it are two different but complementary issues.

The next topic covers the broad category of electrical techniques. The presentation of topics generally follows the trend from methods with higher levels of resolution to methods of lower resolution.

In going from a discussion of seismic techniques to the broad category of electrical methods, two differences become evident. First, the diversity of electrical methods that exists, and second, the relative complexity in deriving quantitative interpretations from field data compared to seismic methods (Burger 1992).

Electrical methods generally fall into two categories: (1) those in which current is applied to the earth and (2) natural energy sources.

Applied current methods

In **electrical resistivity** (ER) methods, direct or low-frequency alternating current is applied at the ground surface and potential difference (voltage) is measured between two points. **Induced polarization** (IP) is a technique that uses the time-dependent decay of the potential difference to obtain information about the polarization properties of the earth materials. A third method, **electromagnetic surveying** (EM), uses a primary electromagnetic field produced by alternating current to induce a secondary field in subsurface conductive bodies. The difference between the primary and secondary fields provides information on subsurface characteristics (Burger 1992).

Natural current methods

Natural current methods take advantage of naturally occurring currents in the earth. **Tellurics** is the method of measuring the potential differences at the earth's surface resulting from variations in the flow of natural (telluric) currents through earth materials. **Magnetotellurics** is similar but measures the magnetic field as well as the electric field. **Spontaneous potential** (SP) is another method that detects potential differences arising from electrochemical effects in subsurface bodies, for example, ore bodies that lie partly above the water table.

Resistivity

Electrical resistivity methods (ER) can provide moderately detailed information about the subsurface in areas with reasonable resistivity contrasts. Burger (1992) provides a list based on his personal experiences for commonly encountered conditions (Table 4.3).

Another useful list is found in Ward (1990), who presents resistivity ranges for common geologic materials (Figure 4.17). Note the variation in resistivity over six orders of magnitudes for these common materials.

The main advantages of ER methods are their low cost and simple equipment requirements. When current is injected into the ground in an ER procedure using a pair of electrodes, the patterns of subsurface current flow reflect the resistivities of the subsurface. These patterns can be inferred by measuring the variations in voltage at the surface using another pair of electrodes (Greenhouse et al. 1998).

Table 4.3. Resistivities of Common Shallow Subsurface Conditions [After Burger (1992)]

Material	Resistivity (Ω·m)
Wet to moist clayey soil and wet clay	1s to 10s
Wet to moist silty soil and silty clay	Low 10s
Wet to moist silty and sandy soils	10s to 100s
Sand and gravel with layers of silt	Low 1000s
Coarse dry sand and gravel deposits	High 1000s
Well-fractured to slightly fractured rock with moist-soil-filled cracks	100s
Slightly fractured rock with dry, soil-filled cracks	Low 1000s
Massively bedded rock	High 1000s

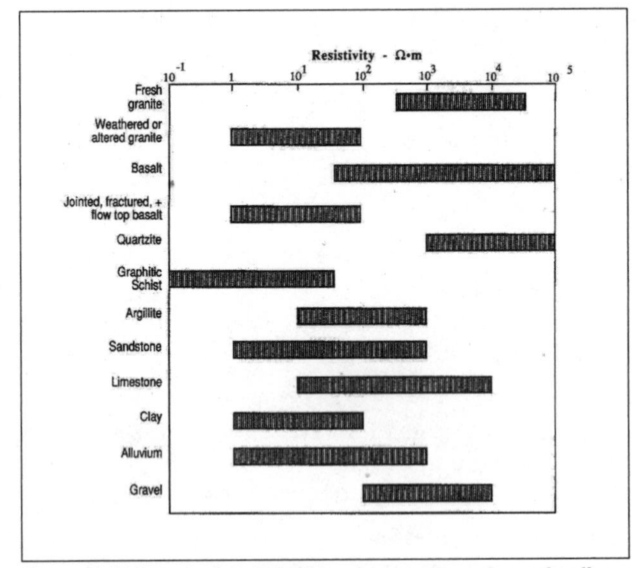

Figure 4.17 Range of resistivities of common rocks and soils. [From Ward (1990).]

Figure 4.18 shows distributions of current lines in a simple model. Using the principle that the measured voltage or potential is proportional to the current density, all else being equal, variations in **current density** (proportional to the spacing of current lines) near the surface will result in variations in **apparent resistivity**. Apparent resistivity is a function of the

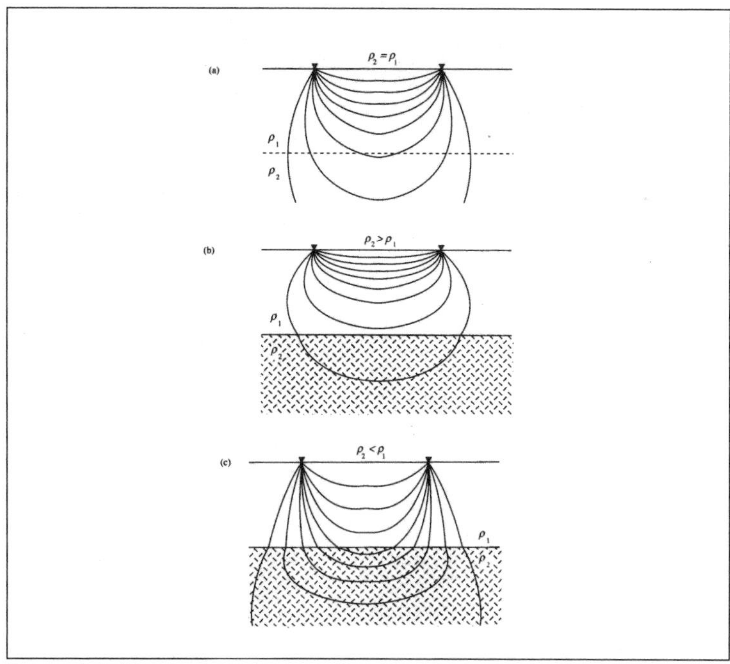

Figure 4.18 Distribution of current lines for a horizontal interface separating two media: (a) homogeneous subsurface; (b) material with greater resistivity beneath the interface—current line spacing increases in the top medium; (c) material with lower resistivity is beneath the interface—current line spacing decreases in the top medium. [From Burger (1992).]

measured potential difference, current injected into the surface, and the spacings of the electrodes.

The spacing of current lines depends on the resistivities of the geologic media. (Burger 1992). Greenhouse et al. (1998) draw a useful analogy between the flow of fluid through a permeable medium and the flow of electricity through a conductive medium. They state that the fluid potential and specific discharge are analogous to voltage and current density, respectively.

Targets and resolution

It is useful to look at an example for a simple two-layer case and a simple electrode arrangement (Figure 4.19). The plot shows modeled apparent resistivity plotted against electrode spacing. Note that electrode spacing is logarithmic. As electrode spacing is increased, more current penetrates to a greater depth; however, the shallow material is also included in the measurement. Therefore, in order for the measured apparent resistivity to approach the resistivity of the second layer, a large volume of the subsurface must be sampled, which is why logarithmic electrode spacing is necessary (Burger 1992).

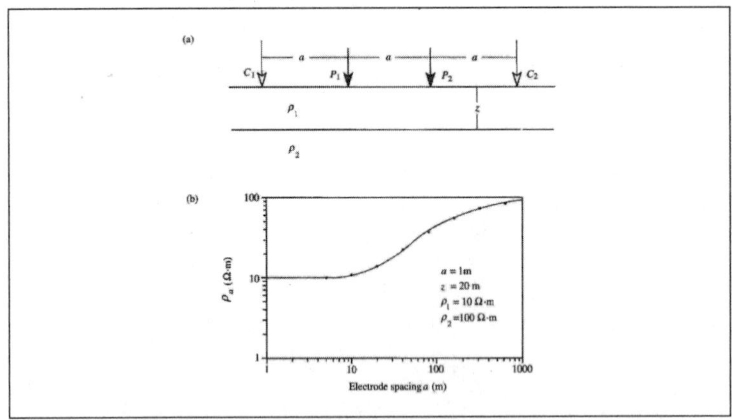

***Figure 4.19** Example of modeled apparent resistivity versus electrode spacing for a simple geologic model. Layer 1 resistivity 10 Ω · μ, layer 2 resistivity 100 Ω · μ, layer 1 thickness 20 m, layer 2 thickness ∞. [From Burger (1992).]*

Also, it should be emphasized that apparent resistivity is a combination of the true resistivities and, as such, requires interpretation to determine the true resistivities of the media.

Field methods

Survey types

ER surveys are usually accomplished in two ways: **profiling** or **sounding**. Many surveys do both simultaneously. In profiling, the lateral distribution of resistivity is studied by maintaining a relatively constant depth of investigation where the depth of investigation is controlled by the electrode spacing. Profiling is used in situations where layer properties vary more horizontally than vertically. Sounding, on the other hand, is used when the zone of investigation varies vertically more than horizontally. In many cases, data are acquired in both profiling and sounding mode (Greenhouse et al. 1998).

Electrode configurations

The electrode patterns used in resistivity surveying are called **arrays**. The commonly used arrays are the Wenner, Schlumberger, and dipole-dipole. Other arrangements are used that are variations of these basic types but these have withstood the test of time. Figure 4.20 is a schematic showing the array arrangements.

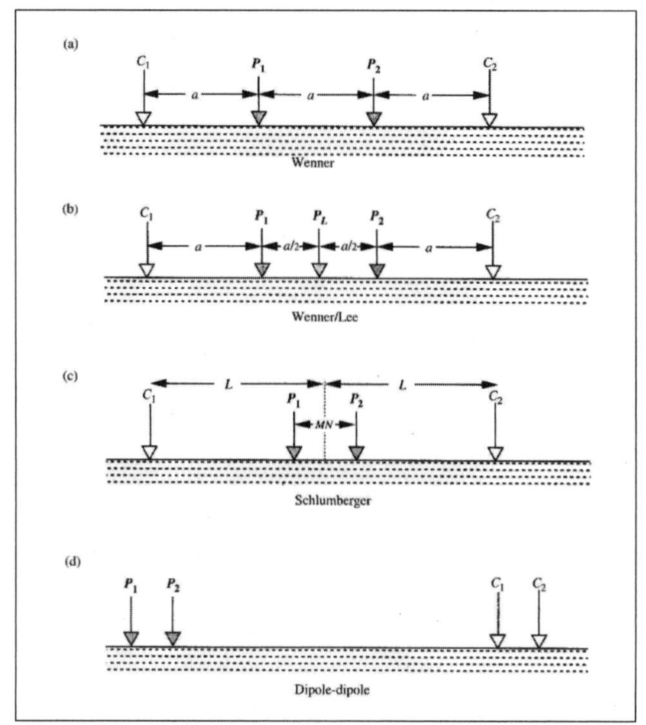

Figure 4.20 Schematic of most common electrode arrays for ER surveying. Electrode spacing is a, current electrodes are C_i, and potential electrodes are P_i. [From Burger (1992).]

Equipment

In its simplest form ER surveying requires relatively inexpensive equipment compared to methods such as seismic. The basic equipment consists of a current source (generator or battery pack), transmitter/receiver (which may be in a single module), wire, and current/potential electrodes (Greenhouse et al. 1998).

In practice, low-frequency alternating current is injected into the ground rather than DC. Manufacturers of equipment include ABEM, Androtex, Iris, Bison, Campus OYO, Soiltest, and Scintrex (Greenhouse et al. 1998). The new generation ER equipment records data digitally and can take advantage of programmable electrodes.

Penetration

In principle, there is no limit to the penetration of the ER method as depth of penetration is controlled by the electrode spacing. Targets can be possibly hundreds of meters below the surface. In practice, penetration is controlled by the types of materials and the strength of the recorded signal compared to background variations or noise. As a rule of thumb, it is difficult to resolve a layer that is thinner than the depth to its upper surface (Greenhouse et al. 1998).

Surveying strategies

As is the case for all geophysical surveys, resistivity survey goals should be specifically defined before fieldwork begins. The appropriate equipment must be selected and geologic control should be available if at all possible. Geologic control becomes critical in the interpretation phase when multiple geologic models can be found to fit the data.

The next step is to decide the type of survey: constant-spread (profiling), expanding-spread (sounding), or a combination. If the goal of the survey is to determine apparent resistivity (ρ_a) information with depth, then sounding using an expanding-spread traverse is the clear choice. An example of this would be to determine the depth to bedrock with an overburden of saturated, silty sands. Another traverse should be run perpendicular to the first to determine lateral variations in both directions (Burger 1992).

If the survey goal is to map lateral variations in resistivity, than a constant-spread survey or profiling is the correct choice. If multiple traverses are recorded and several electrode spacings can be used at each location, then a contour map of ρ_a can be produced for each electrode spacing. This approach would be useful for detection of a gravel bar, for example (Burger 1992).

Concerning electrode spacing, an often used guideline is that depth of penetration is equal to electrode spacing. This is not correct as the amount of current penetrating the subsurface depends not only on electrode spacing but also on the resistivities of the media. If surveying is being conducted in an area for the first time, it is usually advisable to plot ρ_a readings as surveying progresses to determine if the electrode spacings are sufficient to "see" the resistivity of the deepest zone of interest.

Interpretation of results

Curve matching

Curve matching is a relatively straightforward approach to estimating the true resistivities of subsurface materials. This approach works best for the simple case of a single horizontal interface and expanding-spread traverse. The technique is a way to match field measurements of ρ_a and scaled electrode spacing to theoretical curves that have been calculated for various layer thicknesses and resistivities. Figure 4.21 is a plot showing an example of curve matching for a two-layer model with resistivities of 100 and 43 and a top layer thickness of 10 m.

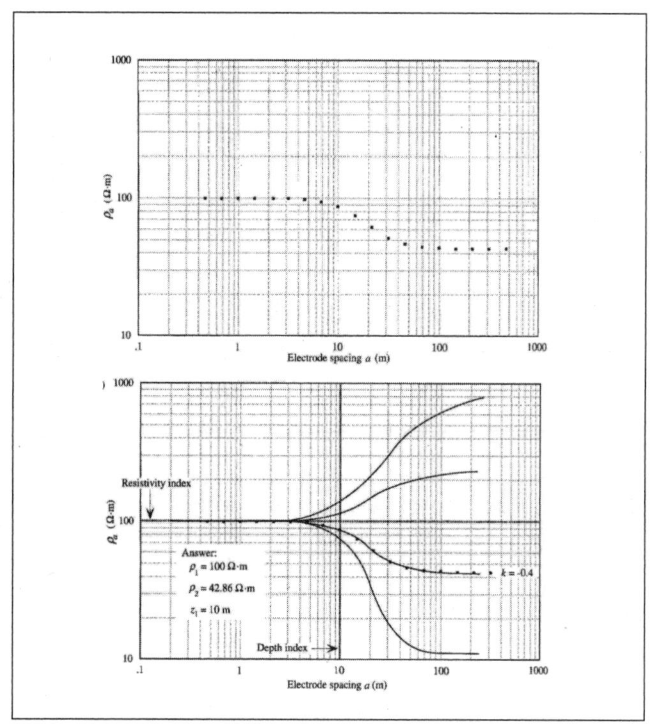

Figure 4.21 Curve-matching example. Top plot shows field measurements. Bottom plot shows field measurements overlaid on a set of theoretical curves. The closest matching curve indicates the best-fit model parameters. [From Burger (1992).]

Analytical methods

With the proliferation of fast, cheap personal computers, analytical methods have largely replaced curve-matching methods. This is typically an inversion approach that requires an initial guess for a starting model and then refines that model by iteratively reducing the error between forward calculations from the current state of the model and the field data. The iterative process is stopped when the error meets a specified criterion.

This is a fast, effective approach that produces a solution in a short time. However, as with any inversion process, the resulting final model is one of many possible models. The result must be compared to existing geologic information to determine its feasibility.

Pseudosections

Sounding data are usually plotted as apparent resistivity (ρ_a) versus electrode spacing, while profiling ER data are plotted as apparent resistivity versus profile distance. It is also possible to plot ER data in a format called **pseudosection**. This format plots ρ_a recorded with a variety of electrode spacings with the spacing as a measure of the depth to the apparent resistivity. The result resembles a geologic cross section, but it must be kept in mind that the depth is an estimate only.

Figure 4.22 is an example of (a) observed pseudosection, (b) computed pseudosection from a finite-element algorithm, and (c) model of the subsurface used to produce the computed results in (b).

Example 4.5

Figures 4.23, 4.24, and 4.25 are examples taken from Burger (1992) illustrating the use of the ER method for two contamination problems and a karst investigation.

Figure 4.23 shows an apparent resistivity contour map (a) generated from constant-spread traverses. The goal of the survey was to map a buried stream channel. The section in (b) shows the interpreted geology from the resistivity survey and borehole information.

Figure 4.24 is a schematic example of how ER can be use to study the extent of leachate plumes leaking from landfills. It is based on patterns observed in studies of several landfills located at various sites throughout the United States. The resistivity contours are determined from variations in apparent resistivities compared to normal values associated with the landfill site (Burger 1992).

Figure 4.25 shows apparent resistivity plotted against offset from an ER survey performed in a karst region in Illinois. The spread used a Wenner array with an electrode spacing of 30.48 m. The bottom plot shows the geologic interpretation. Apparent resis-

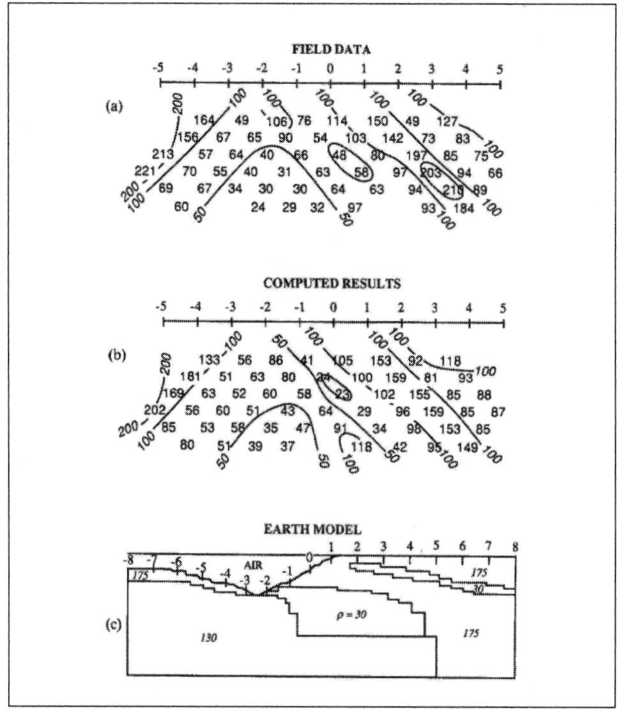

Figure 4.22 Example of (a) observed pseudosection, (b) computed pseudosection from a finite-element algorithm, and (c) model of the subsurface used to produce the computed results in (b). [From Ward (1990), after Hohmann (1982).]

tivity highs correspond to limestone and voids; apparent resistivity lows correlate with clays.

Induced Polarization (IP)

The description of the next four methods will be necessarily brief following the format of Burger (1992). The descriptions will mainly provide a sense of applications, limitations, and advantages and how these methods are similar to or differ from electric resistivity methods. Detailed descriptions of these methods can be found in many books. A particularly useful reference is Ward (1990). A useful tutorial and more information can also be found in the CD-ROM by Greenhouse et al. (1998).

The induced polarization or IP method takes advantage of the fact that when the current applied to a medium is turned off the potential difference does not immediately return to zero. This voltage decay indicates that the

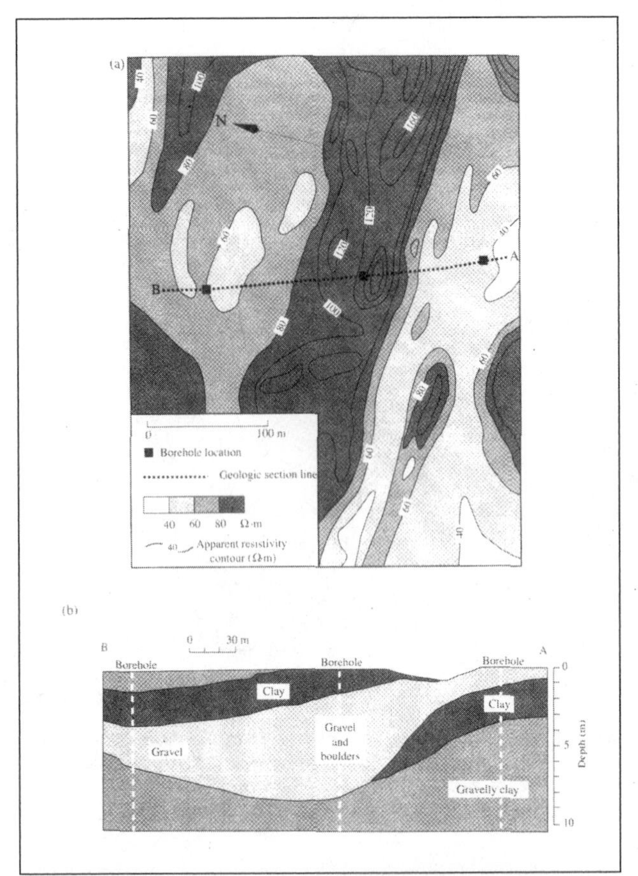

Figure 4.23 Contour map of apparent resistivity values from constant-spread traverses and locations of boreholes (a). Interpreted geologic cross section through the boreholes(b). [From Burger (1992), after Zohdy et al. (1974).]

material has become polarized, analogous to a rechargeable battery. This decay of voltage with time is referred to as an *induced potential in the time domain* (Burger 1992). There is also an approach that uses different frequencies of alternating current. This is called IP in the frequency domain.

A useful measure of the effect of IP in the time domain is chargeability. Chargeability is defined as the ratio of the area under the voltage curve to the potential difference measured before the current was turned off. A theoretical relationship between ρ_a and chargeability allows calculation of relationships between subsurface geometries and apparent chargeabilities (Burger 1992).

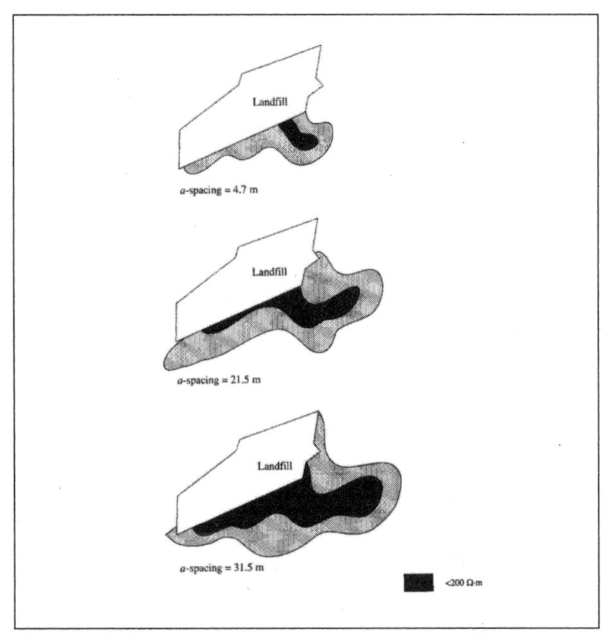

***Figure 4.24** Schematic showing a spreading leachate plume as determined from ER surveys. Each plot is generated at a greater depth. [From Burger (1992).]*

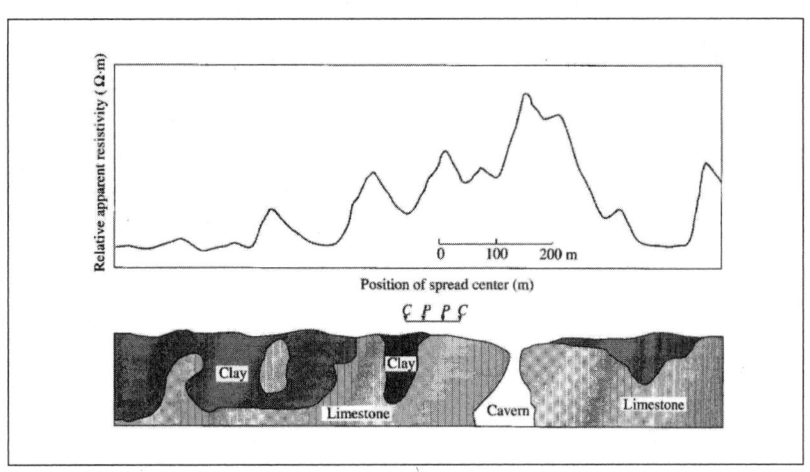

***Figure 4.25** Results from an ER survey in a karst region in Illinois. Apparent resistivity is plotted against offset for constant-spread survey (top) with electrode spacing of 30.48 m (electrode arrangement denoted as CPPC). Geologic interpretation of apparent resistivity where high ρ_a corresponds to limestone and voids and low ρ_a to clays. [From Burger (1992), after Hubbert (1944).]*

Electrode arrays are typically the Schlumberger or dipole-dipole. Apparent chargeability is plotted against electrode spacing, similar to apparent resistivity plotted against electrode spacing. Curve matching with calculated theoretical curves is then used to determine the subsurface model (Burger 1992).

IP applications

IP has probably found more use in the search for base metals rather than the usual ER applications such as groundwater studies. IP tends not to be as cost-effective as ER methods because of the more complex interpretation methods. So it is generally used where the potential payoff is higher.

The equipment for IP is similar to ER with two exceptions: A timing circuit is required, and a recording unit is necessary to record the decay of potential with time. Thus, standard equipment used for ER cannot be used for IP, but IP can be used for measuring apparent resistivities (Burger 1992).

IP could be useful in a situation where ER methods cannot distinguish several layers because of similar resistivities but which might have different chargeabilities (see Ward 1990, Vol. 1).

Spontaneous Potential (SP)

Natural potential differences measurable at the surface may arise from electrochemical reactions at depth that cause currents to flow. For example, in an ore body that lies partly above the water table, oxidation takes place above the water table and reduction below it. These processes are the cause of different charge concentrations that cause a flow of current. Sulfide bodies are probably the most common example of this condition. For this reason, most SP surveys are used for locating ore bodies, especially sulfide ores (Burger 1992).

SP applications

Equipment for SP surveys is minimal. Two electrodes, wire, and a millivoltmeter constitute the basics. Nonmetal electrodes must be used as metal electrodes produce their own self-potential. Porous-pot electrodes are used, consisting of a ceramic container with holes filled with a solution of copper sulfate and a copper tube.

SP surveys are quick and efficient, and the required equipment is minimal. However, quantitative interpretation is difficult and depths of sources are difficult to model (Burger 1992). In addition, SP effects can be masked by other sources such as telluric currents. SP exploration, therefore, is not a common shallow exploration method except in the obvious case of a shallow sulfide body. Corwin (1990) gives a detailed discussion of SP methods (Burger 1992).

Telluric and Magnetotelluric Methods

At any moment, naturally occurring electric currents are flowing in the earth. These currents are induced by the flow of charged particles in the ionosphere. These subsurface telluric currents tend to flow in uniform patterns over large areas. However, variations in resistivity and geometry cause fluctuations in the currents. Thus, measuring the potential differences associated with telluric current flow should give information about the subsurface variations in resistivity (Burger 1992).

Telluric currents vary with time. Thus, the field procedure is to set up a base station at which two pairs of electrodes are placed at right angles. Voltage is then continuously monitored at the base station. Two additional pairs of electrodes are then moved along a grid. Comparison between readings taken at the base station and readings taken at the grid locations yields a potential difference that is caused by the subsurface variation effect on current density (Burger 1992).

Telluric methods can also include measurement of magnetic field intensity variations with voltage. This process is called magnetotellurics. Both sets of data are then used to compute apparent resistivity functions, which are then compared with theoretical curves to determine thicknesses and true resistivities. Magnetotellurics requires a magnetometer capable of measuring weak magnetic field variations that fluctuate frequently (Burger 1992).

The main disadvantage of the natural source magnetotelluric methods is the erratic signal strength. A relatively recent development introduced the principle of an artificial signal source that was dependable and strong enough to speed up data acquisition and improve the reliability of results. The controlled-source magnetotelluric method typically operates in the frequency band 0.1 Hz to 10 kHz, while the natural source magnetotelluric (MT) method uses the range 10^{-3} to 10 Hz. Audio-frequency magnetotellurics (AMT) operates in a still higher range from 10 to 10^4 Hz (Reynolds 1997).

Telluric and magnetotelluric applications

The telluric and magnetotelluric method has been more popular in Europe and the Soviet Union. The natural-source methods are not capable of providing a great level of detail and are thus used in more regional type surveys to determine, for example, basin geometries and other features of that scale (Burger 1992).

Audio-frequency magnetotellurics have been used in groundwater/geothermal resource investigations and in the exploration for major base metal deposits over the depth range from 50 to 100 m to several kilometers. The main application of the MT method, however, has been in hydrocarbon exploration; especially in terrains where the seismic method is difficult to use or ineffective (Reynolds 1997).

Since the 1970s, controlled-source AMT (CSAMT) has been used in an increasing range of applications and, since the 1980s, especially within the geotechnical and environmental community. Reynolds (1997) considers CSAMT to be an underutilized method with many potential applications.

Figure 4.26 shows some plots generated from CSAMT data. It is worth noting that processing CSAMT data is a relatively complicated process accomplished in two phases: preprocessing and interpretative processing (Reynolds 1997).

Electromagnetic (EM) Techniques

Among all the geophysical methods, the EM techniques have the broadest range of different instrumental systems in use corresponding to the broad range of applications of these methods. Recent developments in portability and ease of use have resulted in the increasing popularity of EM methods. Modern EM systems provide a powerful suite of sophisticated instruments. Coupled with major advances in computer interpretation techniques, EM techniques have become increasingly available to engineering and environmental applications (Reynolds 1997).

In its simplest form, EM surveying involves passing alternating current through a wire loop producing a local magnetic field called the primary field. This in turn produces current flow in the earth, which produces a secondary magnetic field that varies in amplitude and phase from the primary. This secondary field is measured by a receiver coil in which current flow is induced by the secondary field. Knowing the primary field, it is possible to isolate the secondary field, which contains information about the subsurface in which it was generated (Burger 1992).

Although the secondary field contains information about the distribution of conducting (resistive) material in the subsurface, quantitative interpretation is fairly complex. However, signatures of good conductors are sufficiently clear that the method is well suited to reconnaissance surveys (Burger 1992).

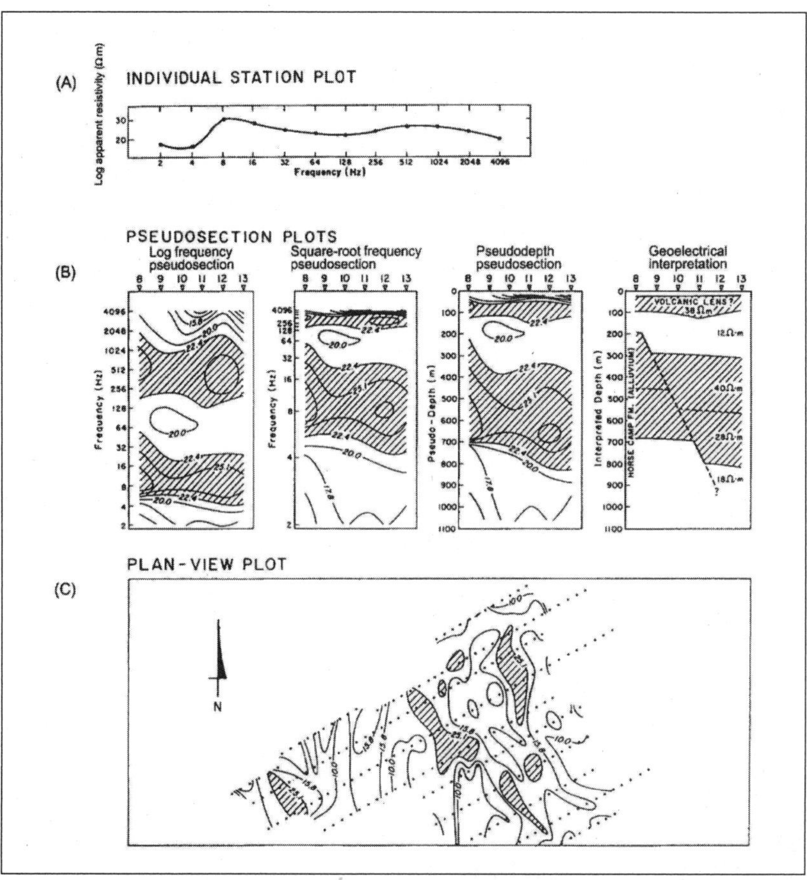

Figure 4.26 Display types for CSAMT data. (A) Individual station plot of apparent resistivity (log) versus frequency. (B) Various pseudosection displays with the corresponding geoelectric/geologic interpretation. (C) Plan view plot of parameter of interest. [From Reynolds (1997), after Zonge and Hughes (1991).]

No electrodes are involved, so fieldwork can be fairly rapid and airborne data collection is common. The EM method is relatively inexpensive, but depth penetration is poor and good conductors are necessary (Burger 1992).

EM applications

The range of applications is large (Reynolds 1997), although Burger (1992) suggests that most of the applications of EM surveying have been for ore exploration. However, Burger (1992) suggests McNeill (1990) for a good, modern review of groundwater applications of EM.

The applications of EM methods depend on the type of equipment being used but can be broadly categorized as in Table 4.4 (Reynolds 1997). Not all EM methods are equally appropriate to the applications listed. For example, ground-penetrating radar (GPR) (see next paragraph) has limited use in the direct investigation of landfills because of the high conductivity levels that are usually present, which attenuate the propagating radio waves rapidly with depth.

Table 4.4 Range of Applications for EM Surveying—Independent of Instrument Type [After Reynolds (1997)]

Applications
Mineral exploration
Mineral resource evaluation
Groundwater surveys
Mapping contaminant plumes
Geothermal resource investigations
Contaminated land mapping
Landfill surveys
Detection of natural and artificial cavities
Location of geological faults, etc.
Geological mapping
Permafrost mapping, etc.

One of the main advantages of EM methods is that the process of induction does not require direct contact with the ground as do the methods that use electrodes. This allows increased efficiency in field operations and permits the method to be used in aircraft and vessels with obvious advantages (Reynolds 1997).

GPR can be considered an EM method but is usually considered separately. Reynolds (1997) suggests Nabighian (1987, 1991) for a comprehensive and detailed discussion of the various EM methods with the exception of GPR (see Ward 1990, Vols. I, II, and III).

Goldstein et al. (1990) present a case study where ground conductivity measurements were used to delineate contaminated water emanating from a series of water storage lagoons in California. The final interpretation of the data is shown in Figure 4.27.

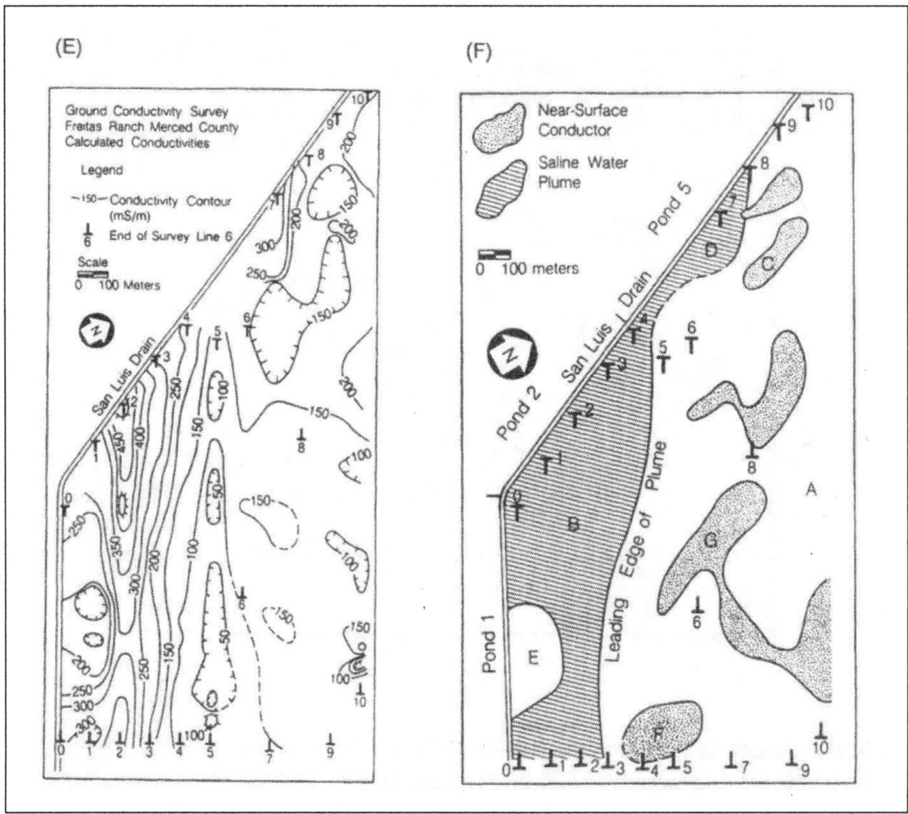

Figure 4.27 Final processed data is conductivity map (E) and final interpretation map from study site at Kesterton, California. [From Reynolds (1997), after Goldstein et al. (1990).]

References

British Standards Institute, 1981. *Code of Practice for Site Investigations, BS 5930*. London, BSI.

Burger, H. R., 1992. *Exploration Geophysics of the Shallow Subsurface*. Prentice-Hall, Upper Saddle River, NJ.

Corwin, R. F., 1990. The Self-Potential Method for Environmental and Engineering Applications, in Ward, S. H., (ed.), *Geotechnical and Environmental Geophysics, Vol. I, Review and Tutorial*. Society of Exploration Geophysicists, Tulsa, OK, pp. 127–145.

Darracott, B. W., and McCann, D. M., 1986. Planning Engineering Geophysical Surveys, in Hawkins, A. B., (ed.), *Site Investigation Practice:*

Assessing BS 5930. Geological Society Engineering Geology Special Publication, No. 2, pp. 85–90.

Evans, B. J., 1997. *A Handbook for Seismic Data Acquisition in Exploration*. Geophysical Monograph Series No. 7, Society of Exploration Geophysicists.

Geological Society Engineering Group Working Party, 1988. Engineering Geophysics. *Quarterly Journal of Engineering Geology*, Vol. 21, No. 3, pp. 207–271.

Gochioco, L. M., and Cotten, S. A., 1989. Locating Faults in Underground Coal Mines Using High-Resolution Seismic Reflection Techniques, *Geophysics*, 54, 1521–1527.

Goldstein, N. E., Benson, S. M., and Alumbough, D., 1990. Saline Groundwater Plume Mapping with Electromagnetics, in Ward, S. H., (ed.), *Geotechnical and Environmental Geophysics, Vol. II, Environmental and Groundwater*. Society of Exploration Geophysicists, Tulsa, Oklahoma, pp. 17–25.

Greenhouse, J. P., Slaine, D. D., and Gudjurgis, P., 1998. *Applications of Geophysics in Environmental Investigations*. CD-ROM.

Hawkins, A. B. (ed.), 1986. *Site Investigation Practice: Assessing BS 5930*. Geological Society Engineering Geology Special Publication No. 2.

Hearst, J. R., and Nelson, P. H., 1985. *Well Logging for Physical Properties*. McGraw-Hill, 571 pp.

Hohmann, G. W., 1982. *Numerical Modeling for Electrical Geophysical Methods*. Proceedings of International Symposium of Applied Geophysics Tropical Region, University do Para, Belem, Brazil, pp. 309–384.

Hubbert, M. K., 1944. An Exploratory Study of Faults in the Cave in Rock and Rosiclare Districts by the Earth Resistivity Method in Geological and Geophysical Survey of Fluorspar Areas in Hardin County, Illinois, U. S. *Geological Survey Bulletin* 942, Part 2, pp. 73–147.

Hunter, J. A., Pullan, S. E., Burns, R. A., Gagne, R. M., and Good, R. L., 1984. Shallow Seismic Reflection Mapping of the Overburden-Bedrock Interface with the Engineering Seismograph—Some Simple Techniques, *Geophysics*, 49, pp. 1381–1385.

Lankston, R. W., 1990. High-Resolution Refraction Seismic Data Acquisition and Interpretation, in Ward, S. H., (ed.), *Geotechnical and Environmental Geophysics, Vol. I: Review and Tutorial*, Society of Exploration Geophysicists, Tulsa, OK.

McNeill, J. D., 1990. Use of Electromagnetic Methods for Groundwater Studies, in Ward, S. H., (ed.), *Geotechnical and Environmental Geophysics, Vol. I: Review and Tutorial.* Society of Exploration Geophysicists, Tulsa, OK. pp. 191–218.

Nabighian, M. N., 1987. *Electromagnetic Methods in Applied Geophysics, Vol. 1.* Society of Exploration Geophysicists, Tulsa, OK.

Nabighian, M. N., and Macnae, J. C., 1991. Time Domain Electromagnetic Prospecting Methods, in Nabighian, M.N., (ed.), *Electromagnetic Methods in Applied Geophysics, Vol. 2a.* Society of Exploration Geophysicists, Tulsa, OK. pp. 427–520.

Press, F., 1966. Seismic Velocities, in Clark, S. P., Jr., (ed.), *Handbook of Physical Constants,* revised edition. Geological Society of America Memoir 97, pp. 97–173.

Reynolds, J. M., 1997. *An Introduction to Applied and Environmental Geophysics.* John Wiley & Sons.

Robinson, E.S. and Coruh, C., 1988. *Basic Exploration Geophysics.* John Wiley & Sons.

Sheriff, R. E., 1991. *Encyclopedic Dictionary of Exploration Geophysics.* Society of Exploration Geophysicists, Tulsa, OK.

Ward, S. H. Ed., 1990. *Geotechnical and Environmental Geophysics, Vol. I, Review and Tutorial.* Society of Exploration Geophysicists, Tulsa, OK.

Ward, S. H. Ed., 1990. *Geotechnical and Environmental Geophysics, Vol. II, Environmental and Groundwater.* Society of Exploration Geophysicists, Tulsa, OK.

Ward, S. H. Ed., 1990. *Geotechnical and Environmental Geophysics, Vol. III, Geotechnical.* Society of Exploration Geophysicists, Tulsa, OK.

Zohdy, A. A. R., Eaton, G. P., and Mabey, D. R., 1974. *Application of Surface Geophysics to Ground Water Investigations.* Techniques of Water-Resources Investigations of the United States Geological Survey, Book 2, pp. 49–50.

Zonge, K. L., and Hughes, L. H., 1991. Controlled-Source Audio-Frequency Magnetotellurics, in Nabighian, M. C. (ed.), *Electromagnetic Methods in Applied Geophysics, Vol. 2, Applications, Part B.* Society of Exploration Geophysicists, Tulsa, OK, pp. 713-809.

Chapter 5

Groundwater Flow

5.1 Groundwater Movement

Groundwater is generally always moving. Movement occurs from higher hydraulic head in **recharge** areas (natural or artificial), where precipitation is generally higher, to **discharge** areas of lower hydraulic head (wells, springs, rivers, lakes, and wetlands). The reason groundwater moves is because there always seems to be a "change in head" or some kind of hydraulic gradient or slope in potentiometric surface of a groundwater system. Gravity is the driving force that moves water. Infiltrating groundwater moves downward until it reaches a horizon with low enough hydraulic conductivity to begin piling up. Groundwater moves so slowly (feet/year to feet/day) that "piles" of water in recharge areas may build up or mound before their effects in the system can be equilibrated. If you picked up one corner of a bathtub full of water, gravity would cause the water to move towards the low corner. If the bathtub were filled with saturated sand and you picked up a corner (you would be very strong!), water would still move, but much more slowly. The quantity of groundwater movement through porous media is defined through **Darcy's Law** (Figure 5.1).

Darcy's Law

In the mid-nineteenth century, Henry Darcy (1856) systematically studied the movement of water through sand columns. He was able to show that the volumetric rate (Q) of groundwater was proportional to the intrinsic permeability (see Chapter 3) of the porous media (k) and the change in head (hydraulic gradient)) over the length of the sand column (Figure 5.1). This can be expressed as Darcy's law:

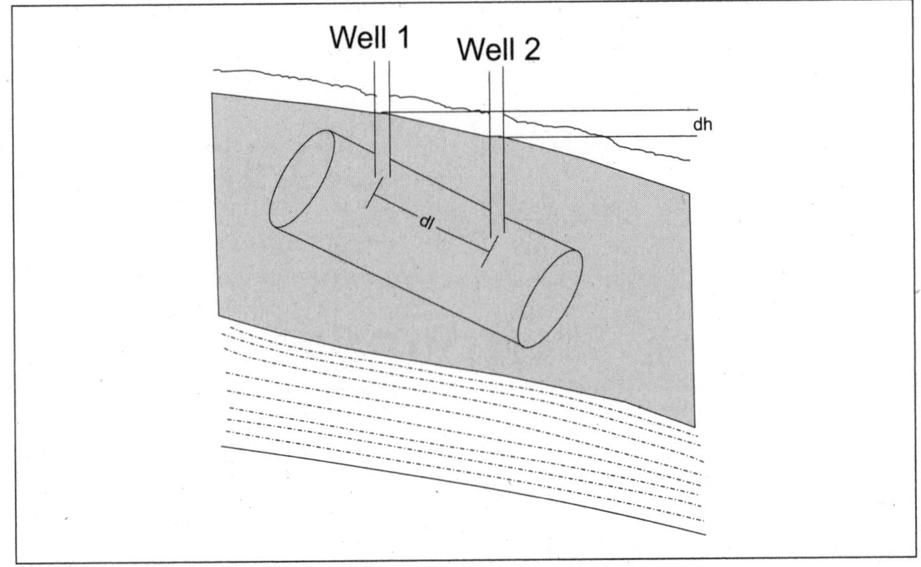

Figure 5.1 Illustration of Darcy's Law.

$$Q = K\frac{\partial h}{\partial l}A = KiA \qquad [5.1]$$

where:

Q = volumetric discharge rate (L^3/t)

$\partial h/\partial l = i$ = hydraulic gradient (L/L) or slope of the potentiometric surface.

A = $b \times w$, the cross-sectional area perpendicular to flow (L^2),(in horizontal flow, the saturated thickness (b) multiplied by the width of the aquifer (w) or a part of the width of an aquifer)

K = hydraulic conductivity (L/t)

There is a significant difference between the hydraulic conductivity (K) and the intrinsic permeability (k), although they tend be to used interchangeably in practice because the fluid properties of most uncontaminated groundwaters are similar at similar temperatures (Chapter 3). Waters with total dissolved solids (TDS) less than 5,000 mg/L have similar enough viscosities and specific weights (Fetter 1994).

Darcy's Law can also be viewed as:

$$Q = q \times A \qquad [5.2]$$

where:

$$q = K\frac{\partial h}{\partial l} = Ki \qquad [5.3]$$

q = the Darcian velocity, also known as the specific discharge.

Even though the Darcian velocity or **specific discharge** (q) has units of velocity (L/t), it does not represent a true groundwater velocity. This is because the sand grains occupy some of the area perpendicular to flow. To account for this, one must include the effective porosity (n_e) of the porous media. The true velocity is really an average linear velocity (V_{ave}) as the water moves around the grains from one point (high head) to another (low head). This value is also known as the **seepage velocity** (V_s), expressed as:

$$V_s = \frac{K}{n_e}\frac{\partial h}{\partial l} = \frac{q}{n_e} \qquad [5.4]$$

Where all terms have been defined previously.

Darcy's Law is valid in the saturated zone, in multiphase flow, and in the vadose zone, although adjustments to Equation 3.1 have to be made. For example, the hydraulic conductivity (K) in the vadose zone becomes a function of the moisture content (Fetter 1993) (Chapter 12).

Hydraulic Head

In Chapter 3 potential hydraulic head and total head are mentioned. Here these concepts will be developed.

Whenever water moves from one place to another, it takes work (W) to accomplish this. The work performed in moving a unit mass of fluid is a mechanical process and is known as the fluid potential (Φ)(Freeze and Cherry 1979). The forces moving the fluid have to overcome the resistant frictional forces of the porous media. The work calculation represents three types of mechanical energy (Freeze and Cherry 1979). The total fluid potential (Φ_T) (the mechanical energy per unit mass) is expressed as the sum of the three components of the Bernoulli Equation, which are velocity potential, elevation potential, and pressure potential (The dimensions of work and energy are the same: ML/t^2.):

$$\Phi_T = \frac{v^2}{2} + gz + \frac{P}{\rho} \qquad [5.5]$$

where:

v = fluid velocity (L/*t*)
g = acceleration due to gravity (L/t^2)
z = elevation above a datum (L)
P = pressure = Force/*A*(ML/$t2$)/L^2 = M(Lt^2)
ρ = fluid density, (M/L^3)

M. King Hubbert (1940) showed that the fluid potential at any point is simply the total hydraulic head multiplied by the acceleration due to gravity (*g*).

$$\Phi_T = gh_t \qquad [5.6]$$

Since *g* is very nearly constant near the Earth's surface, fluid potential and hydraulic head are nearly perfectly correlated (Freeze and Cherry 1979). If the components of the Bernoulli Equation (Equation 5.5) (expressed as energy per unit mass) are divided by gravity, we obtain the three components of total hydraulic head expressed as fluid potential on a unit weight basis.

$$h_t = \frac{v^2}{2g} + Z + \frac{P}{\rho g} \qquad [5.7]$$

Dimensionally, these are all measured in terms of length (feet or meters). The first term is known as the *velocity head* (h_v), the second term is called the *elevation head* (h_z), and the last term is called the *pressure head* (h_p). Since groundwater moves so slowly (feet/year to feet/day), the velocity head (h_v) is considered to be negligible (see Example 5.1). Therefore, the total head (h_t) at any point in a groundwater system (confined or unconfined) can be expressed in terms of the elevation head (h_z) and the pressure head (h_p). Equation 5.7 can be simplified to:

$$h_t = h_z + h_p \qquad [5.8]$$

Example 5.1

The typical hydraulic conductivity of a fine to silty sand is approximately 0.3 ft/day (1×10^{-4} cm/sec). Is the velocity head negligible?

Solution:

The hydraulic conductivity must be converted to a velocity using Equation 5.4 by making some assumptions. If the effective porosity can be assumed to be 25% and a gradient of 0.005 ft/ft is used, then the velocity can be calculated.

By using Equation 5.7 one can calculate the relative velocity head.

$$v = \frac{[(0.3\,\text{ft/day})(0.005\,\text{ft/ft})]}{0.25} = 6.0\times10^{-3}\,\text{ft/day}$$

$$\frac{v^2}{2g} = \frac{[(0.006\,\text{ft/day})(1\,\text{day}/86{,}400\,\text{sec})]^2}{2(32.2\,\text{ft/sec}^2)} = 7.5\times10^{-17}\,\text{ft/sec} = 6.5\times10^{-12}\,\text{ft/day}$$

This appears to be a negligible value.

The total head in an unconfined aquifer is reflected by the elevation of the water table. In a confined aquifer, the total head is reflected by the elevations of water levels in wells that rise above the top of the aquifer. When these elevations are contoured, a surface is generated that is also known as the potentiometric surface. The **potentiometric surface** of an unconfined system is the water table. The potentiometric surface; therefore, is a surface reflecting the total head, which includes elevation head (h_z) and pressure head (h_p).

Example 5.2

A well completed in a unconfined aquifer and a confined aquifer are compared at two points within the well to show the water level represents the total head (Figure 5.2).

Solution:

Equation 5.8 can be used to calculate the heads at A, B, C, D.

$h_A = h_z + h_P = 28\text{ ft} + 0\text{ ft} = 28\text{ ft},$ $\qquad h_B = h_z + h_P = 5\text{ ft} + 23\text{ ft} = 28\text{ ft}$

$h_C = h_z + h_P = 33\text{ f} + 0\text{ ft} = 33\text{ ft},$ $\qquad h_D = h_z + h_P = 20\text{ ft} + 13\text{ ft} = 33\text{ ft}$

Hydraulic Head and Darcy's Law

Between the Bernoulli Equation and Darcy's Law, most groundwater problems associated with fluid flow can be solved. Groundwater models are designed to solve for head distributions, from which Darcy's Law is applied to determine the quantity of flow. Where most people have trouble; however, is in the proper interpretation of the Bernoulli Equation and the correct application of Darcy's Law.

Remember that the total head is composed of the elevation head and the pressure head. The pressure head is created by the weight of the water above a datum. Pressure is often times measured in pounds per square inch (psi). There is a simple relationship between pressure head and equivalent hydraulic head:

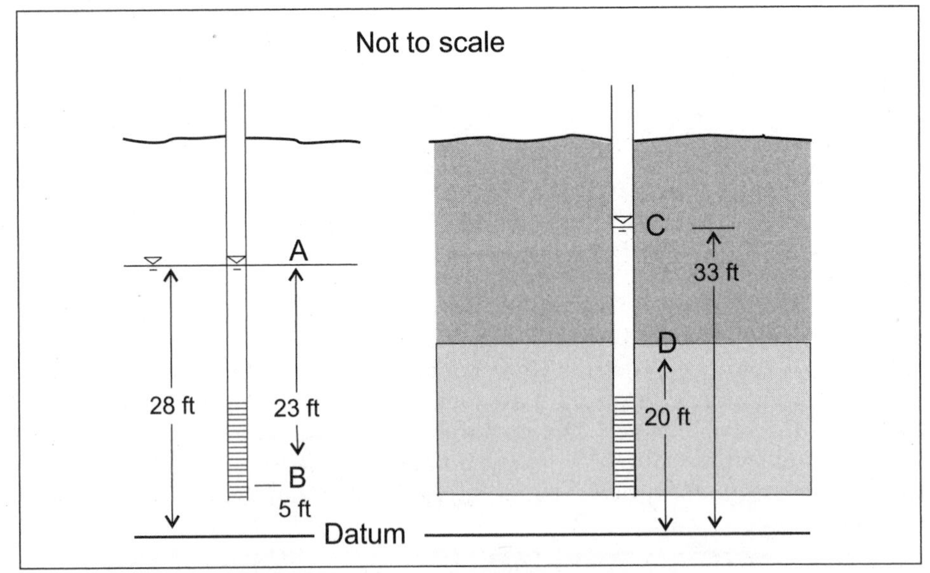

Figure 5.2 Comparison of heads at A, B, C, and D in Example 5.2.

$$1 \text{ psi} = 2.31 \text{ ft of } H_2O \quad [5.9]$$

Let's see where this came from. Recall from the Equation 5.7 that the pressure head can be expressed as:

$$h_p = \frac{P}{\rho g} \quad [5.10]$$

Pressure (P) is equal to force/area, where force (F) is equal to mass times acceleration due to gravity (g). The denominator essentially represents a force over a volume. Therefore; there is an inherent g in both the numerator and the denominator:

$$\frac{P}{\rho g} = \frac{mg / A}{\rho g} = \frac{mg / A}{mg / \text{vol}} \quad [5.11]$$

If it can be assumed that the density of water (ρ) is approximately 62.4 lb/ft^3 and the area (A) is converted from square inches to square feet:

$$1 \text{psi} = \frac{1 \text{ lb} / 1 \text{ in.}^2}{62.4 \text{lb} / \text{ft}^3} \times \frac{144 \text{in}^2}{1 \text{ft}^2} = 2.31 \text{ft} \quad [5.12]$$

This means that if a well is capped and there is a pressure gauge reading at the well cap, the pressure reading could be converted to additional water level height, such as in a stand pipe (Figure 5.3).

Figure 5.3 Well with stand pipe.

Example 5.3

A well completed in Death Valley, California, is capped and has a pressure-gauge reading of 23 psi. The ground surface elevation is –234 ft (using mean sea level as a datum, section 5.3) and the well casing is capped at 2.5 ft above ground surface. What is the total head at this location that would be used to contour a potentiometric surface?

Solution

Using Equation 5.9, we obtain the equivalent pressure head, and using Equation 5.8, we determine the total head:

$$h_p = 23 \text{ psi} \times 2.31 \text{ ft/1 psi} = 53.1 \text{ ft of pressure head}$$

$$h_z = -234 + 2.5 = -231.5 \text{ ft of elevation head}$$

$$h_t = h_p + h_z = 53.1 + (-231.5) = -178.4 \text{ ft, relative to mean sea level.}$$

Example 5.4

As a graduate student in Laramie, Wyoming, I lived in a small house with a basement apartment. In the springtime, the basement apartment would flood from seeping groundwater. The landlord and his son were down there mopping the water up and wringing it into buckets, which they would throw out onto the lawn. After observing them make a couple of trips to the yard, I found out what they were doing and suggested they design a sump-pump system to handle the high water table. If we can assume the walls are impermeable and that all the water is seeping through the concrete slab of the basement, we can use this information to determine the hydraulic conductivity of the concrete slab. Suppose the following information was determined after a brief field excursion, which included placing a few piezometers around the house (Figure 5.4):

Rate of water seepage is 20 gal/ hour.
Average head outside the house is 3 ft above the basement floor.
Basement slab is 4-in. (1/3-ft) thick.
Basement floor area is 24 × 32 ft.

Solution:

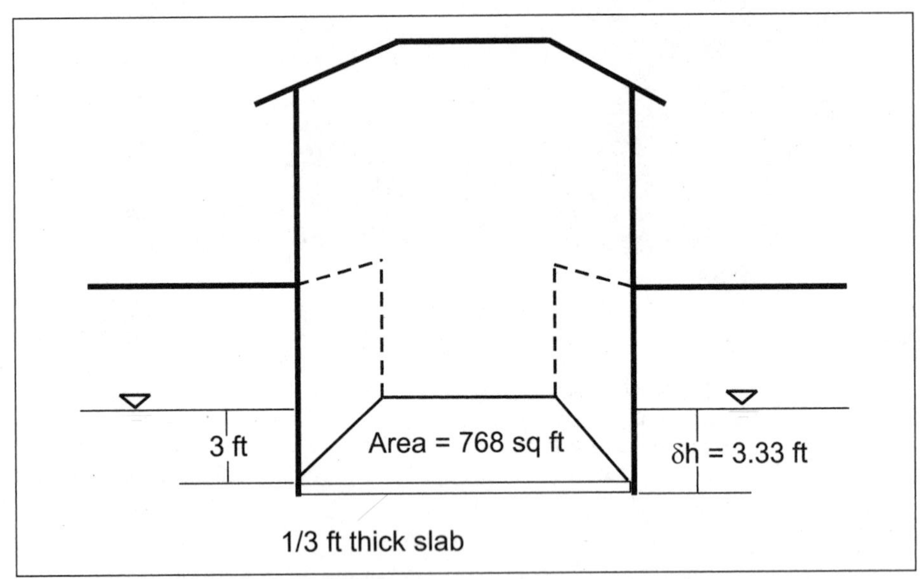

Figure 5.4 Basement flooding problem.

If it can be assumed that the physical properties of the concrete slab are homogeneous and isotropic, Equation 5.1 can be manipulated algebraically to obtain the result.

Q = 20 gallon/hour × 1 ft^3/7.48 gallon × 24 hours/1 day = 64.2 ft^3/day

A = 24 ft × 32 ft = 768 ft^2, the area perpendicular to flow

∂h = 3.33 ft

∂l = 1/3 ft, the distance over which the head change occurs

i = 3.33 ft/0.33 ft = 10

$$K = \frac{Q}{Ai} = \frac{(62.4\ \text{ft}^3/\text{day}}{(768\ \text{ft}^2)10} = 8.3 \times 10^{-3}\ \text{ft/day}$$

Example 5.5

(An example from Dr. Bruce Thompson) Suppose that along the Rio Grande River there is a canal that runs parallel to the river. Because of flooding, the surficial materials are fine grained and clayey. However, there is a connection through a sandy layer between the canal and the river that averages 2-ft thick (Figure 5.5). Suppose that the stage of the river is 2,100 ft and the stage of the canal is 2,113 ft. If the hydraulic conductivity of the sandy material is 10 ft/day and the average distance from the canal to the river is 35 ft, how much **leakance** or seepage loss occurs per mile of canal?

Solution:

Equation 5.1, Darcy's Law, can be used to calculate the result.

Change in head is: $\partial h = 2{,}113 - 2110 = 3$ ft

Flow path length over which the head changes is: 35 ft

Area perpendicular to flow is: 1 mile × 2 ft = 5280 ft × 2 ft. = 10,560 ft^2

Leakance Q = (10 ft/day) (3 ft/35 ft) (10,560 ft^2) = 9,050 ft^3/day per mile of canal.

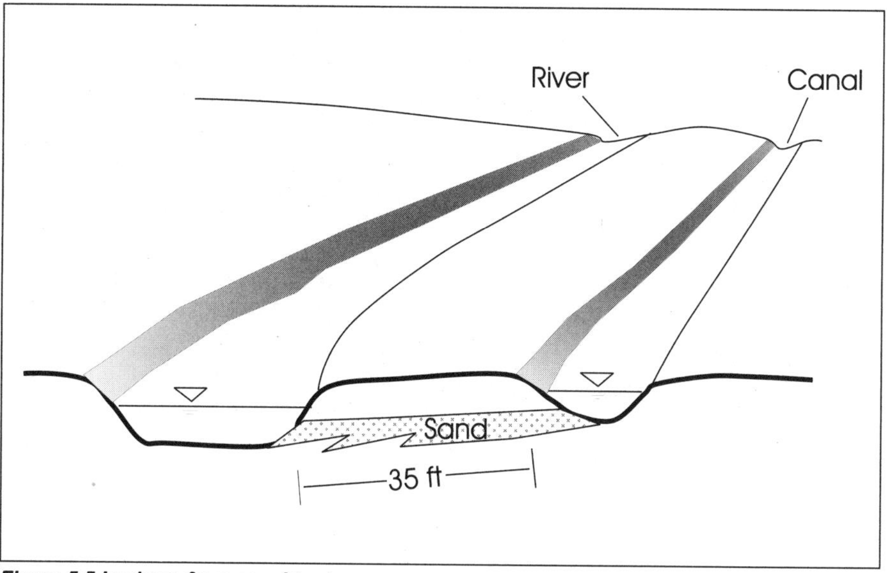

Figure 5.5 Leakage from canal to river.

Example 5.6

One of the best ways to visualize Darcy's Law and gain a perspective of the area perpendicular to flow is by imagining that the room you are in is an aquifer. Pretend that a series of wells drilled from the roof and completed within the aquifer (the room) reveal a gradient of 0.007 from one side of the room to the other. If you assigned a hydraulic conductivity to the aquifer materials, can you determine the amount of flow through the room?

Solution:

Since flow is horizontal from one side of the room to the other, the distance from one wall to the other is the flow length over which the head changes (∂h). The area perpendicular to flow (A) is the height of the room from floor to ceiling multiplied by the width of the wall.

5.2 Flow Nets

One of the ways to graphically represent a 2-D view of groundwater flow, either planar (x,y) or cross-sectional (x,z), is through a **flow net**. The net consists of two sets of lines, flow lines and **equipotential lines**, or contours, of equal hydraulic head. In the case of homogeneity and isotropy flow lines are parallel to groundwater flow and the equipotential or potential lines are perpendicular to flow. The spacing of the equipotential lines reflect the hydraulic gradient (Figure 5.6). The usual spacing of the flow lines and potential lines are selected so that they roughly form a box. This gives the appearance of a net. The quantity of flow (Q) between any two flow lines is the same based on the principle of conservation of mass. If there are sources (recharge contributions) or sinks (discharge areas), the effect would result in the bending (divergence for injection or convergence for discharge) of flow lines. This will be discussed in more detail later on. In a given flow net, the areas of highest transmissivity have the flattest gradients and therefore the largest boxes. Some excellent examples on how to construct flow nets, with wonderful drawings and examples related to flow beneath concrete or earthen structures, are given by Watson and Burnett (1993). It is our experience that flow nets and can be easily create using a variety of software (Waterloo Hydrogeologic, Groundwater Modeling Systems, and others).

The steepening of the hydraulic gradient in Figure 5.6 suggests that the transmissivity of the aquifer has changed. This may be attributed to a region of lower hydraulic conductivity (K) or a thinning of the aquifer (b). When one uses flow nets to estimate flow, it is important to make calculations through flow lines where the hydraulic gradient is similar, rather than averaging the gradient over a larger region. Calculations using Darcy's law are only valid where the hydraulic conductivity or transmissivity is constant. Where this is true, the flow net will show a uniform gradient corresponding to the hydraulic properties.

If the location at hand is not in a recharge or discharge area, the flow is generally considered to be mostly horizontal (horizontal flow). This can oc-

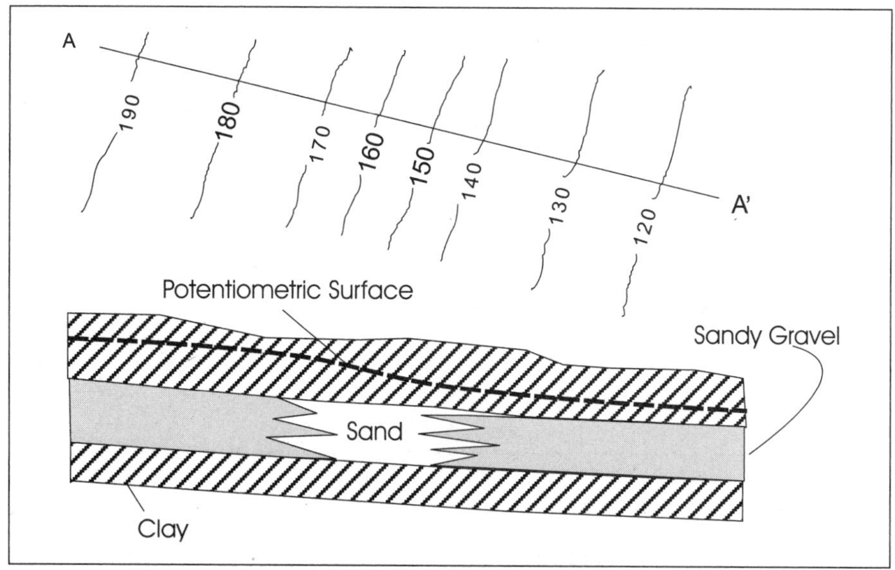

Figure 5.6 Plan view and cross-sectional view of a potentiometric surface map in a confined aquifer. [Modified after Heath and Trainer (1968).]

cur if the material is relatively permeable in flat terrain. Here the potential lines are assumed to be more or less vertical. As a practical matter, this means that wells completed at different depths within the same aquifer at the same location will have the same water level elevation.

Rarely, within a given aquifer unit, is the distribution of horizontal hydraulic conductivity the same. This phenomenon is a function of the depositional environment of the sediments. There are usually lenses of varying hydraulic conductivity or fining upwards or downwards sequences within an aquifer. The coarser units tend to have faster relative velocities than the finer-grained units. This contributes to varying contaminant arrival times than would be predicted using the average linear velocity from Equation 5.4 (Figure 5.7). In plan view flow nets and in 2-dimensional groundwater modeling, however, the vertical distribution of horizontal conductivities (and other characteristics) are averaged over the entire saturated thickness.

Vertical Groundwater Flow

In discharge areas, recharge areas, or undulating topography, there is always a significant vertical component to groundwater flow (Figure 5.8). This means that water levels in wells completed at different depths within the same aquifer unit will be different. The best way to determine this is to

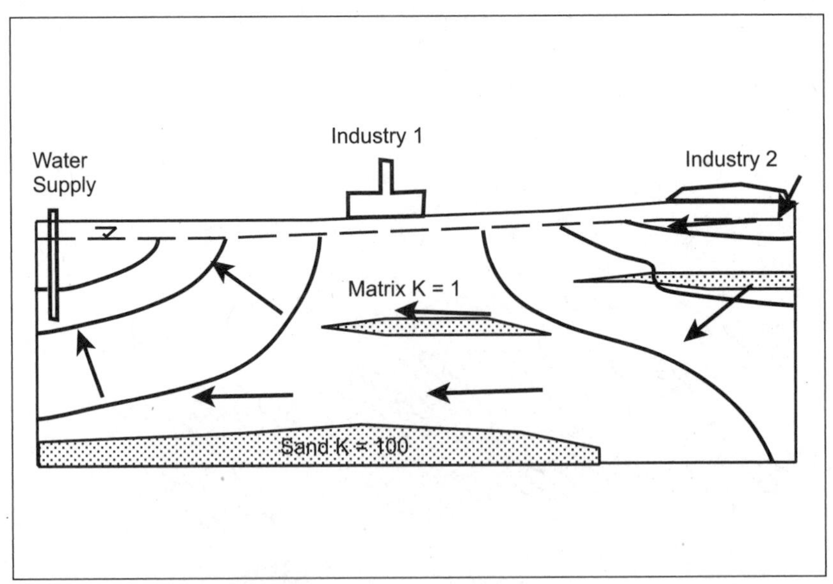

Figure 5.7 Potential lines and flow arrows for a recharge-discharge area with two potential sources affecting a public water supply.

complete two or more wells at varying depths within the same drillhole as a nested set. There should be a trend of downward or upward changes in water levels with depth, depending on whether the location is a recharge or discharge area. Wells completed in the central region of Figure 5.8 will have similar water levels, whereas wells completed deeper in the discharge area will show higher relative heads than wells completed nearer to the surface. The reverse will be true in a recharge area.

At a given contamination site, it is important to know if there is a significant component to vertical flow and which way that component is. For example, wells with long screened intervals may have components of groundwater flow entering the well and components leaving the well (McIluride and Rector 1988). For example, in a discharge area, the higher potential lines are intercepted at the lower part of the well. This results in water flowing into the well in the lower portion of the well screen (Figure 5.9). In the upper portion of the well screen, the water level in the well is higher than the surrounding potential lines within the aquifer. This results in water in the upper portion of the well moving from the well out to the formation. Any contamination that would potentially intercept the upper portion of the well would go around the well and may not be detected.

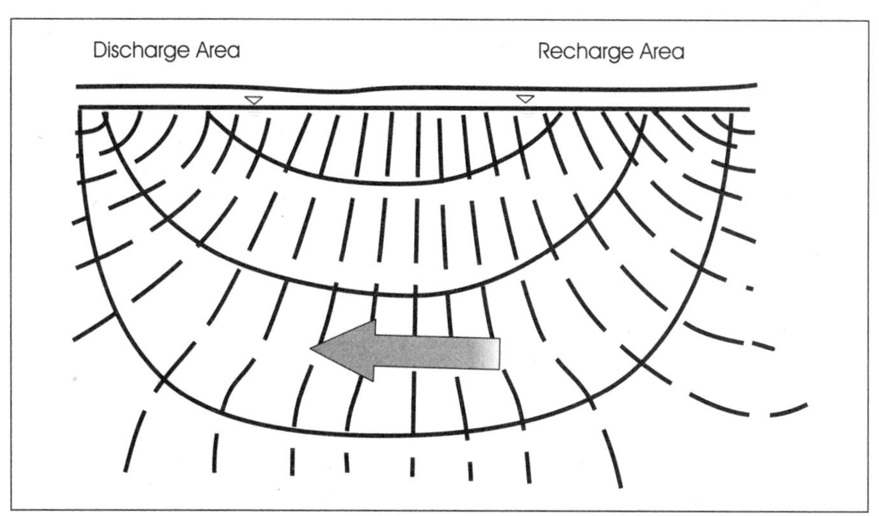

Figure 5.8 Flow net showing recharge and discharge areas with no vertical exaggeration. [Modified from Toth (1962).]

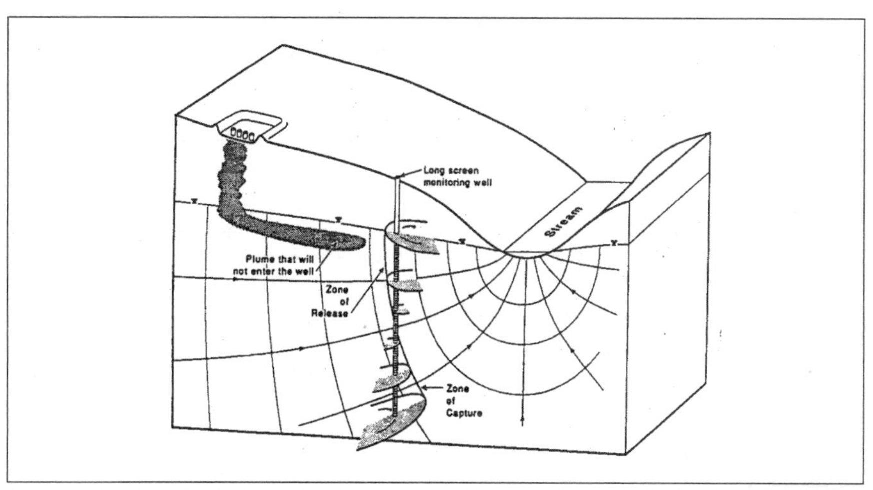

Figure 5.9 Water circulation within fully screened monitoring wells in discharge areas. [From McIlvride and Rector (1988).]

Gaining and Losing Systems

Many perennial streams receive groundwater discharge (gaining stream) along their courses that sustain their flow even during the driest months, even if no tributaries enter. In this case, the river represents the lowest point in a groundwater system. The equipotential lines in a flow net bend such that the flow lines converge on the stream ("V" upstream) (Figure 5.10). Here the water table slopes towards the stream. Although in a 2-D plan view of a flow net, all potential lines appear to be vertical, in a discharge area (e.g., a gaining stream the potential lines actually dip in the direction of stream flow. This means that flow lines are actually curvilinear (Wampler 1998).

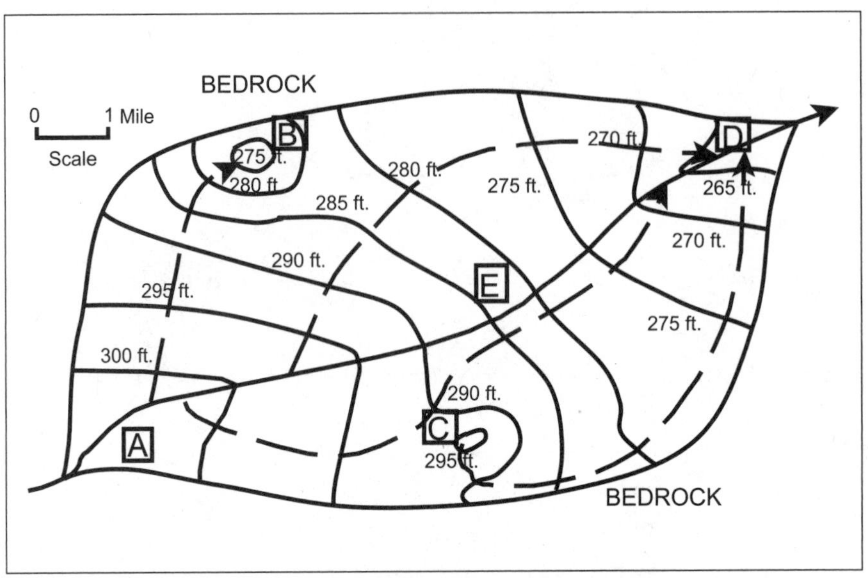

Figure 5.10 Schematic of groundwater/surface water flow in a valley surrounded by bedrock: losing stream (A), discharge area (B), recharge area (C), gaining stream (D), and neither gaining or losing conditions (E). [Adapted from Davis and Dewiest (1966).]

In the arid west of the United States, many streams maintain sustained flow in the mountains, but lose their water into alluvial fans or the sediments of larger valleys. As streams enter areas with lower precipitation, it may be that the water table slopes away from the river or is several feet below the river stage. Here, Darcy's Law would indicate that water will move from the river out to the aquifer. This is a losing stream. In this case, the potential lines bend so that they "V" downstream (Figure 5.10), and also dip away from the direction of stream flow.

A given stream valley can have both components of losing stretches and gaining stretches as shown in Figure 5.10. In the recharge and discharge areas, there are vertical components (3-D components) of flow; whereas, in the areas where groundwater flow parallels the stream, flow is essentially horizontal (2-D). This becomès significant later on in the interpretation of water-level data (Section 5.4).

Refraction of Groundwater Flow

As water passes from porous materials of a certain hydraulic conductivity into a differing hydraulic conductivity, the flow lines will bend or refract according to the principles described by Snell's Law. This phenomenon is observed in a clear glass of water with a straw sticking out of it. The straw seems to bend at the air/water interface. This is due to the refraction of light as light rays pass from the less dense medium of air into the denser medium of water. Snell's Law applied to groundwater flow is described by:

$$\frac{K_1}{K_2} = \frac{\tan(\sigma_1)}{\tan(\sigma_2)} \qquad [5.13]$$

This phenomenon is illustrated in Figure 5.11. What you see is that flow lines or stream lines move vertically through low conductivity layers (at wider spacing) and somewhat horizontally through higher conductivity layers.

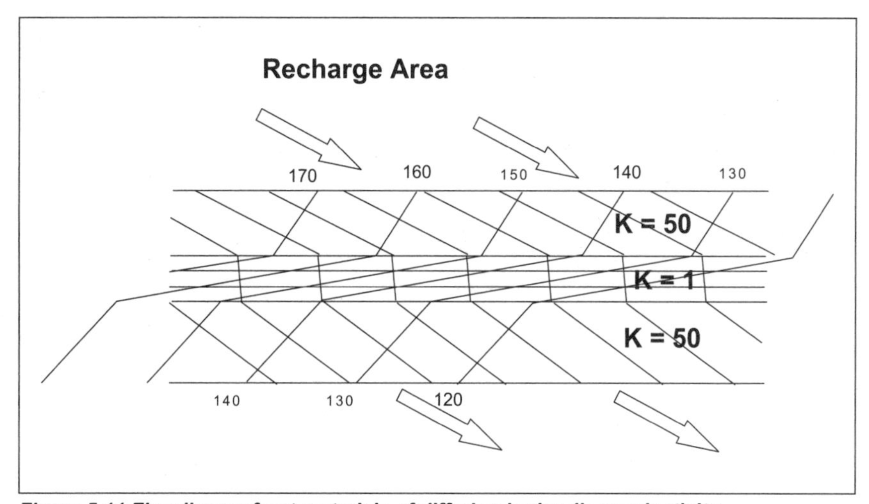

Figure 5.11 Flow lines refract materials of differing hydraulic conductivity.

In terms of pipe flow, if:

$$Q_1 = Q_2 = v_1 A_1 = v_2 A_2 \qquad [5.14]$$

then the closer together (tightly spaced) stream lines represent areas with horizontally faster velocities (moving through flatter gradients), and the wider spaced streamlines represent areas with slower vertical velocities (moving through steep gradients).

Since streamlines will refract when they encounter geologic materials with differing hydraulic conductivities, it is extremely difficult to predict where these streamlines will go. Many depositional environments will create lenses of materials with differing hydraulic conductivities. This becomes significant when one considers locating a landfill or attempts to predict where a source of contamination came from. Depending on the geology and the flow system, a contamination source that appears to be obviously close may not be the source in question at all, but may have come from some more distant source (Figure 5.7).

5.3 Level Measurements in Groundwater Monitoring Wells

The basis for determining the direction of groundwater flow is based upon static water-level data. Water levels can be collected manually with a portable field device or rigged to collect continuous data with a chart recorder or data-logger sentinel. It is possible to telemetrically send these data back to the office, powered by a solar panel (Figure 5.12). Water-level data are of little value without knowing the well-completion information and what the relative elevations are. Grave errors can result if one assumes that all water-level data for a given area are collected from the same aquifer unit, ignores recharge and discharge areas, and fails to survey the measuring points!

The purpose of this section is to present the practical aspects of obtaining level-measurement data and to describe the most commonly used devises to obtain them. It is instructive to present the most common sources of error from field mistakes and idiosyncracies of the equipment to give the reader a better understanding of the pitfalls one can fall into. The pitfalls one can make in interpreting water level data are found in Section 5.4.

Defining Level Measurements

For a given project area or data set, it is imperative that there be some consistency among those who collect the data. One of the first considerations is where the common datum is. Many use mean sea level (MSL) as a datum. This would mean that all data would be reduced to elevations above MSL. For some projects, an arbitrary relative datum may be become the

Figure 5.12 Solar-powered data collector.

base datum. For example, you are out in the field and there is no convenient way to tie into a benchmark of known elevation. You still need to evaluate the relative elevations or may have security reasons for keeping your database as arbitrary elevations. Keeping exact elevations from being known by others may be important if a public presentation is being conducted in a politically sensitive area. Whatever ends up being used as the base station needs to be a relatively permanent feature that is not likely to be disturbed. It would be disastrous to select a large rock or drive a stake that is later excavated and removed during construction or by curious children.

Surveying the well locations and measuring points is critical to proper interpretation of the data. Surveying can be done by estimating on a topographic map, by tape and measure, by simple level surveys, total station systems that give northing, easting, and elevation, or by global positioning systems (GPS). For example, level surveys provide a relative vertical positioning, but are not capable of providing northing and easting positions. GPS systems are very useful for widely spaced monitoring locations, distributed farther than is practical with traditional surveying equipment. The handheld GPS units are becoming more popular with drillers and geologists, but these are only useful for general locating. The absolute vertical positioning is poor and may be off by several tens of feet or even tens of meters. If a more expensive GPS is employed, one with a base station and a rover unit on a tripod, accuracies into the fraction of an inch or centimeter can be achieved. These usually require correction software for changes in drift and atmospheric conditions.

Access to Wells

Taking level measurements in wells is straightforward, provided access to wells is known in advance. In a site where level measurements are being taken from wells constructed by a variety of contractors, there may be an assortment of security devices and locks. Each may require its own key or some special way to remove the cap. For example, monitoring wells that are constructed in playgrounds or in paved roadways may have a flush-mount security plate (Figure 5.13). In this case, a socket or wrench is required to remove the surface metal plate before a subsequent lock can be reached. This flat completion method allows vehicles to run over the well locations or individuals to run around without tripping over something. Without the wrench, wasted time is expended in a fruitless effort.

Figure 5.13 Flush-mount well cap.

Figure 5.14 Typical cap well.

Removing a well cap

To save on costs for more regional studies, monitoring networks are sometimes expanded to include domestic and stock water wells. Nothing is more frustrating in fieldwork than not to have access to a well. This may be something as simple as forgetting the keys to the well caps or the gate to get in. Many stock wells or domestic wells will typically have a well cap attached by three bolts (Figure 5.14). These are usually more than finger tight to prohibit birds or small animals from falling down the hole or knocking off the well caps, and discourage the casual passerby from intruding. An indispensable device is the handy multitool purchased at most sporting goods or department stores. These should be equipped with pliers that can be used to loosen the bolts and free the cap. This is a good excuse to tell your spouse that you need one, and it makes a great safety award.

Well caps that become wedged over a long season from grit, changes in air temperature, or moisture can usually be loosened by taking a flat stick or your multitool and tapping upward on the sides of the well cap until it loosens up. This process will eventually allow access to the well. Tight fitting polyvinyl chloride (PVC) well caps can be modified with slits on the sides of the well cap, to allow more flexibility. Another important point to remember is to place the well cap back on and re-secure any locking device. Well owners can get really mad about such neglect and may not allow additional opportunities.

In areas where gas migration takes place, coal mines, or landfill areas, gas pressures may build up under well caps if they are not perforated with a breathe hole. One field hydrogeologist was measuring water levels in a coal-mining region and had a thought to kick the well cap. When the cap blew 150 ft into the air he contemplated what would have happened to his head. This resulted in a company policy of drilling breathe holes in well caps or in the casing to bleed off methane gas pressures.

Opening tight-strung gates and negotiating fences

Another obstacle in gaining access to wells in the western United States is barbed-wire fences. After one obtains permission to obtain level measurements in rural areas, it is important to leave all fences and gates the way they were found. Some gates are tough to open and are tougher to close. There is actually a technique to opening and closing a tightly strung gate. The gate is usually held at the bottom and the top by a looped wire attached to the fence post. The gate post fits into the bottom loop and then must be pulled toward the fence to achieve securing the upper loop. Both loops in place leave the fence tight. The trick is to place the gate post into the bottom loop first and then pull the gate post toward you by having the fence post (with the loops) against your shoulder in a hugging motion (Figure 5.15). This results in more leverage and therefore being able to close the gate. It is embarrassing to have a co-worker or support vehicle wait long periods of time for you to close a gate. Merely pushing the gate post towards the fence results in safety problems and frustration. The hugging method works much better. In some cases, in extremely tight fences a "cheater bar" will be hanging there along the fence post. In this case, you again place the gate post into the lower loop and then use the "cheater bar" to hug the fence close enough to attach the upper loop. Both apply the same principle.

Figure 5.15 Accessing a well through a barbed-wire gate.

Some fences may have an electrically hot top wire to keep livestock in. In this case if you touch the wire, you will get shocked. The wire looks distinctively different from the other wires and is located near the top of the fence. Test the wire by gently tossing something metal at the suspected wire. Having rubber boots on can be helpful. Hot fence wires can be especially problematic for fences keeping buffalo in. Network mogul Ted Turner has a fence in Montana that puts out 5,000 volts (although the amperage is low) to the unwary animal. It is shocking enough to keep a buffalo away but also strong enough to knock a person down and stop their heart from beating for a couple of minutes.

Measuring Points

Once a datum has been established, data collected in a fieldbook will be brought back to the office. All persons collecting data should clearly indicate whether data were collected from top of casing (TOC) or ground surface (GS). Are the data recorded as depth from TOC or depth from GS or from some other point of reference? For example, a well casing cut off below the floor of a shed may be recorded from the shed floor surface. The distance that the casing extends above the ground surface is referred to as the "stick up." All these become important later on as a level survey is conducted and relative elevations need to be established.

Identifying measuring points (MPs) may not be as straightforward as would seem. Sometimes when many monitoring wells are constructed in a short period of time, simple annoyances, such as not cutting the casing off level, can occur. It may be that a piece of steel casing was cut off with a torch, or a section of PVC casing was cut diagonally with a saw or some other device. This is not only a problem with placement of a well cap, but creates a myriad of problems when different individuals are going out into the field to take measurements. Where do you take the measurement? At the lowest part of the cut, the top of the cut, or somewhere in between? Probably you were not responsible for the well completion, but still need to collect meaningful data.

It is suggested that markings be clearly made on the casing indicating where to take a measurement. On steel casing, it is helpful to take a hacksaw or file and make three nick marks close together. A dark permanent maker can also be used to make an upward arrow pointing to the three marks. Marking MP (for measuring point) is also helpful. A similar arrangement can be used on PVC casing; however, the dark permanent marker works well enough that hacksaw or file markings are usually not necessary (Figure 5.16). This works best if all markings are made inside the casing. Markings on the outside, even with bright spray paint, tend to fade quickly in the weather. If a name plate is not with the well, it is also helpful to indicate well name and completion information, such as total depth drilled (TDD) and screen interval, on the inside of the well cap.

Water-Level Devices

There are a host of water-level devices used to collect level-measurement data. Don't be fooled into thinking one is best; for the different devices each have their own unique applications. Manual devices such as steel tapes and electric tapes, or E-tapes, are discussed first, followed by those with

Figure 5.16 Marking point on a PVC casing.

continuous recording capabilities, such as chart recorders and transducer–data-logger combinations.

How to take a level measurement

The basic procedure of taking a level measurement involves gaining access to the well, removing the well cap, and lowering a device down the well to obtain a reading. Normally, this procedure takes only a matter of a few minutes depending on the well depth, device used, and whether the well is a dedicated monitoring well or a domestic well being used as a monitoring well. As simple as this procedure sounds, if one is not aware of the weaknesses and idiosyncracies of the equipment, serious errors in readings can result that affect the quality and interpretation of the data. These are discussed in this section and in Section 5.3.5.

Most level measurements are taken under static conditions. Dynamic conditions occur during a pumping test. If the intention is to take a static water-level measurement and the well is in use or was recently used, it is important to allow recovery to equilibrium conditions before taking the measurement. For example, a stock well may be on all the time or the pump activated by a timer. The pump can be shut off for a while (until there are no changes between readings), then a reading can be taken, and the pump turned back on or the timer reset. Once again, if timers are not reset or pumps turned back on, then you may not be given a second chance.

Steel tapes

Most data collected before the late 1970s were likely collected with a steel tape. Steel tapes represent the tried-and-true method that still has many important applications today. Water levels from the 1940s on were all measured this way until electric tapes (E-tapes) became the norm. The proper way to use a steel tape is to apply a carpenter's chalk to approximately 5 to 10 ft (1.5 to 3 meters) of tape before lowering it into the well. The tape is lowered to some exact number next to the measuring point, for example, 50 ft (15 meters). The tape is retrieved to the surface, where the "wet" mark on the tape, accentuated by the carpenter's chalk, is recorded (Figure 5.17). A reading like 50 ft minus 4.32 ft would yield a reading of 45.68 ft depth to water below MP. There is usually a historical knowledge of the approximate depth to water to guide this process. Otherwise it is done by trial and error. Usually a good reading can be made even on a hot day, with the wet chalk mark, otherwise the tape dries quickly. Getting a clean marking may require more than one trip with the tape, hence its loss in popularity.

Sometimes the only way to get a water level is with a steel tape. In many pumping wells, with riser pipe and electrical lines, or wells with only a small access port the only way to get a water level is with a thin steel tape. Steel tapes are rigid and tend to not get hung up on equipment down the hole. A 300-ft (100-m) steel tape is an important part of any basic field equipment. It can also be used to measure distances between wells or other useful tasks.

Electrical tapes (E-tapes)

By the early 1980s, E-tapes began to dominate the market. The first devices were usually marked off in some color-coded fashion every 5 ft (1.5 m), with 10-ft markings and 50- or 100-ft markings in some different color. E-tapes now are usually marked every 1/100th of a foot (3 mm) (Figure 5.18). All E-tapes function with the same basic principle. A probe is lowered into the water, which completes an electrical circuit, indicated by a buzzer sound or an activated light or both. The signal from the water level is transmitted up the electronic cable of the E-tape to the reel where the signal occurs. Each has its own appropriate function. During a pumping test with a generator running, it may be hard to hear a buzzer sound, or you may have a hearing impairment from hanging around drill rigs too long without wearing ear protection. In this case a light is helpful. In very bright sunlight, your light indicator may be hard to detect and the buzzer is more helpful.

Figure 5.17 "Wet" mark on steel tape.

Figure 5.18 E-tapes with probes.

Most E-tapes have a sensitivity knob. The sensitivity knob is usually a turn dial that ranges from 1 to 10. Sensitivity is needed for a variety of water qualities. Low sensitivity (1 to 3) will give a clear signal in high total dissolved solids (TDS) waters; such as those found in Cretaceous marine shales. The higher the TDS in the water, the lower the sensitivity needed to detect a water level. Relatively pristine or low TDS waters require that the turn dial be adjusted to a higher sensitivity setting (8 to 10) to get a clear signal. A good general rule is to put the dial in midrange (4 to 6) and lower the probe until a buzz/light signal is indicated. If little attention is paid to the sensitivity setting, the hydrogeologist may obtain false readings. This may be one of the first questions asked to a field technician when trying to explain why a water-level reading does not make sense.

To take a water-level reading using an E-tape, turn the device on by moving the sensitivity knob dial to midrange. Lower the probe until the light goes on or the buzzer sounds. This gets you close to where the reading will be. At this point, lift the E-tape line above the depth that activates the buzzer and gradually lower the probe until the buzz sound is repeated. Hold this spot on the E-tape line with your thumbnail or with a pointer, like a pencil (Figure 5.19). The E-tape line should be held away from the measuring point toward the center of the well and shaken lightly, to remove excess water and repeat the reading. If the light signal is clear and the buzz is crisp, the reading is probably accurate.

Figure 5.19 Taking water level reading using E-tape.

False readings are fairly common if one does not pay attention to the sensitivity knob setting. In some wells, condensation inside the casing can trigger a false reading. The sound of the buzzer may not be clear or the light may flicker instead of providing a clear, bright reading. In this case, the sensitivity should be turned down to the lowest setting that yields a good signal. This will be variable, depending on the water quality involved. Cascading water from perforated sections above the actual water level can also be problematic. This can occur during a pumping test, where one is trying to get manual readings in a pumping well that has been perforated at various depths. In pristine waters, it may be difficult to get a reading at all. In this case, lower sensitivity knob settings may not be discerning enough to detect when the water table has been reached. Detection may or may not sound off until the probe has been submerged well under the water surface. In this case, turn the dial to the high sensitivity range (8 to 10) and the buzzer or light should give a clearer signal. It is always a good idea to lightly shake the line and repeat the process until a clear reading can be made.

E-tapes come in a variety of designs. Some have round or flat marked electronic cable lines and probes come in a variety of sizes. Some are easier to use, but most used by hydrogeologists and field technicians today tend to be marked every 1/100th of a foot (3 mm). Many drillers still have devices marked every 1 ft or 5 ft. Common spool lengths range from 100 (30.5 m) to 700 ft (213.2 m). It is usually handy to have a couple of different depth capabilities. There are also combined temperature, specific conductance, and level measurement E-tapes, Also known as TLC-meters. These tend to be marked every 1 ft (0.305 m) and have finer measurement markings on the back of the spool.

Water levels taken in monitoring wells tend to be uninhibited by riser pipe and electrical lines. These pose the fewest concerns as to probe size. Pumping wells; however, may present some challenges. Here, probes may wind around or get hung up in wiring or spacers placed down the well to hold the riser pipe in the central part of the well. Getting hung up can be a be a real problem, not to mention costly. Here is where steel tapes or the cheaper 5-ft marked E-tapes can be helpful, particularly those with a very narrow probe size. In some cases a service call to a well driller may be cheaper than paying for a new E-tape.

Chart recorders

Many times it is advisable to perform a continuous recording of level measurements. One device that has been used for many years is a chart recorder, such as a Steven's Recorder. A device like this uses a drum system onto which a chart is placed. The position of the water level in the well is tracked by having a weighted float connected to a beaded cable that passes over the drum and is connected to a counterweight (Figure 5.20). As the drum turns forward or rolls backward with the movement of the water level, a stationary ink pen marks the chart. The pen is only allowed to move horizontally to correspond with time set by a timer. Timers can be set for one month or up to three months. Chart paper is gridded where each column line usually represents 8 hours. The row lines mark the vertical water-level changes.

Since the turning of a drum can be significant, control of movement is through setting the ratio of float movement to drum movement of the chart paper. This is known as setting the gear ratio. For example, a 4:1 (four to one) gear ratio means that the float will move four times the distance the drum would turn forward or backward. This helps keep the pen on the chart paper.

Once the chart is replaced with new paper, the information from the chart has to be reduced or converted into numbers. A technician often logs numbers from a chart into a file for data analysis or hydrograph plotting. Corrections for drift are made by noting the time and date at the end of the chart and then applying a linear correction from the beginning time. For example, if the final marking on the chart is 12 hours short (because the timer setting is off) of the supposed time, a correction factor is applied to "stretch" the data to match up with real time. This can be a bit tedious but is done in a few hours by an experienced technician.

Figure 5.20 Chart recorder in the field.

Frequently, wells located near surface water streams are fitted with a chart recorder to compare the groundwater and surface-water elevations or in an area where water-level fluctuations are significant. Other reasons may be for placement during winter months or remote areas when access is difficult.

The chart recorder is positioned directly above the well on some kind of constructed platform. The platform sits above the well casing, with the float attached to a beaded cable lowered down to the water surface. A hole cut in the platform allows the beaded cable to pass up over the drum and then through another hole cut in the platform for the counter weight. The top of the well casing and the recorder can be encased in a 55-gal drum that can be locked.

Errors and malfunctions can occur in a variety of ways. For example, the batteries can go dead on the timer or the pen can become clogged. The beaded cable occasionally comes off the of the drum from lack of maintenance or from rapid water-level changes. Replacement floats can be jury-rigged with a plastic one-liter soda bottle weighted by sand, if one cannot find a replacement float in a timely way. In spite of problems that can occur, the chart recorder is still widely used after many years.

Transducers and data loggers

Transducers are pressure-sensitive devices that "sense" the amount of water above them. The "sensing" is performed by a strain gauge located near the end of the transducer, which measures the water pressure. The water pressure can be converted into feet of head above the strain gauge by Equation 5.9, repeated here for convenience as Equation 5.15:

$$1 \text{ psi} = 2.31 \text{ ft of } H_2O \qquad [5.15]$$

This relationship gives important information, such as the maximum pressure or how deep the transducer can be placed below the static water surface. Typical ranges of pressure transducers are 5, 10, 15, 20, 30, 50, 100, 200, 300, and 400 psi. For example, a 10-psi transducer should not be placed more than 23 ft below the static water surface. If it is, the transducer may become damaged. Another problem may occur when the data logger will not record any meaningful data. Personal experience indicates that the data logger will give level readings of zero. Transducers of 400 psi have been used in deep mine shafts to gauge recovering water levels.

During a pumping test, one should estimate the range of water-level changes that will take place during a test. The pumping well is usually equipped with the highest range followed by lower water-level ranges for observation wells located at various distances away. If the transducer is to be placed in an observation well, it is a good idea to lower the transducer to an appropriate depth and let it hang and straighten out before defining the initial level measurement. A common error is to forget to establish the base level in each well as a last setting in the data logger before the beginning of the test. If this is done as the transducer is placed into the well, especially with the placement of the pump and riser pipe, the water levels may not have equilibrated. It is also a good idea to select a base level other than zero. Fluctuations of the water level may occur above the initial base reading, resulting in negative numbers for drawdown. This can be problematic when trying to plot the data in a log cycle. Placement of the transducer in the pumping well should be just above the top of the pump. The transducer should be secured with electricians tape, with additional tape being placed approximately every 5 ft (1.5 m) as the pump and riser pipe are being lowered into the well.

Data loggers record level-measurement readings from pressure transducers and store these values into memory. The stored data can be retrieved for analysis. Transducers and data loggers are indispensable equipment for taking rapid succession readings from multiple wells during a pumping test or slug test. Loggers can record data in log-cycle mode, which allows hundreds of readings within the first few seconds of turning the pump on or during the initial recovery phase. Data loggers can also record data on a linear time scale determined by the user. In this way, a data logger is similar to a chart recorder; however, the information stored is already in digital format and does not need to be corrected or reduced (Figure 5.21).

Figure 5.21 Data loggers in use in a field setting.

Care must be taken in the field to make sure pressure transducers are connected to the data loggers. Sometimes, even though a proper connection is made, the cable line from the transducer to the data logger may be kinked or damaged, thus prohibiting accurate data collection. Most cables have a hollow tube that allows the strain gauge to sense changes in air pressure. The typical field procedure is to make a loop greater than 1 in. (2.5 cm) with duct tape and attach this to the well casing, also with duct tape. The cable can be protected in heavy traffic areas by placing boards on the ground, with the cable located between the boards. Most field technicians are encouraged to obtain manual backup readings to the data-logger readings using E-tapes in case the automatic system fails.

Well sentinels can be placed into wells to record data over a long time period similar to chart recorders. For example, In Situ Inc. has one called the Troll. They are stainless-steel probelike devices with their own built-in pressure transducers and data loggers. These can be programmed directly via communications software loaded on a personal computer. Well sentinels can also be equipped with a variety of water-quality probes in addition to a pressure transducer. Measurements of specific conductance, pH, dissolved oxygen, and temperature are popular options. These can be very helpful in monitoring changes in water quality over time. For example, during a pumping test, changes in water quality may indicate helpful information that aids in interpretation. A case example occurred during a pumping test in mining impacted waters near Butte, Montana. A steady-state condition was being approached during the test. It was clear to see that a recharge source was contributing cleaner water to the pumping well, indicated by a higher pH and lower specific conductance, possibly from a nearby stream or a gravel channel connected to a nearby stream. The water-quality data nicely augmented the interpretation of the pumping test.

If the well site has a solar panel power source, the level-measurement data can be retrieved directly in the office. Of course, these systems are more costly. If one performs a cost analysis on travel time and labor-hours to obtain the same information, it may show that this technology can be cost-effective on remote sites.

Practical Design of Level-Measurement Devices

As the previous section shows, there are some practical considerations one should be aware of when purchasing level-measurement devices. This section will go into more detail on the designs of probes, cables and battery placement. Most of these have implications for the popular manual E-tapes, although other methods will be addressed.

Probe design

The purpose of the probe is to detect the static water level. Some detection sensors are near the middle, while others are located near the bottom of the probe (Figure 5.22). With an E-tape, when the static-water surface is encountered, an electrical circuit is completed, which activates a light or audible buzzer. Both are desirable. The sensor may be made of copper or some other metal; probes are mostly made of stainless steel. The sensor may be accessible on the outside of the probe, or through a hole in the probe with the sensor in the middle, or, the bottom inside of the probe. In the latter case, the bottom part is somewhat hollow, like a bell housing, with a wire sensor extending across the hollow section. In the case of a transducer, the size needed depends on the application. Several are in the 5/8-in. (1.6-cm) or 3/4-in. range (1.9 cm). These are sufficient to be used in 1-in. piezometers.

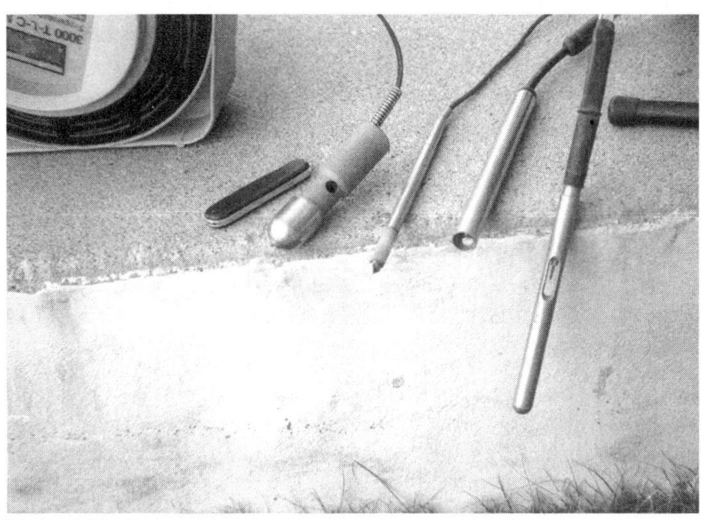

Figure 5.22 Probe sensors.

Sometimes false readings are made when the E-tape probe scrapes moisture down the side of the casing. Condensation is especially problematic in hot weather or very cold weather. The hole in the middle and bell-like probe housing with the sensor near the bottom are designed to avoid this problem. These are common in a probe like those made by Solinst Inc. or the Heron by GeoTech. The only problem is that the diameters of these probes tend to be larger (5/8 inch or larger, 1.6 cm). Whether this is a problem depends on the application. For monitoring wells, probe diameter is not a problem down to a well diameter of about 1 inch (2.5 cm); however, for mini-piezometers or small access ports on well plates at the top of wells, probe diameters may be too large. The shapes of these probes also tend to be more cylindrical and blocky, sometimes with abrupt edges that provide places for them to become hung up on protrusions in the well, such as centralizers or wiring in pumping wells. Newer designs have probes that are more sleek and tapered and have depth capabilities over 500 ft (152.3 m) or metric models to 150 meters.

Smaller well probes, usually equipped with an outside sensor, tend to be tapered and have smaller diameters. The minimum size is down to 1/4 in. (6.4 mm), which easily allows measurement in a ½-in. (1.3-cm) mini-piezometer or access through a port in the top of a well plate Solinst. These can be very effective for shallow wells with depth capacities usually less than 100 ft (30.4 m). Another example of this kind of design is by slope indicator, which has a 5/16-in. (8-mm) probe with depth capacity up to 700 ft (213 m).

Sometimes you may wish to get a water level in a production well that does not have a convenient access port. In this instance, instead of having a solid stiff probe, a string of brass beads covers the water-level sensor (they seem to be carried by drillers more than anyone else). These devices will snake through a difficult access port better than anything else. They are marked every foot (30.5 cm) or every 5 ft (1.5 m), with a brass plate located on the cable for each marking. Once again, these may give false readings from scraping along the casing.

Electronic cable design

Another variation in level-measuring devices is in the cable and reel design. Most E-tapes have a crank handle that allows the cable to spool onto a reel. Some are freestanding, while others require the person to handle the whole thing to maneuver it. Some manufacturers provide both capabilities, with the freestanding design requiring an extra charge. Having some kind of freestanding capability is desirable for taking repeated measurements and wanting to set the device down in between readings. This helps keep the reel, probe, and cable cleaner. Another useful design is a simple braking stop that keeps the reel from turning when it is helpful to "hold" a position. This is usually a plastic or rubber stop that can be twisted in or out or a lever arm that is moved half of a rotation or so to keep the reel from turning.

Some reels and spooling designs have too small a housing for the cable. When the probe is fully reeled in, some of the wraps may slip off, creating a tangled mess. Whenever too much cable is placed on too small a reel, there will be problems. Another aspect of level-measuring devices is the cable design itself. Electronic cables come in a flat, tapelike design, or round, or variations in between. Most of the tape designs have the electronics fully encapsulated in an unbreakable ultra-violet radiation-resistant plastic, with two reinforcing wires or thin metal cables that pass through the outer edges for added strength and resistence to cable stretch. This is a good idea. Markings occur on both sides in hundredths of a foot and fractions of a meter. A drawback to this cable design is in the spooling on the reel. The tapelike design allows easier slippage on the reel.

Round electronic cable designs are also marked to 1/100th ft (3 mm) but have no capability for having metric markings without another separate E-tape. The choice is either in feet or meters but not both. The round electronic cable has a reinforced mesh design around the transmitting signal wire for strength. Round cables tend to not slip, as the flat tapelike cables do, but one worries more about cable stretch with round-cable measurements, especially measurements that are deep.

Other errors in level measurements

Another aspect of any measurement is whether the kinks and bends in the electronic cable lead to inaccurate readings. This can also be argued with steel tapes, too. Generally, as one uses the same device for each measurement, the results tend to be consistent. If minor changes occur in the data that are significant to a project, one should investigate whether there was a change in field technicians or equipment or both.

Other discrepancies could be the result of improper reading of the measuring device. There are fewer errors associated with taking readings from E-tapes marked every 1/100th ft (3 mm) than with those marked every 1 ft (30.5 cm) or every 5 ft (1.5 m). Typically errors amount to reading a 6 or a 9 upside down. In tapes where markings are more widely spaced, readings are referenced either from the next mark above or the one below. The usual procedure is to measure from a given marking on the E-tape to the place held by your thumbnail (the water-level measurement). This is done with a retractable 10 foot (3 m) tape marked every 1/100th ft (3 mm). For example, if the water level is at 18.00 ft (5.48 m), the distance from the thumbnail to the next marking would be either 3 ft (0.91 m) from the 15-foot (4.57 m) mark or 2 ft (0.61 m) subtracted from the 20-foot (6.09-m) mark. The field technician would essentially think "fifteen plus three feet or twenty minus two feet." If one is always consistent at reading from the marking below or above, it is easier to detect errors later on. If one arbitrarily changes how to follow protocol in the field, serious interpretation problems may result. For example, in a high-transmissivity aquifer, 1/10th ft (0.03 m) can make a big difference in interpreting the direction or rate of groundwater flow.

Battery location

As obvious as it sounds, the location where batteries are housed can be a big factor on level-measurement device maintenance. Some new models and many earlier models require that all of the cable be unreeled before the battery compartment can be accessed. Some allow the reel housing to become separated to allow access to the batteries. A good design is where the batteries are quickly accessible through a plate cover on the reel housing (Figure 5.23). This a simple remove-the-plate and change-the-batteries scenario. When purchasing a level-measuring device, accessibility and easy maintenance can be a deciding factor if all other features are the same. Most level-measuring devices are powered with a 9-volt battery.

Level-measuring devices also have a battery or light tester button. One merely presses a button and the light or buzzer goes off or both. Some allow the user to choose whether they want to have the light active or the buzzer

Figure 5.23 Battery access through plate cover on reel housing.

active and can test each separately. If the buzzer signal or light is weak, it is a good idea to replace the batteries. Having extra batteries on hand is always a good idea. The multitool is helpful for removing a few screws, should changes of batteries be required.

Other Practical Applications

Some of the problems associated with level measurements come from well completion rather than device design. For example, pumping wells arranged with spacers and centralizers are more likely to encounter hang-up of the equipment than monitoring wells dedicated to water-level measurements. Once a level-measurement device becomes stuck, how can it be retrieved from the well? Additionally, what should be done about decontamination of equipment before it is lowered into other wells? What do you do if the well is flowing? How do you obtain a measurement on a flowing well?

Retrieving lost equipment

When pumping wells are being used to obtain level-measurement readings, there is always a risk of getting the equipment stuck. This may be during a pumping test, when an operator is trying to follow the pumping level in a well, with an E-tape or simply trying to obtain a level measurement from a domestic well. Extending upward from the top of the pump are the riser pipe and electrical lines to power the motor (Figure 5.24). If the pump is

Figure 5.24 Pump with riser pipe.

fairly deep, plastic centralizers are placed to keep the riser pipe and wiring near the middle of the casing.

In cold climates, pitless adapters are used to make a connection from waterline pipe placed in a trench from the house to the well. An access port is welded to the casing, approximately 6 to 8 ft below the ground surface to prevent freezing of waterlines. An elbow connector from the riser pipe fits into the **pitless adapter**. This way the riser pipe elbow can be removed from the pitless adapter and the pump can be pulled from the well for servicing.

It is amazing how a level-measuring probe can become stuck down a well. When probes become lodged, it is usually because the probe has wound around the riser pipe and electrical lines. Sometimes the probe will slip through the electrical lines and become wedged as it is being retrieved or hung up on a pitless adapter. Some well owners will have a rope or cable attached to the pump, which extends up to the surface for pump retrieval.

If an E-tape probe threads through a gap between the cable and another obstruction, it will become stuck. There are other cases where the probe actually forms a slipknot during the process of obtaining a level measurement. Then the electronic cable line loops outward as an obstruction is encountered, and the probe ends up falling back through the loop.

When an E-tape becomes stuck, the first task is to try and free the probe. Depending on how far down the probe is, one can try different methods. Sometimes success occurs from loosening the electronic cable and striving to jiggle the probe free. If this doesn't work and the probe is within 20 ft (6.1 m), a "fishing" device can be made from one or more sections of 1-in. PVC electrical conduit pipe, which are typically 10-ft (3 m) long. In the bottom piece, a slit or long notch about 6 in. (15.2 cm) long, is cut. The level-measuring cable is hooked into the notch. By sliding the conduit pipe towards the probe, there is more leverage available to move the probe around. In the event the well probe really becomes stuck, it may be necessary to pull the pump and riser pipe to retrieve the E-tape. This can be a real pain, particularly in a production well that has a lot of heavy piping. If you want to save the E-tape, it is cost-effective to call in a service truck with a cable and hoist capability to pull up the piping and pump. The cost of service is a minor percentage of the cost of a new E-tape.

Sometimes if someone pulls really hard on an electronic cable, the line breaks and the cable and probe fall down the well. In this case, after one pulls out the pump and riser pipe it is possible to "fish out" the offending object with a treble hook attached to 30- to 50-lb test monofilament fishing line. Treble hooks may be helpful in hooking onto things that unwittingly fall down a well.

Detangling equipment

When an E-tape cable lines slip off of a reel, a tangled mess can occur. Rather than trying to fight the reel, it is worthwhile to unspool the reel by having someone walk the probe and cable out until the problem area is resolved. Sometimes untangling becomes akin to untangling kite strings or fishing lines. Time and patience can be well worth the effort.

In some cases the electronic cable line may become damaged. It is possible to reattach the probe to a convenient marking on the cable (such as at 100 ft). In this case the repaired device can be used as a backup E-tape. Most manufacturers also have replacement cables. One must compare the economics of purchasing a new probe over the time and effort to repair or replace a new electronic cable line.

Decontaminating equipment

As an operator performs a level-measurement survey of a series of monitoring wells, it is important to decontaminate, or DECON, the equipment. Merely going from one well to the next with the same device can cause cross-contamination. DECON practices vary from company to company. Something as simple as rinsing with an Alconox-water solution to steam cleaning to a submergence bath may be required. It depends on the nature of the site. If the monitoring is for trends in water levels, a simpler DECON step can be taken, compared to monitoring wells completed in a plume of toxic organics. For water-quality sampling it may be better to use dedicated bailers for each well.

Newer level-measuring devices are designed to have the electronic reel built as a module that can be removed for DECON purposes. It is significant to realize that certain chemicals may adsorb onto probes (Fetter 1993). Most probes are constructed of stainless steel, because this tends to be less reactive than other materials, except in mining impacted waters where the pH is low and the amount of dissolved metals is high. It is also helpful to have clean-looking equipment when you go knock on a landowner's door.

Level measurements in flowing wells

In some areas where the recharge area is of a higher elevation than the land surface where a well is drilled into a confined aquifer, the pressure head is sufficient to cause the water level to reach the land surface. The volume of flow is not necessarily an indication of head, rather a function of the casing diameter. Flow can be reduced merely by reducing the diameter of the casing at the land surface.

Some wells can be capped with a pressure gauge attached and read directly (Figure 5.25). In this case the elevation of the cap would be surveyed and the pressure gauge reading converted into feet of head and added to the elevation head. If the pressure head is not great, then a well can be fitted with a standpipe sufficiently high to keep the well from flowing. If the pressure head is too great, sometimes a pressure transducer can be plumbed directly into the well casing where the head reading is recorded on a data logger (Figure 5.26). A section on running pumping tests on flowing wells is given in Chapter 10.

Example 5.7

In Petroleum County, Montana, several wells have been drilled into several confined aquifers, such as the Cretaceous Basal Eagle Sandstone, and the 1st, 2nd, and 3rd Cat

Figure 5.25 Well cap pressure gauge.

Figure 5.26 Well casing with pressure transducer.

Creek (Brayton 1998). These wells had been flowing for decades until a rehabilitation program took place to control free-flowing wells. Some of the wells were controlled by installing flow reducers, while others were redrilled and completed with new casing and insulation and located below the frost zone to keep from splitting from freezing (Figure 5.27). This resulted in saving millions of gallons of water and the recovery of water levels on the order of 5 ft in just one year (Brayton 1998; Weight et al. 1999).

Summary of Level-Measurement Methods

Level-measurement devices are necessary to obtain static water-level data. Each has its own application and design. Generally, it may be helpful to have a variety of designs and depth capabilities.

Summary of manual methods

1. Establish a common datum. It may be worth defining a particular elevation above MSL so that maps of different scales can be constructed later on.

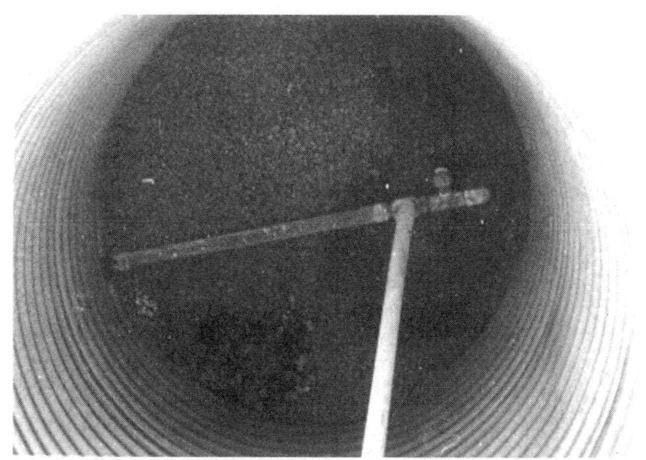

Figure 5.27. Well installed below frost zone.

2. Make sure that the base station is a relatively permanent feature, unlikely to be moved or removed later on.

3. Don't forget your keys, wrenches, and multitool to remove the well cap.

4. Establish a measuring point (MP) physically on the well casing or the location from which all water-level measurements will take place.

5. To prohibit effects from weathering, make all markings on the inside of the well or under the well cap.

6. Use a carpenter's chalk with a steel tape for a clearer reading.

7. E-tapes should have their sensitivity dials adjusted as appropriate to the water quality to get a crisp, clear buzz or light indication.

8. Always move the E-tape cable line to the middle of the well, shake lightly, and repeat, to make sure the reading is the same.

9. Record the depth to water from TOC or GS. The "stick up" should also be measured and recorded. Remember to put the well cap back on after the reading and secure the well.

10. Make sure your field notes are clear, because getting back to some wells again may be difficult.

If wells are locked or inside locked gates, then to gain access one must remember the appropriate keys or tools necessary to gain access. Multitools are handy to remove tightened bolts and loosen wedged well caps. Steel tapes are helpful to gain access to small openings in well caps and can be used to measure field distances. E-tapes are the most common level-measuring devices, and come in a range of designs. Probe size is important for small piezometers and pumping wells, and proper care must be taken to ensure the readings are accurate. Flat electronic cable designs are easy to read, with English and metric markings, and resist stretching, but may slip on the reel. Round cable designs do not slip but have markings on one side and may stretch on deeper readings. Other design features, such as the location of the battery housing, may be important for choosing among devises and maintenance capabilities.

Chart recorders are helpful in monitoring continuous hydrograph data, especially at remote sites. Problems include clogging of the pen, malfunction of the timer, the beaded cable coming off the drum or dead batteries. Data reduction is a more cumbersome step than from a data logger, but chart recorders have been used successfully for years.

Pressure transducers and data loggers are indispensable for rapid succession readings needed during pumping and slug tests. These are more expensive, but are easier to manipulate during the data reduction and analysis stage. This may also be helpful in obtaining levels on flowing wells.

Summary of automated methods

1. Lower each transducer down the well to allow the cable to straighten and spool each transducer cable to the data logger. Allow a 1-inch loop in the cable (to sense air-pressure changes), secured with duct tape and then taped to the well so that it doesn't move.

2. Check the psi range of the transducer and make sure the transducer isn't lowered into a water depth that exceeds its capacity.

3. Attach the transducer in the pumping well above the pump with electricians tape and secure the cable to the riser pipe every 5 ft (1.5 m) to the surface.

4. Make sure each transducer is connected to the data logger.

5. Establish the base level as a last item before starting the pumping test and use a value other than zero.

6. Back up the automated system with manual E-tapes.

Getting level-measurement devices stuck down a well is a real possibility in pumping wells. Fishing techniques are the first step, but it may be necessary to call a service rig to free an E-tape probe. This would only be a percentage of the cost of a new probe. Decontamination of level-measuring devises is essential to prevent cross-contamination. Simple rinsing with an

Alconox-water solution or steam cleaning can do a good job. Having clean equipment is important when showing up at the door of a landowner.

5.4 Misinterpretation of Water-Level Data

The misinterpretation of water-level data is a relatively common occurrence among those just getting started in field hydrogeology. It takes time and experience to think through certain questions to put together a reasonable hydrogeological flow model. Some of the following reasons for making errors in interpreting water-level data can be grouped as topics. The following list is by no means exhaustive, but will hopefully get the hydrogeologist thinking.

- Vertical component of flow
- Misunderstanding the difference between water levels in wells and the elevation of the water table
- Combining shallow and deep wells completed in the same aquifer
- Combining long and short screen lengths for wells completed in the same aquifer
- Combining level data for wells completed in different aquifers
- Combining level data from different times.

Shallow and Deep Wells

The first three topics can be essentially discussed together. Even if the geology of an area is fairly uniform, there is usually a vertical component to flow emanating from recharge to discharge areas (see Figure 5.8). The flow path of water particles is generally rectilinear (Wampler 1998). This means that very few water particles actually move along the top surface of the water table; rather they take a rectilinear path that represent local, intermediate, or regional flow systems. The development of the different systems is a function of the dimensions of the basin. Deeper basins tend to have more systems developed (Fetter 1994). In humid areas, increased topography also affects the local flow regime. More undulations in the surface topography result in greater numbers of local recharge and discharge areas (Fetter 1994). This must be taken into account when deciding on the depth and placement of wells. It also will affect one's interpretation of water-level data. For example, it may mean that wells completed at different depths may be screened at different parts of local and intermediate systems.

This brings up the problem of improper interpretations of water-level data resulting from combining data from wells completed both shallow and deep (Saines 1981). The question arises, how deep is too deep and what difference must there be to have a problem? The answer is complicated by the proximity of wells to the recharge or discharge area, the geology, and dimensions of the valley. One must think about these factors when interpreting the data. It may be that some of these difficulties can be resolved with additional information, such as water-quality data. It may be that there are different water-quality signatures represented by the different flow systems.

Example 5.8

A study was conducted along Blacktail Creek in Butte, Montana, to investigate why local residents were experiencing basement flooding (Figure 5.28). Geology in the immediate vicinity of Blacktail Creek consists of fine-grained sediment lenses mixed with sandy units. The sandy units appear to be laterally connected, with the fine-grained sediments generally constrained within the flood plain area of the creek. Nested wells completed at depths of 15 and 30 ft, respectively, within 100 ft (30.5 m) of the Creek showed water-level differences of 1 ft (0.3 m) between wells. Deeper wells indicated a water level approximately 1ft (0.3 m) higher than shallow wells, indicating an upward gradient. Differences between nested pairs indicated that if all data were combined, a confused interpretation would result.

Figure 5.28 Well drilling along Blacktail Creek, Butte, Montana.

The question exists whether all groundwater within the residential area comes from more than one flow system. Water-quality samples from the area indicate the possibility that waters from mining-impacted areas may be influencing waters discharging into the area. A schematic of this interpretation is presented Figure 5.29.

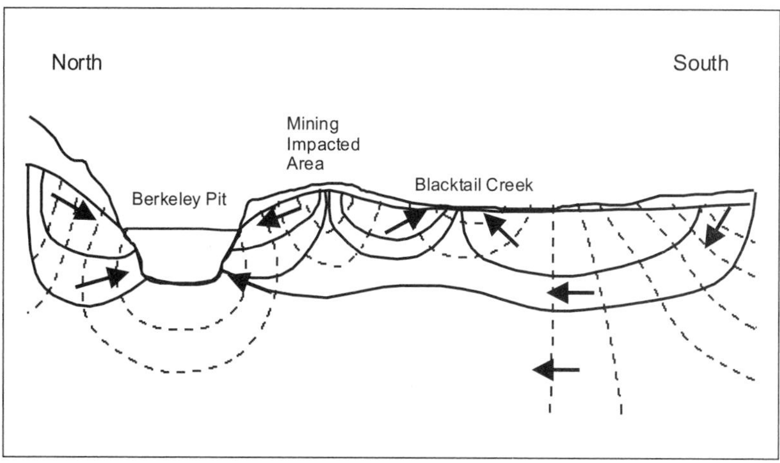

Figure 5.29 Schematic of Example 5.8 interpretation.

Example 5.9

Butte, Montana, is nestled in a valley surrounded by mountains including those of the North American continental divide. Like many intermontane basins of southwestern Montana, the basement rocks of the valley are structurally tilted to the east, resulting in thicker deposits on this side. In this case, the eastern sediments are approximately 900 ft thick and only about 100- to 200-ft thick on the western margin. Sediments are generally coarse grained, with lessor discontinuous clay lenses that locally have a confining affect. The valley slopes to the north in a gradual gradient as does groundwater flow. Shallow wells that penetrate to "first" water indicate a gradient of 15 ft/mi (or 0.003) (Botz 1969).

Two of the local drainages, Little Basin Creek and Blacktail Creek, initiate their headwaters in the Highland mountains to the south and flow northward into the Butte valley. The streams are bedrock controlled as they flow out onto the weathered granitic valley-fill materials, losing most of their water, which results in basin recharge. Deep circulation occurs in the valley, and stream flow picks up again as groundwater begins to discharge to the north. Wells completed near the discharge area to the north at depths of 120 ft (36.5 m) and 35 ft (10.7 m), respectively, show differences of 2 ft (0.61 m) between wells in with an upward gradient, while nested wells in other parts of the valley show the same water-level elevations or in the recharge area a downward component of flow. This essentially the scenario depicted in Figure 5.8.

Short Versus Long Screen Lengths

Another source of interpretation errors may come from wells completed with varying lengths of screen. It may be that wells at a site designed to monitor free product of a light nonaqueous phase liquid (LNAPL), such as gasoline, also have wells completed deeper to monitor cleaner water at depth, all within the same aquifer. Are all of the water levels telling the same story? It may be helpful to evaluate what the water level in a well indicates.

The water level in a well represents the total average head of the midpoint of the screened interval, including elevation and pressure head. The longer the screen length, the more vertical head changes that can be included in the average. Piezometers are distinguished from wells by having a relatively short length of screen (less than 1 ft, 0.3 m). The purpose of piezometers is to obtain hydraulic head data from a point. They are especially useful in distinguishing the vertical distribution of hydraulic head. Sometimes at a given site there are a combination of well completions, ranging from production wells with tens of feet of screen, to monitoring wells with shorter lengths of screen (5 ft, 1.5 m), to driven well points and piezometers with short screen lengths completed near the top of the aquifer (Chapter 6). The hydrogeologist needs to be careful in considering the geology, the hydrogeologic setting, and the well completion when coming up with a conceptual groundwater-flow model. This is particularly important if one is in a recharge or discharge area.

A well completed to detect an LNAPL , such as the example described in Figure 5.9, tends to have long screen lengths extending above and below the water table. This is to be able to monitor during seasons of high and low

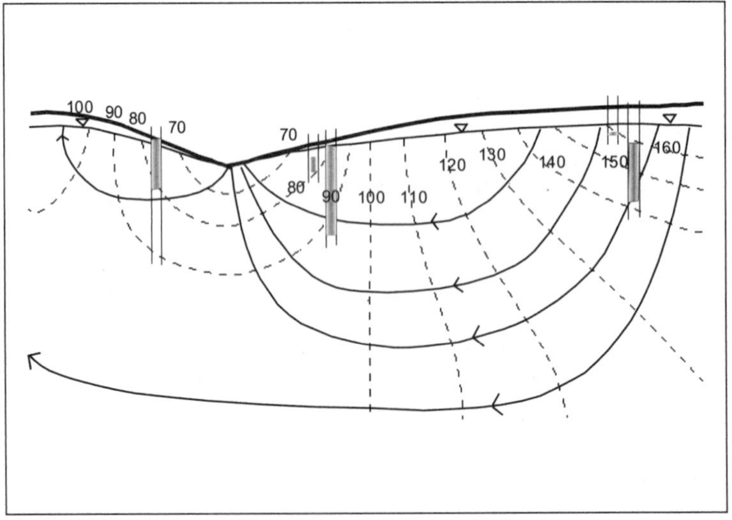

Figure 5.30 Water levels in a recharge area and discharge area, showing an average water level for a longer screen length.

water levels. If the well is completed in a recharge or discharge area, the average head may be below or above the water table. Consider the schematic in Figure 5.30.

Combining Different Aquifers

A woman from a conservation district in eastern Montana called one day to say she had monthly water-level and water-quality data for 24 months. Could I please make some sense out of it and tell whether there were any water-quality trends? My first question to her was, could I please obtain a base map and well completion information, including lithologic logs? One must know how many aquifers and confining units may be involved before proper interpretation can be performed. Understanding this is crucial to interpreting the direction of groundwater flow and contaminant movement.

Example 5.10

A consulting engineer was estimating drilling costs based upon water-level data within a

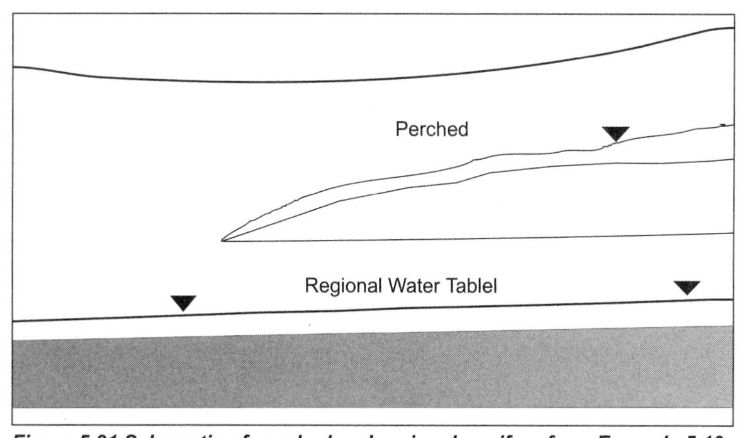

Figure 5.31 Schematic of perched and regional aquifers from Example 5.10.

subdivision area of a couple of square miles (5.2 km^2). Within the area was a perched aquifer and a regional aquifer with wells completed in both (Figure 5.31). The engineer did not understand the geologic setting and averaged the well-completion depths for both aquifers to obtain an estimate for total drilling footage. This resulted in a poor estimate of total drilling costs and surprises in the field when drilling began.

References

Botz, M. K., 1969. Hydrogeology of the Upper Silver Bow Creek Drainage Area, Montana. *Montana Bureau of Mines and Geology Bulletin 75*, 32 pp.

Brayton, M. D., 1998. *Recovery Response from Conservation Methods from Wells in the Basal Eagle Sandstone, Petroleum County, Montana*, Master's Thesis, Montana Tech of the University of Montana, 68 pp.

Davis, S.N., and Dewiest, R.J.M., 1996. *Hydrogeology*. John Wiley & Sons, New York, 463 pp.

Fetter, C. W., 1993. *Contaminant Hydrogeology*. Prentice-Hall, Upper Saddle River, NJ, 500 pp.

Fetter, C. W., 1994. *Applied Hydrogeology, 3rd Edition*. Macmillan, New York, 691 pp.

Freeze, A., and Cherry, J., 1979. *Groundwater*. Prentice-Hall, Upper Saddle River, NJ, 604 pp.

Heath, R.C., and Trainer, F.W., 1981. *Introduction to Groundwater Hydrology*. Water Well Journal Publishing Company, Worthington, OH, 285 pp.

Hubbert, M. K., 1940. The Theory of Ground-water Motion. *Journal of Geology*, Vol. 48, pp.785–944.

McIlvride, W.A., and Rector, B.M., 1988. Comparison of Short- and Long-Screen Monitoring Wells in Alluvial Sediments. *Proceedings of the Second National Outdoor Action Conference on Aquifer Restoration, Ground Water Monitoring and Geophysical Methods*, Vol. 1, Las Vegas, NV, pp. 375–390.

Saines, M., 1981. Errors in Interpretation of Ground-Water Level Data. *Ground Water Monitoring and Remediation*, Spring Issue, p. 56–61.

Toth, J. A., 1962. A Theory of Groundwater Motion in Small Drainage Basins in Central Alberta. *Journal of Geophysical Research*, Vol. 67, pp. 4375–4381.

Toth, J. A., 1963. A Theoretical Analysis of Groundwater Flow in Small Drainage Basins. *Journal of Geophysical Research*, Vol. 68, pp. 4795–4811.

Watson, Ian, and Burnett, A. D., 1993. *Hydrology: An Environmental Approach*. Buchanan Books, Cambridge, 702 pp.

Wampler, J. M., 1998. Misconceptions about Errors in Geoscience Textbooks, Problematic Descriptions of Ground-Water Movement. *Journal of Geoscience Education*, Vol. 46, pp.282–284.

Weight, W.D., Brayton, M.D., and Reiten, J., 1999. Recovery Response From Conservation Methods in Wells from the Basal Eagle Sandstone, Petroleum County, Montana. *1999 GSA Abstracts with Programs*, Vol. 31, No. 4, p. A60.

Chapter 6

Groundwater/Surface-Water Interaction

Another topic a hydrogeologist must be familiar with is how groundwater and surface water interact within rivers and drainage systems. This was briefly addressed in Chapter 5 under gaining and losing streams and in more detail under the discussion of water-level interpretations (Section 5.4). However, in this chapter, an extended treatment of this topic is presented, including concepts of channel geometry and groundwater exchange, the hyporheic zone, and stream health (Meyer 1997; Winter et al. 1998; Woessner 2000).

Traditionally, hydrogeologists considered groundwater/surface-water interaction in terms of water rights and streamflow reduction by evaluating distance-drawdown relationships and pumping wells adjacent to streams (Walton 1970; Sophocleous et al. 1995; Modica 1998; Winter et al. 1998). Flowing streams are complex systems with intricate interactions with living organisms (Hansen 1975; Grimm and Fisher 1984; Dahm et al. 1998), streamflow dynamics (Bencala and Walters 1983; Huggenberger et al. 1998), and groundwater discharge and recharge (Castro and Hornberger 1991). Additionally, streams are situated within a fluvial plain that may be meandering or braided and may have stratigraphically complex relationships, including rapid changes in grain size distributions, facies, and vegetation distribution (Anderson 1989; Mial 1996; Gross and Small 1998). There is a need to perform field work to better conceptualize groundwater/surface-water interaction within the fluvial plain and within channels (Woessner 2000).

Some of the reasons for having a basic understanding of groundwater/surface-water interactions are to better understand groundwater flow in fluvial plains for water-resource management, watershed assessment, watershed restoration, impacts to streamflows including water rights and minimum flows for riparian habitat. More work in the future will need to be performed as companies and regulatory agencies work together on preimpact assessments from mining, housing development, logging, construction, and postconstruction design. The purpose of this chapter is to discuss issues regarding groundwater/surface-water interactions and stream health, and present the basic principles the hydrogeologist will need to know to perform meaningful field studies.

6.1 Fluvial Plain

The **fluvial plain** represents the bigger picture of where field work is performed. The fluvial plain is a fairly planar feature consisting of the active stream channel, flood plain, and associated fluvial sediments, which also include older sediments associate with previous stream positions (Woessner 2000). In intermontane basins, the fluvial plain is bounded by highlands or terraces on either side (Figure 6.1). The fluvial plain system as described by Woessner (2000) is also referred to as an alluvial valley or a riverine valley (Dahm et al. 1998; Winter et al. 1998).

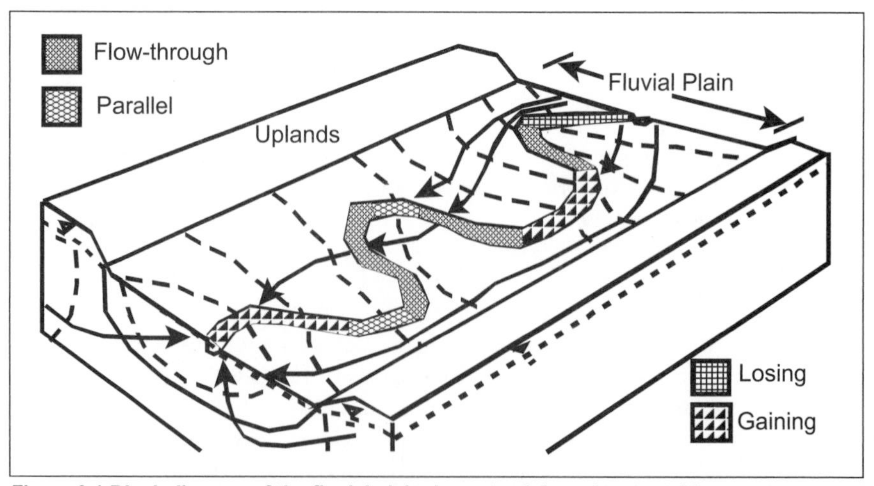

Figure 6.1 Block diagram of the fluvial plain. Isopotentials and water table are represented by dashed lines and groundwater movement by flow arrows; reaches are broken into losing, gaining, flow-through, and parallel stretches. [Adapted from Woessner (2000). Used with permission of Groundwater (2000).]

Characterizing a fluvial plain is difficult, because of the many dynamic processes occurring. Each stream carries a quantity of sediment and flow volume proportional to the gradient of the stream. During spring runoff, stream velocities and flow volumes are great enough to incise certain meandering stretches and may overflow the banks depositing sediments onto the flood plain. The lateral continuity of sediments change rapidly over distances of tens of feet (meters to tens of meters). Geological models of these systems vary significantly as one evaluates the differences between meandering streams (Figure 6.2) with lower gradients to braided stream systems (Figure 6.3) with high sediment loads (Anderson 1989; Mial 1996; Gross and Small 1998; Huggenberger et al. 1998).

Figure 6.2 Meandering stream in southwestern Montana.

Figure 6.3 Braided stream near Anchorage, Alaska.

Example 6.1

A research site along Silver Bow Creek approximately 9 miles (14.5 km) west of Butte, Montana, was selected to evaluate the hydrogeology of a mining-impacted streamside tailings setting. A number of wells and piezometers were placed to evaluate groundwater flow, aquifer characteristics, geology, and water-quality distribution. Over the years, the students from the hydrogeology field camp have added more wells, so well density is quite high. Figures 6.4 and 6.5 show a plan view and cross-sectional view of sediments at the site.

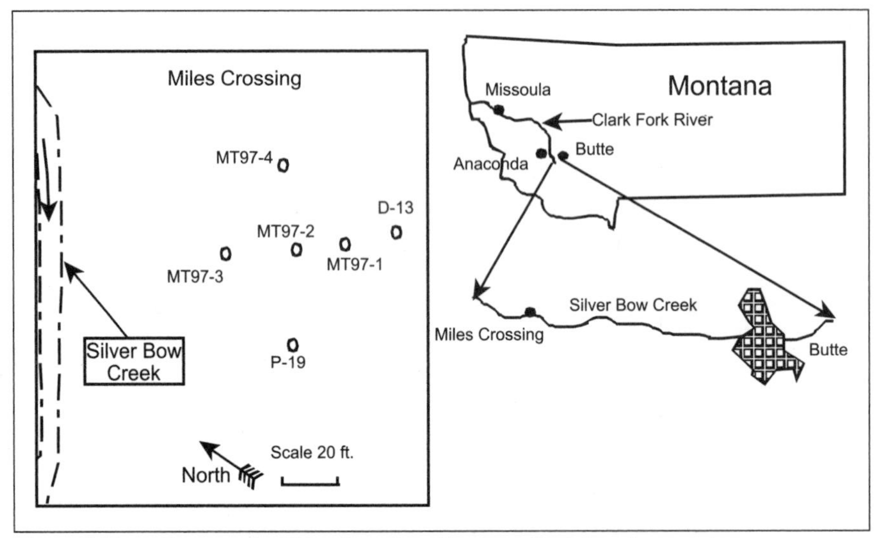

Figure 6.4 Location map for Miles Crossing, Montana, fluvial environment.

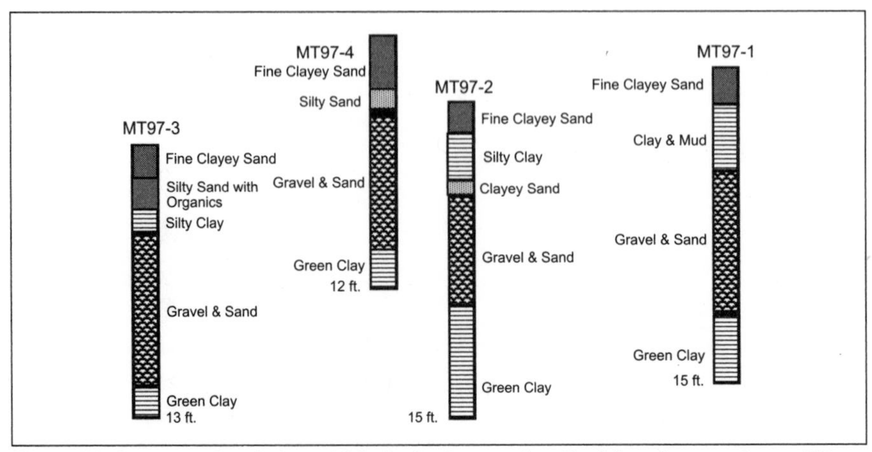

Figure 6.5 Cross-sectional view of lithologic changes in a fluvial environment near Miles Crossing, southwestern Montana.

One general result of the depositional processes within the fluvial plain is that units of higher hydraulic conductivity tend to follow the main axis of the valley, although they can migrate laterally during the meandering process (Winter et al. 1998; Woessner 1998). The trend of groundwater flow is down-plain with fluid transport mechanisms controlled by the higher hydraulic conductivity units (Woessner 2000).

Channel Orientation and Groundwater/Surface-Water Exchange

There are at least four general flow scenarios associated with groundwater/surface-water in the fluvial plain: losing, gaining, flow-through, and parallel (Winter et al. 1998; Woessner 1998, 2000) (Figures 6.1 and 6.6). Streams that lose water (provide recharge to the aquifer) have stages that are greater than the hydraulic head in the sediments underlying and adjacent the stream channel (Figure 6.6A). Conversely, when the head underneath and adjacent to the stream is greater than the stream stage, the stream gains water from the sediments within the fluvial plain (a groundwater discharge area) (Figure 6.6B).

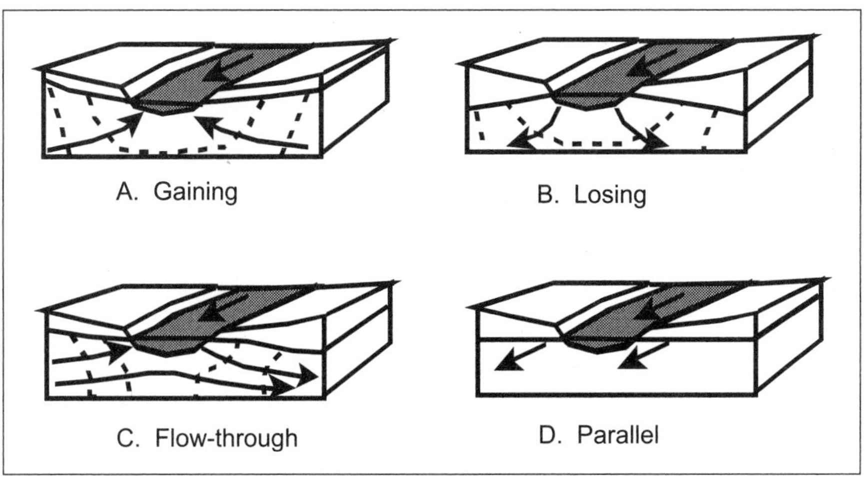

Figure 6.6 General flow scenarios associated with groundwater/surface water in the fluvial plain. [Adapted from Woessner (2000). Used with permission of Groundwater 2000.]

In a flow-through scenario, the head on one side of the stream bank is higher than the head on the other side of the bank (Figure 6.6C). This often occurs when the stream meanders and becomes oriented perpendicular to the fluvial plain and general direction of groundwater flow (Huggenberger et al. 1998; Woessner 1998). When the stream stage and hydraulic head in the adjacent sediments are equal, then parallel flow or "zero exchange" occurs (Figure 6.6D)(Woessner 1998). In this last case, the stream flow and groundwater flow are parallel to each other, although some minor exchanges may be occurring at the streambed scale (Woessner 2000).

Channels can take on a variety of shapes, configurations, and time of flow factors. It is beyond the scope of this chapter to discuss streams that are braided versus anastomosing or meandering and their respective hydrogeologic interactions, but by mentioning these and presenting the general concepts of the fluvial plain an increased awareness of the issues will occur. The discussion is also limited to perennial systems where riparian habitat is present with active hyporheic zones. More work needs to be done on intermittent systems that flow part of the year and develop some of the dynamics mentioned. Ephemeral streams represent channels where flow is only direct connection with snow melt or precipitation runoff. Most of these are losing streams.

6.2 Hyporheic Zone

Surface water and groundwater are considered by hydrogeologists as a single system, since interaction occurs along the total stream length where hydraulic connection with shallow groundwater occurs (Winter et al. 1998; Woessner 2000). This is indicated by the covariation of hydrographs of stream stage and shallow groundwater wells (Stanford and Ward 1993). Although covariation responses can be tracked, this does not necessarily mean that mixing is taking place. Combining the perspectives of stream and riparian ecologists, biologists, geochemists, and hydrogeologists, the **hyporheic zone** can be viewed as the mixing zone of stream water and groundwater (Stanford and Ward 1988; Triska et al. 1989; Winter et al. 1998; Woessner 2000).

Groundwater discharge areas and influxes along meanders from parallel flow in the longitudinal direction may serve as higher nutrient exchange sites that enhance algal growth, the food base for any stream system (Dahm et al. 1998). This may also affect the distribution of macroinvertebrates within active streams (Wright 1995; Wallace et al. 1996). However, macroinvertebrates may be found throughout the floodplain of large alluvial systems in the western United States over a mile (2 to 3 km) distant from the active channel, when high porosity and permeability paleochannels are connected within the flood plain (Sanford and Ward 1993).

At the fluvial plain scale (Figure 6.1), upland waters flow toward the floodplain of a stream system. Discharge occurs in the **riparian zone**. The riparian zone is commonly referred to as the interface between the terrestrial and aquatic zones (Gregory et al. 1991). It is characterized by trees and other large vegetation and generally does not contain surface water except during episodic floods (Dahm et al. 1998). In smaller stream systems, the riparian zone may be characterized by willows or other shrubs (Figure 6.7).

Figure 6.7 Riparian zone characterized by willows and shrubs along the Clark Fork River, southwestern Montana.

Within the fluvial plain system, on the stream side of riparian zones, groundwater and nutrient exchanges are occurring primarily along parafluvial sections, where water enters the gravel at riffles and exits at pools or moves through gravel bars and meanders sections discharging on the other side (Wroblicky et al. 1998).

Channel Geomorphology and Stream Connectivity

The exchange of groundwater and surface water discussed above can be partially explained by understanding the geomorphology within fluvial plain systems. Stream reaches can be categorized as constrained, unconstrained, **aggrading**, or **degrading**. Constrained stream reaches can be defined as fluvial plains less than four times the width of the active channel and unconstrained at greater than four times the active channel (Dahm et al. 1998). Aggrading reaches are those that are depositing sediment, while degrading reaches are where active downcutting and erosion are occurring. Straight channelized sections and canals tend to be areas where groundwater/surface-water interaction becomes disconnected. Additionally, unconstrained degrading streams rarely have losing reaches within them

(Dahm et al. 1998). A summary of nutrient retention studies in groundwater discharge, recharge, and parallel flow scenarios along with various channel configurations was hypothesized by Dahm et al. (1998) and is shown in Figure 6.8.

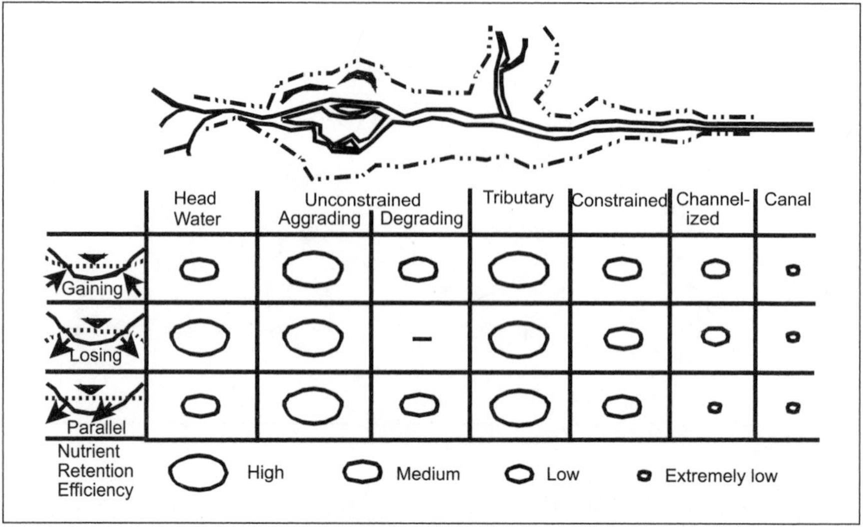

Figure 6.8 Hypothesized nutrient retention efficiencies in various groundwater/surface-water settings and channel configurations. (Losing reaches rarely occur in degrading sections of unconstrained stream.) [Adapted from Dahm et al. (1988).]

Dahm et al. (1998) hypothesize that the geomorphologic and hydrogeologic conditions represented in Figure 6.8 portray the retentive capacities of solutes and their ability to alter the volume and direction of interactive exchange between the active stream, the riparian zone, and the alluvial aquifer. The highest interactions between groundwater and surface water are hypothesized to occur in tributaries and aggrading reaches, while channelized reaches are believed to be disconnected (Dahm et al. 1998).

Nutrient retention can be viewed as solute sources that have sufficient residence times and interactive exchanges between surface waters and groundwater to promote a complex variety of biochemical processes occurring throughout the fluvial plain (Dahm et al. 1998). These processes may also be complicated by an intermittent vadose zone within gravel bars and in the riparian zone, providing an increased oxygenated environment.

A summary of processes within the hyporheic zone of streams can be divided into chemoautotrophic and electron-accepting conditions (Dahm et al. 1998). Stumm and Morgan (1996) present an example of the various processes, including energy yields by taking a pH solution of 7 and using CH_2O as a typical organic substance. These are presented in Table 6.1.

Table 6.1 Chemoautotrophic and Electron Acceptors in the Fluvial Plain [From Stumm and Morgan (1996)]

Chemoautotrophic Processes	Kcal/Equiv	Electron Accepting Processes	Kcal/Equiv
Methane oxidation	-9.1	Aerobic respiration	-29.9
$CH_4 \longrightarrow CO_2$		$CH_2O \longrightarrow CO_2$	
Sulfide oxidation	-23.8	Denitrification	-28.4
$S^{-2} \longrightarrow SO_4^{-2}$		$NO_3^- \longrightarrow N_2$	
Iron oxidation	-21.0	Iron reduction	-7.2
$Fe^{+2} \longrightarrow Fe^{+3}$		$Fe^{+3} \longrightarrow Fe^{+2}$	
Nitrification	-10.3	Sulfate reduction	-5.9
$NH_4^+ \longrightarrow NO_3^-$		$SO_4^{-2} \longrightarrow S^{-2}$	
		Methanogenesis	-5.6
		$CO_2 \longrightarrow CH_4$	

The redox potential within the riparian zone is a function of residence time and metabolism rates within mixing zone of upland waters and hyporheic waters and their chemical characteristics (Dahm et al. 1998). Organic carbon decomposition can be carried out through the reduction of terminal electron accepting processes listed in Table 6.1. The rates and distribution of biogeochimical processes are still not well understood. Additionally, there are a number of metabolic pathways that oxidize dissolved or particulate matter and selected nutrients (Dahm et al. 1998). For example, Hill (1993) has shown that riparian zones are efficient removal sites for nitrate from upland flow areas, with nitrogen removal exceeding 80% from a survey of streams.

At the tens of feet or meters scale, groundwater/surface-water exchange can be partially evaluated through the use of minipiezometers and comparing the head differences between groundwater and surface water (Lee and Cherry 1978; Woessner 2000). A schematic of flow conditions for overall gaining or losing scenarios is presented in Figure 6.9.

Figure 6.9 shows how streambed topography can affect localized water exchanges even though the reach may be overall losing (Figure 6.9A) or gaining (Figure 6.9B). Hydraulic head data can be coupled with basic water-quality surveys (dissolved oxygen, specific conductance, and temperature) to provide a multidisciplinary approach. Additional perspectives can be gained by studying nutrient patterns and macroinvertebrates.

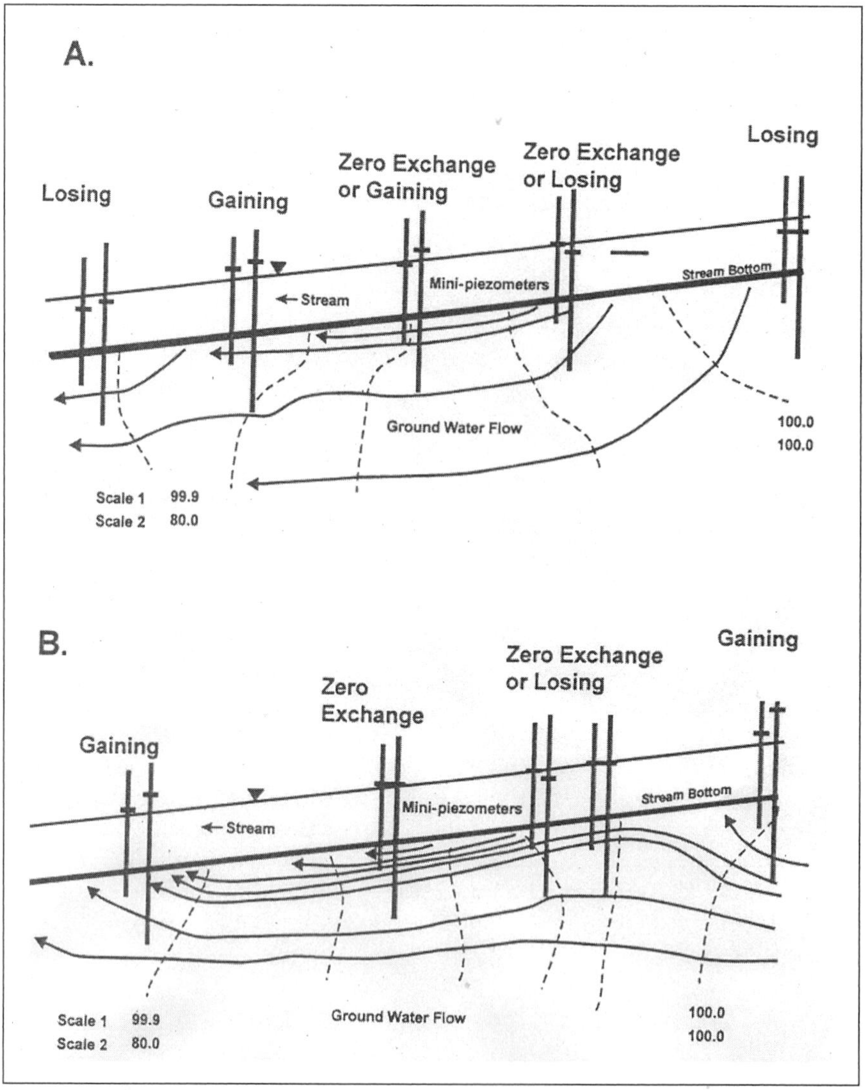

Figure 6.9 Streambed topography and localized water exchanges. [From Woessner (2000). Reprinted with permission of Groundwater 2000.]

One of the keys to understand biochemical processes is tied to an understanding of the fluvial plan hydrogeology. Places of relative constant influx or infiltration and the distribution of macro- and microbial fauna all contribute to gauging the health of a given stream.

6.3 Stream Health

A healthy stream as proposed by Meyer (1997) is an ecosystem that is sustainable and resilient in maintaining its ecological structure and function over time. This structure leads to discussions of stream, hyporheic, riparian zone, and fluvial plain functions and their respective critical components (Woessner 2000). Ecologists and scientists may provide measures of stream health, but defining the respective goods and services expected of a stream requires input of societal values (Meyer 1997). Examples of goods and services valued by policy makers are shown in Table 6.2. As societal values change, so do the policies and laws that are implemented for stream maintenance and restoration.

Table 6.2 Goods and Services Valued by Healthy Streams [After Meyer (1997)]

Goods	Services
Clean water for drinking, washing, and other uses	Cleansing and detoxifying water
Adequate supply for irrigation and industry	Maintaining water supply
Uncontaminated foods (seafood, crayfish, shellfish)	Reducing sediment inputs in coastal areas
Challenging waterways for kayaking	Providing aesthetic pleasure
Sites for swimming	Produce fish for angling
An environment for meditation and renewal	Decomposing organic matter
Unique species and wildlife to observe	Storing and regenerating essential elements

The concept of stream health is analogous to human health, in that the absence of disease in humans can be compared to the absence of pollution, acids, and toxins (Meyer 1997). This can be measured by macroinvertebrate studies, the presence or absence of fish populations, or other biotic indicators (Wright 1995; Wallace et al.1996). Biota are stress indicators but do not give a full picture of the fluvial plain. The focus of modern medicine is more on wellness and maintaining health rather than curing diseases (Meyer 1997). Stream health is conceptually similar but requires a multidisciplinary assessment for a proper evaluation. Both terms are imprecisely defined, but useful.

Another measure of stream health is how quickly it recovers from a disturbance. This is how resilience can be evaluated, although rates of recovery have not been applied to stream health (Meyer 1997). Comparative studies are needed between impacted and natural stream reaches (Woessner 2000). The purpose of presenting this chapter is to heighten the awareness of hydrogeologists and to consider their role in designing field work for stream restoration and evaluating impacts from hazardous waste contamination in the hyporheic zone.

6.4 Field Methods to Determine Groundwater/ Surface-Water Exchange

There are several field methods hydrogeologists use to evaluate the interaction of groundwater and surface water in fluvial plains. A combination of these will usually provide a better understanding of the flow system. Perhaps the most common ones are:

- Stream gauging
- Parshall Flumes
- Crest-stage gauge
- Minipiezometers and wells
- Seepage meters
- Tracer studies
- Chemical mass balance

If a hydrogeologist teamed up with an ecologist, additional insights on riparian functionality and general stream health can be included to obtain a more complete picture (Palmer 1993; Woessner 2000). The scope of this chapter is to include only a cursory presentation of the issues important to stream biota and focus more on the physical phenomenon of groundwater/surface-water interaction and the associated field tasks facing the hydrogeologist.

Stream Gauging

Steam gauging is the field method of quantifying flow along a given stretch of stream. In smaller streams this is done by using a handheld velocity meter and stadia rod to obtain point velocities along a stream cross section (Figure 6.10). This can be done as long as the current flow and depth are not great enough to affect the safety of the field person. In larger and deeper streams, gauging is performed at a bridge or from a cage suspended from a cable pulley system. In deeper, higher-velocity streams, a weighted **Price** meter is used to determine stream velocities per fraction of cross-sectional area (USGS 1980). A Price-type meter has a horizontal wheel approximately 5 inches (12.7 cm) in diameter, with small cups attached. The wheel rotates in the current on a cam attached to the spindle. An electrical contact creates a "click" sound with each rotation. The clicks are transmitted to a headset worn by the field person to obtain a clear signal. The clicks are counted for 30 or 60 seconds or sent to a direct readout meter. The number of clicks is compared with a calibration curve to obtain the velocity.

Figure 6.10 Stream gauging with a handheld velocity meter and stadia rod.

Streams have a parabolic velocity profile that is affected by the bottom and sides of the stream (Fetter 1994)(Figure 6.11). The average velocity is found at approximately the 0.6 depth from the surface, although an average of the 0.2 and 0.8 depths are also used (Buchanan and Sommers 1969). The fastest part of current occurs in the middle and center of the stream. If no velocity-measuring equipment is available, a quick estimate of flow can be obtained by using the **float method**. In the float method, a distance of 10 to 20 ft (3 to 6 m) is paced along the bank of a straight stretch of stream. A stick is tossed in the middle of the stream above the first marking and timed how long it takes to move from the first to the second marking. This yields an average velocity for the center of the stream. The estimated velocity should be multiplied by a correction factor of 0.75 or 0.80 to account for friction along the sides and bottom of the channel. The correction factor is applied according to the uniformity of the profile across the stream. Edges with higher friction factors, such as vegetation, would have the 0.75 correction applied, where the more uniform velocity would have the 0.8 factor applied. The cross-sectional area is estimated by measuring several depths across the stream where the flow "looks" average. The velocity times the area yields a flow in cubic feet per second (ft^3/sec).

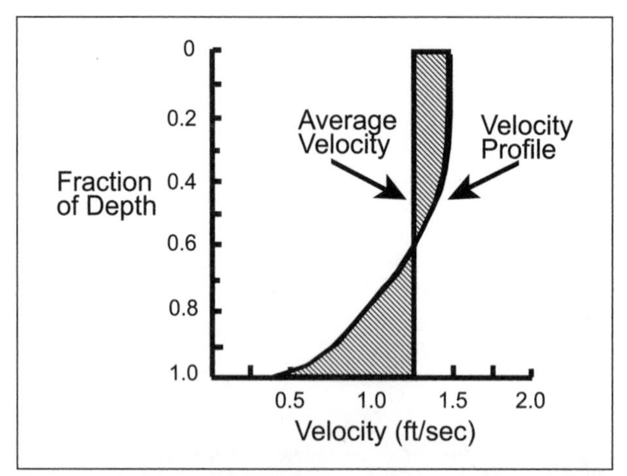

Figure 6.11 Velocity of stream as a function of depth. (The 0.6 depth is near the average velocity.) [After Fetter (1994).]

Larger stream systems may have dedicated gauging stations to evaluate long-term monitoring (Figure 6.12). The distances between stations are typically miles to tens of miles (kilometers to tens of kilometers) apart. In order to establish a gauging station, a detailed cross section of the stream is surveyed during low-flow conditions. Multiple discharge readings are made at different stages to establish a functional relationship between depth and flow, known as a stage-discharge curve (Figure 6.12).

An inlet pipe allows water to enter the stilling well (often constructed of corrugated pipe). The float moves up and down according to the stage, strung by a beaded cable, which passes over a pulley at the recorder with a counterweight hanging on the other side. The shed housing the system is known as the dog house. The stage is recorded via a float-recorder system (described in Chapter 5).

Some watersheds have gauging stations near the watershed outlet, before there is a confluence with another channel. However, to understand the dynamic local groundwater/surface-water interactions occurring within a watershed, it is necessary to take multiple discharge readings along the stream. One must account for every contribution be accurate. For example, there may be minor tributaries, springs, irrigation ditches or pipe discharge occurring along the stream that may not be readily appar-

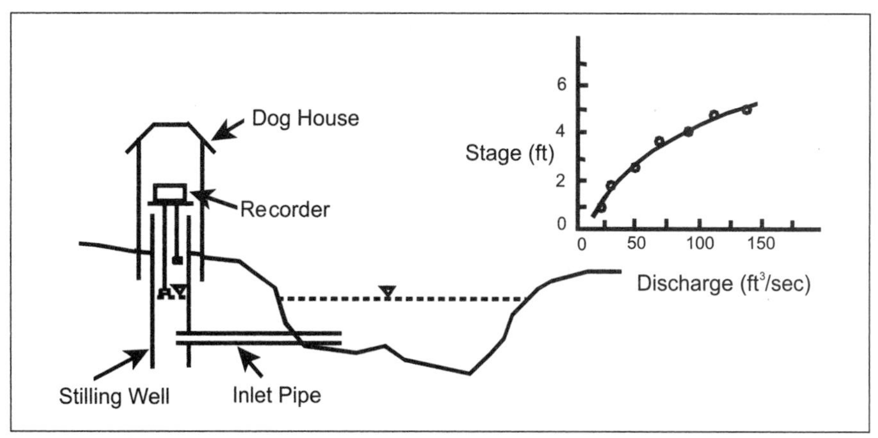

Figure 6.12 Schematic showing a gauging station and associated stage discharge curve.

ent. A detailed evaluation of all gains and losses within a stretch is known as a **synoptic survey**. For the purposes of discussion, it is assumed that these streams and associated tributaries can be gauged with a current meter attached to a wading rod at least some time during the field season.

For smaller streams, the U.S. Geological Survey used or still uses a **pygmy** meter, similar to the Price meter; however, all the components are at a smaller scale. Cups attached to a wheel spin proportionally to the current. It is critical to perform a "spin test" prior to use and properly maintain this equipment. A spin test is conducted by blowing on the cups to allow them to free spin for a minute or so. If they stop within 30 to 40 seconds, there is something wrong that needs to be fixed. It may be that a drop of lubricant is needed or a nick in the spindle tip may be the problem. Maintaining these meters so that they will pass the spin test is probably the most common problem with these devices. For this reason, there is a trend to use meters equipped with sensors to measure velocity.

Another type of current meter creates a magnetic field as the moving water passes the sensing device (Figure 6.13). Changes in velocity are correlated with changes in voltage amplitude. Velocities are directly read from a meter that hangs around the operator's neck. The great thing about this type of device is that it can be used in streams or irrigation ditches with weeds or other debris that would normally clog up a spinning wheel. In comparing velocity results during field camps using both types of current meters, correlations were found to be within the error ranges found with multiple readings using the same device. The following ratings are given when comparing measurements: excellent (2%), good (5%), fair (8%), and poor (over 8 %) (USGS 1980).

Figure 6.13 Flow meter with sensing device to indicate stream velocity.

An ideal site would be a stream stretch that is relatively straight and free from obstructions and vegetation (Sanders 1998). The field operator should test the stream for depth, strength of current, and the bottom for safety and stability before wading out with the equipment. A tag line is stretched across the stream with graduated markings and secured, so that velocity measurements can be accurately divided at the appropriate intervals. A 100-ft (30.5 m) cloth tape works well for many applications. The stream width should be divided into approximately 20 sections, so that less than 5 to 10 % of the total discharge passes through any given section. This provides more repeatable results. The field operator should bring:

- Field book
- Waders
- Graduated tagline or 100-ft (30. 5-m) cloth tape
- 10-ft (3-m) measuring tape, for making shorter measurements
- Writing utensils and extra supplies
- Extra batteries for field equipment
- Calculator

The current meter should be attached to the wading rod, and the operator should stand on the downstream side facing upstream. A velocity reading is taken within each section. Hold the rod out with the cups or current sensor facing upstream set at the 0.6 depth, unless the water depth is more than 2.5 ft (0.75 m) (USGS 1980). The wading rod should be kept vertical. When the water depth is greater than 2.5 ft (0.75 m), readings should be taken at the 0.2 depth and 0.8 depth to obtain an average. The total flow is calculated by summing the individual velocities and multiplying the result by the respective cross-sectional areas (Equation 6.1). A repeat is made of each measurement. An example sample data sheet for stream velocity measurements is also presented.

$$Q = \sum_{i=1}^{n} v_i A_i \quad [6.1]$$

Stream Velocity Measurements Data Sheet										
Date		**Stream**				**Weather**				
Location										
Meter type					**Meter coefficient (vel/rev)**					
Observer					**Stage gauge reading**					
Comments										
Angle Coef.	**Dist. from edge**	**Width**	**Depth**	**Obs. point depth**	**Revs**	**Elapsed Time**	**Vel-city**	**Angle adj.**	**Area**	**Dis-charge**

Example 6.2

Approximately 5 miles west of Butte, Montana, is the Sand Creek drainage basin. A hydrogeologic study was conducted to create a "transitional baseline" in the midst of industrial development and water-use changes (Borduin 1999). A significant area of the drainage basin has been zoned heavy industrial. Stream gauging was conducted along Silver Bow Creek and Sand Creek and other tributaries to document fluxes and gaining and losing stream sections. Later, this information was used to compare numerical modeling results of the basin to compare modeled stream fluxes with field gauged data. Agreement of potentiometric surface data and stream fluxes was an indication that calibration objectives were being achieved. Stream reaches less than 5 ft (1.5 m) wide were divided up every 0.5 ft for velocity measurements. Wider stream reaches were divided into enough widths to represent approximately 10% of the total flow. A direct readout current meter was used to obtain point velocities at the 0.6 ft depth. An example is shown in Table 6.3 for a reading along Silver Bow Creek with associated calculations.

Table 6.3 Example of Stream-Flow Calculations Site 1, Silver Bow Creek, August 27, 1998 [From Borduin (1999)]

Distance (ft)	Depth (ft)	Velocity (ft/sec)	Area (ft²)	Discharge (ft³/sec)
Left bank = 1	0.9	0.16	0.9	0.144
3	0.97	0.26	0.97	0.252
4	1.4	0.36	1.4	0.504
5	1.5	0.48	1.5	0.72
6	1.87	0.76	1.87	1.421
7	2.1	1.04	2.1	2.184
8	2.35	1.05	2.35	2.468
9	2.4	1.15	2.4	2.76
10	2.65	1.09	2.65	2.889
11	2.45	1.05	2.45	2.573
12	2.2	0.99	2.2	2.178
13	2.0	0.94	2.0	1.88
14	1.77	0.74	1.77	1.31
15	1.9	0.68	1.9	1.292
16	1.8	0.48	1.8	0.864
17	1.8	0.43	1.8	0.774
18	1.95	0.34	1.95	0.663
19	1.7	0.26	1.7	0.442
20	1.3	0.14	1.3	0.182
21	0.75	0.0	0.75	0.0
			Total	25.5

Stream gauging is useful when net gains and losses are obvious (flow changes are greater than 10%); however, when performing a synoptic survey within a drainage basin or along a significant length of stream, there are many sources of error (Roberts and Waren 1999). Some of the following can render a stream gauging survey inconclusive:

Gauging during the spring or near the time of a precipitation event when stream flow is changing. Gauging at low-flow when conditions are constant yields improved confidence in results,

Diurnal changes that occur from overnight to early morning to evening may vary significantly if vegetation growth is healthy. In this case, measurements should be taken at similar times of the day over a period when stream flow is stable,

Ignoring or missing small contributions such as minor tributaries, springs, irrigation ditches, or pipe discharge can result in significant water-budget errors.

Parshall Flumes

During a synoptic study of stream discharge, it is very important to identify and gauge all surface water inputs or outputs. This includes irrigation diversions and irrigation return flows. In the semi-arid Western states, irrigation ditches are extremely common and are often difficult or impossible to locate from a topographic map. Ditches are sometimes easier to spot in the field or from an aerial photo, and can be traced upgradient to their origin at the stream source. However, there is no substitute for simply walking down the entire drainage, catching any unmapped diversions and returns. In a synoptic study of Big Lake Creek, south of Wisdom, Montana, a downstream traverse uncovered 10 to 12 irrigation diversions with significant flow over approximately 10 miles of creek. Only a few of these diversions were obvious from preliminary reconnaissance work.

Near the point of divergence, landowners will often place a headgate with a weir or flume to control and quantify the amount of water withdrawn from the stream (Figure 6.14). In Southwest Montana, Parshall flumes are especially common. The Parshall flume is essentially a tapered box with a rectangular cross section, flaring both upstream and downstream, and with a slight constriction in the middle. The total discharge in cubic feet per second (cfs) is computed knowing the depth of the water and the width of the flume at its narrowest point (University of Wyoming 1994). Care should be taken in the field to make sure that the flume is in good condition. Problems include obstructions and making sure that the flow is laminar and that there is no evidence of water passing around or underneath the flume box. If high precision and accuracy is required, it is a good idea to check the flume measurement with your own estimate of flow based on velocity transects. Although Parshall flumes are made to rigid specifications, problems can occur with faulty installation or wear and tear. Because the relationship of flume width, water depth, and stream discharge is nonlinear, there is no simple equation to calculate stream discharge. Instead, the hydrogeologist must consult a Parshall flume table for the appropriate conversion (University of Wyoming 1994).

Figure 6.14 V-notch weir used to estimate discharge from a mine adit.

Crest-Stage Gauge

Sometimes it is helpful to know what the stage was at a flooding event in a remote area or an ephemeral drainage. If no permanent gauging system is available and the location is important to evaluating peak flows, a **crest-stage gauge** can be installed (Buchanan and Sommers 1968). A crest-stage gauge is essentially a calibrated stick inserted inside a clear tube. The tube is constructed so that water can enter at the bottom and allow air to escape at the top. A material such as ground-up cork is inserted in the tube so that when the stage increases, the cork floats along with the rising stage. When the stream reaches a maximum stage (or crests) and begins to drop, the cork will adhere to the calibrated stick and the inside of the tube. The tube and stick can be placed inside a PVC casing to protect the gauge, which is attached with metal strapping to a fence post on the downstream side of an ephemeral drainage (Figure 6.15).

Example 6.3

While performing baseline studies for a work for a prospective coal mine in southwestern Wyoming, it was necessary to obtain peak discharges for several ephemeral drainages that occurred during spring runoff. None of the drainages contained water except in response to spring snow melt and precipitation events. The crest-stage gauge design presented in Figure 6.15 was used at a couple of locations along each drainage to obtain upstream and downstream values. Each post was set approximately 3 ft

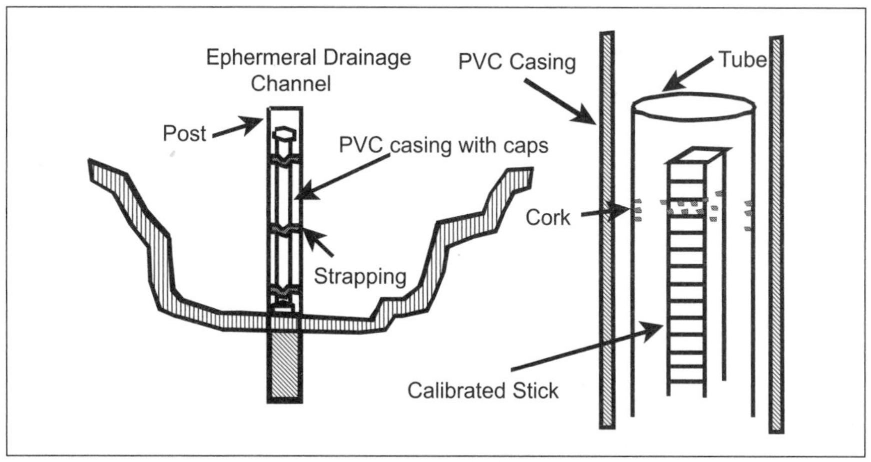

Figure 6.15 Crest-stage gauge attached to a fence post with metal strapping. (Holes drilled around the base of the PVC casing and the tip cap allow air circulation. A detailed diagram on the right shows a clear tube, calibrated stick, and cork material. Cork material floats during high stage and adheres to the tube and stick as the stage decreases.

(0.9 m) into the stream channel in concrete. Each crest-stage gauge location was surveyed to obtain a cross-sectional area. An estimate of the velocity was made by using the Manning Equation. The Manning Equation uses four parameters to estimate the velocity (Equation 6.2).

$$V = \frac{1.49 R^{2/3} S^{1/2}}{n} \qquad [6.2]$$

where:

V = average estimated velocity, ft/sec

R = hydraulic radius, ft^2/ ft and R = Area/WP

WP = wetted perimeter (Figure 6.16)

S = estimated slope of the water surface

n = Manning roughness coefficient

The Manning equation, when length is measured in meters and time in seconds is:

$$V = \frac{R^{2/3} S^{1/2}}{n} \qquad [6.3]$$

The Manning roughness coefficient ranges from 0.025 for streams that are straight, without brush and weeds or riffles and pools to 0.15 for channels containing dense willows (Chow 1959). Some of the more common values for streams with grasses, rocks, and pools range from 0.035 to 0.045. See Chow (1959) for tables and photographs of all types of roughness coef-

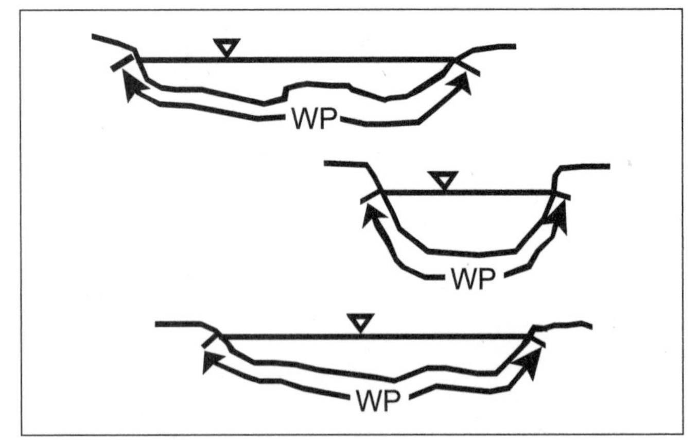

Figure 6.16 Schematic showing the wetter perimeter (WP), the bottom surface from edge to edge. (Note that the WP is wider than the width of the channel.)

ficients. To make a reading, the PVC cap is removed and the tube with the calibrated stick is pulled out. After a reading is taken, the cork material is washed to the bottom of the tube, to be ready for the next event.

For areas with established gauging stations, a crest-stage gauge would be installed next to the bank at average flow. In this case, the crest-stage gauges would have to be checked more frequently. The ground-up cork material will only record the peak stage height. If there were multiple peak events, only the highest one can be measured. It will be hard to distinguish which peak event was recorded by the crest-stage gauge.

Minipiezometers

Elevations of hydraulic head for local and intermediate groundwater flow systems are usually determined from water levels from wells (Chapter 5). Wells tend to be more permanent features installed using power equipment or drill rigs contrasted with minipiezometers that are manually placed and retrieved (Lee and Cherry 1978). In evaluating groundwater/surface-water interactions or in determining the head differences at a smaller scale (meters to tens of meters), **minipiezometers** are great tools. Minipiezometers can be constructed with a variety of construction materials including stainless steel, different types of tubing, or PVC. Minipiezometers are screened over a short length (a few inches or centimeters) and therefore provide "point" estimates of hydraulic head at a particular depth.

The mixing of surface water and groundwater takes place within the near-channel sediments (Woessner 2000). The mixing process can be very complicated (discussed in Section 6.4) and at many different scales because of the dynamics of surface-water movement and heterogenieties within the near channel sediments (Bencala et al. 1983, Huggenberger et al. 1998). Minipiezometers can be installed along the stream bank or in the streambed to partially evaluate the different flow regimes, as depicted in Figure 6.6.

Installation of minipiezometers into stream sediments devoid of cobbles is pretty simple; however, stream beds armored with rocks and cobbles are nearly impossible to install manually. A couple of installation methods are presented. The examples will assume we are using ½-in. (1.2-cm) stainless steel electrical conduit. Electrical conduit is typically sold in 10-ft (3-m) lengths at minimal cost. These may be cut to any shorter lengths with a hack saw. Perforations can also be made with a hack saw, staggering the cuts for increased strength. A loosely fitted bolt is "scotch" taped to the bottom of the piezometer, to keep the tube from crimping (Lee and Cherry 1978). At the top end, another bolt is used for support and in keeping the top open during pounding with a mallet or heavy hammer. The author has found reasonable success pounding on a block of wood placed on the top of the minipiezometer (Figure 6.17). Once the piezometer is driven 1 ft (30 cm) or more into the sediments, the minipiezometer tube is twisted and pulled back slightly to free the bolt, which is left in the sediments (Lee and Cherry 1978). This process could be referred to as the **bolt method**.

When a significant fraction of gravel is found in the sediments, the bolt method may not provide sufficient support for the piezometer tube. In this case, a metal support rod similar in length to the piezometer can be used effectively (the **support-rod method**). An ideal material to fit inside the thin piezometer tubing material is a steel grounding rod. The smoother the support-rod material, the better (English 1999). Good results can also be obtained with a heavy gauge copper wire. Steel rebar turns out to be a poor choice because the ribbed edges pull sediments into the piezometer when retracting, and will bend relatively easily when encountering cobbles and rocks. The technique is to insert the support rod inside the piezometer tubing and pound it in with a hammer. The support rod should only protrude an inch (2.5 cm) or so beyond the end of the piezometer and extend a short distance above the top end. The stress of the pounding is all taken up by the support rod. Once retraction of the rod has taken place, some water can be poured into the piezometer to flush the screened zone and evaluate the connection with the stream sediments. The screened zone can be constructed by cutting slots with a hacksaw or drill holes with the perforations wrapped with 0.2-mm nylon mesh or fiberglass cloth (Lee and Cherry 1978). The support-rod method is the most efficient method in gravelly conditions, but may not work if the stream-bed rock density is too great.

Figure 6.17 Installing a minipiezometer.

The water-level difference between the piezometer and the stream stage indicates the change in head with depth. The respective levels can be found by lowering a small diameter E-tape down the piezometer to get a reading and then dangling the E-tape from the top of the piezometer. To obtain a clearer understanding of the vertical gradient profile in gaining streams, it may be helpful to have a group of piezometers placed at graduated depths (1, 2, and 3 ft; 30, 60, and 90 cm).

Something should be said about where to place a minipiezometer. A minipiezometer should not be placed in the middle of the stream where current is high and difficult to read levels. Near the edge of a stream or just next to the channel will allow installation and readings without fighting current effects. If multiple minipiezometers are being placed as a nest, it may be helpful to place adjacent to the channel, however placing up on the bank away from the stream will require greater depth of penetration. It is also helpful to have the tops of nested piezometers to be level one with another so that rapid comparisons of water levels can be made, even though they may be driven to different depths. Vertical gradients from minipiezometers are measured by comparing the heads in the respective piezometers with the perforation zones (Figure 6.18).

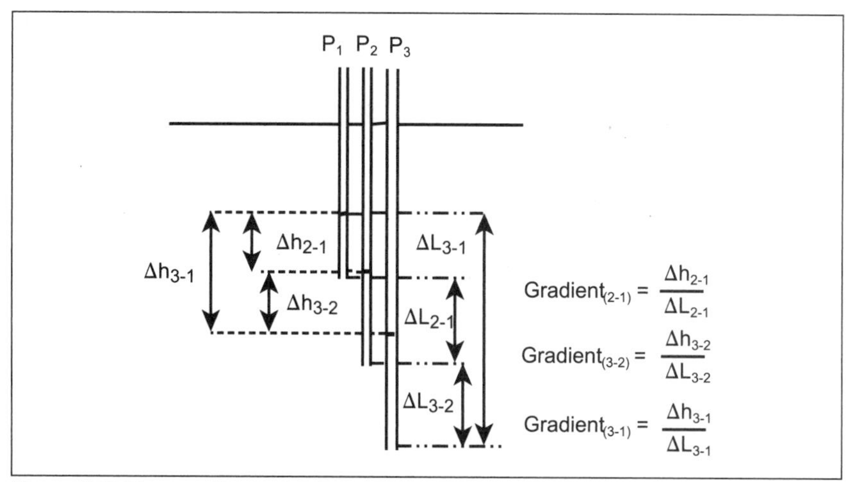

Figure 6.18 Diagram illustrating how to calculate vertical gradients in nested mini-piezometers. (In this case, all gradients are negative, indicating downward movement.)

Another way of measuring differential hydraulic head readings between stream stage and minipiezometers is from a method described by Lee and Cherry (1978). This method is also practical for obtaining water quality samples from stream sediments. The bolt method, described above, is used to drive a piece of nonperforated piezometer tubing (casing) into the stream sediments. Once the desired depth has been reached, a smaller-diameter plastic tube with a perforated tip is inserted into the bottom. The tubing (casing) is carefully pulled out and the sediments collapse against the plastic tube. The bottom bolt is again left in the sediments near the tube tip. A measuring scale can be held against the tube as it is extended vertically to compare differences. If the stream is gaining and the tube is bent close to the stream elevation, water will flow from the tube and water quality parameters can be measured. If the stream is losing or under parallel flow conditions, a bag can be attached to the tubing and held under the stream surface to obtain a sample (Figure 6.19).

Example 6.4

Near Silver Gate, Montana, a study was conducted to understand the relationship between surface water and groundwater in the Soda Butte Creek drainage before entering Yellowstone National Park (Figure 6.20). As part of the study, groundwater flow paths were evaluated using domestic wells in the area. Groundwater/surface-water interactions were evaluated by installing nested minipiezometers, stream gauging and comparing groundwater and surface-water data (English 1999). Over 20 nested pairs of minipiezometers were installed to evaluate vertical gradients and compare seasonal trends (Table 6.4 and Figure 6.21).

Figure 6.19 Seepage meter on Box Elder Creek, Petroleum County, Montana. The bag is downstream of the stick.

Table 6.4 Vertical Hydraulic Gradients for Paired Minipiezometers near Sliver Gate, Montana [From English (1999)]

Piezometer ID	Vertical W.L Difference (ft.)	6 Aug 1998 Gradient	6 Oct 1998 Gradient	2 Dec 1998 Gradient	18 Mar 1999 Gradient
19s &19d	4.7	0.019 up	0.021 up	0.023 up	0.021 up
20s & 20d	4.0	flat	0.008 up	0.005 up	0.005 up
21s & 21d	3.5	flat	0.006 down	0.006 down	0.006 down
22s & 22d	3.5	flat	flat	flat	0.020 down

Seepage Meters

The seepage flux between groundwater and surface water can be measured directly using a **seepage meter**. A seepage meter is a constructed by inserting an open-bottomed container into the stream sediments and then measuring the time it takes for a volume of water to flow into or out of a bag connected to the container (Lee and Cherry 1978). The bag can be protected by a #10 can, (such as a coffee can) held in place with a rock. A hole is drilled into the container and fitted with a plastic barbed fitting. Tygon or plastic tubing is inserted to provide an extension to which the bag can be attached (Figure 6.22). Another tube can be attached to the top of the container, near one edge, to serve as a vent for any gases released from the sed-

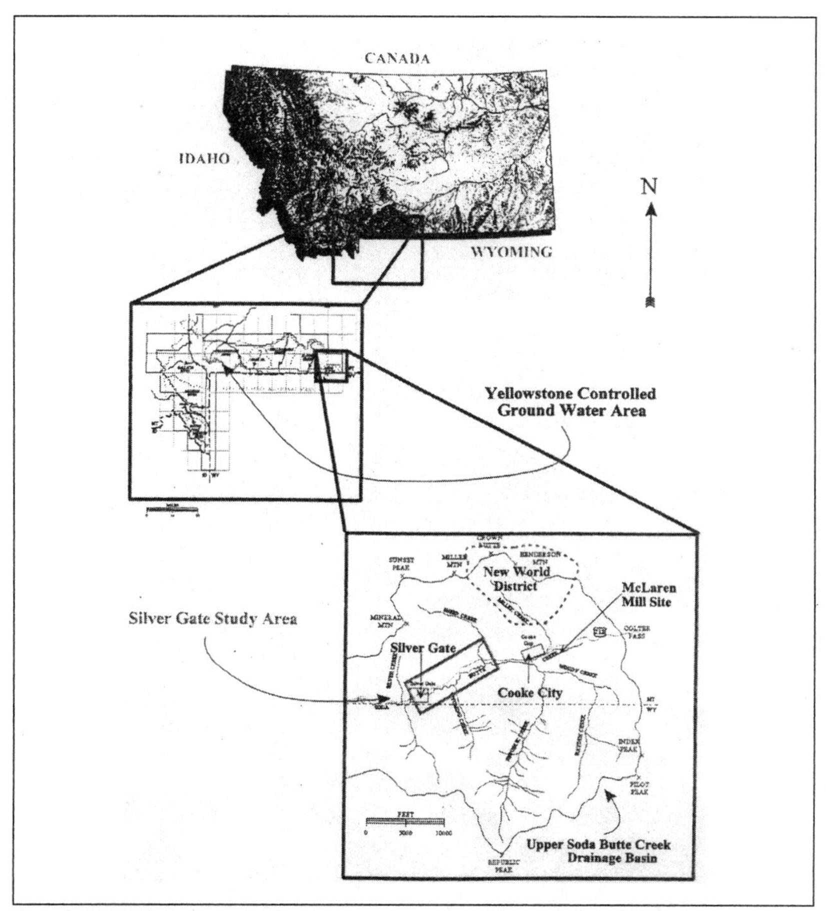

***Figure 6.20** Silver Gate study area location map in south-central Montana, within the upper Soda Butte Creek drainage. [From English (1999).]*

iments. Metal or stiff sided containers work best. Good results can be obtained by cutting a 5-gal bucket (18.9 L) in half horizontally or using a 55-gal (0.208-m^3) drum with sides cut to approximately 1 ft (30 cm) in length. The larger the surface area of the container, the more representative the averaged seepage rate is. Seepage-meter installation has the highest success in lakes or surface streams with low current velocities that do not have significant gravel or rocky armor fraction.

The seepage meter is installed by pushing the container slowly into the sediments with the container slightly tilted. Tilting allows the vent tube to work better (Lee and Cherry 1978). It is essential that the container cavity be filled with water with no air bubbles. The bag is attached to the plastic tubing and clamped until ready for timing. If the flow is upward, the empty bag does not need to be preweighed (Figures 6.21 and 6.22). A Mylar balloon works well because it is strong, lightweight, and has a narrow filling access port that is easily clamped to the tubing. If the stream is losing, the bag needs to be filled or partially filled with water and weighed before attaching and clamped to the tube extending from the container (not the vent tube). When the operator is ready, the tube is unclamped and the time

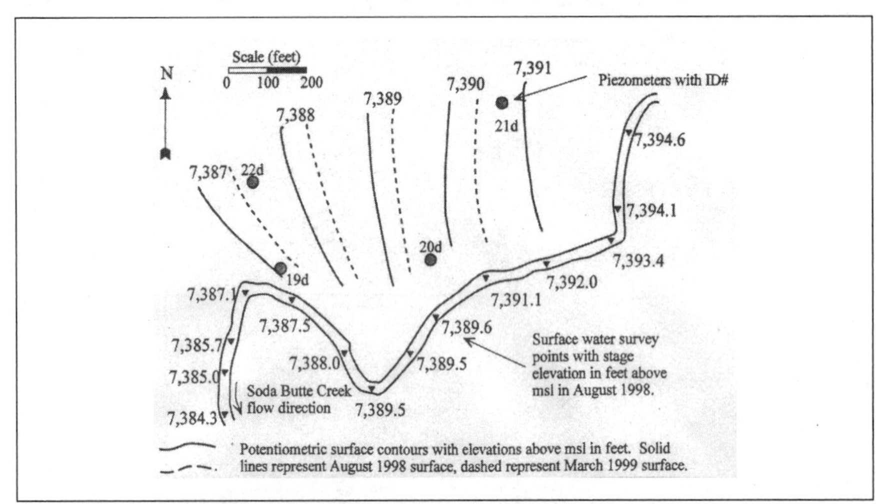

Figure 6.21 Shallow groundwater flow in the marshy area by Silver Gate is toward the west and does not appear to be strongly influenced by Soda Butte Creek until it reaches the western edge of the marshy area. [From English (1999).]

Figure 6.22 Seepage meter in Flathead Lake evaluating discharge. (Photo courtesy of Bill Woessner.)

noted. Once the bag empties or fills it is reweighed and the time noted once again. The bag can be protected from the current in the stream by punching a hole through the top of a #10 can (coffee can size) and running the tube from the container into the can before attaching the bag. The can/bag configuration can be held down with a rock (Figure 6.23).

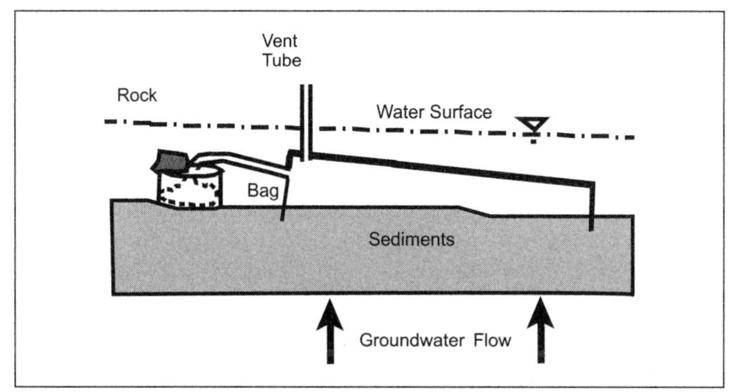

Figure 6.23 Schematic of installed seepage meter (gaining scenario).

The volume of water represents the flow (Q) of Darcy's Law (Chapter 5). The hydraulic gradient (I, change in head per distance) could be determined by installing a minipiezometer to the depth of the seepage meter and reading it directly. The area (A) affected by seepage is the cross-sectional area of the container. From this a hydraulic conductivity (K) can be calculated from Equation 6.4.

$$K = \frac{Q}{IA} \tag{6.4}$$

The hydraulic conductivity (K) is not a seepage velocity. The average linear seepage velocity (v_s) is determined by multiplying K by the hydraulic gradient (I) and dividing by the effective porosity (Equation 6.5) of the unconsolidated sediments (25 to 35%) (Chapter 3).

$$V_s = \frac{K}{n_e} I$$

Example 6.5

In east-central Montana, a study was conducted to evaluate the recovery of hydraulic head from rehabilitated flowing wells completed in the Basal Cretaceous Eagle Sandstone (Brayton 1998). Artesian wells had been flowing since before the 1960s and through rehabilitation and conservation efforts, recoveries of hydraulic head in excess of

5 ft (1.5 m) was achieved. As part of the study, seepage meter tests were performed in streams in the outcrop areas of the Eagle Sandstone to evaluate recharge (downward gradient) from surface flow.

A 5-gal (18.9-L) bucket was fitted with a nylon nipple on the bottom (Figure 6.19). A 500-ml bag full of water was attached and allowed to sit for 1 hour. At that time, the quantity of water was remeasured to obtain a volume lost per time (Table 6.5).

Table 6.5 Seepage Meter Results at Streams Crossing the Eagle Sandstone Outcrop [From Brayton (1998)]

Location	Volume Lost, mL/hr
NW 1/4, NE 1/4, SW 1/4 , Section 25, T14N, R27E	60
SE 1/4, SW 1/4, NE 1/4, Section 11, T14N, R27E	30
SW 1/4, SW 1/4, SW 1/4, Section 6, T14N, R28E	72
NE 1/4, NE 1/4, NW 1/4, Section 15, T15N, R27E	230

Care was taken to ensure that the stream sediments were well connected to the Eagle Sandstone aquifer. Minipiezometers were also installed near the seepage meter sites and all indicated a downward gradient.

Tracer Studies

Tracer studies have been used for years in determining "time of travel" in streams (Wilson 1968; Kilpatrick 1970). Methods include injecting chloride or bromide or a flourescent dye (Smart and Laidlaw 1977). Tracer studies conducted for aquifer velocities and aquifer properties are presented in Chapter 13. However, separate discussion of tracer-study applications for groundwater/surface-water interactions is needed because of the trends shown about storage (Bencala and Walters 1983) and dynamic movement within stream sediments (Castro and Hornberger 1991).

Time-of-travel studies are useful when stream velocities cannot be measured from ice cover or steep rocky channels, or when one wishes to evaluate dispersal patterns from waste or thermal pollution (Kilpatrick 1970).

A detailed comparison of eight fluorescent dyes was presented by Smart and Laidlaw (1977). Their evaluation yielded a recommendation of three: rhodamine WT (orange), lissamine FF (green) and amino G (blue). Evaluation criteria such as pH of waters, temperature effects, salinity (for estuarine studies), resistance to adsorption, and toxicity were used in the analysis. Kilpatrick (1970) found that rhodamine WT could be used at a lower concentration (20% solution compared to 40% solution) than rhodamine BA and at a lower cost.

For time-of-travel studies, the Equation 6.5 from Kilpatrick (1970) is helpful.

$$V_d = 3.4 \times 10^{-4} \left(\frac{Q_m L}{V} \right)^{0.93} C_p \qquad [6.5]$$

where:

V_d = Volume of dye in liters

Q_m = Discharge in the reach in ft^3/sec

L = Length of reach in miles

V = Mean stream velocity in ft/sec

C_p = Peak concentration and micrograms per liter (μg/L) desired at the lower end of the reach

Equation 6.5 represents the case where C_p = 1 μg/L at the lower end of the reach and for a stream velocities that range between 0.2 and 5.0 ft./sec. To increase the peak concentration at the lower end of the reach (C_p), a proportionately greater volume of dye is needed. The U.S. Geological Survey does not recommend exceeding 10 μg/L as a peak concentration (Kilpatrick 1970, Kilpatrick and Cobb 1985). Equation 6.5 is presented graphically in Figure 6.24 to evaluate dye volumes directly.

Tracers are also helpful in determining the movement of groundwater in near-channel sediments. This may have implications for understanding the hyporheic zone (Section 6.4) and stream ecology (Stanford and Ward 1988, 1993; Meyer 1997; Dahm et al. 1998).

A stream reach can be selected that has a riffle zone or shallowing area where the tracer can readily enter the streambed sediments (Figure 6.25). Ideally, there is a place on the downstream end where most of the tracer influenced water can be accounted for (a discharge point). The sampling points should be distributed along the stream reach to be able to obtain a representative distribution of the tracer throughout the stream sediments and within the channel.

Example 6.6

A tracer study was conducted in the North Fork of Dry Run Creek in Shenandoah National Park in Page County, Virginia, by Castro and Hornberger (1991) to evaluate groundwater/surface-water interaction within fluvial sediments in a steep mountain stream (gradient 0.057). A 225-m reach was selected where the spatial variability of the channel characteristics was great. The width of the alluvial materials varied between 22 m and 77 m (Figure 6.26). The surficial flow pattern was dominated by pools and riffles and was conducted at low flow.

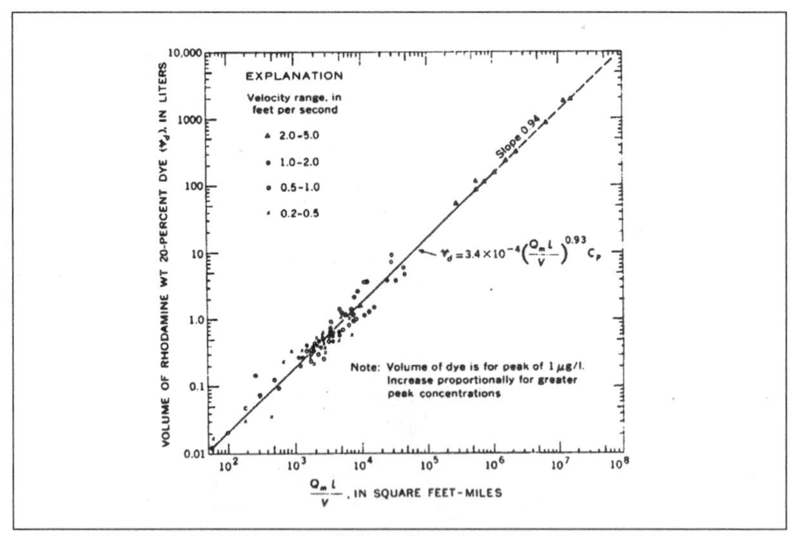

Figure 6.24 Quantity of rhodamine WT 20% dye required for slug injection to produce a peak concentration of 1µg/L at a distance downstream, L, at a mean velocity, V, and a maximum discharge, Q_m, in the reach [From Kilpatrick (1970)].

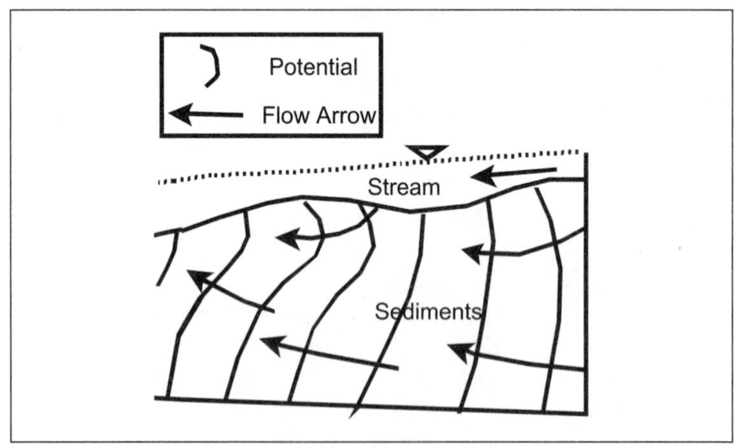

Figure 6.25 Longitudinal profile of a stream indicating where surface water enters stream sediments. [Adapted from Lee and Cherry (1978).]

The (KBr) tracer was mixed with stream water in a 378 L (100 gal) polyethylene tank. On the day before the test, the solution was mixed via a small pump for 8 hr. During the test, the tank was stirred manually twice a day during the injection period. The tracer concentration in the tank was 48.6 g/L. The tracer was introduced into the stream at a riffle zone at a rate of 50 mL/min over the 93-hr injection period (approximately 0.05 to 0.1% of the total discharge). The injection rate was checked using a bucket and stopwatch several times during the injection period.

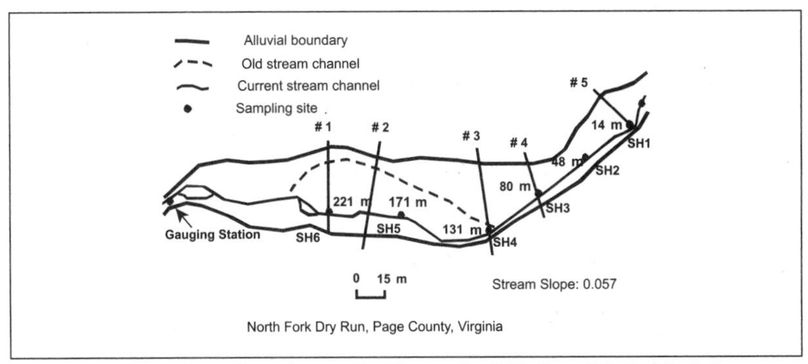

Figure 6.26 Plan view of study area in Shenandoah National Park, Virginia. [Adapted from Castro and Hornberger (1991). Used with permission of American Geophysical Union (1991.)]

Six surface water sampling sites (SH1 through SH6) and a series of transects were established to evaluate the distribution of tracer in the sediments (Figure 6.26). Groundwater samples were obtained from hand-dug wells along each transect near the active channel and from an old abandoned channel within the alluvial plain. Each hole was dug deep enough to fill each hole with 0.15 to 0.3 m of standing water. Additionally, a few cased wells were drilled with completion depths ranging from 0.75 to 1.75 m. These were up to 24 m from the active stream to provide sampling stations from deeper sediments (Figure 6.27).

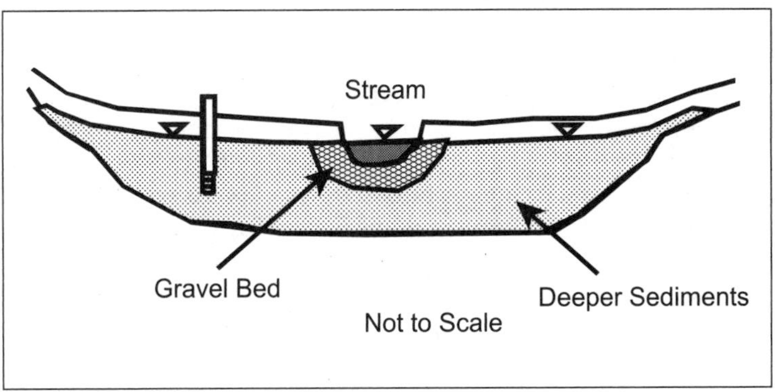

Figure 6.27 Schematic cross section showing mixing zones in the stream, gravel bed, and deeper sediments.

Samples were collected from the stream and gravel bed every 0.5 to 1.0 hr until the arrival of the leading edge of the tracer. Subsequently, samples were collected every 4 hr while concentrations were relatively constant during injection of the tracer. Following tracer injection, samples were collected at 0.5 to 1.0, 4, 8, and 24 hr, and ultimately weekly (Castro and Hornberger 1991).

A specific ion electrode (accurate to 0.25 mg/L) was used to measure bromide concentrations (peak concentrations were near 30 mg/L). This was used in conjunction with a pH/conductivity/temperature meter. Because of the low ionic strength of the stream water, all samples and standards, prior to analysis, were mixed with an ionic strength adjuster (ISA), 2 *M* potassium nitrate, to the ratio of 1:10 (ISA:sample) (Castro and Hornberger 1991). Standards were made with upstream water and compared with standards made with deionized water with no differences found.

Significant tracer concentrations were found in wells at distances of 8 m and 16 m from the active stream channel after concentrations within the stream had cleared, supporting the concept of long-term storage. Concentration tails (abrupt concentration arrivals followed by gradual decreasing trends; see Chapter 13) at surface-water sites waned after 4 to 6 days, while similar tails in wells lingered approximately 20 days before clearing to background levels. This study showed how mixing within the hyporheic zone is active at significant distances away from the main channel and that movement away from and back into the channel through the bottom sediments does occur.

Chemical Mass Balance

The purpose of this section is to give a brief presentation on water quality indicators at groundwater/surface-water settings and how a mass balance approach can be used to compare general groundwater and surface-water qualities in accounting for gaining streams. A more complete presentation on water chemistry is given in Chapter 7.

If minipiezometers are installed in a gaining stream stretch and cut close enough to the stream surface, they will flow like mini-artesian wells. These can then be tested for water quality parameters such as specific conductance, temperature, and dissolved oxygen to distinguish water-quality differences between groundwater and surface water. In losing stretches, minipiezometers would have to be completed differently. After the minipiezometer casing is placed, plastic tubing is inserted to the bottom of the casing. The minipiezometer casing is carefully pulled out, leaving the tube as the piezometer. A plastic bag is attached to the tubing and held under the surface of the water sufficiently low to induce flow into the bag (Lee and Cherry 1978). The bagged water can then be tested for water quality parameters. Redox potential and nutrient dynamics can also be studied within the hyporheic zone (Dahm et al. 1998). The usual practice, however, is to insert a small tube and pull a water-quality sample with a paristaltic pump.

Sometimes a mass-balance approach can be used to account for in-fluxing groundwater into a stream. It is necessary to obtain stream water-quality samples from places where stream gauging data has also been collected. The groundwater quality must be characterized from "representative" well samples collected near the active stream. It is a good idea to identify whether significant changes with time are occurring, as this may complicate the analysis. If the two water chemistries are distinctive enough, it is possible to make a comparison between the two and perform an estimate of the influx of groundwater. The approach is to use Equation 6.6:

$$Q_{sw,\text{in}} \times WQP + S_{gw,\text{in}} \times WQP = Q_{sw,\text{out}} \times WQP \qquad [6.6]$$

where:

$Q_{sw,\text{in}}$ = stream flow into the area, in L^3/t

WQP = water quality parameter of choice (consistent with flow term)

$S_{gw,\text{in}}$ = groundwater flow into the area, in L^3/t

$Q_{sw,\text{out}}$ = stream flow out of the area, in L^3/t

The objective is to use the streamflow and water-quality data to evaluate the groundwater flow into the area ($Q_{gw,\text{in}}$). It can be tested using several parameters for comparison purposes.

Example 6.7

In the study conducted by English (1999) near Silver Gate, Montana, a mass balance approach was taken to estimate the influx of groundwater into Soda Butte Creek as it flows through the study area (Example 6.4). Stream gauging stations 7, 11, and 20 were used in the analysis along with their respective water chemistries, along with local shallow wells (less than 65 ft below ground surface) (Figure 6.28). Station 7 is on Soda Butte Creek just downstream from the confluence of a tributary (Sheep Creek). Station 11 is on Wyoming Creek, a tributary in the middle of the study area, just before it merges with Soda Butte Creek. The water quality and gauging data only reflect this contribution. Station 20 is on Soda Butte Creek at the exit point of the study area (Figure 6.24). The respective water qualities and gauging data are shown in Table 6.6.

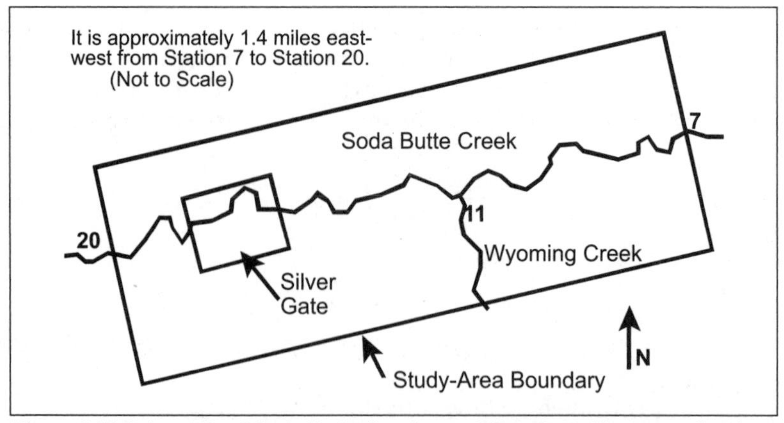

Figure 6.27 Schematic of Soda Butte Creek near Silver Gate, Montana, showing surface water gauging station and water-quality sites used in a mass-balance analysis of groundwater influx into the study area.

Table 6.6 Water Quality and Stream Gauging Data near Silver Gate, Montana [From English (1999)]

Well	Depth (ft bgs*)	SC (μmhos/cm²)	Ca (mg/L)	Mg (mg/L)	Na (mg/L)	Cl (mg/L)
5-well ave.	35–65	270	32.68	7.35	4.23	0.66
SW Station	**Flow (L/s)**	**SC (μmhos/cm²)**	**Ca (mg/L)**	**Mg (mg/L)**	**Na (mg/L)**	**Cl (mg/L)**
7	192.86	189	22.1	5.84	3.93	0.52
11	15.86	81.1	5.37	1.96	8.1	0
20	277.82	218	26.5	6.02	3.87	0.6

*Below ground surface

Using the following form of Equation 6.6 and calcium as an example, the groundwater inflow ($Q_{gw,\text{in}}$) was estimated. This same process was repeated for the other five water-quality parameters to obtain an average estimate for ($Q_{gw,\text{in}}$). The results are tabulated in Table 6.7.

$$192.86\,\text{L/s} \times [22.1\,\text{mg/L}] + 15.86\,\text{L/s} \times [5.37\,\text{mg/L}] + Q_{gw,in} \times [32.68\,\text{mg/L}] =$$

$$277.82\,\text{L/s} \times [26.5\,\text{mg/L}]$$

Table 6.7 Summary of Estimated Groundwater Influx in L/sec, Based on Groundwater Quality Data near Silver Gate, Montana [From English (1999)]

Specific Cond.	Calcium	Magnesium	Sodium	Chloride	Average
82.1	93.4	70.8	45.3	99.1	78.1

The average estimate of groundwater influx from stream gauging was 76.5 L/sec. (2.7 cfs).

6.5 Summary

Groundwater/surface-water interactions within riverine systems are complex. A variety of flow scenarios occur at a many scales. Interactions of groundwater/surface-water within the hyporheic and riparian zones, coupled with biotic distributions and chemical gradients suggest the need for hydrogeologists to team up with other professionals to understand these systems. The field methodologies presented will assist the hydrogeologist in collecting meaningful data to better manage the mass balance components occurring within a watershed.

References

Anderson, M.P. 1989. Hydrogeologic Facies Models to Delineate Large-scale Spatial Trends in Glacial and Glaciofluvial Sediments. *Geological Society of America Bulletin,* 101, pp. 501–511.

Bencala, K.E., and Walters, R.A., 1983. Simulation of Solute Transport in a Mountain Pool-and-riffle Stream: A Transient Storage Model. *Water Resources Research,* Vol. 19, No. 3, pp. 718–724.

Bencala, K.E., 1993. A Perspective on Steam-catchment Connections. *Journal of North American Benthological Society,* Vol. 12, No. 1, pp. 44–47.

Borduin, M.B. 1999. *Geology and Hydrogeology of the Sand Creek Drainage Basin, Southwest of Butte, Montana,* Master's Thesis, Montana Tech of the University of Montana, 103 pp.

Brayton, M., 1998. *Recovery Response from Conservation Methods in Wells found in the Basal Eagle Sandstone, Petroleum County,* Montana. Master's Thesis, Montana Tech of the University of Montana, 68 pp.

Buchanan, T.L., and Sommers, W.T, 1968. *Stage Measurement at Gauging Stations.* Techniques of Water-Resources Investigations of the U.S.

Geological Survey, Book 3, Chapter A7. U.S. Geological Survey, Washington, DC.

Buchanan, T.L., and Sommers, W.T, 1969. *Discharge Measurements at Gauging Stations.* Techniques of Water-Resources Investigations of the U.S. Geological Survey, Book 3, Chapter A8. U.S. Geological Survey, Washington, DC.

Castro, N.M, and Hornberger, G.M., 1991. Surface-subsurface Water Interaction in an Alluviated Mountain Stream Channel. *Water Resources Research,* Vol. 27, No. 7, pp. 1613–1621.

Chow, V.T., 1959. *Open Channel Hydraulics,* McGraw-Hill, New York.

Dahm, C.N., Grimm N.B., Mamonier P., Valett H.M., and Vervier P., 1998. Nutrient Dynamics at the Interface between Surface Waters and Groundwaters. *Freshwater Biology,* Vol. 40, pp. 427–451.

English, A., 1999, *Hydrogeology and Hydrogeochemistry of the Silver Gate Area, Park County, Montana.* Master's Thesis, Montana Tech of the University of Montana, 131 pp.

Fetter, C. W., 1994. *Applied Hydrogeology, 3rd Edition.* Macmillan College Publishing Company, New York, 691 pp.

Gregory, S.V., Swanson F.J., McKee W.A., and Cummins K.W. 1991. An Ecosystem Perspective of Riparian Zones. *Bioscience,* Vol. 41, pp. 540–551.

Grimm, N.B., and Fisher, S.G., 1984. Exchange Between Interstitial and Surface Water: Implications for Stream Metabolism and Nutrient Cycling. *Hydrobiologia, Vol. III,* pp. 219–228.

Gross, L.J., and Small M.J., 1998. River and Floodplain Process Simulation for Subsurface Characterization. *Water Resources Research,* Vol 34, No. 9, pp. 2365–2376.

Hansen, E.A., 1975. Some Effects of Groundwater on Brown Trout Redds. *Transactions of American Fisheries Society,* Vol. 104, No. 1, pp. 100–110.

Hill, A.R., 1993. Nitrogen Dynamics of Storm Runoff in Riparian Zones. *Journal of Environmental Quality,* Vol. 25, pp. 743–755.

Huggenberger, P., Hoehn, E., Beschta, R., and Woessner W., 1998. Abiotic Aspects of Channels and Floodplains in Riparian Ecology. *Freshwater Biology,* Vol. 40, pp. 407–425.

Kilpatrick, F.A., 1970. Dosage Requirements for Slug Injections of Rhodamine BA and WT dyes. In *Geological Survey Research*, U.S. Geological Professional Paper 700-B, pp. B250–B253.

Kilpatrick, F.A., and Cobb, E.D. 1985. Measurement of Discharge Using Tracers. Techniques of Water-Resources Investigations of the U.S. Geological Survey, Book 3, Chapter A16.

Lee, D.R., and Cherry J.A., 1978. A Field Exercise on Groundwater Flow using Seepage Meters and Mini-piezometers. *Journal of Geological Education.* Vol. 27, pp. 6–10.

Meyer, J.L., 1997. Stream Health: Incorporating the Human Dimension to Advance Stream Ecology. *Journal of the North American Benthological Society*, Vol. 16, No. 2, pp. 439–447.

Mial, A.D., 1996. *The Geology of Fluvial Deposits: Sedimentary Facies, Basin Analysis, and Petroleum Geology.* Springer-Verlag, New York, 582 pp.

Modica, E,. 1998. Analytical Methods, Numerical Modeling, and Monitoring Strategies for Evaluating the Effects of Ground-water Withdrawals on Unconfined Aquifers in the New Jersey Coastal Plain. *USGS Water Resources Investigation Report,* 98-4003, 66 pp.

Palmer, M.A., 1993. Experimentation in the Hyporheic Zone: Challenges and Prospectus. *Journal of the North American Benthological Society*, Vol. 12, No. 1, pp. 84-93.

Roberts, M., and Waren, K., 1999. North *Fork Blackfoot River Hydrologic Analysis.* Montana Department of Natural Resources and Conservation, Helena, MT, 36 pp.

Sanders, L.L. 1998. *A Manual of Field Hydrogeology.* Prentice-Hall, Upper Saddle River, NJ, 381 pp.

Smart, P.L., and Laidlaw, I.M.S., 1977. An Evaluation of Some Fluorescent Dyes for Water Tracing. W*ater Resources Research*, Vol. 13, No.1, pp. 15–33.

Sophocleous, M., Koussis A., Martin, J.L., and Perkins S.P., 1995. Evaluation of Simplified Stream-aquifer Depletion Models for Water-rights Administration. *Ground Water*, Vol. 33, No. 4, pp. 579–588.

Stanford, J.A., and Ward, J.V., 1988. The Hyporheic Habitat of River Ecosystems. *Nature*, Vol. 335, pp. 64–66.

Stanford, J.A., and Ward, J.V., 1993. An Ecosystem Perspective of Alluvial Rivers: Connectivity and the Hyporheic Corridor. *Journal of the North American Benthological Society*, Vol. 12, No. 1, p. 48–60.

Stumm, W., and Morgan, J.J., 1996, *Aquatic Chemistry, 3rd Edition.* John Wiley & Sons, New York, 1022 pp.

Triska, F.J., Kennedy, V.C., Avanzino, R.J., Zellweger, G.W., and Bencala, K.E., 1989. Retention and Transport of Nutrients in a Third-order Stream in Northwestern California: Hyporheic Processes. *Ecology,* Vol. 70, No. 6, pp. 1893–1905.

University of Wyoming, College of Agriculture, 1994. *Irrigation Water Measurement, Irrigation Ditches and Pipelines.* Cooperative Extension Service Bulletin 583 R, University of Wyoming Laramie, WY, 70 pp.

U.S. Geological Survey (USGS), 1980. *National Handbook of Recommended Methods for Water-data Acquisition (updates).* Office of Water Data Coordination, U.S. Geological Survey. Reston, VA, U.S. Department of Interior.

Wallace, J.B., Grubaugh, J.W., and Whiles, M.R., 1996. Biotic Indices and Stream Ecosystem Process: Results from an Experimental Study. *Ecological Applications,* Vol. 6, pp. 140–151.

Walton, W.C., 1970. *Groundwater Resource Evaluation.* McGraw-Hill, New York, 664 p.

Wilson, J.F., 1968. An Empirical Formula for Determining the Amount of Dye Needed for Time-of-travel Measurements in Geological Survey Research. *U.S. Geological Survey Professional Paper 600-D,* pp. D54–D56.

Winter, T.C., Harvey, J.W, Franke, O.L., and Alley, W.M., 1998. Ground Water and Surface Water: A Single Resource. *USGS Circular,* 1139, 79 pp.

Woessner, W.W., 1998. Changing Views of Stream-groundwater Interaction. In Van Brahana, J., Eckstein Y., Ongley L.W., Schneider R., and Moore, J.E., (eds.), *Proceedings of the Joint Meeting of the XXVIII Congress of the International Association of Hydrogeologists and the Annual Meeting of the American Institute of Hydrology.* American Institute of Hydrology, St. Paul, MN, pp. 1–6.

Woessner, W.W., 2000. Stream and Fluvial Plain Ground-water Interactions: Re-scaling Hydrogeologic Thought. *Ground Water,* Vol. 38, No. 3, pp. 423–429.

Wright, J.F., 1995. Development and Use of a System for Predicting the Macroinvertebrate Fauna in Flowing Waters. *Australian Journal of Ecology,* Vol. 20, pp. 181–197.

Wroblicky G.J., Campana M.E., Valett, H.M., and Dahm, C.N., 1998. Seasonal Variation in Surface-subsurface Water Exchange and Lateral Hyporheic Area of Two Stream-aquifer Systems. *Water Resources Research*, Vol. 34, pp. 317–328.

Chapter 7

Water Chemistry Sampling and Results

Many projects will require that groundwater samples be collected and analyzed as an integral part of the project. This chapter concentrates on how to collect samples correctly, get all of the necessary field measurements, determine and order the appropriate analyses from the lab, check the laboratory results, and utilize the laboratory and field data. It is intended as an initial guideline document with questions and a checklist for you to use as an aid. However, to get the right laboratory results, you really have to know the purpose of the sampling and what regulatory constraints your client or employer will have.

At the conclusion of this chapter, you should be able to go to a source (well, spring, lake, or creek) and, after asking a few questions about the setting and the purpose of sampling, determine what methods to use and what constituents should be analyzed.

It may be useful to start by looking at what can be done with the data, provided that the sampling and analyses were done properly.

7.1 How Can Groundwater Chemistry Be Used?

Cementing materials in granular porous media may be growing (precipitating) or dissolving, altering the effective porosity, and affecting the water chemistry. The common precipitated cementing agents are silica (SiO_2) and calcium carbonate ($CaCO_3$). Also, in carbonate aquifers, secondary porosity is often the main control on the permeability of the aquifer; the water chemistry can be used to evaluate dissolution and precipitation along the flow path and so the long-term impact upon porosity and permeability can be predicted. Alternatively, we may try to interpret what has happened up-gradient from a sampling point based upon the chemistry of the water sample. This is normally done on computers with geochemical speciation programs, but simply looking at the trend of ion ratios (such as Ca/Mg) can be quite useful in carbonate systems. The carbonate system is often discussed because of its wide usefulness in understanding ongoing processes.

Helpful Theory

The chemistry classes that you took need to be recalled (mainly pH definition, solubility products, solution speciation, and ion exchange principles; the books by Drever and Hem listed at the end of the chapter are good places to review or learn the basics, while the books by Appelo and Postma, Langmuir, and Stumm and Morgan deal with these in a more advanced manner).

Charge balance

One of the basic characteristics of water is that the sum of the positive and negative charges for dissolved species should equal zero (principle of electrical neutrality). This can be used as a quick-and-dirty check on the completeness of the analysis. Your analysis will probably give constituents (analytes) in units of concentration (mg/L or g/L) except for pH. pH is the negative logarithm of the activity of the hydronium ion (H^+ ion) using base-10 logarithms. Neglecting the correction for activity coefficients, at a pH of 7, the hydrogen ion concentration is 10^{-7} moles/L or about 10^{-7} grams/L (10^{-4} mg/L or 0.1 g/L).

Unless the pH was extreme (<5 or >9), we can probably skip the hydrogen and hydroxyl ions and just add up the charge information (concentration of the species multiplied by its presumed valence; typically this will give you milligrams/liter [mg/L] times valence divided by the atomic weight of the element or compound (grams/mole), yielding the charge in milliequivalents/liter [meq/L]).

If the sum of cation (positive ions) charge minus the anion charge divided by the average charge is less than 5%, you've got a good analysis (in terms of ionic balance). If the difference is greater than 10%, you've got a problem; either the lab data have an error, something present in significant concentration wasn't analyzed, or you are dealing with very pure water (when the total dissolved solids [TDS] are less than 50 mg/L, laboratories typically have problems getting an acceptable ionic balance; snow samples almost always drive the laboratory quality assurance person a bit crazy) or very high TDS waters (TDS $\geq$ 5,000 mg/L). (See Table 7.1.)

Table 7.1 Example of How to Evaluate Electrical Balance from Laboratory Data

Cation				Anion			
Analyte	**mg/L**	**Valence**	**meq/L**	**Analyte**	**mg/L**	**Valence**	**meq/L**
Ca	40.1	+2	2.0	HCO_3	122.	-1	2.1
Mg	12.2	+2	1.0	SO4	48.0	-2	1.0
Na	11.5	+1	0.5	Cl	10.6	-1	0.3
K	3.9	+1	0.1	NO_3	1.4	-1	0.1
Mn	0.05	+2	0.001	F	1.9	-1	0.1
Fe	0.05	+2	0.001	—	—	—	—
Sum	—	—	3.602	Sum	—	—	3.60

This approach will be used to evaluate laboratory results later in this chapter.

pH and alkalinity

As kids, we've probably all shaken a warm bottle of pop and opened it, squirting the pop with glee. Gases are more soluble in cold water than warm water (this is why a warm pop goes flat faster than a cold pop and why boiling chlorinated tap water gets rid of the chlorine taste). Also, at any given temperature, the higher the partial pressure of the gas, the larger the absolute amount of the gas that is dissolved into the water. When sampling, we don't want to lose gases dissolved in the water from the sample being sent in to be analyzed; this will be discussed in the sampling methodology section. However, we can also use geochemical speciation models to calculate gas partial pressures from existing chemical analyses as a quick check on whether gas loss may be a problem, before going out to collect new samples. These factors are important in obtaining an accurate alkalinity analysis, and the reason that pH and alkalinity should be determined in the field.

As an easy example, let us evaluate the pH and TDS of "pure rainwater." This rainwater should have anions balancing cations for electrical neutrality. Your first assumption might be that the pH is 7, but the water has interacted with the atmosphere and atmospheric carbon dioxide gas. Consequently, some CO_2 has dissolved in the water, as indicated by the reaction:

$$CO_{2(g)} + H_2O = H_2CO_3{}^0 \quad K_0 = 10^{-1.47} = \alpha(H_2CO_3)/P(CO_2)\alpha(H_2O) \quad [7.1]$$

The carbonic acid (H_2CO_3) acts as a relatively weak acid with a first dissociation constant of about $10^{-6.35}$ as indicated by:

$$H_2CO_3^0 = H^+ + HCO_3^- \; K_1 = 10^{-6.35} = \alpha(H^+)\alpha(HCO_3^-)/\alpha(HCO_3^0) \quad [7.2]$$

and a second dissociation constant of about $10^{-10.33}$ as indicated by:

$$HCO_3^- = H^+ + CO_3^{-2} \; K_2 = 10^{-10.33} = \alpha(H^+)\alpha(CO_3^{-2})/\alpha(HCO_3^-) \quad [7.3]$$

The positive charge comes only from H^+, which must balance the charge of OH^-, HCO_3^-, and CO_3^{-2}. (See Table 7.2.)

Table 7.2 Rain Equilibrated with Atmospheric Pco₂ ($10^{-3.5}$ atm)

Cation				Anion			
Analyte	**mmole/L**	**Valence**	**meq/L**	**Analyte**	**mmole/L**	**Valence**	**meq/L**
H^+	2.21E-03	1	2.21E-03	OH^-	4.57E-06	-1	4.57E-06
				HCO_3^-	2.19E-03	-1	2.19E-03
				CO_3^{-2}	4.71E-08	-2	-9.42E-08
Sum			2.21E-03				2.19E-03
Equilibrium pH = 5.659							

For comparison, water that infiltrates into the soil picks up additional carbon dioxide in the vicinity of the plant roots. Studies of soil P_{CO_2} conducted in Alabama showed typical soil partial pressures ranging from 10^{-3} to 10^{-2} atmospheres (greater P_{CO_2} for longer periods since the last rainfall (recharge) event). Table 7.3 presents these values.

Table 7.3 Rain Equilibrated with P_{CO_2} = 0.01 atm

Cation				Anion			
Analyte	**mmole/L**	**Valence**	**meq/L**	**Analyte**	**mmole/L**	**Valence**	**meq/L**
H^+	1.24E-02	1	1.24E-02	OH^-	8.17E-08	-1	8.17E-07
				HCO_3^-	1.24E-01	-1	1.24E-02
				CO_3^{-2}	4.77E-08	-2	9.53E-08
Sum			1.24E-02				1.24E-02
Equilibrium pH = 4.910							

In these examples, the pH of the water drops from 7 to 5.659 when exposed and equilibrated with roughly the atmospheric partial pressure of carbon dioxide (this is a moving target because atmospheric concentrations of CO_2 have been rising for over a hundred years); when exposed and equilibrated with an approximated soil vapor CO_2, the pH is lowered further to 4.91.

The TDS content is essentially just that of the bicarbonate in either example. The rainwater example has 0.00219 mm/L HCO_3^- and bicarbonate has an atomic weight of about 61 milligrams per millimole, so the TDS is about 0.134 mg/L, while the soil-water example has 0.0124 mm/L bicarbonate and a TDS of about 0.756 mg/L.

We will come back to these and compare each of these when also equilibrated with calcite as an example of speciation, and we will see much larger TDS values associated with those cases.

Oxidation and reduction

The oxidation/reduction (redox) state of the system is also of importance. This value may be calculated as the **pe** or measured as the **Eh** of the water. If not determined directly with a platinum electrode, it can be calculated from elements, compounds, or species with multiple oxidation states, if two of those states are present in quantities large enough to be analyzed.

1. The pe is defined as the negative log of the activity of the electron; this is analogous to the definition of pH. Unfortunately, this cannot be measured directly, so we measure the "effective concentration" of electrons using a platinum electrode and then correct that to what the value would be using a hydrogen gas electrode.

2. The Eh is defined as:

$$Eh = Eh^0 + (2.303RT/nF)\log_{10}[\pi_i\{ox\}^{n_i}/\pi_j\{red\}^{n_j}] \qquad [7.4]$$

where n is the number of electrons involved in the reaction, F is the Faraday's constant, R is the gas constant, T is the temperature in degrees Kelvin, $\pi_i\{ox\}^{n_i}$ is the product of the activities of the products of the reaction (as written, by convention the oxidized form of the element is on the product side of the reaction and the reduced form is on the reactant side; note that all components *except* the electrons are present in these terms) and $\pi_j\{red\}^{n_j}$ is the product of the activities of the reactants in the reaction as written. E_h^0 is the potential measured (in volts) when all of the species in the reaction are present at unit activities (the reference state for the reaction) and the rest of the terms are the correction for constituents not being present at unit activities. If all products and reactants are present at unit activities ($a_{i,j}$= 1) the right-hand term becomes the log of 1, which is zero, and the terms other than E_h^0 disappear.

3. Typically used for back calculating redox are ferric and ferrous iron [Fe (III) and Fe(II)] or As(V) and As(III) species. It should be noted up front that almost all textbooks discussing these relationships discuss

the theory using activities (i.e., $a_{Fe^{+3}}$) as is required for the half reactions as shown in Equation 7.5, and then shift in calculations to the dissolved concentration of the oxidized or reduced form of the element. This "mixed" approach has historical roots from the analytical results giving the total dissolved concentration of, for example, Fe (III) and Fe(II), and the questionability of the complexing constants used to generate the free ion (i.e., Fe^{+3}) activities, especially in the times before computer programs to assist these calculations were widely available. As long as the student understands that for these calculations the value used for the activity of Fe^{+3} is commonly the concentration of Fe(III), we can both follow the conventions used in most textbooks and understand their example calculations (see Stumm and Morgan 1996, pp. 432–433, example 1).

As an example, using dissolved iron at 25°C:

$$Fe^{+2} \rightarrow Fe^{+3} + e^{-} \quad K = (\alpha_{Fe}{}^{+3}\, \alpha_{e}{}^{-} / \alpha_{Fe}{}^{+2}) = 10^{-13.0} \qquad [7.5]$$

$$pe = -\log_{10}(\alpha_{e}{}^{-}) = -\log_{10}(K) + \log_{10}(\alpha Fe^{+3} / \alpha_{Fe}{}^{+2}) \qquad [7.6]$$

$$= 13.0 + \log_{10}(\alpha_{Fe}{}^{+3} / \alpha_{Fe}{}^{+2})$$

$$Eh = Eh^{\circ} + (2.303\, RT / 1F)\log_{10}(\alpha_{Fe}{}^{+3} / \alpha_{Fe}{}^{+2}) \qquad [7.7]$$

$$= 0.77 + (2.303\, RT / 1F)\log_{10}(\alpha_{Fe}{}^{+3} / \alpha_{Fe}{}^{+2})$$

where α is the activity of the subscripted species, R is the universal gas constant, T is the temperature in degrees Kelvin, F is the Faraday constant.

Obviously, pe is dimensionless, and Eh will depend upon units used but is normally expressed in volts or millivolts. When having to convert Eh to pe for some geochemical programs, you use the relationship

$$Eh = \left[(2.303 \times R \times T)/F\right] \times pe \qquad [7.8]$$

Or, this can also be expressed

$$(Eh \times 5.0395) / T = pe \qquad [7.9]$$

where Eh is in millivolts and T is in degrees Kelvin.

It should be pointed out that while we use pe in most of the geochemical models for calculation purposes, our field measurement is an Eh value. This measured value is a "mixed" value in that there may be several reactions involving electrons going on (ferric iron and sulfate reduction, oxidation of organic matter, etc.) and several metals or metalloids (Fe, Mn, As,

Se, etc.) participating. Concentration of these elements and the reaction kinetics relating to oxidation and reduction will affect the measured Eh value. Generally, if iron concentrations are significant and conditions are acidic, we presume that the ratio of Fe^{+3} to Fe^{+2} is controlling the measured Eh. This is based on the fact that you rarely get the same ferric/ferrous ratios in the lab analysis that you get with an immediate measurement at the well head because ferrous ions oxidize rapidly, whereas Mn and As ratios tend to be much more stable, with reasonably good agreement between field determinations, rapid (same-day) lab analysis, and lab analysis of stored samples (Cherry et al. 1979; Lindberg and Runnells 1984; Barcelona et al. 1989; Kempton et al. 1990).

Table 7.4 Data from the Berkeley Pit, Sampled 10/16/1987

Analyte	**depth 100 m**	**depth 130 m**
Analytical	**mg/L**	**mg/L**
Fe(III)	14	24
Fe(II)	944	962
As(V)	0.768	0.807
As(III)	0.087	0.101
Calculated		
Fe-pe	9.84/9.93	10.05/9.90
Fe-Eh	0.560/0.565	0.572/0.563
As-pe	6.89	6.88
As-Eh	0.392	0.392
Measured		
Eh	0.47	0.46
pH	3.15	3.14
Temp	13.5	13.7

As an example of this, Table 7.4 presents data from the Berkeley Pit (Davis 1988), an inactive open pit copper mine in Butte that has been filling with water since 1982. Note that we have the Eh measured using a platinum electrode, plus the Eh and pe values calculated from the iron and arsenic data. Note that the calculated values (PHREEQC and MINTEQA2, respectively, for iron with minor disparities, MINTEQA2 for arsenic) bracket the measured Eh value, showing a rather wide (0.27-volt) spread. The key point is that the disparities between geochemical program calculations are minor, while the calculated values from different elements may be

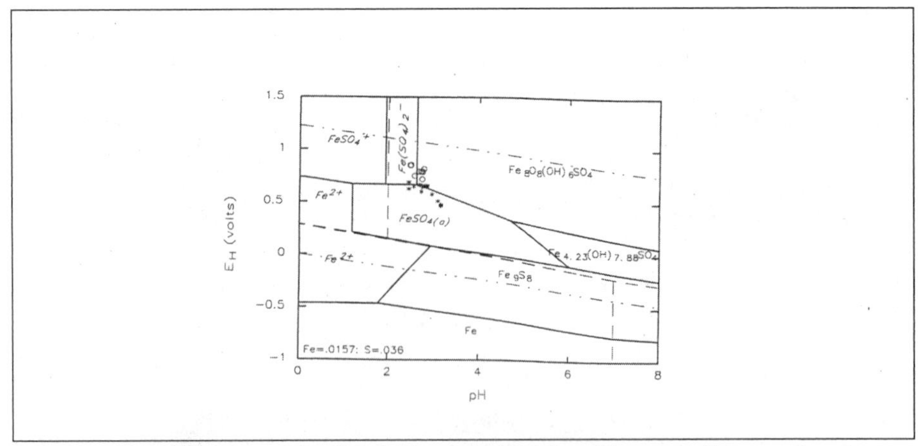

Figure 7.1 Plot of Eh versus pH with total iron concentration. (Concentrations used are in molality and are equivalent to 877 ppm iron and 3458 ppm sulfate. The sulfate number is about half the concentration typically reported.) [From Robins et al., with permission.]

quite large. This is not atypical of redox data, as disequilibrium conditions appear to be common.

The redox data are normally used in the preparation of Eh–pH diagrams as an aid to understanding geochemical processes. Robins et al. (1996), utilized data collected in the Berkeley Pit from 1984 through 1985 to construct an Eh–pH diagram which strongly suggests that the mineral phase schwertmannite ($Fe_8O_8(OH)_6SO_4 \cdot 5H_2O$) controlled iron content in the water chemistry. Figure 7.1 shows that the data cluster along and within the schwertmannite boundary with the aqueous ferric and ferrous species [$Fe(SO_4)_2^{-1}$ and $FeSO_4^0$, respectively], which supports this conclusion.

Speciation

Speciation is simply the assigning of the analytically determined values to the species containing that element/constituent. For example, the analytical calcium is normally divided between Ca^{2+}, $CaHCO_3^+$, $CaCO_3^0$, $CaSO_4^0$, and possibly $Ca(OH)^+$ and $Ca(OH)_2^0$ (only important under high pH conditions). This is normally done using computer programs such as PHREEQC or MINTEQA2. Both programs are public domain, available from the USGS and EPA, respectively. The USGS URL is:

http://water.usgs.gov/software/geochemical.html

When downloading a copy of the program, the GUI-based version [PHREEQC] is strongly recommended (it runs under NT4 and Windows

2000). The EPA program (v. 3.11) is DOS based. The EPA URL for version 4.02 of the MINTEQ program is:

http://www.epa.gov/ceampub/softwdos.htm

This program provides a DOS version in the 2.5 megabyte file, INSTALMT.EXE; note that the file attempts to directly access the hard drive during decompression and file extraction (not permitted by Windows 2000) and needs to be run in DOS mode under Windows 98.

To show speciation and a simple use of this program, an example in which calcite (the main mineral in limestone or caliche) is equilibrated with our soil water (Pco_2 of 0.01 atm.) is presented in Table 7.5.

Table 7.5 $CaCO_3$ Equilibrated with Pco_2 = 0.01 atm

Cation				Anion			
Analyte	**mmole/L**	**Valence**	**meq/L**	**Analyte**	**mmole/L**	**Valence**	**meq/L**
H^+	5.403E-05	1	5.403E-05	OH^-	2.137E-04	-1	2.137E-04
Ca^{2+}	1.591E+00	2	3.182E+00	HCO_3^-	3.224E+00	-1	3.224E+00
$CaHCO_3^+$	4.894E-02	1	4.894E-02	CO_3^{-2}	3.741E-03	-2	7.482E-03
$CaCO_3$	5.559E-03	0	0.000E+00				
Sum			3.231E+00				3.232E+00
Equilibrium pH = 7.297							

However, this example is a bit too simplistic to be realistic. In soil systems the water is migrating downward and the soil atmosphere reservoir is not infinite in size and volume. For a somewhat more realistic model, the connection to the soil atmosphere is "broken" before equilibrating this water (Table 7.2) with the caliche, with the result as shown in Table 7.6.

Table 7.6 $CaCO_3$ Equilibrated with Water Previously Equilibrated with Pco_2 = 0.01 atm

Cation				Anion			
Analyte	**mmole/L**	**Valence**	**meq/L**	**Analyte**	**mmole/L**	**Valence**	**meq/L**
H^+	3.075E-06	1	3.075E-06	OH^-	3.501E-03	-1	3.501E-03
Ca^{2+}	3.613E-01	2	7.226E-01	HCO_3^-	6.974E-01	-1	6.974E-01
$CaHCO_3^+$	2.775E-03	1	2.775E-03	CO_3^{-2}	1.230E-02	-2	2.460E-02
$CaCO_3$	5.563E-03	0	0.000E+00				
Sum			7.254E-01				7.255E-01
Equilibrium pH = 8.527							

Note that this revised approach decreased the amount of calcite or caliche dissolved from 1.6455 millimole/liter (66.0 mg/L dissolved calcium or 165 mg $CaCO_3$/L) to 0.3696 millimole/liter (14.8 mg/L dissolved calcium or 37 mg $CaCO_3$/L) and is a bit more in order with the data from shallow groundwater systems and vadose zones.

Kinetic factors

Kinetics will be touched upon only briefly, but you should know that: (1) complexation reactions between dissolved species normally occur rapidly (seconds or less); (2) reactions between species in aqueous and nonaqueous phases are much slower; and (3) many redox reactions are extremely slow to equilibrate, unless biologically catalyzed. Equilibration of the water with gases normally takes hours to days unless you have an experimental apparatus with bubbling frits and stirring which provides large numbers of small gas bubbles (giving a large surface to volume ratio) and mixing of these bubbles throughout the container. Solids dissolve at varying rates, with equilibrium usually not reached for times ranging from days to hundreds of years.

Minerals Forming/Dissolving, Predictions, and Flow Paths

If you pour a small amount of table salt into a glass of water and stir it, the salt will dissolve and you still have clear water. Table salt is derived from the mineral halite, which normally only occurs as a result of strongly evaporating conditions. However, all minerals have some (even if it is very, very small) solubility in water. Typically, if we follow a recharge event and track the water down gradient in the flowpath, you will find that the total dissolved solids in the water increase as it migrates down gradient.

One of the classic examples of this increase was presented in a paper by Back and Hanshaw (1971) using the Tertiary carbonate aquifer system of the Florida peninsula, pretty well-defined flow paths, carbon-14 dating, and chemical analyses. Unfortunately, contouring is in terms of entropy rather than moles of calcite (or equivalent) dissolved, but the entropy contours do approximate the TDS increase down gradient in this flow system. A point noted by the authors is that the energy loss associated with the down-gradient movement is approximately balanced by the entropy energy gain associated with the calculated mineral reactions (Figures 7.2, 7.3, and 7.4). Of particular interest is the fact that there is a close correlation between the amount dissolved and the time of residence in the aquifer (Figure 7.5). In other words, the distance from the recharge area varied with the relative permeabilities of the facies along different flow-path arrows, but the water chemistry roughly indicated the age of the water and shows us

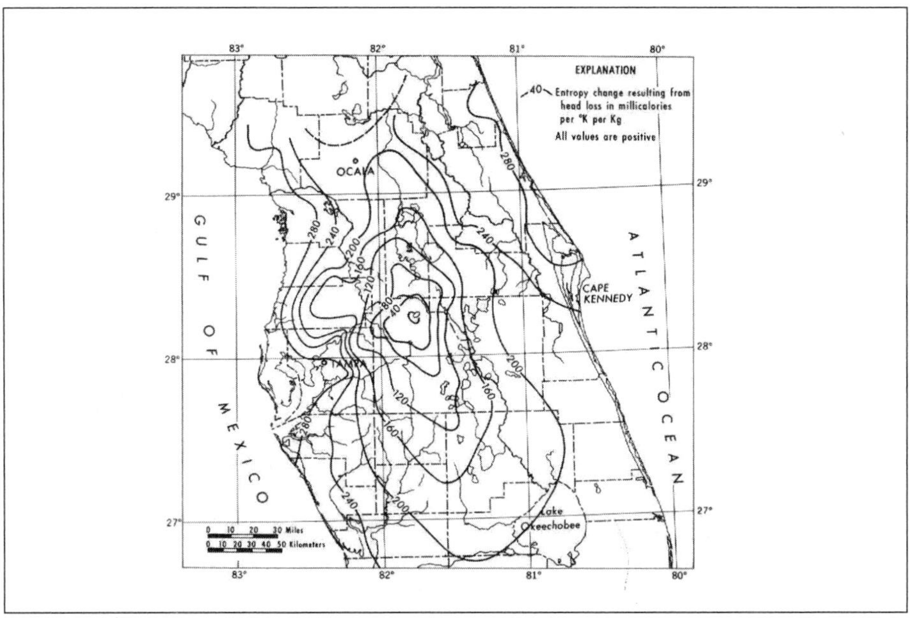

Figure 7.2 Calculated entropy resulting from head loss in the Tertiary-age Principal Artesian Aquifer of central Florida, positive values. [From Back and Hanshaw (1971).]

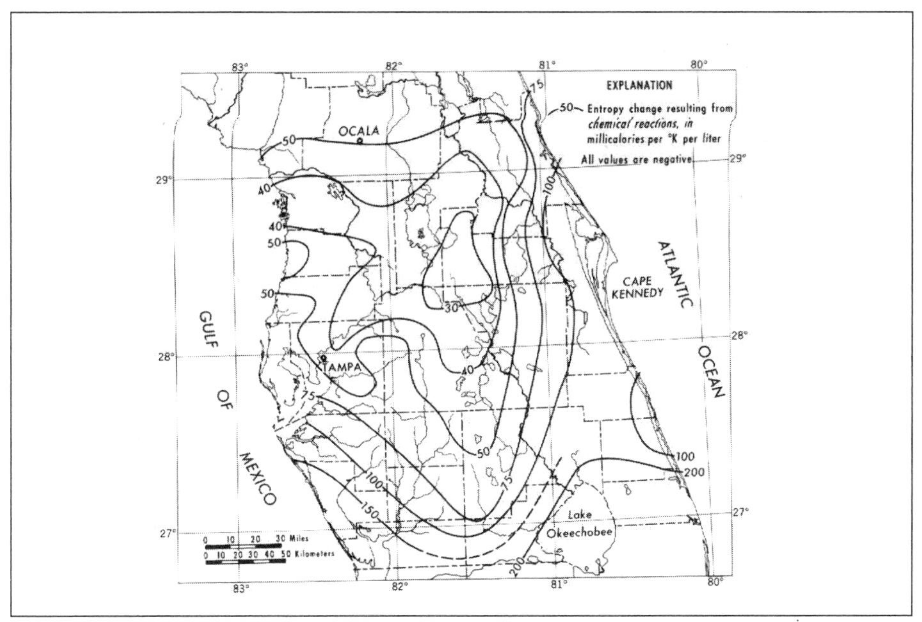

Figure 7.3 Calculated entropy resulting from head loss in the Tertiary-age Principal Artesian Aquifer of central Florida, negative values. [From Back and Hanshaw (1971).]

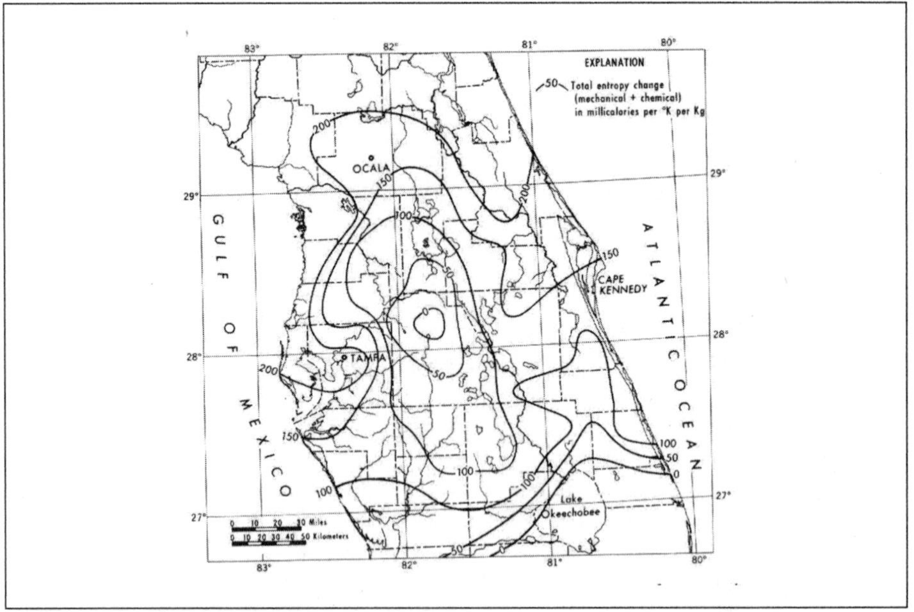

Figure 7.4 Distribution on entropy change resulting from combination of physical and chemical processes. [From Back and Hanshaw (1971).]

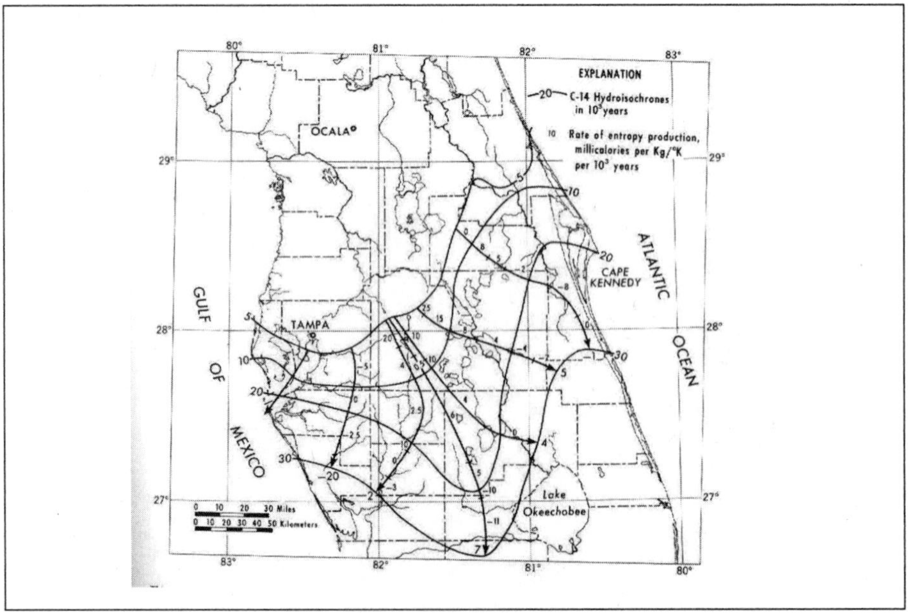

Figure 7.5 Rate of entropy production from both chemical and physical processes. (Values at arrow tips are averages for the entire flow path. [From Back and Hanshaw (1971).]

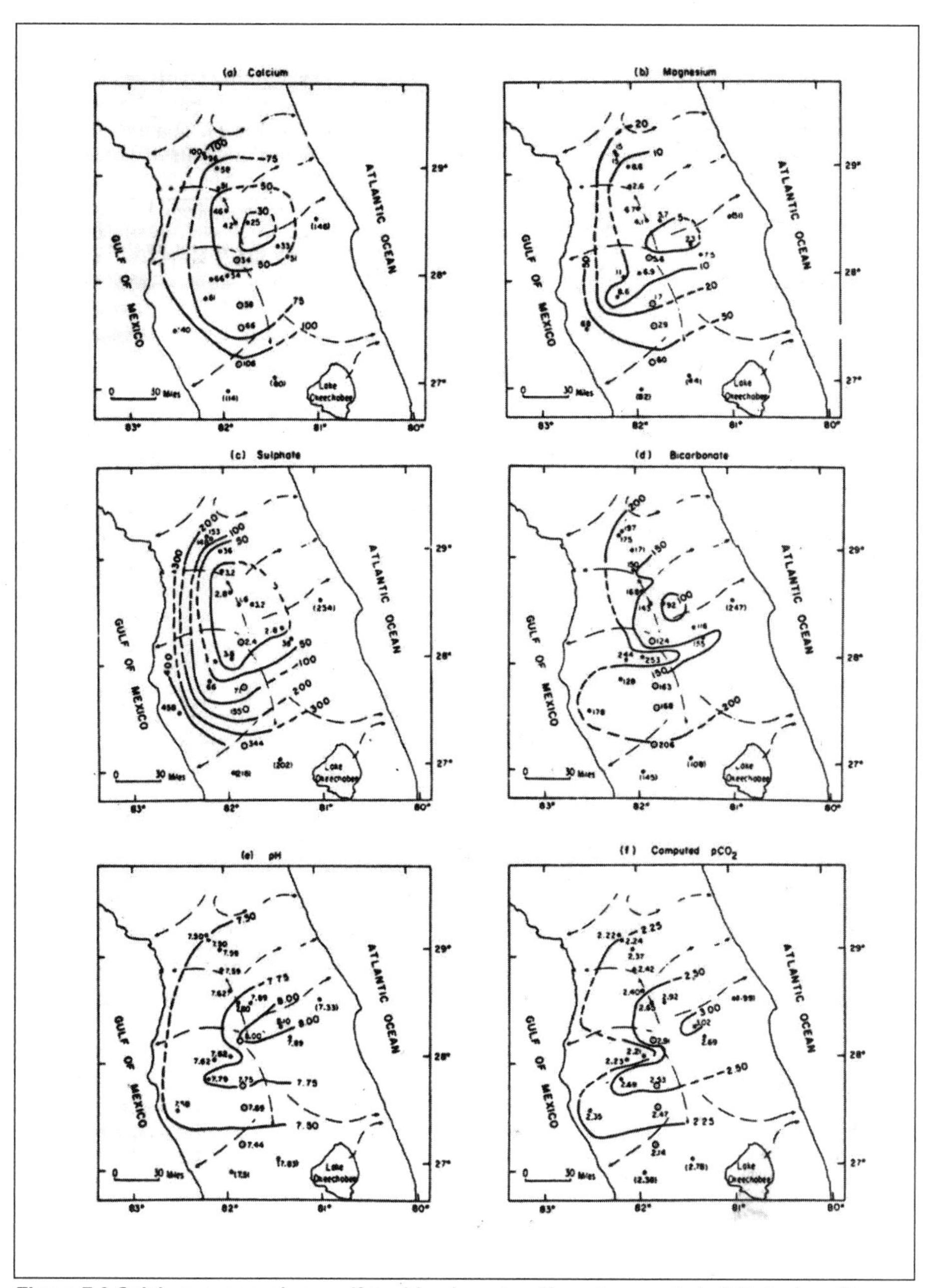

Figure 7.6 Calcium, magnesium, sulfate, bicarbonate, pH, and computed pCO_2 (= $-\log_{10}P_{CO2}$ concentration) maps of the limestone aquifer in central Florida. [From Mercado and Billings (1975). Used with permission.]

that the dissolution kinetics were roughly the same throughout the system being studied. While this paper is not for beginners, it is one of the few papers in the literature that actually looks at the change in potential energy related to elevation and tries to tie that back into the chemical potential energy.

A follow-up study by Mercado and Billings (1975) using Back and Hanshaw's data expands this approach and explains the chemical calculations far more clearly, and is strongly recommended as an aid to understanding this approach. Their figures for dissolved constituents (Figure 7.6) present a more conventional approach of presenting this type of data. The key point of this is that if we know the mineralogy of the rocks, and the system isn't too complicated, we can use the water chemistry to check or refine the potentiometric flow maps.

Predictions about Direction of Processes (Groundwater Evolution)

Fortunately, the thermodynamic calculations are less complicated than most flow systems. Admittedly, we will use ideal mineral formulas with little or no substitution, and (normally) well-crystallized lattices. As an example, let us assume a limestone aquifer containing some interbedded dolomite. Rain falls in the recharge areas, seeps into the soil and acquires some additional carbon dioxide gas (typically $P_{CO_2} = 10^{-2}$ atm. compared to the partial pressure in air of about $10^{-3.5}$ atm.) and reacts with the limestone and dolomite. As we follow the recharge down gradient, both Ca and Mg increase. Because there is more calcite than dolomite, and because calcite is more reactive (both affect dissolution fluxes, remember, you had to scratch the mineral dolomite [making a fine-grained powder on the mineral surface] to get it to "fizz" with the same strength HCl acid that vigorously attacked calcite in freshman geology or mineralogy labs), the concentration of Ca is greater than that of Mg.

Geochemical models will show that both calcite and dolomite are approaching saturation as we evaluate groundwater in the down-gradient direction. If we continue looking at samples from further down-gradient they will show that the amount of calcium in the water reaches a maximum and then may actually decrease a bit, while the amount of magnesium continues to increase. The geochemical models will show that calcite is saturated or slightly supersaturated, and that dolomite is significantly supersaturated. This suggests that calcite is precipitating, and has precipitated, in this latter part of the flow path, but dolomite is not precipitating. What is going on? Is this thermodynamic approach invalid?

In a word, yes. If you remember being presented with what was classically called the "dolomite problem," you will remember that dolomite does not commonly form under earth-surface conditions, and that it has only been found forming in highly evaporative sabka environments. There is a kinetic barrier to the formation of dolomite. It requires what the chemists call a large "activation energy" (considerable supersaturation before precipitation will occur, estimated to be about 77 times the saturation value based upon some unpublished data of salinized groundwater in the Montana high plains where dolomite is only found in the very evaporative discharge area of the local flow systems).

However, because the mineral dolomite does not precipitate easily, the Mg/Ca ratio can be used to indicate travel down the flow path. The higher the ratio, the further the water has come or the slower the water has traveled (kinetic factors), assuming that the abundance of limestone and dolomite is homogeneous. Even in real (nonideal, varying carbonate mineralogy percentages and "impurities" such as low-Mg calcite and the presence of clays, which will react with Mg systems, this generalization works. This is shown in Figure 7.6, where in the recharge area (roughly the center of the bull's eye created by the contours) the Mg/Ca ratio is about 5/30 or 1/6. Following the flow to the south, we find the ratio changing and increasing to about 50/100 or 1/2, demonstrating this relationship.

There is basically a kinetic barrier preventing the "rapid" (in terms of the rate of groundwater flow and the available equilibration time) equilibration of many minerals. The silicates are good examples; a number of minerals (such as quartz and the feldspars) that form at relatively high temperatures are commonly supersaturated in groundwater analyses.

As an example, Table 7.7 depicts the saturated (first two columns) and undersaturated mineral phases from a 1997 sample from a shallow (less than 15 ft deep) monitoring well in valley-fill alluvium on the upgradient end of the Butte Reduction Works below and downstream of the main mineralization zone in the Butte Quartz Monzonite bedrock. This water chemistry is largely the result of weathering of the granitic intrusive and the outer zone mineralization associated with the copper deposit, but accompanying extrusive tuffs which are partially preserved help explain the high silica content (dissolution of volcanic glass).

While most of these are common minerals, the cadmium and zinc carbonates (otavite and smithsonite) have rather obscure names and are listed by formula. Note that with the exception of chlorite, typical aluminous, silicate, and aluminosilicate minerals are supersaturated. The saturation index [SI = log (IAP/Ksp)] is the log to the base 10 of the ion activity product divided by the solubility product. Thus at saturation it has a value of zero

[$\log_{10}$ (1) = 0.0], and the value of 1.08 for quartz means that the water contains about 12 times more dissolved SiO_2 (36 mg/L) than it would if in equilibrium with quartz at a water temperature of 6°C. Disequilibrium will be more pronounced (the absolute value of log (IAP/Ksp) is large) with clays and other phases containing multiple hydroxyl units because a small change in pH yields a significant change in the saturation index. More complex minerals than simple "salts," such as feldspars will be intermediate in disequilibrium magnitude because of the exponentiation of some components (such as $H_4SiO_4^0$).

Table 7.7 Selected Saturated and Undersaturated Mineral Phases in Monitoring Well 8A at the Butte Reduction Works, Sampled April 8, 1997

Mineral	Log (IAP/Ksp)	Mineral	Log (IAP/Ksp)
Alunite	4.31	Calcite	-1.47
Chalcedony	0.51	Chlorite	-8.82
Gibbsite	1.17	Dolomite	-3.56
Quartz	1.08	Fluorite	-0.34
CdCO3	0.56	Gypsum	-0.99
Illite	4.12	ZnCO3	-0.55
Kaolinite	8.20		
Low Albite	2.56		
Microcline	3.35		
Montmorillonite	4.79		
Muscovite	10.80		

History of Flow Path, Can It Be Deduced from the Chemistry?

This is the reverse process of what was just described. Given the water analysis from a point in a flow system, can we deduce something about the flow path that the water took and/or the mineralogical influences upon the water chemistry? Commonly the answer is yes, but only to a limited extent. The more knowledge that we have about the geologic setting (the groundwater "plumbing system"), the easier it is to get a meaningful interpretation. From the water analysis we have a list of elements and their concentrations; these had to have been in the original precipitation, or have been derived from minerals that the water contacted along its flow path. However, how many minerals are viable sources of the first element? What minerals are potential sinks (precipitates that remove or limit the amount of an element in the water)?

In general, high-temperature minerals will dissolve but not precipitate. For example, feldspars are attacked and dissolve, but aren't formed; instead, one or more clay minerals and chert (microcrystalline quartz) are the common limiting phases/precipitates in near-surface groundwater flow systems. Classic papers on this topic by Garrels (1967; Garrels and MacKenzie 1967) are listed in the references at the end of the chapter should you wish to read more about this.

Despite these limitations, some inferences can be drawn using the minerals known to occur or probably present in the aquifer(s) in question. For example, "dirty" sandstones commonly contain feldspars and amphiboles that will weather and form secondary clays. In turn, these clays may have significant ion-exchange capacity. Frequently, groundwaters in the near-recharge area are "hard" (Ca plus Mg as $CaCO_3$ 60 mg/L), while down-gradient groundwaters are "soft" due to ion-exchange reactions. This is a common situation in the Tertiary "fill" materials of the western Montana intermontane basins.

7.2 Collecting Samples

A sample should accurately reflect the composition of the groundwater in the aquifer. To achieve this, one needs to measure some parameters in the field. In some cases, a down-hole determination of temperature, pH, specific conductance, and redox may be needed in conjunction with down-hole sample collection (Chapter 5, Section 5.3). Why might that be?

The pressure at the sampling point will be approximately atmospheric plus 0.433 pound per square inch (psi) per foot of water above the sampling point (Chapter 5, Section 5.3). Hence, any pressure-dependent reaction that will affect the water will result in different values for samples collected in situ and at the well head (atmospheric pressure). The most common example of this is carbon dioxide degassing in samples brought to the land surface and the concomitant loss of alkalinity and rise in the measured pH resulting from the reaction:

$$HCO_3^{-1} + H^{+} = H_2O + CO_{2(gas)} \qquad [7.10]$$

If the partial pressure of carbon dioxide in the aquifer at the sampling depth is not equal to the atmospheric partial pressure, carbon dioxide will migrate out from or into the sample. Ideally, we would get in situ measurements of pH, SC, Eh, and temperature, and a pressurized sample for alkalinity determination. If that is not possible, we attempt to measure these variables at the well head, commonly using a flow-through cell with a restricted outflow (to minimize pressure loss), but, if not, with a small (1- or

Figure 7.7 Hach titration tool containing a 0.01639 N acid cartridge for an alkalinity titration.

2-gal bucket) container receiving the well discharge. In this last case, it is preferred that the discharge be transferred using a hose to the bottom of the container to minimize aeration of the sample. Alkalinity is determined as rapidly as possible with field titration using a burette or tools like Hach's field alkalinity kit (Figure 7.7). This is normally done with a beaker or a large-mouthed Erlenmeyer flask, inserting a rapid-response glass combination pH electrode into the water and gently swirling the solution while titrating acid into the beaker or flask until the end-point pH (typically 4.5) is reached. Because bicarbonate is the major ion determined in most alkalinity titrations, the original equation and acid strength chosen were set up to simplify the calculation of the bicarbonate concentration in milligrams per liter:

$$\text{Alk(as mg/L } HCO_3^{-1}) = 1000 \times (V_a/V_s) \qquad \textbf{[7.11]}$$

where V_a and V_s are the volumes of 0.01639 normal sulfuric acid and the sample, respectively. Typically, a 50-mL sample volume is used. A more generalized equation for alkalinity as $CaCO_3$ is:

$$\text{Alk (as mg/L } CaCo_3) = (1000/V_s) \times (V_a) \times (0.820/0.01639) \times (N_a) \qquad \textbf{[7.12]}$$

where V_s is the volume of the sample in mL, V_a is the volume of acid added in mL, 0.820 is the conversion term to change two bicarbonates to one calcium carbonate, 0.01639 is the preferred normality of the acid and is equal to 1 divided by the mass of the bicarbonate ion, and N_a is the normality of the acid actually used.

Typical problems encountered in the field are overshooting the end point, error in measuring the sample volume accurately (25- and 50-mL graduated cylinders are very handy; glass is preferred but plastic has a longer useful life under most field conditions), and recording error of the initial or final acid volume. One of the causes for overshooting the end point (adding too much acid) is a slow pH electrode. Never try to do alkalinity determinations with gel electrodes as they are quite slow to respond. When you are getting to the end point, the buffer capacity of the water is essentially used up and a small acid addition has a large effect on the pH of the sample, so use small addition steps.

Alkalinity is evaluated on an unfiltered sample. However, samples going to a laboratory are normally filtered, unless the determination is for EPA "total" values. There is a long history of discussion regarding the use of filtered versus unfiltered samples. The purpose of the use of your analytical data should control how the samples are collected and preserved. However, one valid generalization is that surface waters are more likely to have unfiltered samples submitted to the laboratory. Routine filtration employs an initial 0.45-micron (45-angstrom) filter opening; as sediment and organic matter collect on the filter, the effective diameter of the openings is decreased. The author recalls having to use five 0.45-μm filters to get 750 mL of sample from an irrigation ditch during the rising limb of the spring runoff event.

The trend has been toward the use of smaller filter openings, both to minimize the content of colloids and finely ground clays and to prevent clogging of analytical system "plumbing." Even 0.45-micron filters will plug fairly rapidly when filtering "dirty" water (such as irrigation ditch water or undeveloped/poorly developed or constructed wells). Monitoring wells (particularly sub-2-inch casing diameters) may be difficult to develop and in some cases (due to requirements to dispose of development water as a hazardous material) have not been developed. Frequently, prefilters that utilize a larger opening are used to remove the coarser material suspended in the sample. However, be aware that filters can contaminate your sample. Have blanks run on the field equipment (analyze your deionized or distilled water used in the field to clean the equipment and also filter this water with a volume equal to the smallest sample size being collected; if the filtered water contains elements not found or in higher concentrations than the unfiltered water, the difference is attributed to leaching from the filter and filtration equipment). At one point in time, the Montana Bureau of Mines and Geology was utilizing a hypodermic disposable filtration system that contributed as much as 0.7 ppm zinc to a 250-mL sample. Also, the amount of zinc released varied from lot to lot of these filters. Shifting to a different brand reduced the problem, both in terms of the amount of zinc released and the lot to lot variability of the filters. Figure 7.8 shows several types of filtration equipment.

Figure 7.8 Atila positive pressure filtration device (A), vacuum filtration system (B), and hypodermic syringe with disposable filter (C). A and B use replaceable 47-mm diameter filters from companies such as Millipore.

The discussion of sample contamination by the filtration equipment was to demonstrate why it is important to discuss the reason for sampling with the laboratory to ensure that you are using equipment that will not significantly impact the analytical results needed for your task and to be certain that you are collecting the necessary types of samples to allow the lab to give you the best results. This naturally leads to sample preservation.

Samples are commonly preserved by cooling (putting the samples into an ice chest that is already cooled down and in which ice is maintained until the samples are delivered to the lab), freezing (common with core samples in which interstitial water is to be analyzed), or acidification. Nitric acid is commonly used for metals preservation, but it is an oxidant and will convert ferrous iron to ferric iron. If the Eh is to be determined from analytes, hydrochloric acid is the common preserving agent. The laboratory should provide the acid used, so that it can discount the trace amount of metals in the acid (even "spec pure" grade acid contains some contaminants). In general, the filtered acidified samples are considered "good" for several months for total metals, alkalis, and alkaline earths, but not for redox states of metals.

Nutrient samples used to be preserved by adding mercuric chloride; however, it was found that enough mercury escaped as vapor to contaminate the other samples. More recently, nutrient samples are typically preserved by adding sulfuric acid. Anion samples are typically filtered and chilled, but not preserved other than by being chilled. Thus, there can be fairly rapid changes in things like orthophosphate ($H_3PO_4^0$ + $H_2PO_4^{-1}$ + HPO_4^{-2} + PO_4^{-3}) if a separate nutrient sample isn't also collected and used for the orthophosphate determination. A study done in Georgia showed a logarithmic decay due to holding time problems in getting the chilled but not preserved samples to the lab for orthophosphate analysis. Generally, a 48-hour maximum holding time is quoted as a guideline for anion samples, but we were finding 50 to 60 % loss of orthophosphate 48 hours after sampling.

Again, the important concept is to discuss the sampling objectives with the laboratory staff and to be sure that the field personnel understand and follow the sampling and preservation protocols needed to meet your objectives.

Inorganic Constituents

Analytical determination of the inorganic, dissolved content of a water is an old "art form" now set up for routine automated analysis by modern instrumentation. If you are using data from before about 1955, check the data for an "Na + K" component. This is because the determination of sodium and potassium by classic "wet chemistry" methods was exceedingly difficult. Commonly the alkaline earths and metals were analyzed, as were the common anions. A charge balance was calculated and the surplus (if any) of anions was attributed to the sodium and potassium content in the water. Actually, many of these old analyses are surprisingly good, but you should be aware of potential problems with the data.

Modern analyses using atomic absorption spectrophotometry (AA) or inductively coupled argon plasma spectrometry (ICAP or ICP) are normally good, for what was reported. Often more constituents were determined than are reported, because the customer didn't request all of the constituents that the automated analytical equipment routinely determines with the newer (since the late 1980s) equipment. One of the potential pitfalls here is that something present in significant quantities is not reported. For example, anion chromatography generates curves for both nitrate and nitrite, yet labs commonly report nitrate plus nitrite. From a health perspective, significant values of nitrite are very important as it commonly indicates recycling of septic effluent into the water supply system. Sulfate is now determined by ion chromatography (IC), but older data (from before 1980 or so) were usually based upon gravimetric analysis or spectroscopy and may be less precise. ICP units (the optical emission spectroscopy [ICP-OES] or mass spectroscopy [ICP-MS] units) typically analyze up to 40 constituents at once, but only those requested are reported. Yet values for titanium can be very useful as high values are commonly associated with failure of the filtering membrane or membrane installation. The mass spectrometry units give very low detection limits (parts per trillion or lower), which may be needed for environmental work.

From the perspective of the data user, it is recommended that when negotiating a contract with a laboratory, you try to get all of the constituent determinations reported. If the lab will not add these for a modest increase in cost, at least have the complete "laundry list" run on a few samples representing the range of conditions in the study area. This should help you determine if anything important is being omitted from the standard samples.

Because the analytical methods for the analysis of typical inorganic constituents are well established, there are rarely problems with the lab results. Probably the most common error is forgetting to record a dilution at the lab. Dilutions are often needed because as the detection levels have dropped, the upper range of the methods have tended to decrease, also. You can do a quick ballpark check using the specific conductance and the laboratory total dissolved solids calculation:

$$\text{TDS (mg/L)} \cong (0.67) \times \text{SC at 25 °C}(\mu\text{mho/cm or } \mu\text{Siemen/cm}) \qquad [7.13]$$

There are a number of good books available on using and interpreting aqueous geochemical data. One of the oldest but still one of the better ones, even for novices, is John Hem's *Study and Interpretation of the Chemical Characteristics of Natural Water* (1985). For more chemically advanced students, Drever's *The Geochemistry of Natural Waters*, Stumm and Morgan's *Aquatic Chemistry* (1996), Langmuir's *Aqueous Environmental Geochemistry*, Krauskopf and Bird's *Introduction to Geochemistry*, or Appelo and Postma's *Geochemistry, Groundwater and Pollution* (1993) may be better choices. This last book is excellent for hydrogeologists with a reasonable chemistry background, as it emphasizes what happens to water from falling as precipitation through recharge and transport to discharge or deep "stagnant" reservoirs.

Organic Constituents

Both natural and man-made organic constituents may be found in water samples. Most laboratories offer determination of "organic carbon," either total (from an unfiltered sample) or dissolved (from a filtered sample). Some laboratories will refuse to do totals because of frequent plugging of the instrumental plumbing. For groundwater studies, a dissolved organic carbon is the more appropriate in nearly all cases, so this shouldn't be a difficulty. However, you should know that these compounds can be subdivided into volatile and nonvolatile subgroups. The nonvolatile organic compounds are relatively easy to sample for, whereas the volatile constituents are much harder to obtain representative samples, and special procedures and equipment are normally called for. See Chapter 8 for a discussion of in situ gas chromatographic analyses of soil vapor.

Natural nonvolatile constituents include organic compounds such as fulvic, humic, and tannic acids. Waters containing organic acids are commonly colored (pale yellow to brown) as any tea drinker would know. Swamps are the areas where surface waters commonly contain significant amounts of these acids, but small amounts are typically found in groundwaters affected by less extensive wetlands.

Man-made nonvolatile compounds may be the sparingly soluble DNAPLs (dense, nonaqueous phase liquids) such as TCE (trichloroethylene), PCP (pentachlorophenol) or PCBs (polychlorinated biphenyls). These dense liquids consist of "heavy" molecules with relatively low vapor pressures (hence nonvolatile) and larger octanol-water partition coefficients (hence, less soluble or "water loving").

If you are involved with emergency response activities related to transportation spills, you won't know what was carried until the DOT (U.S. Department of Transportation) manifest can be found or a copy acquired. First responders have to presume the worst and wear appropriate personal protective equipment (PPE) for the worst case. Sampling typically comes after a first response when the contents that were spilled are known and risks should be less.

Probably the most important thing that can be stressed in this section is to always follow the PPE guidelines for the materials that you may be sampling, and if there are any uncertainties, err on the side of caution. Oftentimes, the effects of exposure don't show up for a number of years and may be difficult to prove (just look up the history of Agent Orange on the World Wide Web if you need an example), and no amount of insurance can buy back your health.

7.3 Evaluating Laboratory Results

There are a number of methods for evaluating laboratory analytical results. Ion balancing, spike recovery, sample splits (actually duplicates) and third party lab results, and comparison with field parameters as indicators of the "adequacy" of the analytical results will be explored. However, you should be forewarned that all of these approaches are methods of improving quality control but do not guarantee the accuracy (versus the precision) of the laboratory results (discussed in Section 7.4). Different labs using the same methods and equipment can have good agreement and still be "wrong" if the method is not completely appropriate for the sample. Since sample collection has just been discussed, a logical approach is to start with comparing the field parameters with the laboratory result.

Comparison with Field Parameters

As pH and SC (specific conductance or electrical conductivity) were measured in the field, comparing those field results with the laboratory results can indicate potential problems of sample stability or poor equipment or technique. The field sample had determinations made at or near the in situ water temperature. Laboratory analyses may be done on water just out of the refrigerator, or at room temperature. Most modern SC meters give results in microsiemens/cm or millisiemens/cm; these units are equivalent to the older terms micromhos/cm and millimhos/cm in the literature. There is a temperature dependence in the conductance measurement that becomes more pronounced as the temperature approaches the freezing point. A reasonable approximation of the temperature correction to convert the specific conductance of a water sample at temperature T to the specific conductance at 25°C is to multiply the measured (uncorrected) value by the ratio of the viscosity of water at temperature T divided by the viscosity of water at 25°C (see Table 7.8). Laboratory values of conductance are reported at 25°C. If your field meter makes the correction to 25°C, the values can be compared directly. The magnitude of the correction terms for temperatures ranging from 0 to 35°C are 2.007 to 0.808, so you can see that failure to convert the uncorrected field value can lead to significant disparities between the field and lab values.

Table 7.8 SC at 25°C = SC at Field Temperature × Correction Factor

Temperature	0	5	10	15	20	25	30	35
Correction Factor	2.007	1.702	1.466	1.279	1.125	1.000	0.896	0.808

The earlier discussions of CO_2 degassing may suggest why the sample might be unstable and the laboratory pH measurement, while accurate on that sample at that time and under the current sample conditions, is not representative of the water when sampled. For reporting, the field value is preferred; the best way to do this is to have two adjacent columns giving the field and laboratory pH values. This will let you look for systematic differences between the two as well as making it easy to note the unstable samples (depending upon equipment, etc., a shift of more than 0.2 to 0.3 pH units indicates an unstable sample). Because pH measurements are temperature dependent, there will be some shift resulting from that if the water is not at neutral pH (about 7, neutrality is also temperature dependent) and the field measurement was not temperature corrected.

With older pH meters, we showed the students in the University of Montana's Montana Tech Hydrogeology Field course how significant this could be by having them do a temperature shift from ambient (20 to 25°C in the summer) to a 5 or 10°C water temperature when they were buffering their pH meters in the field. This is best done using a buffer near pH = 10, despite the slow electrode equilibration because the shift is significant; we then had them recheck the 7 buffer and see that the temperature adjustment caused no change in pH (because the voltage is very close to zero). The importance of measuring the temperature of the buffer solution was also demonstrated by this "exercise." Comparison of several meters will show that while the shifts are in the same direction and of about the same magnitude, they are rarely the same.

Ionic Balances

Water has to have its negative charges from dissolved anions balanced by the positive charges of dissolved cations; this is often referred to as the "principle of electrical neutrality" and is implicit in some calculations of ionic strength. From your laboratory results, adding up the sum of the positively charged analyte concentrations (mole per liter or per kg solution) multiplied by their respective charge should be equal to the sum of the negatively charged analyte concentrations multiplied by their respective charge, i.e.,

$$\sum C_i \times z_i = -\sum A_j \times z_i \qquad [7.14]$$

where C_i is the concentration of the ith cation, z_i is the charge on the ith cation, A_j is the concentration of the jth anion and z_i is the charge on the jth anion. Using the data from Table 7.5, we have the sum of cations times charge as 3.231 meq/L, and the sum of anions times charge as -3.232 meq/L (the negative sign was left off in the table, by convention).

Normally, the multivalent metals are calculated at their lower valences state (i.e., iron and manganese as Fe^{2+} and Mn^{2+}), while some anions have to be converted to their dominant species (i.e., phosphate reported as PO_4 must be converted to HPO_4^{-2} at pH values between 7.2 and 12.4 or to $H_2PO_4^{-1}$ for 2.1 <pH < 7.2). Generally the concentrations of multivalent species are small, relative the total concentrations of cations or anions and the error for charge on minor species is minor. When dealing with quite acidic mine drainage (pH < 3.5), the trace metals may have a significant concentrations and a sizeable fraction of the metal may be in an oxidized form (higher valence state), and field or laboratory determination of the metal speciation may be required.

In reality, these numbers almost never balance because of analytical error and species not determined. If you get results for a number of samples that have virtually identical cation and anion charge sums, your laboratory is possibly "massaging" its results, and it should be viewed with suspicion (Table 7.5 balanced because it was a theoretical calculation). The balance is commonly calculated as:

$$\text{Balance} = {[(\sum C_i \times z_i) + (\sum A_j \times z_j)]} \Big/ {(0.5[(\sum C_i \times z_i) - (\sum A_j \times z_j)]} \qquad \textbf{[7.15]}$$

where a positive value indicates an excess of cations, and a negative number indicates an excess on anions (remember, $\mathring{a}A_j \times z_j$ is a negative value). Ideally the balance value should be no greater than ± 0.05; however, values as large as ±0.10 are common and considered acceptable. Balance values up to ± 0.15 may be accepted when high TDS (total dissolved solids) waters (>5000 mg/L) are analyzed. As an example of a difficult water, Table 7.9 shows the 100-m depth sample collected on October 16, 1987 (used for the Eh discussion, second column of Table 7.4).

This balance may be artificially poor, a construct of the standard assumption of the lower valence state for "groundwaters," as almost all of the copper was actually in the +2 state. One offsetting factor is the incompleteness of the analysis (or, more accurately, of the presented data); fluoride and silica were not included in the analytical results. The silica won't effect the balance at low pH values, but the missing 8.4 mg/L fluoride (0.442 mmole/L and meq/L) would have made the balance a tad worse, but still well less than the plus or minus 10 % that is generally considered acceptable. For comparison purposes, a speciated version (using PHREEQC to generate the species and the balance) is presented.

Description of solution:

$$\text{Balance} = (134.935 - 142.472)/(0.5 \times (134.935 + 142.472))$$
$$= -7.537/138.703 = -0.054$$

pH	3.15
pe	8.228
Activity of water	0.998
Ionic strength	1.484E-01
Mass of water (kg)	1.000E+00
Total alkalinity (eq/kg)	-1.786E-03
Total carbon (mol/kg)	0.000E+00
Total CO_2 (mol/kg)	0.000E+00
Temperature (degrees C)	13.5
Electrical balance (eq)	-3.420E-03
% Error = 100*(Cat-|An|)/(Cat + |An|)	-2.23
Iterations	10
Total H	1.110142E+02
Total O	5.578762E+02

Note that the error has dropped from 5.4 to 2.2%, from having the copper oxidation state calculated (an additional 3.19 meq/L of positive charge); however, this may be partially due to the 1.87 milliequivalents of alkalinity calculated by the program and the 0.86 meq/L reduction from HSO_4^{-1} based upon the speciation. The missing fluoride was not added to the speciation calculation in order to be consistent. The output from this calculation is included as Appendix C. The speciated balance is not normally needed, but if you have the needed field data (temperature, pH, alkalinity, and Eh), the information can help you understand more complex waters. If you are really perplexed with an analysis containing considerable oxidizable species but lacking an Eh measurement, try a range of Eh values in a speciation program and see if that helps you understand why the conventional balance calculations (with dissolved metals in their lower oxidation states) seem “out of whack.”

Table 7.9 Berkeley Pit Water from 100-m Depth, Collected October 16, 1987, Dissolved [Data from Davis (1988)]

Cation					Anion				
Analyte	mg /L	Mmol e /L	Valence	meq /L	Analyte	mg /L	Mmole /L	Valence	meq /L
H^+	0.94	0.933	1	0.933	OH^-	0	0	-1	
Ca	482	12.119	2	24.238	Cl	22.1	0.628	-1	-0.628
Mg	280	11.606	2	23.212	SO4	6760	70.919	-2	-141.838
Na	70.8	3.104	1	3.104	As	1.15	0.002	-3	-0.006
K	18.7	0.482	1	0.482					
Al	193	7.209	3	21.627					
Cd	1.87	0.017	2	0.034					
Cu	203	3.219	1	3.219					
Fe	1020	18.406	2	36.812					
Pb	0.522	0.003	2	0.006					
Mn	162	2.972	2	5.944					
Zn	497	7.662	2	15.324					
Sum				134.935					-142.472

pH = 3.15
T = 13.5°C
Eh = .468-v

Spikes and Spike Recovery

A spike is a controlled addition to a sample of an already analyzed liquid or solid. The reason for this is to ensure that higher concentrations of elements are not "lost" due to reaction with the sample or within the analytical process (results significantly off scale because of limited calibration, insufficient reagents, etc.; this was a greater problem when using colorimetric techniques); the method is most frequently used for internal processing control in industry and to validate analytical results for regulatory agencies. For our purposes, we will be looking at aqueous solution additions to water samples. The laboratory will typically make up the spikes in elemental ratios that approximate the sample characteristics for your site samples, using the elements that are of human health or aquatic life concern, depending upon the purpose of your investigation. Thus in a mining district, one would expect that a typical spike might include metals such as Cu, Cd, Pb, ±Ag; these solutions would have nitrate, chloride, or acetate as the balancing anion, chloride cannot be used if silver is in the spike because silver chloride is relatively insoluble. The spike volume should not be less than 1 mL in volume and a larger volume would be preferred (at least

1% of the sample size if the sample aliquot is greater than 100 mL) to minimize measurement error.

For this quality-control approach, the sample and the spiked sample are analyzed, and the percent recovery (amount of the spike detected compared to the calculated concentration of sample plus the amount added with the spike) is reported. For example, assuming that your water sample had 10 μg/L Cd, the spike volume was 1% of the sample volume, and the spike concentration was 5 mg/L Cd, the ideal (100% recovery) analysis of the spiked sample should be 60 mg/L (10 + [0.01×5000] μg/L); or 10 + 50 μg/L), neglecting the small dilution effect). "Good" spike recoveries range from 90 to 110% (10 + 45 to 10 + 55 μg/L), while 80 to 120% are normally considered acceptable.

Blind Controls

The purpose of blind controls is to evaluate precision and "accuracy." The term "accuracy" has to be qualified, because it is often estimated by statistical means using the results from a large number of laboratories (round robin results) on a standard, very stable, sample sent to the labs; this is done by the EPA and USGS. Precision is simply reproducibility of the analysis. This is commonly accomplished by giving the laboratory splits of a sample with different sample numbers and comparing the laboratory results. For this to be meaningful, the samples must be as identical as possible. There are a few commercially available sample splitters for water, but taking the same type of samples (unfiltered, filtered, filtered and acidified, unfiltered but preserved or acidified) sequentially from a bucket usually works okay for the filtered samples, but is a disaster for unfiltered samples that will be analyzed for suspended sediment (not a common groundwater analysis; for surface-water samples you really must use a device that mixes the sample and delivers equivalent aliquots). Typically, the reported values for these splits will be within 1 or 2% of each other. Significant discrepancies may result from poor analytical procedures or poor splits in the field. It then becomes the responsibility of the team leader to ensure that a second suite of blind splits are as identical as possible when submitted to the laboratory to see if the lab is the problem.

Instead of sending blind duplicate (or triplicate) samples to your lab, you may choose to send these to alternative lab(s) for comparison. While this doesn't give you a handle on your laboratory's precision, it does provide a check on its accuracy, especially if it uses different equipment or methodology. Major elements should agree within ±5%, and trace elements within ±10%.

7.4 Tips and Tricks

What follows are really common-sense approaches to sampling. Because you never have a big enough budget to do everything that "might be useful," you need to think about why the work is being done, what possible approaches might work for this particular project, and what you will need for a final report.

Why Are You Sampling?

Can you skip sampling? Collecting samples properly takes time and costs money. If you have to "waste" part of your budget collecting water samples for analysis (or doing field determinations), what do you hope that these data will tell you? Will statistical evaluation of the data be crucial to the study?

This is the point at which the old USGS *Suggestions to Authors,* 5th ed., 1958, strongly stressed as detailed a preliminary outline of the final report as was possible, before starting the work. The real purpose of this is to help you determine what your data needs will be in order to have the necessary information to draw the conclusions expected from the study. Such advice is still as relevant today as it was in 1958.

When the goals or objectives of the study are laid out, most of your sampling decisions will come out of that as logical byproducts of this analysis. Typically, sampling is done to assess the suitability of a water source for human consumption, aquatic life, or to meet some regulatory standard. Frequently such objectives can be met with limited sampling (nutrients plus *e-coli* analyses, etc.). However, if a geochemical interpretation of will be part of the report, some complete analyses will be needed.

Ways to Stretch the Budget

Reconnaissance techniques can help you select samples for more detailed analysis. Part of the initial stages of most studies include an inventory stage. Getting temperature and SC data at that time (plus possibly pH and/or simple colorimetric or capillary tube concentrations) will give you an idea of the range of dissolved constituents, letting you select a limited number of the sources for further sampling. Use of silver nitrate capillary tubes to get field values for chloride (accuracy about ±10%) successfully permitted the determination of oilfield brine contamination plumes in eastern Montana for graduate students with limited analytical resources. This same type of approach can be used prior to drilling wells for shallow volatile hydrocarbon monitoring by using a soil-vapor sniffer and shallow hand-probed holes to determine where the higher concentrations are located; but, don't forget to go sufficiently far up-gradient from suspected sources to close off your anomalies.

The follow-up sampling can then implement a selected number of complete analyses, if needed, while the remainder of the analyses may contain fewer analytes—just what is thought to be needed for the major objective. Care should be taken to ensure that the more detailed samples encompass the full range of water-quality conditions. Typically, one can break the total sampling population into subgroups based upon the reconnaissance parameters, and then select random wells or sites from within each of the subgroups for the more detailed analyses.

An alternate approach that can be used is to randomly select a number of sampling sites for an initial, comprehensive analytical suite. Based upon the results of that initial sampling, the analytes can be selected that will be determined on the "routine" sampling of larger population of the available sampling sites.

Inexpensive standards for checking specific conductance meters can be made up using table salt and stored in 5-gal (20-L) jugs with dispenser spouts. A piece of plastic wrap over the top openings (under the cap) will minimize evaporative loss. Approximate the SC that you want (2.3 g in 20 L yields an SC of about 245 mS/cm; 25 g in 20 L yields an SC of about 2,300 mS/cm; and 300 g yields an SC of about 25,000 mS/cm) and don't worry about drying the salt before weighing it, as you will send a sample of the solution to your lab for precise measurement.

When dealing with existing field and analytical data, you may wish to give extra scrutiny to data from urban or densely populated suburban areas. When doing a screening for geothermal resources in Montana, anomalies (higher temperatures) were found in several urban areas that had no geological justification. Discussion with samplers at the U.S. Geological Survey and Montana Bureau of Mines and Geology determined that frequently it was not possible to run water long enough to thoroughly purge the pressure tanks and lines and get samples at near aquifer temperatures. The only "safe" data came from outdoor hydrants plumbed into the supply line between the well and the pressure tank.

Carrying spare batteries for all equipment used in the field, plus a spare probe for pH meters may sound excessive, but if you work "out in the country" or utilize equipment with special batteries (Lab Line graphite cup SC meters come to mind due to a battery fiasco near Yellowstone Park), the time saved for one case of equipment failure will justify the cost of these items.

Folks working with acid-mine drainage (AMD) may want to carry plastic bottles of acid and base to "play" with the water. With a couple of beakers, you can add base and look at the color of precipitates as the pH is increased with the base. The acid can be used to bring the pH back down if you use too much base. If you do this, make sure that you review the amphoteric nature of the metals that you may be looking at. This approach has been found to be useful to get a quick and dirty guess at the aluminum concentration in quite acidic waters.

Ice chests and bus transport used to be adequate to get samples to a "local" lab within the mandated holding times. However, you do have to set this up with your laboratory. Also, don't trust the bus station to call the lab; give it a call telling it when the samples should arrive if the bus is on time.

Asking the residents if their well water makes decent coffee or tea may sound like voodoo science, but metal-rich waters tend to degrade the "taste" of these drinks, and even if the taste is okay, comments about having to run vinegar through a drip coffee maker warn you about the hardness of the water. It really does pay to talk with the landowners about their well and the suitability of the water from their well(s).

Statistical Analysis of Data

There are two basic reasons for using statistics: (1) to help you set up the project with "justifiable" conclusions when finished, and to develop a better project approach before doing the work; or (2) to attempt to justify your conclusions after the work has been done. The former is the approach that you should be using, but it isn't uncommon to be handed a project in the middle or near its end, and have to try to salvage all that you can from the prior work. What follows is a quick overview of the approaches open to you.

More important questions might be how many samples are needed from a specific source, and how frequently should they be sampled, if you are involved with environmental remediation. Obviously, this goes back to the purpose of the sampling and the regulatory framework. For example, public supply well fields are normally sampled quarterly maintain state approval, but if a problem develops, weekly or monthly sampling may be mandated until it can be shown that the problem constituent(s) is(are) no longer present in the product water in significant quantities. Once the problem is noted, it is probable that each well is sampled to evaluate the source well(s) with the intention of cutting the polluted wells out of the supply system and to facilitate remediation. In each stage, time-concentration plots may be used to estimate aspects of the problem (rate of plume migration to additional wells, for example, and where to install monitoring wells to warn the water supplier before additional supply wells produce the contaminant), but because prior samples were well-field composites, and the new samples are individual well samples, comparisons may fall into the apples and oranges category.

Alternatively, in doing an areal study of the groundwater resources, statistical tests (such as the Students T Test) comparing the sole analyses for all of the wells sampled is a reasonable first cut, but grouping the analyses by aquifer and analyzing these groups individually will give you a better idea about the "uniformness" of the groundwater in each aquifer. The usual problem here is that you may have so few samples from some units that the statistical analysis is virtually meaningless because of the small sample population. As a crude guideline, you need about five samples from groundwater source, with seasonal variation, to indicate un-impacted variability from which a significant shift in composition can be determined.

The necessary samples in surface water settings (especially streams because of their seasonal variability in flow) will be much larger. This is because of daily and seasonal variations in biological activity and sediment loads, etc.

7.5 Field Equipment Use Guidelines

The proper care and maintenance of your equipment will result in better quality field data and let you feel more confident about interpreting that data. What follows is somewhat colored by personal experiences and is bound to be a bit biased. The one "rule" that should be stressed, is that you really must read the manual and familiarize yourself with the instrument before attempting to use it, if you expect to get meaningful results.

pH

When dealing with pH meters, probe care and proper calibration are the most significant factors. Almost all fieldwork employs the use of a combination electrode (which contains both the "reference electrode" and the potential-measuring, indicator electrode in one body), and that is what is described here. The probe tip (typically a glass "ball" at the bottom of the probe) must remain hydrated; if the tip dries out, it will cause erratic and irreproducible results until it is rehydrated. To avoid this problem, you should have a flexible plastic "cup" that slips over the bottom of the probe. This cup should be nearly filled with the desired aqueous solution (pure water, buffer solution, or acidified water) and slipped over the probe tip after use to keep the tip wetted. This will cause a little bit of leakage as the cup is slipped on. Some cups are pretty firm (harder plastic), and it is more difficult to allow easy exit of surplus solution in the cup. For liquid-filled probes, you don't want to force liquid from the cup to migrate into the reference solution through the liquid junction frit or wick. This problem has been lessened by the development of higher flow and flushable liquid junction electrodes.

Gel pH electrodes have improved, with better junctions that speed up their response a bit; but their main value is the lower maintenance level and (usually) sturdier construction. These are typically viewed as "student grade" electrodes, but are fine if rapid response is not needed. However, when sampling waters that may release carbon dioxide, or performing alkalinity titrations, a faster electrode is needed.

Refillable electrodes have a variety of reference solutions. For example, the standard Beckman electrode uses a 4M KCl solution saturated with AgCl, while their Star™ electrode uses 1M KCl saturated with AgCl and is preferred for low ionic strength waters (to minimize the liquid junction potential difference). Alternatively, Orion's Ross™ electrode can use a variety of filling solutions, but the standard is a 3M KCl solution without silver or mercuric chloride at saturation; this electrode has proven very satisfactory under a variety of difficult field conditions.

The refillable electrodes need to have the filling solution maintained at close to full. Typically, they have a soft plastic sleeve that slides over the filling hole to keep the filling solution in the body of the electrode. The sleeve is pushed down to "top off" the electrode (do this before calibrating the meter) with filling solution and to make a measurement. Failure to lower the sleeve (which allows gravity drainage of the solution through the liquid junction) before obtaining a measurement is the most common student mistake with these electrodes.

The current crop of field pH meters are quite sophisticated with automatic temperature compensation and electronic buffer settings, which tends to mask how the meters work. Some of these meters will require the ATC (automatic temperature correction) electrodes to function properly, or a separate ATC probe to use with electrodes that don't contain the ATC feature, while others will permit manual temperature input by the user. Older meters used two-point buffer calibration (you would use a 7.00 buffer and either a 4.01 or 10.01 buffer, depending upon whether your water samples were more or less acidic than neutral) generating a straight-line correction for probe nonideal behavior. The newer, computerized meters can store up to (currently) five buffer points, providing a more accurate result over a wider pH range if all five points are used and the buffers are appropriate for the waters being sampled. This is particularly valuable when dealing with extreme waters; for the Berkeley Pit conditions we would want to add a pH = 2 buffer to the calibration procedure.

These field meters almost universally provide digital readout values and many store data points with a time stamp. One advantage of the digital meter output is that you can see the changes as the pH value settles in (provided that you haven't programmed the meter with a long time averaging factor; using the average for 2 or 3 seconds, definitely not longer than 5 seconds, aids you in seeing the change). The best advice is to read the manual for your meter and probe brochure first, and then walk through all of the setup and calibration steps in the laboratory or office before starting to use it in the field. For extended fieldwork, make sure that you have spare solutions, batteries for the meter, and possibly a spare probe, making sure that the spare probe's storage cup is filled with the chosen wetting solution. Also, we have found that keeping the original manual filed, and putting photocopies of the manual with the meter helps to ensure that there are manuals still available for the older meters.

Being from the "old school," the author's recommendation is to also have two thermometers (to allow for breakage) that can be used to check the ATC probe occasionally. You can check the thermometer calibration by putting it in a glass of ice water, being sure that it is submerged sufficiently. Some varieties require complete immersion; this design is not recommended, but you may have to use them until they can be replaced with ones that only require immersion of the lower third of the thermometer. They will also let you check the S.C. probes that have temperature readouts, which is mandatory if you don't have temperature correction.

Specific Conductance Meters

Specific conductance meters are normally pretty stable. For extended use we have found that taking homemade calibration solutions (made with table salt and run through the laboratory for their values) and checking these daily (may become weekly once you are satisfied with the meter and probe stability) should be sufficient. Waters that have caused us "drift" problems (and require probe cleaning once you get back from the field) have either been rich in metals or organics.

Basically, all these meters consist of is a suite of resistor pots (you select the pot by selecting the "range" of the sample) and a readout based upon the percent of the range of the selected pot required by the sample. In old meters, this was done with a Wheatstone bridge (Robinson and Stokes 1959, Figure 5.2), but in modern field units this is all done with solid-state electronics. Consequently, selection of a meter and probe is mainly a function of probe design relating to sturdiness and ease of cleaning and/or replatinizing as well as multiple function usage. We have had good luck with the basic Yellow Springs units, both for downhole measurement of water level, S.C. and temperature (SCT™ meters), and the "tailgate" meter. The early SCT™ reel meters did not have a quite big enough spool and if the cable wasn't wound on tightly, it tended to escape from the spool and make unwinding at the next site a pain.

Alkalinity

As discussed in Section 7.2, alkalinity really should be determined in the field. The acid can be added with an "eye dropper" or from a buret or titrator. The hassle with storing buret stands and burets, and emptying the buret after use at each sampling site, explains the popularity of the titrator devices. As an example, the Hach Digital Titrator offers higher levels of precision and accuracy than drop count procedures, using titrants in interchangeable cartridges.

Hach claims that accuracy is ±1% for titrations requiring over 100 counts of reagent; however, accuracy is normally a bit lower because of the pH being slightly below the desired end point. This can be avoided by performing Gran plots (Stumm and Morgan 1996), but in practice it is rarely done by laboratories, much less in the field.

The key factors in doing field alkalinity determinations are accurate recording of the titrant volumes, not using too strong an acid (i.e., using 0.1639 N acid on a low alkalinity (also referred to as "poorly buffered") sample) such that the titrant volume is quite small and measurement error becomes significant, and determining if your pH electrode has significant "shearing potential" (change of pH when water is stirred, normally about 0.1 pH unit lower when being stirred rapidly). When using battery-powered magnetic stirbars, you do not get the same speed every time, but you can be fairly consistent; check the stirred versus unstirred pH in the lab at the endpoint pH (usually 4.50) to see what measured pH you should titrate to get the proper unstirred pH value. If you are stirring the solution with the pH probe itself, the stirring rates are lower and the effect is usually minimal, but check this also.

Eh

Eh measurements are pretty hard to foul up; however, you must do the correction to the standard hydrogen electrode (SHE, used in all of the early lab work on oxidation-reduction reactions; virtually all reference tables for redox half reactions are keyed to the SHE). Topping off the reference solution and checking the measured Eh against freshly made up ZoBell's or

Quinhydrone solution in the lab before going into the field will let you get the offset from the hydrogen electrode at the ambient temperature and compare the result with Table 7.10. We have mainly used Orion Eh electrodes in the past because of the paper by Nordstrom (1977) characterizing this electrode and giving the correction factors to bring the measured voltage to that which one would get with the standard hydrogen electrode. However, that electrode design has been changed and other brands perform equally as well. Store the rest of the ZoBell's or Quinhydrone solution in a brown plastic bottle and take it with you for extended field work, so that you can do a morning check on the meter and probe before starting sampling. These solutions degrade with time, so you can do a repeat check of the meter and probe with freshly made-up solution upon returning from the field if you have been out in the field for some time.

Table 7.10 Correction to Add to Measured Eh Value Using a Platinum Eh Electrode Containing a Silver/Silver Chloride Reference Electrode with a Saturated KCl Solution

Temp. (°C)	add XXX mV
0	231
5	221
10	212
15	203
20	194
25	184
30	175

You can correct any platinum combination electrode using a saturated KCl, Ag/AgCl reference by adding the factor in the table (calculated from Nordstrom 1977) to your instrument measurement to get the SHE value.

Dissolved Oxygen

Dissolved oxygen (DO) can be determined by titration or by use of a DO probe. Most direct titration approaches use the Winkler or Modified Winkler (Alsterberg azide) methods. The latter method involves creating an $Mn(OH)_2$ floc, which reacts with the dissolved oxygen to form $MnO(OH)_2$, followed by acidification and titration with KI to release iodine gas that is titrated with sodium thiosulfate and a starch indicator. Needless to say, the process is sufficiently involved to generate interest in the membrane probes, which are currently the most common method of obtaining DO values.

The membrane probes require a filling solution and membrane that used to be sealed with an O-ring. Some of the newer versions use threaded screw-on membrane caps, which typically provide much more stable sealing. The best accuracy cited for top of the line meters is ±0.2 mg/L, so you will probably only need the meter for stream or lake work and shallow groundwater sampling. There are several pitfalls to be aware of when working with these meters.

1. The values are dependent upon air pressure. Thus if you change elevation or the barometric pressure changes, you need a correction or recalibration.

2. The reading is only as good as the condition of the membrane and the reference solution. Solution volumes must be kept within the accepted working range. Membranes can be punctured, which rapidly ruins the results. Also, membranes cane become "fouled" or plugged by organic matter attaching to the membrane. We have had problems with both algae and ferric iron flocs preventing oxygen passage through the membrane.

3. Most of the probes consume oxygen, and the water must be stirred to get fresh (without O_2 loss) water in contact with the probe, or you will get a reading lower than the true value for the water.

4. If hydrogen sulfide (H_2S) gas is present, it will give erroneous positive DO values. This should not be a problem as the H_2S odor (smells like rotten eggs) can be noted by most people at concentrations (0.1 to 0.3 ppm) below most analytical detection limits. However, you need to know that if you can smell H_2S, the meter reading is invalid.

Yellow Springs Instruments has a new model with a micro electrode array (MEA™) that supposedly eliminates the need for stirring. Some of the meters (e.g., Orion) now come with automatic barometric compensation. Consequently, getting reliable DO measurements is becoming easier. However, DO measurements are still probably the most difficult field measurement that you will commonly be making. If you are going to be doing much of this, try to have a meter permanently assigned to you (to minimize maintenance problems) and always carry spare membranes, solution, and batteries.

Batteries

Some instruments require batteries that you can't find at the closest store or gas station. You should always have a spare or spare set (if the instrument requires more than one battery) of batteries for that instrument. If you are working "out of town" (meaning you don't get into town for lunch) you should carry the enough of what you need to replace the "common" batteries used by your instruments.

We had a pair of beautiful Lab Line conductivity meters at the Montana Bureau of Mines and Geology in 1974 (very accurate Wheatstone bridges, circular carbon electrodes in a cup that were easily cleaned). The meters took a very odd battery, which could only be found in one or two towns in Montana (a fact learned the hard way). When replacement batteries were finally obtained, their date of manufacture was three years prior to our obtaining them. Needless to say, the meters were retired from fieldwork once suitable replacements (that used common batteries) could be obtained.

Test Strips, Capillary Tube, and Colorimetric Determinations

Many firms now offer test strips for determining everything from pH to chloride. Some of these use color matching while others function as chromatographic indicators. One of our graduate students with a limited analytical budget used silver nitrate capillary tube strips in conjunction with his well inventory to get a handle on a shallow groundwater chloride plume associated with oilfield drilling. Then a limited number of samples were collected for laboratory analysis. Most of his samples were within 10% of the field value, with the larger errors being associated with the high chloride waters.

While the accuracy of these methods is generally poorer than lab work, these test strips can be used to get an initial data set showing the range of conditions within a study area. However, you do have to be aware of the potential interferences with the method employed with the type of strip being used.

Most of the test kits (Hach puts out a variety of single parameter kits and several multiparameter kits) are a bit better than test strips but use "eyeballed" color comparisons or an inexpensive spectrophotometer to measure loss of light transmission through the sample. Many of these determinations have severe interferences. For a number of years we drove this point home by having students use a colorimetric hardness kit on two wells in a tailings setting. The up-gradient well had very little metals content and gave reasonable results with the colorimetric kit. The down-gradient well had high levels of iron, copper, and zinc, and the kit gave horrible results. These kits can be very useful, but you have to be aware of their limitations when using them.

References

Appelo, C. A. J., and Postma, D., 1993. *Geochemistry, Groundwater and Pollution*, Balkema, Rotterdam, Netherlands, 536 pp.

Back, W., and Hanshaw, B. B., 1971. Rates of Physical and Chemical Processes in a Carbonate Aquifer. In *Nonequilibrium Systems in Natural Water Chemistry*, R. F. Gould (ed.), American Chemical Society, Advances in Chemistry Series, 106, pp. 77–93.

Barcelona, M. J., Holm, T. R., Schock, M. R., and George, G. K., 1989. Spatial and Temporal Gradients in Aquifer Oxidation-reduction Conditions. *Water Resources Research*, Vol. 25, No. 5, pp. 991–1003.

Cherry, J. A., Shaikh, A. U., Tallman, D. E., and Nicholson, R. V., 1979. Arsenic Species as an Indicator of Redox Conditions in Groundwater. *Journal of Hydrology*, Vol. 43, pp. 373–392.

Davis, A., 1988. *A Preliminary Analysis of Aqueous Geochemistry in the Berkeley Pit*, Manuscript of work performed under EPA contract no. 68-01-6939.

Davis, A., and Ashenberg, D., 1989. The Aqueous Geochemistry of the Berkeley Pit, Butte, Montana, U.S.A. *Applied Geochemistry*, Vol. 4, pp. 23–36.

Deutsch, W. J., 1997. *Groundwater Geochemistry: Fundamentals and Applications to Contamination.* Lewis Publishers, Boca Raton, FL.

Drever, J. I., 1997. *The Geochemistry of Natural Waters: The Surface and Groundwater Environments, 3rd ed.* Prentice-Hall, Upper Saddle River, NJ, 436 pp.

Garrels, R. M., 1967. Genesis of Some Ground Waters from Igneous Rocks. In *Researches in Geochemistry*, Vol. 2, pp. 405–420, P. H. Ableson (ed.), John Wiley & Sons, New York, 633 pp.

Garrels, R. M., and MacKenzie, F. T., 1967. Origin of the Chemical Composition of Some Springs and Lakes. In *Equilibrium Concepts in Natural Water Systems*, pp. 222-242, American Chemical Society, Advances in Chemistry Series, No. 67, Washington, D. C.

Hem, J. D., 1985. *Study and Interpretation of the Chemical Characteristics of Natural Water, 3d ed.* U.S. Geological Survey Water- Supply Paper 2254, 263 pp.

Kempton, J. H., Lindberg, R. D., and Runnells, D. D., 1990. Numerical Modeling of Platinum Eh Measurements by Using Heterogeneous

Electron-kinetics. In *Chemical Modeling of Aqueous Systems II*, pp. 339–349, American Chemical Society, Adv. Chem. Ser. 416, Washington, D. C.

Krauskopf, K. B., and Bird, D. K., 1994. *Introduction to Geochemistry, 3rd ed.* McGraw Hill, New York, 640 pp.

Langmuir, D., 1996. *Aqueous Environmental Geochemistry.* Prentice-Hall, Upper Saddle River, NJ, 600 pp.

Lindberg, R. D., and Runnells, D. D., 1984. Ground Water Redox Reactions: An Analysis of Equilibrium State Applied to Eh Measurements and Geochemical Modeling. *Science*, Vol. 225, pp. 925–927.

Mercado, A., and Billings, G. K., 1975. The Kinetics of Mineral Dissolution in Carbonate Aquifers as a Tool for Hydrological Investigations, I: Concentration-time Relationships. *Journal of Hydrology*, Vol. 24, pp. 303–331.

Nordstrom, D. K., 1977. Thermodynamical Redox Equilibria of ZoBell's Solution. *Geochimica et Cosmochimica Acta*, Vol. 41, pp. 1835–1841.

Robins, R. G., Berg, R. B., Dysinger, D. K., Duaime, T. E., Metesh, J. J., Diebold, F. E., Twidwell, L. G., Mitman, G. G., Chatham, W. H., Huang, H. H., and Young, C. A., 1996. *Chemical, Physical and Biological Interaction at the Berkeley Pit, Butte, Montana.* Paper presented at the January 13–17, 1997, Tailings and Mine Waste 97 meeting at Colorado State University, 13 pp.

Robinson, R. A., and Stokes, R. H., 1959. *Electrolyte Solutions.* Butterworth & Co., London, 571 pp.

Schwarzenbach, R. P., Gschend, P. M., and Imboden, D. M., 1993. *Environmental Organic Chemistry*, John Wiley & Sons, New York, 681 pp.

Stumm, W., and Morgan, J. J., 1996. *Aquatic Chemistry, 3rd ed.*, John Wiley & Sons, New York, 1022 pp.

U.S. Geological Survey, 1958. *Suggestions to Authors of the Reports of the United States Geological Survey, 5th ed.*, 225 pp.

Chapter 8

Drilling and Well Completion

Perhaps one of the more common tasks a hydrogeologist will be involved with is obtaining subsurface information, which can be found through a variety of methods, including geophysical (Chapter 4), hand tools, and drilling methods. This chapter presents the common drilling methods and explains how monitoring and production wells are installed. A hydrogeologist does not need to know how to run a drill rig but should be familiar with drilling methods, terms, and well-completion strategies so that meaningful subsurface information can be obtained. Also, the hydrogeologist with some guidance will be better informed to make decisions in the field.

Hydrogeologists need to be able to work safely with the drill crews in obtaining subsurface information. It is helpful to know the basics of drilling methodologies, how to describe drill cuttings, and what is involved in well completion. The approach taken here is to familiarize entry-level hydrogeologists or professionals unfamiliar with drilling operations with what they should know to safely proceed.

Understanding the geology of an area (Chapter 2) is helpful in knowing which drilling methodologies would be most productive and result in obtaining the best subsurface information. Examples of drilling in different geologic settings are presented throughout to show applications of the appropriate drilling method in a given geologic environment.

8.1 Getting Along with Drillers

Drillers are an interesting breed and can make your life highly productive and successful or exceedingly miserable. The biggest factor, generally, is you and your attitude. A friendly, helpful, congenial attitude on your part can go a long way to getting the best information possible. During any given day, things can and will go wrong, but it is up to you to decide how you will let your circumstances govern your actions.

Drillers are generally very professional and knowledgeable about what they do. They know when "first" water has been found or what the conditions are like at depth (Figure 8.1). Nothing is more irritating to them than to have a young inexperienced person tell them how to do their job. If you get on their bad side, watch out. It has been the author's experience from being involved with a variety of drilling conditions in various parts of the world that disgruntled drillers do unpleasant things to smart aleck geologists or hydrogeologists. For example, while you are away, your briefcase may be rearranged, your vehicle sabotaged, or your person may be adorned with lubrication grease. It is better to be polite, act interested, and ask

***Figure 8.1** Discussing drilling conditions during monitoring well installation. (Photo courtesy of O'Keefe Drilling.)*

questions: for example; Did we just go through a gravel layer? Are we making a bit more water? Was that last drill rod 120 ft or 140 ft?

There is a fine line between the drillers making footage and your obtaining complete information. It is your responsibility to obtain good subsurface information. If the drillers are going too fast, you need to politely ask them to slow down. How the job is bid is an important factor. Is payment by the job, by the foot, or by the hour? Being cognizant that a job needs to be done and that drillers need to make a living, too, leads to a cooperative relationship. The bottom line, however, is that you, the hydrogeologist, are responsible for the subsurface information.

Other factors that help or hinder a working relationship are the little things. Do you collect your cuttings and then sit in the vehicle until the next rod is ready to go in? Or do you always sit in your vehicle while the drill rods are being coupled together and going into the hole ("**tripping in**") or being pulled out of the ground ("**tripping out**")? Simple gestures like helping shovel cuttings out of the way, helping guide a **sand line**, or retrieving a tool helps the process run smoother and contributes to positive working relationships (Figure 8.2). While the driller helper has gone for water do you assist or stand around? This is something each professional has to decide

Figure 8.2 Driller and geologist in conference. (Photo courtesy of O'Keefe Drilling.)

for him- or herself, but can make a big difference in how a job gets done. Company policies may address these issues, but being helpful is a good way to be.

Sometimes breakdowns occur, the equipment needs attention, or something may be inadvertently dropped down the hole and a "fishing" expedition is under way (see Example 8.12, Section 8.6). In some cases, there is nothing you can do but stay out of the way. Being patient and understanding that "things happen" is more helpful than yelling at your help.

8.2 Rig Safety

In the above discussion, no information is required at your peril. Always think safety. There are few pieces of equipment with more moving parts than a drill rig. If you are within the mast length of the drill rig, then you need to be checking and looking around. You need to look up as often as you check your rearview mirror while driving a vehicle (every few seconds) because a hardhat cannot save you from a falling drill pipe. Equipment can break loose, cables can break, and conditions can change in a heartbeat. Always have a sense of caution and safety in mind.

Figure 8.3 Minimum safety equipment: hardhat, steel-toed boots, safety glasses, and gloves. (Photo courtesy of O'Keefe Drilling.)

The minimum appropriate attire at the drill site is steel-toed shoes, a hardhat, gloves, and safety glasses (Figure 8.3). If you are working at a hazardous waste site, other protective clothing will be required. Loose clothing is susceptible to becoming snagged by a rotating machine, pulling you in or getting torn up.

The safest place to stand near a drill rig is on the driller side (Figure 8.4). The driller is in charge of the controls and is the most knowledgeable person about what is going on. He or she is the person you need to have an ongoing dialog with and will be most helpful in answering your questions. The "helper" side is generally the direction where more things fall and more moving items are located. The helper's job is to help make connections, place pipe into position, or place "slips," wrenches, or other equipment in place.

***Figure 8.4** The driller's side of the rig is the safest place to stand. (Photo courtesy of O'Keefe Drilling.)*

If you want to help, observe what the helper does and then ask at the appropriate time. Never put your hands, fingers, or other where you can be pinched, trapped, or will compromise safety. No drill hole is worth an injury. If you don't feel comfortable helping, then stand back and get out of the way. Another obvious, but important point is, there is a fair amount of welding and grinding at times. Do not stare at the bright lights, it can damage your eyes (Figure 8.5). Grinding wheels can throw hot, sharp metals pieces, so please stand back.

Figure 8.5 Welding on another section of casing.

If you need to write something down that will take longer than the few seconds needed to keep looking up, move well away to the driller's side or walk over to your vehicle. If you have to write something down that will take time, tell the drillers and have them wait. You may miss something important. Work methodically, conscientiously, and safely. Remember, your vehicle should have been parked farther than the length of the drill mast away from the rig. The best thing to do is ask the driller where to park, because he or she may have some maneuvering plans for the water truck or other equipment.

Before you drill, it is a good idea to call a toll-free number to have the utility lines located. This has to be arranged 48 hours in advance, so plan ahead. A person will be dispatched to locate water (marked blue), power (marked red), gas(marked yellow), phone or TV cable (marked green or black) in spray paint in the ground surface (Figure 8.6). Contacting the local courthouse or city government can be helpful in locating sewer or other utility lines. Installing monitoring wells and making a spark from encountering a gas line while drilling can be interesting.

Example 8.1

An elementary school in Butte, Montana, was having basement flooding problems year after year, each spring. School district maintenance officials were interested in knowing why they had a problem and not just how to solve the problem. The biggest problem appeared to be that a 72-in. (1.8-m) culvert was directing runoff water from a drainage originating at the Continental Divide under an Interstate highway towards the school.

Figure 8.6 Drill site with utility lines marked on the surface.

Another significant problem was that the school was located on fill materials in a floodplain (Figure 8.7). Surface water and groundwater would move towards the school and back up against the fill materials and basement wall as the water table rose each spring. By summer time, the groundwater levels dropped more than 10 ft (3 m) below the basement slab. A dewatering system was designed to alleviate the problem until the various parties of county and state could work out disputes on how to correct the surface drainage.

Figure 8.7 Elementary school site (bottom center) built on fill placed in the middle of a drainage. (Photo courtesy of Hugh Dresser.)

As part of the investigation, a series of monitoring wells were installed. Before drilling, the toll-free telephone number was called and the utility lines were located. While drilling one hole near the front of the school, a light-colored powder came up with the cuttings at a depth of 7 ft (2.1 m). Smelling the power revealed that we were drilling into concrete. A decision was made to abandon the hole, as it was thought it might be an unmarked sewer line. Later on, it was discovered from an oldtimer that it was more likely that we were drilling into a concrete block that was part of the fill material. This illustrates the need to use all your senses, be cautious, and make good use of the available resources in finding out information about a site.

Drill rigs make good lightning rods. If stormy conditions are possible, keep a wary eye open. Often, you can watch clouds approaching from a particular direction. The sound of thunder is a sure sign that lightning is not far away. A simple conversation with the driller will result in a mutual waiting period for the storm to pass over. Drill rig masts can also extend up into power lines. In conversations with other drillers, it is the author's understanding that a rig should be more than 15 to 30 ft (4.6 to 9.1 m) away from power lines or arcing may occur. A phone call to a power utility may result in it placing insulated covers over the power lines so that drilling can proceed safely. Ask your driller what he or she would suggest.

Drilling operations attract people. Drilling in remote areas often draws curious onlookers (Figure 8.8). Flag off your safety area so that onlookers are kept back. People getting too close can slow down working operations and increase the potential for hazzards. A smile and firm demeanor usually gets the message across. One of the biggest problems drillers encounter is clients that do not follow the safety rules. As a hydrogeologist on site, make sure the clients are wearing the minimum safety equipment or politely tell them to back over to a safer place.

Figure 8.8 Curious children from the Peoples Republic of China.

Summary of Safety Points

- Always look up, as you would check a rearview mirror.
- Park your vehicle the drill mast length away from the rig or ask the driller where to park, as he or she may have a particular method of maneuvering equipment.
- Stand on the driller's side of the rig.
- If you have to write a long time (more than approximately 3 seconds), move away from the rig to the driller's side.
- Don't watch people weld.
- Never drill without locating utility lines or identifying other hazards, such as power lines.
- Thunder usually means lightning, so stop what you are doing until this danger has passed over.
- Use caution when drilling near power lines.
- Rope off your area to keep curious onlookers back.
- Prior to donning personal protective equipment or respirators, work out a communication protocol, so that signals are clear.
- Never, never, never compromise safety, it can be fatal.

Other Considerations

Common sense is an important part of your tool box. Most of the time it is the hydrogeologist's responsibility to tell the driller where to drill. The direction of the wind is a significant factor. If you have a serious wind blowing from the front of the rig towards the back, you are going to eat the dust and cuttings swirling around you. This is especially unpleasant when drilling with air and a forward rotary rig in a coal bed.

Have plenty of layers of clothing to put on or take off. Nothing is more miserable than freezing while you are trying to collect data. Your hands do not write well when the fingers are stiff and your lips are blue. At the same time, dripping perspiration on your log book can damage what you have written down. Layers allow flexibility in staying comfortable. A person too cold or too hot is more inclined to make mistakes. This affects safety and the quality of information collected.

Drilling near hazardous waste sites and other conditions require an extra dose of common sense. Dressing out in level A personal protective equipment (PPE) can affect your performance (Figure 8.9). Heat stroke is of-

Figure 8.9 Performing field tasks dressed out in level A personal protective equipment (PPE).

ten as dangerous as the hazards you are being protected from. Communication and frequent breaks will often be necessary. A particular communication protocol needs to be worked out in advance with all parties involved, along with step-by-step procedures before drilling begins. Sometimes methane or other gases may be present in certain situations. In this case, keeping a detection devise operating to identify these gases will be necessary. Sometimes if conditions are poor, safety can be inappropriately compromised.

Example 8.2

In eastern Montana, a number of coal mines are operating in the Fort Union Formation. Drilling is part of the ongoing exploration and production process. During a very cold day -10 °F (-23 °C), a propane torch was being used to thaw out some of the fluid lines on the rig. The TIP gas detector was inadvertently shut off because of the use of propane. This was a coal well, and methane gas was emitting from the well. The torch ignited the methane and the drill rig burned for 3 days before it could be pulled off the drill site!

It is not the intention of this section to provide a treatise on safety, however pointing out several of the more common hazards found near drill rigs may prove helpful. To obtain more information and discussion about the principles of safety, the reader may consult Spellman and Whiting (1999).

8.3 Drilling Methods

A variety of drilling methods have been developed over time to account for the many geologic conditions. Some formations are very hard, such as granite, while others are soft or unconsolidated such as the sands and gravels found in an alluvial setting. Drilling and subsequent well completion should be considered together when deciding which drilling method is most appropriate. Some methods may be faster than others but may result in disturbing the aquifer to the point of reduced yield. For example, if the production zone is in unconsolidated sediments, high-pressure drilling fluids may actually cause damage to the aquifer. In this case, production wells need to be drilled using a slower, less-disruptive method so that production-zone disturbance is minimized. This chapter will explore the most common truck-mounted drilling methods. For a more complete discussion of drilling applications in the water-well industry, the reader is referred to Driscoll (1986).

Monitoring wells provide a point of access to the aquifer. They need to be constructed so that the hydraulic head and water-quality information collected is representative of the aquifer under investigation. Again, the drilling method will depend on the geologic conditions. The well-completion materials need to be appropriate for the water quality of the site, which may include such factors as pH, presence of organics, water temperature, or depth.

Cable-Tool Method

Perhaps one of the oldest drilling methods used is a percussion approach known as the cable-tool method. These technologies were developed by the Chinese some 4,000 years ago (Driscoll 1986). In this drilling method, a heavy drill bit is repeatedly lifted and dropped by a walking beam or spudding beam attached to a spooled cable via a pitman arm (Figure 8.10). The motion of the bit is up and down in a strokelike fashion, so that the bit strikes the bottom of the hole each time it is dropped. The spudding beam is also the reason they are known as "spudder" rigs, although the author also knows of the spudding bean being referred to as a walking beam or rocker arm (from the up-and-down rocking motion). The lifting and dropping of the drill tool results in the breaking up and crushing of geologic materials. The drilling action is usually followed by driving casing down the hole as the drilling proceeds. It is an effective drilling method in unconsolidated and softer materials, but most effective in brittle formations such as shales and limestone. It can be used in hard formations but may be very slow. The well casing associated with this method usually ranges from 4 in. (10.1 cm)

Figure 8.10 Cable-tool drill rig.

to 24 in. (0.61 m), with depth capacities up to 2,000 ft (609 m) (NGWA 1998).

After the materials are broken up or crushed, they are bailed to the surface via a separate cable and spool. If the hole is dry, the bailing operation is done by first pouring several buckets of water down the hole or running a hose attached to a water truck to facilitate mixing with the cuttings. A heavy bailer equipped with a closing valve is lowered to the bottom of the hole through an auxiliary cable. One common name for an auxiliary cable is the sand line (Figure 8.11). The bailer is a long tube with an opening/closing valve. A **dart valve** bailer is an example. When the bailer is lowered to the bottom of the hole, the dart valve is compressed and the water/cuttings mixture gushes into the bailer. When the bailer is lifted off, bottom the valve closes, thus trapping the mixture. The sand line retrieves the bailer where it is brought to the land surface. This can be lowered into a discharge chamber. When this is done, the dart valve once again compresses, and the water/cuttings mixture is released out through the discharge chamber (Figure 8.12). The cuttings can be directed away from the working area by digging a shallow trench.

Figure 8.11 Dart valve bailer on a cable-tool drill rig.

Figure 8.12 Discharge chamber to direct cuttings away from the drilling.

Once the hole has been drilled a few feet (a meter or so), casing is forced or driven to the bottom of the hole to keep the hole from sloughing. Depending on the casing and tool-handling capabilities of the rig, the length of casing driven and added varies. The spudding wheel turns proportionally to the drilling rate. One can observe whether the formations are hard or soft by how much the spudding wheel rotates. It has been the author's experience for smaller diameter wells (6 to 8 in., 0.15 to 0.2 m) that 10 ft (3 m) is welded on each time. This results in an overall drilling rate of approximately 10 ft (3 m) per hour. Rates may vary significantly, but this rate is given for comparison with other methods.

The cable-tool method is ideal when time is not of the essence, the geologic materials are relatively soft or brittle, and the formation disturbance be needs to be minimized. Attempting the drill in hard materials may result in drilling rates too slow to be cost effective, unless the rig is capable of handing a much larger heaver bit. Cable-tool rigs are also useful in drilling shallow, larger-diameter holes, 18 to 24 in. (0.46 to 0.61 m), used for construction supports, such as caissons.

The author has found this drilling method especially useful in drilling unconsolidated materials that may have **heaving sands**. Heaving sands are loosely consolidated sand zones with relatively high confining pressures. When the drill hole breaches one of these zones, the sand gushes into the casing. The remedy is to drill out the intruding sand and attempt to proceed with the hole. This is very disruptive to the formation and is exacerbated using the forward rotary method.

Example 8.3

An irrigation well capable of 250 gpm was desired for irrigating a cemetery in Butte, Montana. The sediments consisted of unconsolidated weathered granitic materials, alternating beds of fine gravel, sand and clay (Figure 8.13). The clay beds confined or semiconfined some of the sandy zones. Recharge to the sandy zones were from elevations sufficiently high to produce heaving sands or sands subject to high confining pressures. These are usually found immediately under a clay horizon. The cable-tool method was used to drill an 8-in. (0.2-m) cased well. Heaving sand events occurred on several occasions while drilling the 170-ft (52-m) well. The sands would intrude approximately 15 to 20 ft (4.6 to 6.1 m) into the casing. The fine gravel zones were stable enough to complete a production zone in the well.

Forward (Direct) Rotary Method

The forward rotary or direct rotary method is perhaps the most versatile drilling method for smaller-diameter drilling (less than 12-inch diameter wells). Drilling depths can be greater than for other methods, and it is well suited for drilling in harder geologic materials. Part of the versatility comes from the variety of drilling additives that enhance the drilling environment.

In the forward rotary method, a bit and drill pipe string are rotated by means of a power drive system. The power drive can be at the top of the drill string (a top-head drive)(Figure 8.14) or near the working level of the drillers, a table-drive system.

Figure 8.13 Split-spoon core sediments from Example 8.3.

Figure 8.14 Forward rotary rig with top-head drive.

In a table-drive system, the top of the drill string is a **kelly** bar, which is turned inside the table. The kelly slides through drive bushings in the rotary table. When these are in their proper place, the table turns the kelly and the drill string bit configuration and feeds it down hole (Driscoll 1986). The shape of the kelly bar is often square or hexagonal and approximately 3 ft (1 m) longer than the drill pipe (Driscoll 1986). Connected to the bottom of the kelly is a short (less than 1 m) removable section used to connect the first drill pipe to the kelly. This removable section or sub is there to save on wear and tear of the threads at the end of the kelly. A similar sub is used to connect the bit to the lower most drill pipe. The drill string is lifted up the mast via a heavy cable with the top section laterally connected to a heavy chain collectively known as the **drawworks**. The driller can raise and lower the drill string by the drawworks while the kelly slides past the kelly bushings in the table. The chain moves in a circular fashion within the mast. The hydrogeologist usually observes the rate of drilling by observing chalk markings made on the chain by the driller or numbered increments made down the mast.

The diameter of the drill pipe is smaller than the diameter of the borehole cut by the bit. The drill pipe is thick but hollow enough to allow the flow of drilling fluids. Drilling fluids are injected via a hose connected to the top of the drill string from a pumping system. The fluids exit ports in the bit that help lubricate it and keep it cool. The cuttings are forced up the annulus between the borehole wall and the drill pipe to the land surface (Figure 8.15).

During mud drilling, a pit is excavated (or a portable pit with baffles is used when cuttings must be removed from the site). The pump intake hose is placed on one side of the baffles where drilling additives can be easily mixed. The returning fluid and cuttings are discharged back into the opposite side of the pit. The baffles filter or separate the cuttings from the intake side. When there is a continuous movement of fluids through the drill pipe to the bit and a return of fluids and cuttings up the annulus of the borehole to the land surface, this is known as circulation. The drilling fluids added during forward rotary drilling are designed to "maintain" circulation.

This a hierarchy of drilling fluids in order of hole stability:

- Air
- Water
- Foam (a detergent)
- Stiff foam (foam with polymer added)
- Mud

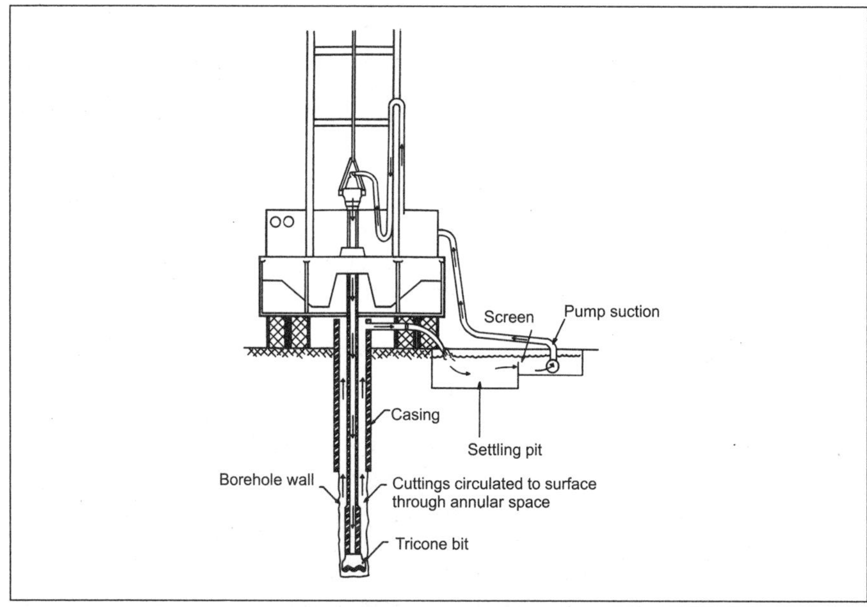

Figure 8.15 Direct rotary circulation system. [From NWWAA (1984).]

An air compressor on the rig is designed to produce sufficient airflow to lift the cuttings to the surface. The driller stands near the control station and monitors the flow volume of air or fluid. The drawworks and other controls are also handy from this position.

The author has observed holes drilled with air in softer layered sediments achieve drilling rates of 100 ft per hour (30 m/hr.) In unconsolidated sediments and in clays drilling with air is not viable. Unconsolidated sediments will collapse against the drill pipe and clays will ball up and not reach the surface. By adding water or mist, the circulation may once again be restored. Drilling with air is also viable in very hard formations until "first water" is encountered.

When drilling with water or other fluids, some of the cuttings coat the borehole wall, thus providing stability. If the borehole caves or washes out, the cuttings may not return to the surface even with a thick additive like stiff foam. At this point, the driller may have to drill with mud to provide a wall cake needed for a stable hole. Once the driller uses mud, then there is no turning back, since water or other fluids would cut through the wall cake and the hole may lose circulation once again. Drilling with mud should also be avoided to keep from sealing off production zones.

When hole diameters exceed 12 in. (0.3 m), most air compressors are not able to supply sufficient air volume to lift the cuttings to the surface.

This can be overcome by using an auxiliary compressor. Sometimes two drill rigs are "plumped" side by side to use the combined strength of both compressors. This is often necessary to complete 10- or 12-in. (0.25- or 0.3-m) wells. Wells with even larger diameters become impractical using this drilling method unless the drill rigs are very large; even then maximums are reached at diameters of approximately 22 in. (0.56 m) (Driscoll 1986).

Another versatile thing about forward rotary drilling is the variety of bit types available for use. Two of the most common bits are the tri-cone and the drag bit. A tri-cone bit has bearings that turn as the bit is rotated. Each cone has patterned cutting teeth or is studded with bumps or buttons that accommodate the hardness of the geologic materials (Figures 8.16, 8.16a). Generally the shorter the teeth or buttons, the harder the materials. Short buttons may be studded with tungsten carbide while larger teeth may be constructed of hard surfaced alloy steel or other materials resistant to abrasion. Drag bits are fashioned with nonrotating metal cutters that are used for rapid penetration, such as through soft shales. Ask your driller about bits and their function.

Figure 8.16 Button bit, tri-cone bits, and drag bit, from left to right.

Figure 8.16a Tri-cone bits; the larger the teeth, the softer the formation.

Forward rotary drilling is ideal for harder geologic materials and faster drilling scenarios. Although forward rotary drilling methods can drill through most geologic materials, there are at least four general conditions when this method is not recommended:

1. Rocks materials that do not allow circulation, such as highly fractured basalts or karstic limestone (Chapter 2)
2. Soft unconsolidated materials with high confining pressures
3. Zones that change from soft to hard and back to soft once again
4. Large diameter wells (greater than 22 in. or 0.56 m)

In highly fractured rocks or limestones with large void spaces, drilling fluids do not remain in the borehole because they are lost out into the formation. If there are heaving sands or other unstable conditions, forcing drilling fluids up the borehole may actually result in decreasing the hydraulic conductivity of the formation. Another problem occurs when the geologic zones alternate between hard and soft horizons. This is illustrated by Example 8.4.

Example 8.4

A family in southwestern Montana was interested in building a home in a topographic saddle between two hills. They called a drilling company to get their well drilled. The drillers were hopeful they could complete the well before the Labor Day weekend. A forward rotary drilling method was employed. The drillers went through some fairly hard zones and then soft zones. Worried that they might get the bit stuck in the hole, the drillers advised the property owners that if they continued drilling they would have to pay for the bit. Were they prepared to pay if it became stuck (approximately $5,000 U.S. at the time)? The owners said no. The drillers pulled off the well and left with no explanation.

Shortly thereafter the author received a call from the property owners seeking help on what to do next. After receiving an explanation of the problem and being advised of the location of the property, the geology of the site was researched and plotted on a topographic map (Figure 8.17). The well site, chosen by the property owners, was near one of the hills to protect their home from the prevailing winds. As it turned out, the geology of the hill near the well site consists of Cambrian sedimentary rocks and the saddle area consists of basalt (Chapter 2). The author reasoned that the well was drilled in a “chill” zone near the contact between the sedimentary and igneous rocks. The chill zone resulted in layers of basalt alternating between soft weathered zones and hard unweathered zones. This was confirmed in the field by looking at the cuttings and other field evidence. Figure 8.18 illustrates the layering that occurs in rock formations in igneous rocks. Under the ledges, the softer materials can accumulate and bunch up around the drill pipe, closing off the diameter of the hole. When the drill string is pulled upward, the bit may become stuck at the ledges.

The author suggested that if they could drill away from the chill zone, perhaps the rock would be brittle or hard all the way down and the hole diameter would remain uniform. Since an E-tape was not available (Chapter 5), the depth of the water table was estimated by dropping a small rock and timing when it hit water. This was repeated three times with consistent values. The depth was estimated to be approximately 100 ft (30 m). With this information, we drove to the top of the hill to examine the property area. It was noticed that a linear drainage extended from the top of the saddle area and seemed to correspond with a fault in the geologic information. By drilling near the fault, it was hoped that increased fracture permeability would be encountered. A site was recommended, and the author later sought more information on the thickness of the basalt. Fortunately an oil well had been drilled within the property indicating that the basalt was approximately 400-ft thick and underlain by lacustrine shales (Chapter 2). The author reasoned that recharge waters from precipitation would “pile up” on the shales and that a drill hole with a depth of 250 ft (76 m) or so would do the job.

The property owners called another driller to perform the task. This driller wouldn't drill at that high an elevation unless he consulted with a water witch (Section 8.7) . Interestingly enough, the "witcher" picked the same place that was recommended by the author (the site was flagged) and a successful well was drilled with a depth of 200 ft (61 m).

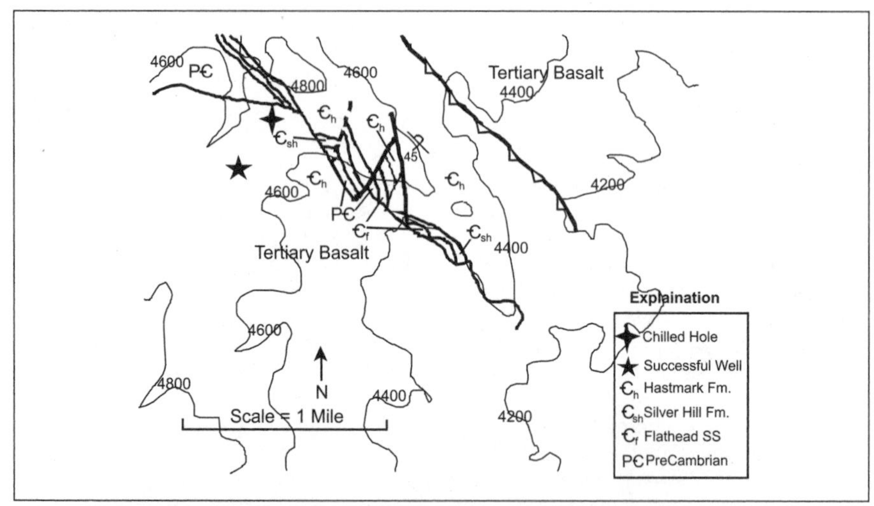

Figure 8.17 Surface geology and elevation contours from Example 8.4.

Figure 8.18 Layering in granite.

Reverse Circulation Drilling

Unlike forward rotary methods where continuous circulation is required before drilling can continue, reverse circulation drilling avoids the problems of losing cuttings and drilling fluids into the formation by taking up all cuttings and drilling fluids up through the drill pipe. This is a reverse drilling process from forward rotary methods. Drilling fluids are introduced between the borehole wall and the drill pipe and then drawn up at the bit. The discharge hose is located at the top of the kelly where water and cuttings may be diverted out through a cyclone and drop to the ground or are directed back into a mud pit (Figure 8.19).

Figure 8.19 Author collecting cuttings from the bottom of a cyclone.

Perhaps the most common application of reverse circulation methods is for drilling larger diameter wells from 24 to 72 inches (0.61 to 1.8 m) (NGWA 1998). The cost for drilling larger diameter wells is not much greater than for smaller diameter ones in unconsolidated materials and the well-completion applications are very straightforward. A summary of useful applications for reverse circulation drilling of larger diameter wells in unconsolidated or soft formations is given by Driscoll (1986):

- This method causes very little disturbance of the natural porosity and permeability of formation materials, an important consideration for production zones when other methods may disrupt the aquifer characteristics.

- Large diameter wells can be drilled quickly and economically.
- No casing is required during the drilling operation.
- Well screens are easily installed as part of the well completion process.
- Borehole wall erosion is less likely because down-hole fluid velocities are low.

There are also some conditions that must be met in order for this method to be effective (Driscoll 1986):

- Formations should be soft sedimentary rocks with no large boulders. Drawing boulders up through the center of the drill pipe is a problem.
- The static water level should be more than 10 ft (3 m) below the land surface to be able to maintain a sufficiently high hydrostatic pressure along the outside of the borehole. Problems also arise if the static water level is high and adequate water for drilling is lacking.
- Drilling using this method requires a high volume of available water with requirements ranging up to hundreds of gallons per minute (tens of liters per second). If this cannot be obtained, this method will not work well.
- Drill rigs, equipment, and components are very large and expensive and require more room to work.

Another reverse circulation drilling method that has very successful applications is known as the dual-wall reverse circulation method. Instead of fluids moving along the outside of the drilling pipe, there are two fluid passageways through the drill pipe (Figure 8.20). The drilling fluids pass down the outer passageway, and the cuttings and fluids are drawn up through the center passageway. The inner sleeve is sealed by an "O" ring. During this method, the only place the drilling fluids contact the formation is near the bit. This allows a continuous sampling of the formation with minimal disturbance and reduces the chances of cross contamination.

The author first became aware of dual-wall reverse circulation drilling in placer gold exploration applications. Any cuttings within 3 inches (7.6 cm) of the bit are drawn up through the pipe. Drilling recovery rates using this drilling method are in excess of 95%. Heavy minerals, like gold, would not likely make it to the surface using another drilling method. This is the method of choice for exploration test drilling. Representative samples can be obtained using this method. Although drill pipe outer diameters (OD) range from 3 1/2 in. to 9 5/8 inch (8.9 cm to 24.4 cm), the most common drill pipe size is 4 1/2 OD with a 2 1/5 ID (11.4 cm OD and 6.4 cm ID) (Driscoll 1986).

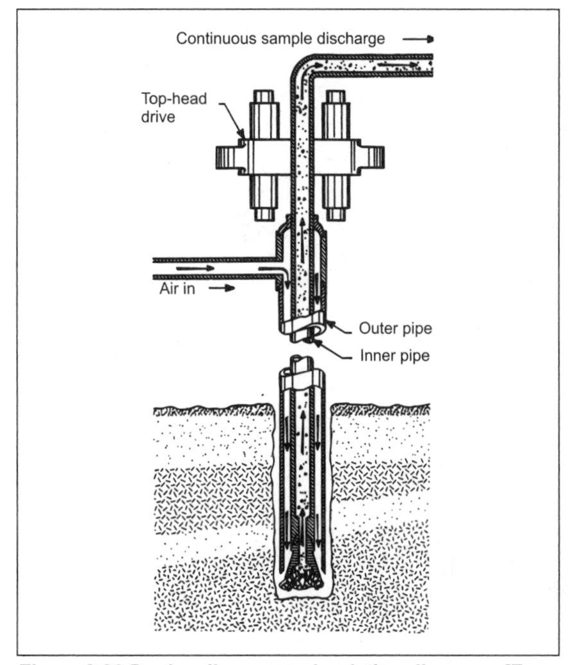

Figure 8.20 Dual-wall reverse circulation diagram. [From Driscoll (1986).]

Casing Advancement Drilling Methods

In cable-tool drilling, casing advancement is an integral part of the drilling process. Two other casing advancement methods will be discussed in this section: top-head forward rotary and wire-line drilling. Casing advancement methods are helpful when the borehole conditions are unstable or there may be a danger of contamination of materials from up hole to down hole. In top-head forward rotary methods, the casing driver is usually at the top of the mast. The drillpipe and casing are suspended together below. At the bottom of the first piece of casing is a thick forged drive shoe that is welded to the bottom to keep the casing from crimping as it is driven. The drill bit and casing advance simultaneously or the drill bit may advance a few feet (a meter or so) ahead of the casing before it is driven.

The casing driver is a piston percussion-driven system activated by air pressure. The bit grinds up the cuttings, as in forward rotary drilling, and the cuttings are blown up through the casing to a discharge tube near the bottom of the casing driver (Figure 8.21). When the bit advances ahead of the casing, the bit can be retracted inside the casing before driving the casing. There are many advantages to this method over traditional forward rotary methods in spite of the additional cost of the casing hammer and the noise of operation. Most drillers who use this method wear hearing protection attached to their hardhats (Figure 8.22). It is also advisable for the hydrogeologist logging the hole to wear hearing protection.

Figure 8.21 Collecting cuttings from the discharge tube.

Figure 8.22 Hearing protection on hardhats; driller measuring flow rate.

Several advantages of this method have been summarized by Driscoll (1986):

- Unstable formations that would result in too much disturbance using forward rotary methods can now be attempted with a good rate of success.

- The borehole is stable through the entire drilling operation.
- Penetration rates are rapid in most drilling environments.
- Loss of circulation is no longer a problem since cuttings are blown up through the casing.
- Specific zones can be tested for water production and sampling without up-hole contamination.
- This method can be used in all weather conditions.

In wire-line drilling, the casing serves as the drill string (NGWA 1998). The bit is attached to the bottom of the first piece of casing and advances along with drilling. As the hole advances, new casing is added to the drill string. All tools can be withdrawn upon completion of the hole. Wire-line drilling is ideal for continuous coring applications at depths where tripping in or tripping out can be cumbersome. The core barrel can be retrieved via a sand line by slinging a brass messenger down the sand line to the top of the core catcher. The sand line can then be pulled up with the core to the surface. A new core barrel can be sent back to the bottom of the hole and reattached to the bit already in place. In this way continuous coring can occur.

Example 8.5

The author spent a couple of months in the Peoples Republic of China performing wire-line coring on a coal property a couple of hundred kilometers west of Beijing. The drilling was unstable with soft overlying formations and coal thicknesses in excess of 50 ft (15 m). The surface consisted of 80 to 100 ft of silty loess (Figure 8.23). This was underlain by a few tens of feet of unconsolidated fine gravel before we encountered stable layered sediments of Pennsylvanian age. A surface casing was set to isolate the soft overlying materials and continuous coring of the Pennsylvanian rocks occurred to depths in excess of 300 ft (100 m). The wire-line coring was very convenient to obtain continuous core at relatively deep depths below very unstable drilling conditions.

Auger Drilling

Two methods of auger drilling are presented, solid-stem and hollow-stem auger drilling. The applications are more common for monitoring well completion rather than production wells, although shallow (approximately 100-ft, 30-m) production wells can be completed in soft formations (NGWA 1998) using this method. The chief reason for auger drilling is that it requires no drilling fluids when drilling the hole. This is especially significant for contamination sites that require monitoring wells. Drilling depths can reach 200 ft or more depending on the geologic formation and capabilities of the drill rig.

Figure 8.23 Wire-line drilling in China; canyons reveal silty loess.

Solid-stem auger drilling is often used when the well can be completed "open hole." This means that all of the auger flights can be retrieved from the hole before placing the casing. Solid-stem drilling can usually achieve greater drilling depths than hollow-stem augers (HSA) because the shafts on auger flights are narrower. The auger flights are rotated via a rotary head drive with a hydraulic-feed mechanism that allows either downward pushing or upward pulling (Driscoll 1986). The kelly bar is hexagonal and is attached to the first flight with two rod bolts that are often turned finger tight. As additional flights are added, they are usually tightened with a wrench. As one driller put it "anything below ground gets wrenched."

As the augers turn, the geologic materials spiral upward to the surface. It is confusing at first to tell which depth the cuttings came from, so it is a good idea to communicate with your driller. Cuttings depths can be correlated between the sound and "feel" of the drilling and the associated materials observed. An experienced driller will be able to tell you what you are in. For example, "We hit that gravel about a foot ago," even though the gravel may not be observed at the surface for another 10 ft (3 m) of drilling. Sometimes cuttings do not make it to the surface and cannot be sampled until the auger flights are pulled.

Many auger flights are 5 ft (1.5 m) long except for the first flight that has a cutter head attached, extending the length an additional foot or so (Figure 8.24). This, of course, depends on the diameter of hole being drilled and size of the drill rig, etc. The author's experience is primarily with smaller-diameter wells (less than 4 in., 0.1 m). Additional discussion on larger-diameter wells can be found in Driscoll (1986). Completing monitoring wells is discussed in Section 8.5.

Figure 8.24 First flight with cutter head.

Direct-Push Methods

Another method of obtaining subsurface information with or without completing a well is through direct-push methods. These are smaller track or van-mounted units that are equipped with a hydraulic system to push a 4-ft (1.2-m) core barrel into the ground (Figure 8.25). The 2 3/8-in. (6-cm) diameter stainless-steel core tubing captures a 1 7/8-in. (4.8-cm) core in a plastic tube. The tube is laid in a tray and sliced in half with a knife for inspection when performing geologic site evaluations (Figure 8.26) or may be sealed for laboratory soil-gas work. Newer systems can be altered to an augering system in the field within minutes. This allows flexibility of continuous coring followed by installing a 3/4-, 1-, or 2-in. (1.9-, 2.5-, or 5.1-cm) or larger diameter monitoring wells.

Figure 8.25 Direct-push method.

Figure 8.26 Direct-push core split in half.

The strength of these devices is in the ability to perform soil-gas surveys or delineating plumes with volatile organic carbon sources or to define bedrock in relatively shallow aquifer systems. There is a capability to extract soil gases directly into a gas chromatograph on the rig or pull a groundwater sample via a peristaltic pump or suction pump. A check ball enhances the ability of lifting fluids from deeper depths. It is a relatively quick way to obtain several point samples in a short period of time with minimal surface impact, since no well completions are needed. This is ideal for investigative and site evaluation work. The track-mounted vehicles maneuver well at small sites.

Direct-push methods are only effective in unconsolidated materials or softer sediments. Depth capacities are usually in the 10- to 30-ft (3- to 10-m) range, although successful stories of drillers exceeding depths of 100 ft are documented by Geoprobe Systems' 100 club (Geoprobe 1999).

Horizontal Drilling

An older technology that has expanded rapidly during the 1990s is horizontal drilling. Horizontal drilling has been used in placing utility lines for years but has found new applications in the environmental field. Horizontal drilling has found many applications in drilling and completing wells under permanent structures such as landing strips, airport fueling areas, roadways, and other areas that would be problematic for remediation if there were a host of vertical wells.

Drilling is accomplished by a rotating disklike bit that is directed at a low angle to the surface. Control is maintained by rotating the disk in a continuous motion. Changes in the direction are made via the orientation of the bit. Often times a "foot" person with a surface-detecting device works with the driller to identify exact positioning of the bit, particularly if one is drilling along existing utility lines (Figure 8.27). There are an increasing number of drill rig companies that perform horizontal drilling services and would be happy to provide additional information. One source of information is the annual rigs issue of *Water Well Journal* (July issue). For additional reading, it is suggested that one refer to various contractors that advertize in *Ground Water Monitoring and Remediation* or consult drilling magazines.

8.4 How to Log a Drill Hole

Before newcomers begin logging subsurface information, it is helpful to be familiar with the different drilling methodologies described above and the section on rig safety before attempting to log your first hole. You need to know how the cuttings arrive at the surface and the objectives of the drilling project. In addition to logging a hole, you are also responsible to see that

Figure 8.27 Horizontal drilling rig performing utility line work.

the drilling project is completed. There needs to a balance between making good footage for the driller and also obtaining the needed subsurface information. The hydrogeologist is responsible for the data that are collected.

It is typically your responsibility to locate where to drill. How will you decide? Will the locations be premarked? Were they located by someone else, which means that you have to be able to read a map and locate yourself? Do you know where the site is and where to begin? Is it necessary to call a utility line locator? Remember that you need 48 hours advanced notice (Section 8.2). Are there any issues regarding personal protective clothing? These may sound like obvious questions, but for someone new they are a significant source of stress and point out the need for a predrilling meeting with the driller and other involved parties.

In addition to knowing the drill-site locations (or enough sites to get started), there are a number of tasks that one can prepare before going into the field. Does your company have a standard form they use for drill logs? Maybe you are using a field book from which you will be transferring information to a standard form. Consult Chapter 1, Section 1.8, on field note taking. Will you be taking samples that will be stored? Will you collect them every 5 ft (1.5 m) or at a major lithology change? Will these be chip samples, core samples, or something else (Figure 8.28)? Do you have whatever containers, bags, or boxes prepared so the drillers don't have to wait for you?

Figure 8.28 Chip samples in compartmentalized container.

Make a list and be sure you are prepared to go into the field. Some common items needed are:

- Site location map
- A topographic map, if it is not included on the site map
- Geologic map, if possible, or at least have well logs from other wells in the vicinity
- Field book, well log forms
- Hand lens, grain-size chart, acid bottle, or hardness tester
- Containers for samples, or materials needed for samples, core boxes, etc.
- Chain-of-custody information if needed
- Writing utensils, with extra backup writing capability
- If using a laptop, extra batteries or a power source
- Appropriate clothing, rain gear, etc.
- Drinking water and food
- Reading material for breakdowns or delays
- Cell phone or other communication devise for emergencies

This is not an exhaustive list, but serves to jog one's memory. It will be essential to have a cursory idea of site geology, even if you have never been to the site before. Which drilling method is most appropriate for the conditions? Is forward rotary better or auger drilling?

You will need to know how to tell how deep you are in the ground. When the driller sets up at the hole and raises the mast, he or she will level the rig. When the bit rests at the ground surface, this is surface or zero drilling depth on your log. If there are preexisting numbers painted down the mast, you need to note where the zero marking coincides. Another way to keep track of footage is to mark the drawworks chain, every 5 ft (1.5 m) or so with a bright colored grease marker. Another way is mark directly on the drill casing with chalk (Figure 8.29). It is best to ask the driller the most appropriate way to keep track of drilling footage.

Figure 8.29 Depth markings on the steel casing.

The example of logging a drill hole that follows illustrates using the forward rotary method, initially drilling with air and then switching to water and then foam.

Example 8.6

The driller asks you if you are ready and the deafening sound from the air compressor is directed down the kelly into the drill pipe. Air begins gushing out the ports at the end of the bit and dust begins to blow. The driller engages the power drive system and the rotary table begins to turn. The cuttings begin to blow around at the base of the drill rig and pile up. The driller helper begins to shovel the cuttings to channel the fluids and cuttings away from the back of the rig.

You have prefilled out most of the header of your logging form and stand near the driller with your sieve ready to catch cuttings. You do so frequently and look up frequently to make sure you are aware of what is going on. In your log book you write down descriptions for every 5 ft (1.5 m) or at major lithology changes. Generally, lithology changes more frequent than 3 ft (1 m) will require too much writing to keep up with drilling. You start piling representative cuttings sequentially on the ground to be able to refer back to them, or place them in premarked bags. After 50 ft or so you notice the samples are getting moist and you ask, did we hit water yet (Figure 8.30)?

Figure 8.30 Describing drill cuttings.

The driller nods and then starts pumping water along with the air to maintain circulation. It will be necessary to fill a 5-gal bucket (19-L) from the back of the water truck to rinse your cuttings now. What you observe in your sieve may represent a mixture of cuttings from the full length of the hole. What are we in? You communicate with the driller and get back on track. The hole seems to making an increase in water, you ask the driller to stop at the end of the next joint an blow on it (only inject air) to see how much it is making (Figure 8.22).

Another 50 ft and the cuttings are not coming to the surface like before. The driller adds detergent to a mixing tank and begins to pump foam down the hole for added stability. It gets pretty messy around the base of the rig. The sieve fills up with mostly foam. You vigorously rinse the cuttings, but the operation is a bit more challenging.

Example 8.6 presents some issues worth discussing in more detail. How do you know what you are drilling in when the cuttings represent a mixture from the total length of the hole? As you drill deeper it takes longer for cuttings to reach the surface. How does one tell what the conditions of the subsurface are like?

Describing the Cuttings

When drilling with air and at relatively shallow depths, cuttings arrive almost instantaneously at the surface. As drilling proceeds, cuttings from near the bit become mixed in with cuttings from up hole as they are continuously blown to the surface. This yields a mixture of cuttings in your sieve. Generally, about half of what you see is from the current drilling depth. This needs to be evaluated along with the sound of drilling, drilling rate, and drilling depth. Shales and other softer materials will drill more quickly and quietly, while sandstone and limestone may be much louder and slower. Once again, consult with the driller in determining where a formation change took place or when you drilled through fractured rock.

In describing cuttings, it is helpful to find a representative large piece in the sieve and break it in half to obtain a fresh surface. This will help you distinguish among the different lithologies, and more specifically among fine-grained materials. Harder cuttings will not break in half. A quick decision must be made when assessing the materials before writing something down. It becomes difficult at times to distinguish differences among mudstones, claystones, and siltstones. This is best evaluated from a field test, described in Table 8.1. Of course, don't try this if there is any danger of contaminated soils. This is mostly an exploration drilling technique, but can be very helpful.

Table 8.1 Field Methodology for Distinguishing Finer-Grained Sediments

% Silt	Lithology	Taste test (rub against teeth)
0–33% Silt	Claystone or clay shale	Cuttings smear on teeth with little or no grit
33–67% Silt	Mudstone or mud shale	Very gritty on teeth, grit observable
67–100% Silt	Siltstone	Mostly grit observed

Lag Time

The time it takes cuttings to reach from the drill bit to the surface is known as lag time. When drilling with air or foam, the fluids and air from the compressor must be able to lift all cuttings to the surface to maintain circulation. There is turbulence during this process and cuttings may take quite a while to reach the surface after exceeding a depth of 300 ft (>100 m) or so. The best way to gage lag time is to notice a distinctive formation and glance at your watch when you first hear it. Note the time and continue to log cuttings as they are blown out at the surface. Coal or some other marker bed is especially helpful. When the cuttings you noted appear at the surface, look at your watch again; this is the lag time. At depths greater than 100 meters, the sound and chatter of the rig will be as helpful as the actual observed cuttings. By being able to have a marker bed or two every so often, the drill log can be adjusted to match the depths better. Adjustments can also be made on lithology logs if geophysical methods are also available.

Example 8.7

While performing coal exploration in Wyoming, each coal bed had a particular signature. It was easy to tell when you were in coal because of the chatter of the rig and rapid penetration. When drilling at depths of 500 ft or so, the lag times ran from between 15 and 20 minutes. The time depended on the drill hole diameter, volume of cubic feet per minute (cfm) of air being injected, and fluids being added.

As another example, lag times for cuttings mixed with mud from a 1,000-ft hole at the Big Sky Montana ski resort area required 30 minutes or so to reach the surface.

How Much Water Is Being Made and Where It Came From

One of the objectives of drilling is to know when you first hit water and which zones the water was likely coming from. It may be that there is a significant difference between first water and where the static water level is. Confining units may be keeping the water from rising to its unrestrained levels. You can learn a lot about the hydrogeology of an area by observing which zones produce water and where the static water level is each day before you continue drilling.

You should make it a practice to measure the water level in the hole at the beginning of each day and after periods of breakdown where the hole has sat unperturbed for a significant length of time. This along with your notations of producing zones all tell part of the story.

If you suspect that the hole is producing water, ask the driller to stop at the end of the next joint or rod and inject air only. After 5 to 10 minutes, the water being produced from the hole will be coming from the formation. Realize that this may represent minimum production, since not all of the water makes it out of the hole. Some may be going out into the formation. It is the author's experience that a hole will produce approximately 25 to 30% more than what is blowing out at the surface. This can have important implications for water production depending on whether water is being produced from the length of the hole or from a specific zone, because the upper zones are cased off. Table 8.2 gives a rough idea of the difference between volumes blowing from the hole and what could be produced with a pump.

Table 8.2 Production Estimates Based on Water Exiting the Hole

Production observed (gpm)	Possible from pump production
3–5	5 +
5	7
7–8	10
10–12	12–15

Table 8.2 may be important because if drilling has proceeded for a couple of hundred feet (66 m) and total production is on the order of 2 to 3 gpm (0.13 to 0.19 L/sec) one must evaluate the probability of success for additional production at deeper depths and the additional drilling costs.

Estimating the volume takes practice and can be solved a variety of ways. If the volume of water can all be channeled through a single place, the chances are greater of getting an accurate estimate. Using a bucket and timer, a weir, or flume are common methods (Figure 8.22) (see also Section 6.5). Most experienced drillers can estimate flow rates just by looking. If flow rate is an important issue, attempts to measure the volumes accurately are warranted.

8.5 Monitoring-Well Construction

Monitoring-well construction is ideally performed to provide a window into the aquifer at a given completion depth. The goal is to complete the well in such a way that you can have confidence in the hydraulic-head information and water-quality samples. Monitoring wells are usually not designed to evaluate aquifer hydraulics, although they are often used to do so. They are usable as observation wells, but not designed as production wells. Production wells, discussed in the next section, are designed to be pumped at a level that stresses the aquifer to evaluate its hydraulic properties.

Objectives of a Groundwater Monitoring Program

Monitoring-well programs require significant thought and planning to be able to accomplish their intended task. Table 8.3 lists some of the objectives associated with monitoring-well designs.

From Table 8.3 it is apparent that the objectives must be well defined prior to achieving a successful program. Merely placing wells haphazardly is not an effective use of budget funds. Some of the objectives in Table 8.3 require that wells be placed with a certain component of randomness or the assumptions inherent in the analytical methods to be used will be violated. It is also important to have a knowledge of some of the following aquifer properties:

- Are there multiple aquifers?
- Are there perched zones?
- Is there contamination in the vadose zone?
- Is there more than one flow system? Local and regional?
- Recharge and discharge areas?
- What is the nature of a contaminant? Its solubility or density?

All of these things complicate the monitoring design, since it may be necessary for multiple sampling ports at various depths. Are the wells being placed in the groundwater flow path and at the appropriate depths to capture what is being monitored? Another consideration that seems to take precedence is what is the budget and how soon does this monitoring program need to be in place? Is the study being conducted in house or is it under public scrutiny? All these things tend to complicate the process. In a public setting, there may be a negotiation process among regulators, the public, and the company, before the monitoring program can be implemented.

The method used to actually install monitoring wells depends on the geologic setting, the depth to be completed, diameter of borehole, and so on. A common method for drilling and completing monitoring wells in unconsolidated materials with hollow-stem augers (HSA) is presented next.

Installing a Groundwater Monitoring Well

This example will be for a 2-in. (5.-cm) monitoring well to be used for hydraulic-head measurements and water-quality samples (Figure 8.31).

Table 8.3 Monitoring System Objectives

Objective	Discussion
Reconnaissance	This may be a regional study to see what is there. It may consist of a few sparsely placed wells. Shallow/deep pairs may be needed for head and quality data.
Fixed station	Long-term monitoring. Each station may have greater expense, such as a more secure access or a protective shelter.
Research	Temporary monitoring design that may have materials that need to be retrieved later on after research is completed.
Cause and effect	This is where monitoring is established to evaluate whether a process or procedure is effective. Usually there is a control area where the standard approaches or status quo is occurring, compared to a site where a treatment is being applied.
Compliance	These may be wells placed to make sure compliance requirements are being met according to some minimum or maximum standard.
Quality control	Monitoring may be occurring to check whether a process or remediation strategy is working properly.
Trend analysis	A monitoring system may be designed to determine whether a decreasing or increasing trend may be occurring.

Figure 8.31 Water-quality testing from a monitoring well.

1. The casing materials and supplies for building the well have been laid out near the well site or are easily accessible from a vehicle nearby.

2. The bottom of the bit is fitted with a knock-out plate or a removable plug that slips down flush with the bit. The knock-out plate or plug serves to keep cuttings from passing up through the hollow stem and getting bound into the hole. This allows the cuttings to spiral upward, similar to a solid-stem auger system. Once the desired depth has been reached, the knock-out plate can be dislodged so that the well can be constructed. The materials the knock-out plate is constructed with vary, but the author has seen wood to polytetrafluoroethylene (PFTE, Teflon) used.

3. The drill hole penetrates through the first 5 ft (1.5 m). The helper loosens the two rod bolts, and the driller hoists up the kelly and drive sleeve up the mast. The helper slings over the next auger flight, fitting it on top of the one in the ground. Two bolts are pneumatically wrenched into place to secure the flights together (Figure 8.32). The driller lowers the kelly and drive sleeve over the new auger flight and the rod bolts are twisted on finger tight once again. The next 5 ft of drilling continues. As a hydrogeologist, you carefully describe the cuttings that spiral upward along the hollow stem. You record the information in the borehole log so that you know what all the critical depths are when constructing the well.

4. To meet the objectives of this study, you decide to penetrate the aquifer an additional 10 ft below the water table. Once there, you make ready to construct the well. A very important point will be made here. Before dislodging the knock-out plate it is important to **load the hole**. There may be a difference between the water level inside the bore hole and inside the hollow stem. Loading the hole means that water is added to the inside of the hollow stem to equalize the two levels. If the hole is not "loaded," then muddy water from the borehole gushes up into the hollow stem chamber. This may produce a "skin" along the inside of the borehole that is difficult to remove during well development (Chapter 10). This is such a small step, but can make a great difference in well development. The author has seen wells become unusable by forgetting to "load the hole." Care must be taken to introduce clean water. You must know the source.

5. It is now time to remove the knock-out plate. This is done by lifting a heavy metal rod with a sand line and allowing it to "free-fall" inside the hollow stem to the bottom. This procedure is repeated several times. The driller checks the depth by lowering a cloth tape weighted

Figure 8.32 Adding additional auger flights.

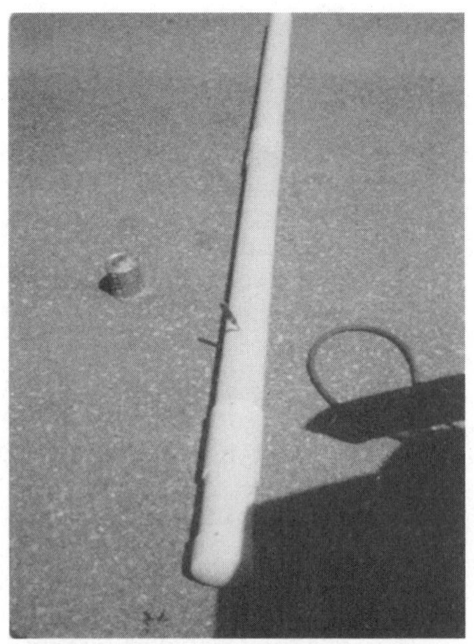

Figure 8.33 Monitoring well bottom cap, short sump, and well screen (knife in screen).

with a heavy metal piece, so that it will fall freely to the bottom. The driller can also "feel" if the knock-out plate has been dislodged by distinguishing the difference between the bottom and the hard plate.

6. It is time to start building the well. A bottom cap is secured onto the bottom of the well screen. It is important to make sure that when the well is constructed the screen is placed exactly where desired (Figure 8.33). Therefore, if the well penetrated into a confining layer, a sump consisting of blank casing will need to be added to the bottom so that the screen is positioned within the aquifer. Blank casing is added to the top of the screen until it extends a couple of feet (0.6 m) above the top of the hole, unless it is located in a street or pathway where one could trip on the casing. In this case, one should complete the well with a flush mounting (Chapter 5). A top cap is placed to keep the inside clean during the rest of the operation. Preferably this is all done with threaded pipe. Screens and pipe come manufactured in a variety of materials and lengths. This will be discussed later on.

7. Now that the well is in the hole, it is time to complete the well. Typically, the next step is the placement of a filter pack around the screen. The packing material is poured down through the hollow stem

pipe around the casing. The filter packing materials should be larger than the screen slot size (<10% passing through the slots). Sieved silica sand is a good choice and these come in 50- and 100-lb (22.7- and 44.3-kg) bags (Figure 8.34). The 50-lb (22.7kg) bags are much easier to handle if used directly, but the 100-lb (44.3-kg) bags may be more cost

Figure 8.34 6 to 9 mesh sieved sand.

Figure 8.35 Stainless-steel screen.

effective. The sand size should not be confused with the slot size. Sand sizes represent sieved intervals, for example, 10 to 20 sand was sized through a number 10 to 20 sieve (2 to 0.83 mm, very coarse sand). And a 20 screen slot size means that the openings are 20 thousandths of an inch (0.51 mm). This is a commonly used scenario. Another example would be to use 6 to 9 sieved (3.35 to 2.16 mm, very fine gravel) sand with a 40 slot (1.02 mm) screen (Figures 8.34, 8.35).

8. One effective way to add sand to the hole is to pour it from a 5-gal (18.9-L) bucket with a notched flap cut in the bottom and side. A 1-in. (2.5-cm) cut is made on the side of the bucket, and then a 1-in. (2.5-cm) cut is made to the bottom to make an L-shaped flap. This piece can be fitted into place while sand is being added to the bucket and flipped down to allow the sand to pour in a controlled manner in the annulus between the hollow stem and the casing (Figure 8.36). This is a good way to keep sand or filter pack materials from spilling all over the place, and it allows the worker to use the more cost-effective 100-lb (44.3-kg) bags. While the sand is being added, the auger flights are being lifted upward to avoid creating a sand lock between the casing and the auger. This is checked by the driller with the weighted cloth tape in a tamping motion to "feel" the placement of the sand and make sure there is a space between the bottom of the auger and the top of the sand. This procedure is continued until the sand is at least a foot or two above the top of the well screen (this is different for a production well).

9. Once the sand is in place, depending on the stability of the hole, the rest of the auger flights may be pulled out for easier well completion. As each flight is raised up to where you can see the bolts, the helper shoves in a plate to suspend the drill string below ground surface, while the connecting bolts are loosened with an air wrench. Once loosened, the auger flight is pulled upward and the helper guides the auger flight away from the rig and lays it on the ground for cleaning. This process is repeated until all auger flights are out of the ground. If the hole is not stable, then the bentonite sealing materials are added in a similar way, being poured through the annulus of the hollow stem and casing.

10. A bentonite seal is placed on top of the sand-packing materials to isolate the zone that the monitoring well is sampled from and to protect this zone from contamination from above. There are a variety of methods to accomplish this. One method is to gradually pour 3/8-inch (1-cm) bentonite chips around the casing. Care must be taken not to pour too quickly or a bridge can form that creates a void in the completion zone. Bentonite chips will fall through a water column if poured slowly (Figure 8.37). Deeper well completions may require the use of a tremie pipe. A tremie pipe is a smaller-diameter pipe that passes down

Figure 8.36 Adding sand using a bucket with a notch cut in the bottom.

Figure 8.37 Pouring bentonite chips to grout a monitoring well.

the hollow stem to the desired depth. Bentonite also comes in pellet form. Both chips and pellets partially hydrate when they come in contact with water. The pellets can be placed with greater success if water is added to them in a bucket to make a slurry and then poured down the hole. If larger volumes are required, then the grouting materials may need to be pumped into place with the rig. The length of the seal depends on regulations on a state-by-state basis. An expansive grouting material can be made by mixing cuttings with bentonite chips and then backfilled within approximately 2 ft (0.6 m) from the land surface.

11. A top plug is made with concrete with framed sides to form a pad with sides sloping away from the casing. A metal locking security devise can be shoved into the concrete to hold it tightly into place. This way the well could have a slip cap over the casing, with a locking painted metal security cover (Figure 8.38).

12. A drainhole should be drilled into the outer security casing to allow for condensation drainage (Fetter 1994). It is also a good idea to drill a hole below the cap of the monitoring well casing to allow for pressure changes to equilibrate easier, unless the site is for monitoring gases. This can be very important for widely fluctuating water levels (Chapter 5).

Figure 8.38 Well with metal security cap removed to observe water depth using reflected light.

Well Completion Materials

Monitoring wells are supposed to provide hydraulic head and water-quality information representative of the aquifer. One of the most important criteria for selection of construction materials is water quality. For example, there is a tendency for chlorinated solvents to sorb onto polyvinyl chloride (PVC) and polytetrafluoroethylene PFTE (Teflon) from aqueous solutions (Reynolds and Gillham 1985). Stainless steel 304 and 316 may sorb heavy metals from aqueous concentrations as low as 50 to 100 μg/L (Parker et al. 1990). PVC glue, used for bell couplings, may also be leached into the monitoring well. This is one reason threaded couplings are desired over fitted couplings that require an adhesive.

Ranney and Parker (1997) performed tests on six different casing materials to see how they would react to 28 different pure "free product" organic chemicals and some acids and bases. PFTE and fluorinated ethylene propylene (FEP) casing showed good resistence to all chemicals, and stainless steel showed good resistence to the organic compounds. Stainless steel is less affected by organic compounds and is suitable for most geologic settings, except in the natural setting of acid rock drainage (ARD) sites where heavy metals may be present or where contamination with strong acids and bases has occurred. A reasonable casing material at an ARD site is PVC.

Some materials may be the very best for chemical resistence but also the most costly. The various pros and cons of different construction materials involve chemical resistence, weight, strength, and cost. A compromise may be required depending on the budget and site requirements. Table 8.4 lists some of the more common monitoring-well construction materials and lists respective advantages and disadvantages from Driscoll (1986) and Fetter (1993).

Table 8.4 Well Construction Materials Comparison

Type	Advantage	Disadvantage	Relative Cost
PVC	Lightweight, great for ARD acids, alkalies, alcohols, and oils	May adsorb VOC organics, react with ketones and esters	1.0
Stainless steel 304,316	Least reactive to organic materials, high strength, and temperature ranges	May corrode in heavy metal or ARD waters, higher cost than plastic, heavy to use	6.0, 11.2
Teflon	Resistant to many chemicals, lightweight	Lower wear resistence and tensile strength, expensive	20.7
Mild steel	Inexpensive, strong, lower temperature sensitivity	Heavy to use, not as chemically resistant as stainless steel	1.1
Polypropylene	Lightweight, resistant to acids, alkalies, alcohols, and oils	Weak, not very rigid, temperature sensitive, hard to make slots for screen	2.1

Well Development

One of the problems with monitoring wells is that they are often not properly developed. If the well-construction process goes as it should, and the water level recovers quickly after bailing the well a couple of times, these wells are often deemed completed, and the crew moves on to the next location. This may be adequate for establishing a connection with the aquifer to verify the hydraulic-head data but may not be adequate for water-quality sampling. The sediments from the drilling operation have not been properly mobilized from the borehole and may affect water-quality results. This can be kind of a dilemma because there is a tendency to limit well development at contaminated sites because of the need to handle the development water. There may also be a concern about disturbing the configuration of the plume. Decisions made about well development must be weighed with the objectives of the study.

If the monitoring well is to used for slug testing (Chapter 11) or the evaluation of aquifer properties (Chapters 9 and 10), then proper well development is a must. This is done by pumping and surging the well to the get the finer sediments to move back and forth through the screened interval. This may take an hour or more, but greatly improves the confidence in obtaining reliable water-level and water-quality data.

The are various techniques that can be used for developing monitoring wells. Many wells are simply overpumped until the water runs clear. Using low-volume water-quality sampling pumps is a poor choice for well development because the impellers become clogged quickly and may cease the pump. It is better to use some form of surging action to mobilize the fines prior to using a pump designed to handle high sediment content. One good choice for development yields in the range of 10 to 15 gpm (0.63 to 0.95 L/sec) is a converted chainsaw engine pump produced by Homelite. The pump needs to be primed, but does a great job of producing dirty water (Figure 8.39). Generally, it is *not* a good idea to add external water to the well, particularly if it to be used for water-quality sampling. A couple of nonconventional well-development methods are given next.

Figure 8.39 Using a converted chainsaw engine pump.

Example 8.8

At one site a garden hose was within reach of a 2-in. (5-cm) monitoring well. A successful well-development method was devised by using the garden hose with duct tape wrapped around it approximately 2 ft (0.6 m) below the nozzle to form a surging tool. The hose contributed clean water to the well as the surging tool was raised and lowered. The dirty water was lifted out of the well during the process for a successful completion.

In an another application a commercial hand pump was not working properly, so a water-bed venturi was attached to a hose connected to the handpump. This caused water to be produced from the well at a rate equal to production from the hose (Figure 8.40). These two examples violate the recommendation of not adding water to a well, but were justified because in both cases the wells were only being used for water-level and hydraulic property data.

Figure 8.40 Using a garden hose and water-bed venturi to make a surging tool.

Example 8.8 was not given as an example of standard practices, but as illustrations of the principles of surging and overpumping.

8.6 Production-Well Completion

Production-well completion could easily comprise another chapter, but is included here as a comparison to monitoring-well completion. Proper production-well completion allows well yields to be maximized while maintaining efficiency. This also leads to confidence in the results from aquifer testing (Chapters 9 and 10).

A general question regarding well completion is: Are we testing the aquifer or the well? An aquifer capable of producing a fair amount of water may be hindered by drilling practices, choice of well-completion materials, or well-development technique. This may result in poor well performance.

Another type of problem arises when well-completion materials are preselected before there is a proper knowledge of the aquifer properties. A slot size or packing material is used that is similar to other wells in the area when they may not be appropriate. This is an inefficient use of resources. Just because a well may be designed to yield a particular quantity of water, the aquifer may be incapable of doing so.

The best way to tell what the aquifer is like is to drill a small diameter pilot or test hole. Larger-diameter production wells can then be drilled with a good knowledge of where the production zones are. If the cable-tool drilling method is used and the target aquifer is known to be productive, production diameter casing may be used from the beginning. This approach can also be used for "drill and drive" rotary methods. Samples from the production zone can be collected and analyzed to select the appropriate screen- and gravel-packing materials. The production zone can later be exposed by jacking up the casing to expose the screen.

Williams (1981) suggests that well-design criteria should include approach velocity and turbulence concepts rather than entrance velocity alone. This process begins with selecting the appropriate packing materials around the screen and selecting the right screen slot size. This is usually done by performing a sieve analysis on the formation materials. Well development around screens can be for naturally developed or artificial materials. Naturally developed wells are less expensive than gravel packed wells because of the additional steps involved in well completion. If the best production zone has a significant amount of stratified fines, then gravel packing allows for greater flexibility (Williams 1981).

Sieve Analysis

Assuming the well has been drilled to a desired depth and is cased to the bottom, the hydrogeologist can take the cuttings to a lab for analysis or perform a sieve analysis. If the cuttings or formation materials have been bagged every 5 ft (1.5 m), a sieve analysis should be performed over every potential zone to be included within the screened interval. A way to evaluate the grain-sizes distribution is to plot them on a graph. Some graphs forms are prepared for this purpose and can be provided by any screen manufacturer (Figure 8.41). A general evaluation of grain-distribution or sorting is through the uniformity coefficient (C_u), Equation 8.1.

$$C_u = {}^{d_{60}}\!/_{d_{10}} \qquad [8.1]$$

where:

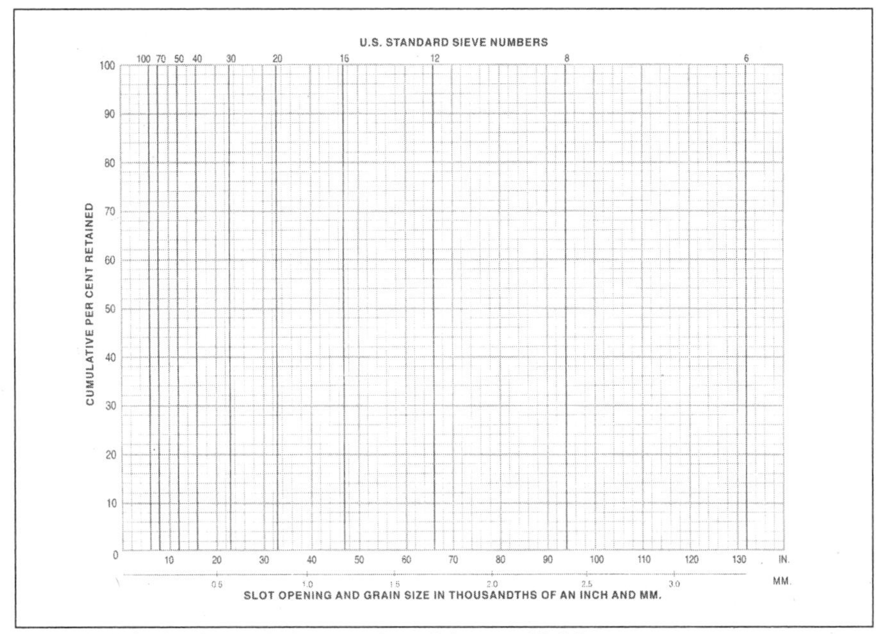

Figure 8.41 Grain-size analysis paper from Johnsons Well Screens.

d_{60} = diameter where 60% of the grain sizes are finer

d_{10} = diameter where only 10% of the grain sizes are finer, also known as the **effective grain size**

If this ratio is less than 4 it is considered to be well sorted, and if it is greater than 6 it is considered to be poorly sorted (Fetter 1994). Walton (1962) suggests that in heterogeneous aquifer materials having a C_u greater than 6 the screen should retain 50% (allow 50% passing) if the overlying materials are unconsolidated and collapsible, or as little as 30% retained (70% passing) if the overlying materials are firm. In homogeneous materials that are well sorted, if the overburden is soft, the screen should retain 60% (allow 40% passing), and if firm allow 40% retained (60% passing). The reason for indicating percent retained or percent passing is because the graph paper is presented both ways. It is just as easy to plot the data using a spreadsheet program. A comparison of grain sizes versus sieve size is shown in Chapter 2.

Example 8.9

In southwestern Montana, an 8-inch (0.2-m) irrigation well was drilled, and the driller wanted to know what size screen slot to recommend. He had collected representative

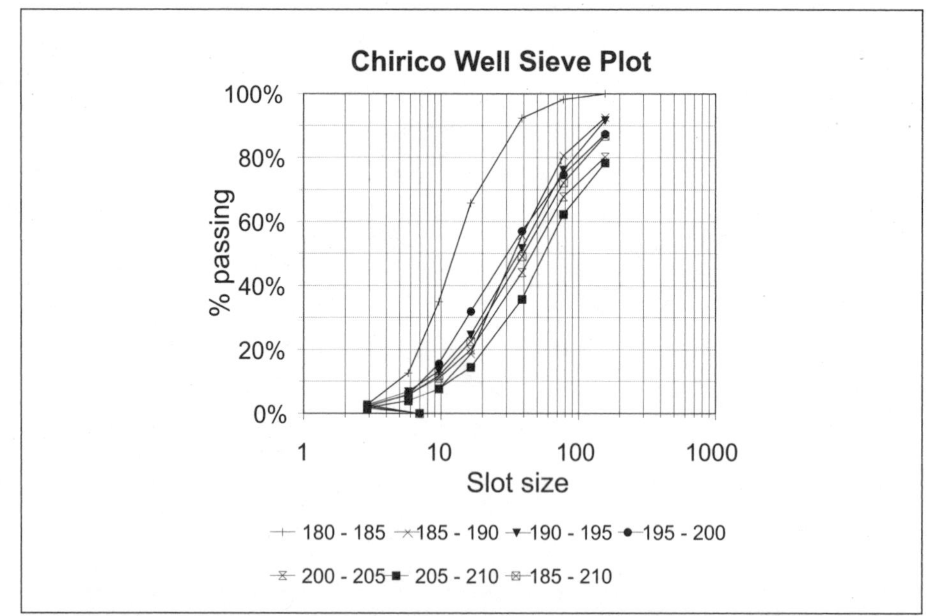

Figure 8.42 Graph of percent passing grain sizes versus slot size to evaluate the appropriate screen slot size, for an 8-in irrigation well in southwestern Montana.

samples every 5 ft (1.5 m) between 180 and 210 ft (56.1 and 64 m). Approximately 150 to 200 g of sample were placed in tins and cooked at 105°C until dry. Each sample was weighed and ran separately through a series of sieves in a ro-tap machine for 15 minutes. The results are shown in the partial spreadsheet of Table 8.5 and plotted in Figure 8.42.

The grain sizes for the sediments between 185 and 210 ft were more uniform, and the 180 to 185 sample was too small to be included in the production zone. The d_{60} ranges between 1.1 and 1.9 mm (coarse to very coarse sand) and the d_{10} ranges between 0.2 to 0.3 mm (fine to medium sand) with a C_u around 6. By evaluating Figure 8.42 at a 50% passing, a slot size of 40 was chosen.

Grain sizes with a d_{10} smaller than 0.25 mm and are fairly well sorted should probably have a gravel packing (Williams 1981). The well in Example 8.9 was successfully developed without a gravel packing because of the coarser fraction present.

When artificial gravel packing is introduced, a different philosophy is employed. Normally, the filter packing materials should be uniform in size with a uniformity coefficient less than 2. The aquifer grain-size distribution curves are evaluated to identify the finest grain-size interval in the screened zone. The d_{30} of the aquifer material is multiplied by a factor to obtain the d_{30} of the gravel packing materials (Williams 1981). The slot size is selected so that 90% of the filter pack material is retained (only 10% passes).

Table 8.5 Sieve Analysis of Samples from an 8-in. Irrigation Well Production Zone between 180 and 210 ft, from Figure 8.42

Screen #	Size mm	Size inch	180 - 185	185 - 190	190 - 195	195 - 200	200 - 205	205 - 210	185 - 210
5	4	0.157	0	6.22	10.548	10.862	27.272	23.673	10.887
10	1.981	0.078	1.354	9.94	19.605	10.866	17.045	17.618	11.987
18	0.991	0.039	4.426	20.96	31.14	15.104	32.77	29.175	19.004
40	0.425	0.0165	20.06	30.95	34.398	21.628	33.169	23.21	21.833
60	0.25	0.0097	23.38	9.31	14.365	14.02	12.387	7.502	8.831
100	0.149	0.0058	16.84	2.91	8.205	8.485	7.049	4.09	5.031
200	0.074	0.0029	7.568	1.783	5.233	3.184	4.727	2.578	2.851
<200			2.06	1.65	3.472	1.765	3.497	1.767	1.82
		Total	75.688	83.723	126.966	85.914	137.916	109.613	82.244
	Sample weight		76.76	86.13	127.16	85.85	138.04	110.16	82.36
		Screen #	**Cum. %**	**Cum. %**	**Cum. %**	**Cum. %**	**Cum. %**	**Cum. %**	**Cum. %**
		5	0.00	7.43	8.31	12.64	19.77	21.60	13.24
		10	1.79	19.30	23.75	25.29	32.13	37.67	27.81
		18	7.64	44.34	48.28	42.87	55.89	64.29	50.92
		40	34.14	81.30	75.37	68.04	79.94	85.46	77.47
		60	65.03%	92.42%	86.68%	84.36%	88.93%	92.30%	88.20%
		100	87.28%	95.90%	93.14%	94.24%	94.04%	96.04%	94.32%
		200	97.28%	98.03%	97.27%	97.95%	97.46%	98.39%	97.79%
		<200	100.00%	100.00%	100.00%	100.00%	100.00%	100.00%	100.00%
	Slot size	**Screen #**	**% passing**	**% passing**	**% passing**	**% passing**	**% passing**	**% passing**	**% passing**
	157	5	100.00	92.57	91.69	87.36	80.23	78.40	86.76
	78	10	98.21	80.70	76.25	74.71	67.87	62.33	72.19
	39	18	92.36	55.66	51.72	57.13	44.11	35.71	49.08
	16.5	40	65.86	18.70	24.63	31.96	20.06	14.54	22.53
	9.7	60	34.97	7.58	13.32	15.64	11.07	7.70	11.80
	5.8	100	12.72	4.10	6.86	5.76	5.96	3.96	5.68
	2.9	200	2.72	1.97	2.73	2.05	2.54	1.61	2.21
		<200	0.00	0.00	0.00	0.00	-0.00	-0.00	0.00

The thickness of gravel-packing materials measured between the borehole wall and the well screen should be at least 0.5 in. (1.3 cm) thick, but not greater than 8 in. (20 cm) (Williams 1981; Driscoll 1986). Greater thicknesses may actually hinder aquifer development because the energy required to develop the aquifer must be able to reach through the gravel packing to the aquifer. For most production wells, gravel pack thickness between 2 and 6 in. around the well screen is a good target. Since well casing may not be placed dead center in the well bore, this can usually be accomplished.

Well Screen Criteria

The well screen is designed to allow entry of groundwater from the aquifer into the well bore. One of the greatest sources of well problems is from overpumping (Williams 1981). Overpumping causes fines from the aquifer to entrain and creates turbulence near the well bore. Turbulence reduces well efficiency, enhances encrustation, and reduces well life. There is a limitation of how much water can be delivered from an aquifer, so the construction design should complement this maximum capacity. Sufficient funds should be invested for any production wells that are intended to be used for a long time. Temporary wells or monitoring wells may not require as high a level of design or expense. A list of the screen criteria and functionality is listed in Table 8.6 (modified from Driscoll 1986).

Table 8.6 Criteria and Function of Well Screen [Adapted from Driscoll (1986)]

Criteria
Large open area
Nonclogging slots
Resistant to corrosion
Sufficient strength
Function
Easily developed
Minimizes incrustation
Low head loss
Control of sediment entry

The slot size used in wells is measured in thousandths of an inch. For example, a 40-slot screen has a slot opening of 0.040 in. (1 mm). A table of slot size versus millimeter equivalents is shown in Table 8.7.

As water moves towards a well, its velocity increases. If the velocity exceeds a critical limit, finer particles from the aquifer will become entrained and move into the gravel packing (Williams 1981). This critical limit is known as the **approach velocity** (**V_a**). This is not to be confused with the entrance velocity. The entrance velocity is measured where water passes through the well screen into the well, and the V_a is measured at the borehole wall at the damage zone (Chapter 10). Additionally, the V_a is a specific discharge and not a velocity where the specific discharge is divided by the effective porosity (Chapter 3). It can also be viewed as the discharge divided by circumferential area. For design purposes a conservative V_a (Huisman 1972) can be expressed as:

$$V_a = \sqrt{K}/30 \quad [8.2]$$

where:

V_a and K are both in units of m/sec.

This can also be expressed in terms of grain size (Huisman 1972) from the aquifer materials as:

$$V_a < 2d_{40} \quad [8.3]$$

with similar units.

Table 8.7 Slot Size in Inches and Millimeters

Slot Size	Inches	Millimeters
10	0.010	0.246
16	0.016	0.417
20	0.020	0.495
25	0.025	0.635
30	0.030	0.762
35	0.035	0.889
40	0.040	0.990
60	0.060	1.52
80	0.080	2.03
100	0.100	2.54
120	0.120	3.05
150	0.150	3.81
180	0.180	4.57
210	0.210	5.33
250	0.250	6.35

This assumes that the grain-size distribution is representative of the aquifer materials in the production zone (Williams 1981). Once the V_a is known, it can be multiplied by the circumferential area of the filter's outside area to estimate the maximum yield capacity (Q_c). If the distance (r_d) at which fines are removed during development can be estimated, the discharge (Q_d) necessary for proper development can also be estimated from the following relationship (Williams 1981).

$$\frac{Q_d}{Q_c} = \frac{V_a A_d}{V_a A_p} = \frac{r_d}{r_p} \quad \text{or} \quad \frac{Q_c r_d}{r_p} \qquad [8.4]$$

where:

Q_c = capacity at the maximum approach velocity

Q_d = discharge during development

r_p = distance to outside edge of filter pack or screen diameter in naturally developed wells

r_d = distance of effective development

A_d = area at the outer effectiveness of development

A_p = area at the outside edge of filter pack or the outside the screen for naturally developed wells

Williams (1981) points out that although the approach velocity cancels out in Equation 8.4, it is necessary for deriving the maximum yield capacity (Q_c).

Example 8.10

Referring to Example 8.9, the d_{40} (40% passing) is nearly equivalent to slot size 30. This yields a diameter of approximately 0.76 mm. From Equation 8.3 and Table 8.7 the maximum approach velocity is approximately:

$$V_a \leq 2 \times 0.000762\,\text{m} = 1.5 \times 10^{-3}\ \text{m/sec}$$

From pumping-test data, the hydraulic conductivity (K) is estimated to be 10 m/day. If this K value is plugged into Equation 8.2 and converted to m/sec, another estimate of the maximum approach velocity is calculated to be:

$$V_a \leq (1.2 \times 10^{-4}\,\text{m/sec})^{0.5} / 30 = 3.7 \times 10^{-4}\,\text{m/sec}$$

These are both within a half order of magnitude of each other. By using an 8-inch telescoping screen with outer diameter of 7.5 inches (19 cm) and dividing by 2 to obtain the (r_p) distance, assuming the development distance is an additional 4 in. (10.2 cm) and the screen length is 25 ft (7.6 m), these can be plugged into Equation 8.4 to obtain a range of discharge values (Q_d). The two areas are first converted into meters squared:

$$A_d = 2 \times \pi \times r_d \times \text{Screen Length} = 2\,\pi\,(0.197\text{ m})(7.62\text{ m}) = 9.43\text{ m}^2$$

$$A_p = 2 \times \pi \times r_p \times \text{Screen Length} = 2\,\pi\,(0.095\text{ m})(7.62\text{ m}) = 4.56\text{ m}^2$$

$$Q_c = V_a \times A_p = 1.5 \times 10^{-3}\text{ m/sec} \times 4.56\text{ m}^2 = 6.9 \times 10^{-3}\text{ m}^3\text{/sec}$$

$$Q_d = \frac{Q_c r_d}{r_p} = \frac{(6.9 \times 10^{-3}\text{m}^3\text{ /sec})(0.197\text{ m})}{0.095\text{ m}} = 1.43 \times 10^{-2}\text{m}^3\text{ /sec} \quad (230\text{ gpm})$$

Using the other approach velocity, a value of 3.5×10^{-3} m^3/sec (55 gpm) was obtained. Actual production rates were approximately 180 gpm (1.1×10^{-2} m^3/sec), which suggests the grain-size approach of Equation 8.3 may give a better approximation, *if* the production zone samples are representative. This analysis can be performed if production-zone samples and sieve analyses will be performed prior to a pumping test or well development.

Screen Entrance Velocity

Although the approach velocity is important for evaluating production rates for well development, the screen entrance velocity is probably one of the most frequently used criteria for sizing well screen. The volume of water pumped from a well is a function of percent open area and the entrance velocity. If the desired well discharge (Q) is known and a minimum entrance velocity is fixed in the design criteria, the only variable is the percent open area. The most commonly used entrance velocity for design purposes is 0.1 ft/sec (0.03 m/sec) (Williams 1981; Driscoll 1986); however, other screen manufacturers report acceptable entrance velocities as high a 3 ft/sec (1 m/sec) (Roscoe Moss 1994). A variety of different slot types are available, continuous, slotted, bridge-slot, and louvered (Figure 8.43), in a variety of materials. The continuous-slot screen has the largest percent open area per length, because a continuous wire spacing is welded to vertical supporting rods. The functions of well screen are listed in Table 8.6. As entrance velocities increase, the potential for turbulence and head losses also increases. This results in higher electrical costs, promotes encrustation, and enhances sediment mobilization.

The percent open area is increased by lengthening the screen or increasing its diameter. For thin aquifers the only choice may be increasing the diameter; however, only percentage gains are made by increasing the diameter. This can be evaluated by comparing some of the continuous slot values listed in Table 8.8. In unconfined aquifers, lengthening the screen may effectively reduce well yield by reducing the available drawdown. The

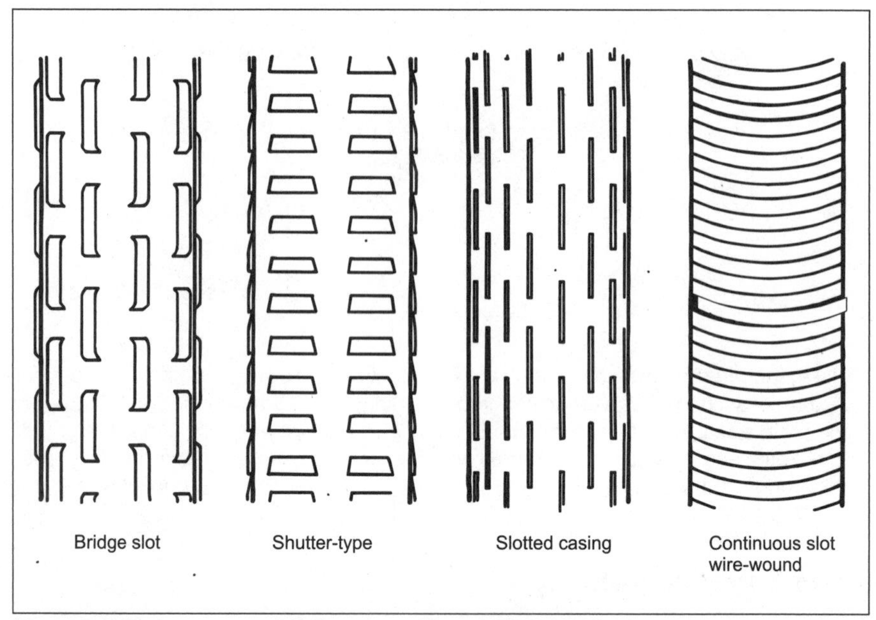

Figure 8.43 Various types of well screens. [From NGWA (1989).]

opposite may be true for a confined aquifer, as long as the hydraulic head remains above the top of the aquifer.

Table 8.8 Percent Open Area for Stainless-Steel Continuous-Slot Screens

	Slot Size	in² /ft	%Open Area	Screen Diameter	Slot Size	in² /ft	%Open Area
4" ID	20	44	25	**12" ID**	20	69	14
	40	72	41		40	99	21
	60	90	52		60	135	28
	100	112	64		100	189	39
8" ID	20	45	18	**16" ID**	20	68	11
	40	77	31		40	124	21
	60	100	40		60	169	28
	100	131	53		100	238	40

Bridge slotting runs one-fifth to one-third less percent open area than continuous slot screen for similar casing diameters. Louvered slot screen runs less than one-tenth to one-fifth of continuous slot screen (Driscoll 1986). Vertical slot screens generally yield a maximum of 3% open area.

Other perforations that are handmade may be from torch-cut slots. These are usually used for monitoring wells but are known to be used in domestic well completions. This results in sediments coming into the well over the life of the well. Some perforations are made in situ in steel casing using a mill knife. This is a device placed downhole that hydraulically punches a hole outward in the casing like a can opener. This is commonly done on 6-in. domestic wells, but once again may allow sediments to pass into the well.

Domestic wells may have their "pump liner" (described in the next section) perforated by a 1/16 -in. (1.6-mm) or 1/8-in. (3.2-mm) slot cut with a circular saw. Generally this done cutting a slot approximately 6 in. (15.2 cm) to 1 ft (30.5 cm) long, spaced every 1 ft or so. The casing is turned approximately 120° in successive rows that are staggered to make three rows that have a spiral pattern of perforations. This maintains a level of strength. The author recommends a maximum depth of 400 ft using this process for schedule 80 PVC screen. He has witnessed the collapse of vertically slotted screens at greater depths. Deeper wells should be completed with steel.

Example 8.11

This example will refer to the well discussed in Example 8.3, an 8-inch production well designed to irrigate a 30-acre cemetery. The production yield was supposed to yield 250 gpm (15.8 L/sec). The drilling method used was cable tool. Drilling would proceed 10 ft before welding on another section of casing. During drilling, a clay zone was encountered at approximately 90 ft (27.4 m) with "first water" at 25 ft (7.6 m). Drilling was hampered by heaving sands. The decision was made to continue looking for the next coarse-grained zone. A fine gravel zone was encountered between 148 and 165 ft (45.1 to 50.3 m) before encountering another clay zone. A sieve analysis was performed on samples collected from the production zone, assuming a natural gravel pack. The analysis indicated a 46-slot was appropriate for a 50% passing grain-size (Figure 8.44). If a 15-ft section of 46-slot screen was used in the production zone, would the entrance velocity be less than 0.1 ft/sec.? There are two ways to check, (1) either the entrance velocity calculated to be less than 0.1 ft/sec using the desired Q or (2) the area and entrance velocity can be used to calculate a Q. As long as the Q exceeds 250 gpm, then the design would be adequate. The percent open area from the chosen manufacturer's 46-slot continuous screen was 33%.

The surface area for the 15-ft screen was first determined

$= \pi \times d \times 12$ in./ft (per ft of screen)

$= \pi \times 7.5$ in. $\times$ 12 in./ft

= 282.7 in.^{2}/ft (for 1 ft of screen)

Multiplying this by 15 ft yields: 283 in.^{2}/ft $\times$ 15 ft of screen = 4,240 in.^{2}

The percent open area is 33%, so:

Figure 8.44 Stainless-steel, v-wire wrapped production screen.

$4240 \text{ in.}^2 \times 0.33 = 1399 \text{ in.}^2$, or

$1,399 \text{ in.}^2 / 144 \text{ in.}^2/\text{ft}^2 = 9.7 \text{ ft}^2$

$Q = 250$ gpm, or 250 gpm/(448.8 gpm) per ft^3/sec = 0.557 ft^3/sec

From the relationship: $Q = V \times A$

We can estimate the velocity moving through the slots, where:

Q = flow in ft^3/sec

V = velocity in ft/sec

A = screen open area in ft^2

The velocity is calculated by: $V = Q/A = 0.557 \text{ ft}^3/9.7 \text{ ft}^2 = 0.057$ ft/sec, which is well below 0.1 ft/sec.

The maximum yield at 0.1 ft/sec Is given by:

$Q = V \quad A = 0.1 \text{ ft/sec} \times 9.7 \text{ ft}^2\text{/sec} = 0.97 \text{ ft}^3\text{/sec}$

$0.97 \text{ ft}^3\text{/sec} \times 448.8 \text{ gpm}/ 1 \text{ ft}^3\text{/sec} = 435 \text{ gpm}$

The well was over-designed, but the well owner wanted a well that would last more than 50 years, or many years beyond his lifetime. In this case, 10 ft of screen may have been adequate.

Well Completion and Development

Wells are either completed open hole or through a packing procedure similar to the monitoring-well process. Domestic wells, if drilled in indurated materials, will be drilled open hole and later have a plastic "pump liner" placed to protect the pump from materials that may slough off from the borehole wall. This is not desirable because of the increased potential for downward migration and cross contamination. Most well drillers realize this, but the competition of drilling by the foot results in well completions that are less than desired.

Larger production wells that are completed open hole and require that the screen be the same diameter as the casing may have a pipe-size screen diameter (Driscoll 1986). They should be welded to the casing for appropriate strength, along with a sump or tailpipe section and bottom plate.

Wells that have been drilled and cased to the bottom have their production zones exposed later by jacking up the casing. These are completed with telescoping screens that slide through the casing to the bottom. A tailpipe and bottom plate are welded to the bottom of the screen, with a piece extending upward that has a packer placed to form a seal between the inside of the casing and the extension piece (Figure 8.45). The whole package is dropped out of sight and followed in with the drillpipe to be held in place while the casing jacks pull up the casing. The hydrogeologist is responsible

Figure 8.45 Neoprene K-packer indicated by a pencil.

for making sure that the screen has been exposed properly. If the telescoping screen package is not in the right place, it must be "fished" out and the hole cleaned out, and the process repeated.

Example 8.12

The 8-inch well from Example 8.11 was drilled to a depth of 170 ft (51.8 m), having penetrated 5 ft (1.5 m) of clay. The telescoping package was designed with a 5-ft tailpipe, 15 ft of screen and a 5-ft extension equipped with a neoprene K-packer. The tailpipe would be located in the lower clay and the screen would fit in the 17 ft of available fine gravel. The package was dropped out of sight and came to a stop 5 ft short of the goal.

Figure 8.46 Fishing tool next to well screen.

A steel fishing tool was constructed by the driller with two metal pieces that could rotate on a bolt that would flange outward to a maximum of ½-in. (1.2cm) smaller than the inside diameter of the telescoping screen (Figure 8.46). It could collapse inward but only flange outward to this maximum distance. This fishing tool was placed inside the well screen to hook onto the lip at the top of the screen. It worked beautifully on the first try.

The hole was cleaned out, and the telescoping package was dropped out of sight again. This time the tailpipe was within 1 ft of the bottom, so the 15-ft screen was fitted exactly within the 17-ft gravel zone.

Well development in production wells is performed to correct the damage done during drilling and enhance the flow characteristics near the screen. With gravel-packed wells, the concepts of approach velocity should be used to determine maximum pumping rates (Williams 1981). Correct pumping rates will mobilize the fines at an appropriate distance from the packing materials. In naturally packed wells, a significant portion of the fines near the screen are washed out during the development process to create a higher transmissivity zone next to the well screen. There are different methods used to accomplish this.

Overpumping is probably the simplest method of mobilizing fines within the screened zone. Pumping until the sediments cease to entrain works for pumping rates lower than the development rate. The problem with this method is that some fines will become trapped from one-way movement. When the pumping stops, these sediments may become loosened and then mobilize again when the pump is turn on. It is better to employ a back-and-forth motion, such as was described earlier in the monitoring-well section.

One effective method of obtaining high volumes of water in a back-and-forth motion is known as **backwashing**. (This term is not to be confused with sharing a soda with a 2-year old). Using a forward rotary rig, the drill pipe is lowered to depth a few feet above the well screen. Air is injected into the well and displaces the water up to the surface. This is done for 15 minutes or so, and then the air is shut off, allowing the water column to rush back out through the screen into the formation. This process is repeated until the water rinses clean (Figure 8.47). This creates a back-and-forth motion in the production zone. It may take several hours to clean up the well, but it is effective. It should be mentioned that when injecting air from the drill pipe, it should not be within the screened interval because the air flow is too intense. If this occurs, the aquifer materials may become more damaged than from the drilling process.

8.7 Water Witching

If there is to be section on water witching (or dowsing), it seems appropriate to be placed in this chapter since many well drillers and property owners use this service prior to drilling. There are individuals who claim to have incredible success rates and must therefore have certain divining powers. It seems appropriate that the author make a few observations that represent his opinion.

The technique involves using welding rods (coat hangers or whatever) approximately 15 to 8 in. (38 to 46 cm) long bent approximately 2 in. down on one end. Forked hickory or willow sticks are also often used. The short lengths of the welding rods are held loosely in the hands while the long ends are allowed to rotate freely. In the case of the sticks, the two forked

Figure 8.47 Clean water using the backwashing development method.

ends are held in the hands while the single end protrudes outward. The author's understanding is that when "water" is encountered the rods turn inward or outward and the hickory stick points downward. This has been used by many people to locate septic tanks or buried pipes on properties. There are mixed reviews on this process compared to technical methods (Mellet 2000). Some individuals use this for locating water on properties for a fee. The author is often asked by students and others what he thinks about this. A fairly complete history on this topic is found in U.S. Geological Water Supply Paper Number 416.

Almost anywhere on the surface of the earth one will encounter water at depth. In an alluvial valley, almost anywhere in the valley would yield a productive well. The challenge comes when evaluating locations in solid rock formations or where the formations may be structurally tilted or disturbed. Even then, water will be eventually encountered; however, the yield may be very low. In hard-layered sedimentary, igneous, and metamorphic rocks, significant well yields may only be derived from secondary fracturing (Chapter 2). It may be that experienced water witchers use similar geologic principles as a scientist would. They may recognize structural trends and the orientations of fracture systems and have experience with other drillholes in the area. When this approach is taken, that is one thing; but when water witchers predict exact depths and yield rates, no matter how sincere and dedicated these folks may be, it is the author's opinion that this appears to be akin to consulting an Ouija board.

Example 8.13

The following example took place in July 1993 in southwestern Montana. At the time, local ranchers had hired a driller to perform some wildcat drilling on the property at an hourly rate, prospecting for water. They owned 2,000 acres of land on either side of a small perennial stream. The property extends towards the mountains on either side of the valley. The thought was that they could run more cattle if there was more water available. Alluvial fans coalesce and slope towards the center of the valley. Drill site locations were constrained to be within 0.25 mile (0.16 km) of a power source or a main road. A couple of drilling sites were selected and the lithologies encountered were alternating silty and gravely materials, with yields only in the 30 to 40 gpm (1.9 to .5 L/sec) range instead of the order of magnitude higher that was hoped for. We were just finishing the second, and last, hole when Elmer the neighbor came over.

Elmer was walking along the alluvial fan in a direction parallel with the mountains. As he walked the welding rods would move and cross at times. He was a larger man who said he couldn't always do this. He had worked for the U.S. Forest Service for about 20 years when an accident occurred that affected his neck. He was on disability and didn't charge for his services. His is a nice enough person with an interest in helping. He acknowledged that most people did not believe in "witching."

Elmer asked the rancher if he would like to try. The rods were placed in his hands and Elmer put his hand to the man's neck and wrist. The rancher exclaimed that no amount of squeezing his thumbs could keep the rods from moving. He was amazed and ran to get his daughter. It did not work for me. The daughter came and it worked for her. Elmer said, "ask how deep it is." The daughter said, "ask who?" (A significant question). "The rods," Elmer replied. At that point the daughter handed the rods back to him.

Elmer proceeded to demonstrate his skills. He walked in a straight line, and the rods crossed where the water "started." They would cross and move back and forth where the "highest" flow was and then swing around pointing behind him when he walked "past" the water. He turned around and repeated the process. The "water" was approximately 20 ft (6.1 m) wide, the possible width of a gravel channel. He asked the rods,"How deep, how many gallons per minute, and how thick?" The rods crossed as he sequentially called out numbers. This was all pretty eerie, but what happened next was especially strange and didn't seem to have much to do with locating water.

Elmer asked, "Where does he water come from?" and the rods pointed towards the mountains, and when asked, "Where is the water going?", they pointed downslope towards the center of the valley. The final question was, "Where is [the rancher's name]?"

and the rods swung around and pointed to him. About that time, I was making tracks for my pickup truck.

The author wishes to apologize to those who may have great faith in this process and may become offended by the above example, but he believes that there is a point where witchers take this process way too far. After all, they do refer to themselves as witches. Following geologic, hydrologic, or geophysical principles are still the only methods the author chooses to estimate drill sites for water production. People will have to decide what they will have faith in for themselves.

8.8 Summary

Selection of the appropriate drilling process requires an understanding of the geology and the objectives of the project. Rig safety and getting along with the drilling contractor can be done in a cooperative environment if the hydrogeologist shows common courtesy and common sense. Well construction and development requires that accurate samples be collected in the production zone. If it is a monitoring well, then the construction materials need to be resistant to the contaminants involved. For production wells, a sieve analysis, screen design, and development strategy can be chosen that effectively produces a maximum yield.

References

Driscoll, F.G., 1986. *Groundwater and Wells.* Johnson Screens, St Paul, MN, 1108 pp.

Fetter, C.W., 1993. *Contaminant Hydrogeology, 2nd Edition.* Prentice-Hall, Upper Saddle River, NJ, 500 pp.

Fetter, C.W, 1994. *Applied Hydrogeology, 3.* Merrill, New York, 691 pp.

Geoprobe Systems, 1999. *The Probing Times*, Vol. 8, No. 1, Summer 1999, Salina, KS, www.geoprobesystems.com.

Huisman, L., 1972. *Groundwater Recovery.* Winchester Press, New York.

Mellet, J., 2000, GPR 5. Dowsers Zero. *Fast Times—The EEG Newsletter*, May 2000, p. 10, The Environmental and Engineering Geophysical Society.

NGWA, 1989. *Handbook of Suggested Practices for the Design and Installation of Ground-Water Monitoring Wells, EPA 600/4-89/034.*

National Water Well Association, Dublin, OH and Environmental Monitoring Systems Laboratory, Las Vegas, NV, p. 398.

NGWA, 1998. *Manual of Water Well Construction Practices, 2nd Edition.* National Ground Water Association, Westerville, OH, pp.1–1 to 13–4.

NGWA, 1999. *Water Well Journal*, July. Westerville, OH, pp.56–58.

NWWAA, 1984, *Drillers Training and Reference Manual.* National Water Well Association of Australia, St. Ives, New South Wales, 267 pp.

Parker, L.V., Hewitt, A.D., and Jenkins, T.F., 1990. Influence of Casing Materials on Trace-level Chemicals in Ground Water. *Ground Water Monitoring Review*, Vol. 14, No. 2, pp. 130–141.

Ranney, T.A and Parker, L.V., 1997. Comparison of Fiberglass And Other Polymeric Well Casings. Part I: Susceptibility to Degradation by Chemicals. *Ground Water Monitoring and Remediation*, Vol. 17, No. 1, pp. 97–103.

Reynolds, G.W. and Gillham, R.W, 1985. *Absorption Of Haloginated Organic Compounds by Polymer Materials Commonly Used in Ground-Water Monitors.* Proceedings of the Second Canadian/American Conference on Hydrogeology, Banff, Alberta, Canada, National Water Well Association, pp. 125–132.

Roscoe Moss, 1994. *Handbook of Ground Water Development.* John Wiley & Sons, New York, 493 pp.

Spellman, F. and Whiting, N., 1999. *Safety Engineering: Principles and Practices.* Government Institutes and ABS Group Company, Rockville, MD, 459 pp.

USGS, 1917. *The Divining Rod—A History of Water Witching.* U.S. Geological Survey Water Supply Paper No. 416, 59 pp.

Walton, W.C., 1962. *Selected Analytical Methods for Well and Aquifer Evaluation.* Illinois State Water Survey Bulletin 49, 81 pp.

Williams, E.B., 1981. Fundamental Concepts of Well Design, *Ground Water,* Vol. 19, No. 5, pp. 527–542.

Chapter 9

Pumping Tests

As a young person in high school, my friends and I would come to a stop at a red light, put the vehicle in park, open the doors, run around the vehicle, and then get back in before the light turned green. This silly activity was fun because of the chaotic flurry of confusion that took place, reminiscent of the Keystone cops, plus the bemused expressions of other drivers as they looked on. It was always a challenge to get everyone around the vehicle and back in so that traffic was not held up. This activity will be referred to as a wacky fire drill (WFD). The first few times one runs a pumping test, the chaotic flurry at the beginning of the test of synchronizing watches, obtaining manual readings every few tens of seconds, the management of a fairly long list of equipment, and the stress of being responsible to obtain meaningful data is much like a WFD.

Example 9.1

I can still remember my first experience in being involved with a pumping test back in the fall of 1980 in northern Wyoming. My job was to continue manual readings for the night shift and to radio in if there was trouble. At about 2 o'clock in the morning the pump started making a terrible noise and the radio had gone faint because the vehicle battery had run low from not restarting the engine to keep things charged up. I felt foolish and scared and wondered how long it would be before anyone found me. To my rescue came the hydrology supervisor, Jim Bowlby, who immediately realized the water-level had drawn down to the pump intake, creating a potential cavitation situation. He also helped give me a jump-start to get the truck and the much needed heater going. We attempted to salvage the test, but the data were messed up. This was a WFD gone bad.

9.1 Why Pumping Tests?

There are many methods of obtaining hydraulic information from aquifers, but perhaps the most common and best is the pumping test or aquifer test. Please note that there is an industry-wide error in referring to these as "pump tests." Hopefully it will be the aquifer we are testing, *not* the pump. Without an estimate of the hydraulic properties of an aquifer, calculations of groundwater movement and contaminant transport cannot be performed with any level of confidence. Saying that pumping tests are the best method of obtaining hydraulic information presumes that the test has been designed properly and was run for an adequate period of time. This also includes setting the pump discharge at a sufficient rate to stress the aquifer and supposes that there is at least one observation well to obtain storativity values. By stressing the aquifer for a sufficient amount of time, a glimpse of the aquifer properties away from the pumping well can be obtained, including boundary conditions that may affect groundwater flow. Although less expensive point estimates can be obtained from specific capacity tests and slug tests, the insights gained from pumping tests can be far more meaningful. This, of course, depends on the objectives of the study. (Specific capacity tests and the evaluation of aquifer-test data are discussed in Chapter 10, and the field methodologies for slug testing an analysis are presented in Chapter 11).

Perhaps the most important reason for conducting a pumping test is to obtain a picture of the general hydraulic properties of an aquifer. Well-hydraulic theory assumes that an aquifer is homogeneous and isotropic (Chapter 10), when in fact this is rarely the case. Another assumption that is often violated is that there are no boundary conditions, meaning that the aquifer is of infinite areal extent. To satisfy the condition of horizontal radial flow, the aquifer is supposed to be screened over the entire thickness of the aquifer. Geologic conditions and well completion costs often make this condition not practical. How can a pumping test be designed that will yield meaningful results? One that perhaps would reveal spatial variations in aquifer properties and the effects of boundary conditions either perceived or not? This is the topic of the next section.

9.2 Pumping-Test Design

Pumping-test design begins with establishing the objectives and conditions of the study. The objectives are usually to obtain hydraulic information; however, conditions correspond to situations that may exist in the field, for example:

- Is the pumping test being conducted near or within a contaminant plume? If so, where will the discharge water be stored and treated and how much will be produced? Or will the pumping test affect the configuration of a plume?
- Will pumping affect or impact other existing water users' wells?
- Will pumping occur near a construction site or downtown area where dewatered sediments may result in compaction and foundation stabilization problems?
- What size of pump will be necessary to adequately stress the aquifer? Will a sampling pump be adequate, or will a submersible or even a turbine pump be required? Think of equipment and costs.
- Is this well being tested for production purposes, such as for irrigation or public water supply, that would require long-term pumping? What impacts, if any, are expected to occur, and what additional personnel and equipment might be necessary?
- Is the aquifer being tested to check the requirements for a subdivision, or to perform dewatering estimates at a construction site or mining property?
- Is the test going to take place near a recharge or discharge area (Chapter 5)? Are vertical gradients or components of flow important?
- Are multiple aquifers being affected?
- What are the potential boundary conditions in the area? Are there streams, faults, constructed barriers, or confining layers or discontinuities that may affect the shape of the cone of depression (Chapter 10)?

Sometimes there aren't any special conditions; it's just that a pumping test is needed to estimate the hydraulic properties for use in calculations for groundwater flow or for a groundwater-flow model or transport model. Maybe initial estimates are needed to estimate distance-drawdown relationships.

Geologic Conditions

When designing a pumping test, it is imperative that a general background of the geology be known. Will this be a porous media test in unconsolidated or lithified sediments? What depositional environments are probably represented? Will this take place in fractured rock or karst conditions? Is this a dual or triple porosity system? For a general presentation on geologic conditions, see Chapter 2.

General geologic information can be obtained from other wells drilled in the area. State agencies have well logs that are accessible by Section, Township, and Range (Chapter 1). For larger production wells, a small diameter pilot hole is advisable to anticipate the well construction design. In a monitoring-well field, it may be that all wells will basically be constructed the same. The pipe and screen will be fairly uniform. Details on monitoring-well construction and design are found in Chapter 8.

For the purposes of pumping tests and the analysis of results, the following geologic conditions are helpful to notice in advance of the pumping test:

- Do the sediments fine upward or downward? This may affect the rate of vertical contribution. Layered sediments will have an effect on the results in terms of potential delayed yield or other sources of recharge. These may also constrain the lateral contribution of water from course-grained sediments that may be confined or semi-confined above and below. This is especially important in determining the saturated thickness contributing water to partially penetrating wells during the pumping test (Chapter 10). To check the vertical connection of sediments, wells completed at various depths will be needed to observe any changes.

- What is the potential for lateral boundary affects? Are there surface water bodies, such as lakes streams or ponds, that are hydraulically connected with the groundwater system (Chapter 6)? Or, could the aquifer be expected to thin or thicken in a particular direction? Are either recharge or barrier conditions anticipated? If so, it is advisable to locate an observation well away from the pumping well in the direction of anticipated impacts.

- Is the aquifer of low or high transmissivity? This will affect the spacing and distance of the observation wells. Cones of depression are steep in low-transmissivity aquifers and do not extend as far as in higher-transmissivity sediments.

- In fractured systems, what are the orientations of the major fracture zones? Observation wells drilled nearby may show little or no response if not connected to the main system. This is especially important in a forced gradient tracer test if a nearby monitoring well is to be used as the source well (Chapter 13).

Example 9.2

During the 1992 Montana Tech field camp a pumping-test site was selected with five observation wells (Figure 9.1). The students were to perform a 24-hour pumping test and interpret the results of a preexisting site. The pumping well was sounded to check for depth and was found to be silted in within the entire 10-ft screened zone. The slot size of the screened interval during well completion was too coarse. Through bailing, the pumping well was cleaned out. Each of the three existing observation wells was also cleaned out, and two new ones were drilled and completed. A configuration is shown in Figure 9.2. One thing that surprised the students was the rapid response of an observation well located much farther away than wells completed closer, until they considered the geologic conditions. After some thought, they realized that the completion depth and lithologies of the more distant well were very similar to the pumping well (Figure 9.3). Wells completed closer to the pumping well were completed more shallowly and screened in finer-grained sediments. Partial penetration effects were also a factor (Chapter 10).

Figure 9.1 Pumping test with five observation wells.

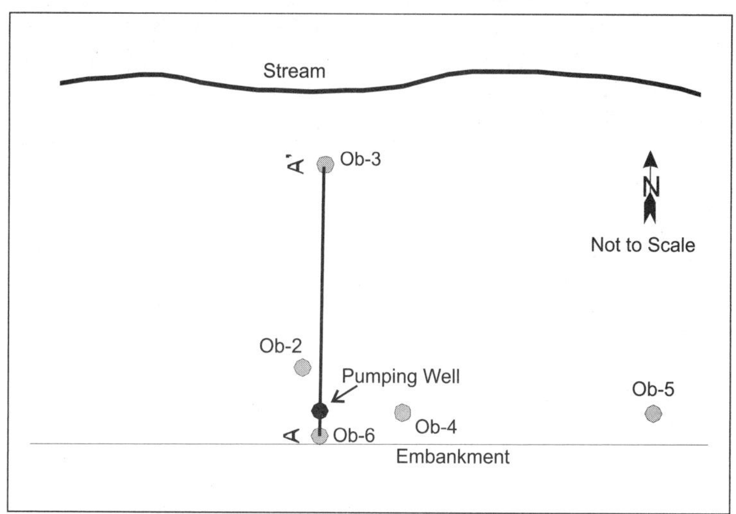

Figure 9.2 1994 Field camp pumping test layout. (Wells Ob-3 and Ob-5 are about 49 and 53 ft, respectively, from the pumping well.)

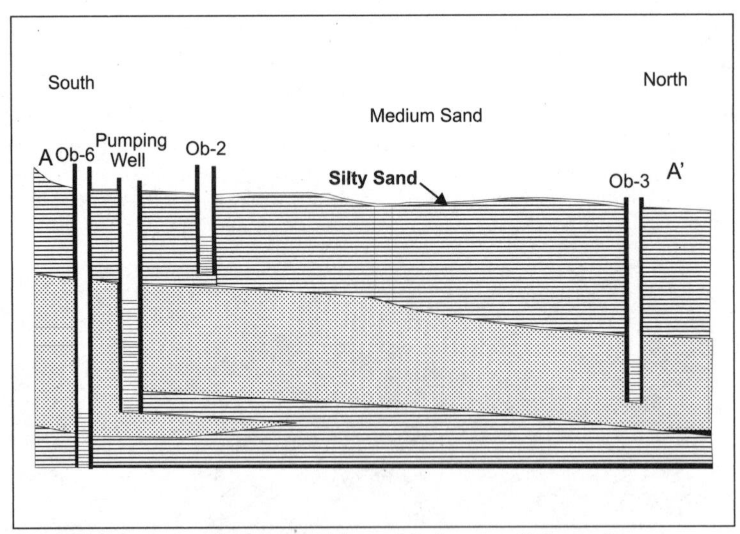

Figure 9.3 Cross-section view of Figure 9.2, from south to north.

Observation well Ob-3 in Figures 9.2 and 9.3 responded sooner than Ob-2. Both the pumping well and northernmost observation well were completed at approximately 30 to 35 ft depths, while Ob-2 was only completed at 12 ft in finer-grained sediments. In time, all wells were impacted by drawdown. A schematic cross section is shown in Figure 9.3.

During a pumping test, the cone of depression expands in response to the requirements of the pump and the discharge constrictions. Observation wells detect the aquifer response in the direction of the expanding cone. If a boundary exists, the observation well closest to the boundary will pick up the change reflected in the time/drawdown data first (Chapter 10). With more than one observation well picking up the same boundary, the distance to the boundary can be estimated (Heath and Trainer 1968).

Distance-Depth Requirements of Observation Wells

The depth of completion of observation wells should reflect the conditions of the aquifer to be developed or the hydraulic properties of the aquifer where calculations will be applied for the intended purpose of the study. If there are additional objectives, such as vertical connections or gradients between shallow and deep systems, then nested pairs are desirable.

The distance of observation wells from the pumping well should also reflect a basic understanding of the geologic conditions already mentioned above and whether the aquifer is confined or unconfined. Confined aquifers can have observation points literally miles (kilometers) away. It will also be a function of the time of pumping. Short-duration tests may require that monitoring wells be placed closer than those being tested at longer pumping times. If delayed yield is expected in the time/drawdown response curve, an observation well close enough to the pumping well to observe this will be necessary. This is usually on the order of 10 ft (3 m) or so, but distances of approximately 10 times that much have been observed. Once again it depends on the duration of pumping and aquifer properties.

Example 9.3

In the Little Bitterroot Valley in northwestern Montana is an elongated N-NW trending intermontane basin (Figure 9.4). Before homesteaders arrived in the valley, water was only utilized at springs (Donovan 1985). The first developable water came from the Lone Pine aquifer.

Lone Pine aquifer consists of very permeable unconsolidated glaciofluvial gravels and sands of early Pleistocene age. These are overlain by about 200 ft (61 m) of silts and clays from Pleistocene Lake Missoula. A shallow surficial aquifer consisting of Pleistocene sand and gravel deposits and Holocene fluvial terrace gravels sits on top of the lacustrine sediments (Donovan 1985).

Drilling was accomplished through a jetting technique that cut through the soft lacustrine deposits but was only able to penetrate a few feet into the Lone Pine gravels. The pressure head within the aquifer was sufficient to cause flowing artesian wells, with yields in excess of 1,000 gpm (3,790 L/min). The main use of developed water is for irrigation of approximately 3,000 to 3,500 acres, although some is used for domestic and stock watering purposes (Donovan 1983). Controversy arose over declines in flow from an increase in the number of wells. Additionally, the valley has warm water 25 to 53°C over approximately 600 acres of the Camp Aqua area (Figure 9.4). This resulted in problems with irrigation applications and with bath houses being developed (Donovan 1985).

Pumping tests were conducted during 1980–1983 to determine the aquifer characteristics of the Lone Pine aquifer. Measurable drawdown was observed in observation wells as far away as 9 miles (14 km)!

Ideally, any pumping test should have two observation wells. The two wells should be placed at different radial distances from the pumping well and at different compass orientations. The closer well would detect near-well conditions and be sure to pick up drawdown, and the other well would be helpful in evaluating the distance the cone of depression expands. It is also extremely important to have an understanding of the direction of groundwater flow. Drawdown will be detected in wells up gradient at farther distances than wells placed down gradient (Fetter 1993). In fact, wells down gradient may be out of the cone of influence of the radius of depression regardless of how long pumping takes place (Figure 9.5).

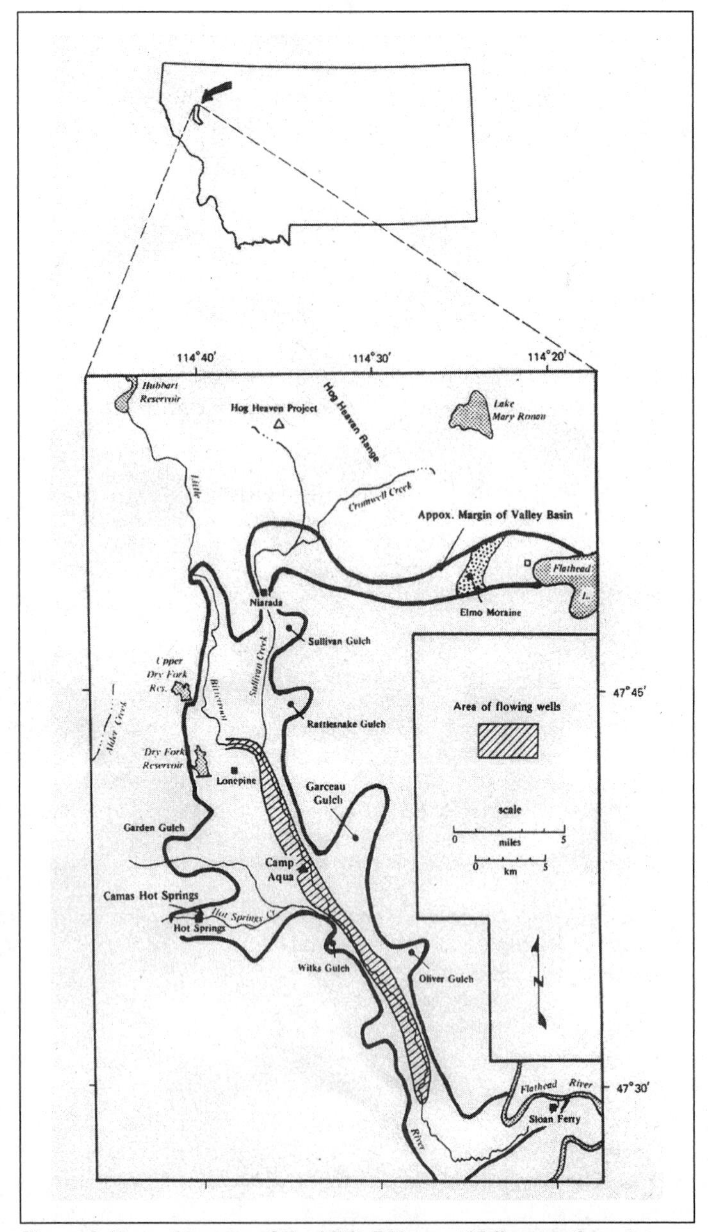

Figure 9.4 Location map of the Little Bitterroot Valley, northwestern Montana. [Reprinted with permission from the MBMG (1985).]

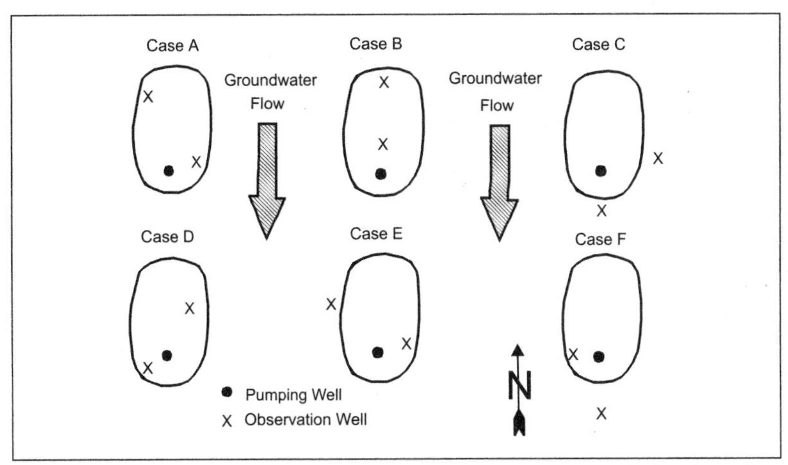

Figure 9.5 Configurations of pumping and observation wells with direction of groundwater flow at some time t. (Cases A, B, and D will detect drawdown in all wells, and Cases C, E, and F will not detect drawdown in one or more wells at time t.)

Figure 9.5 represents six cones of depression from pumping wells over the same duration. In each case there are two observation wells with one placed close to the pumping well and one located farther away. However, in Case C, E, and F, one or more wells do not detect drawdown at a given time *t*. It appears that the duration of pumping would have little effect on whether drawdown would be detected or not. In case B, although both wells would detect drawdown, little information would be gained as to changes occurring east or west of the line of wells. There are not many advantages to a design like this. A schematic cross section of case B is shown in Figure 9.6 to illustrate the concept of a sloping water table and its respective cone of depression shape. This concept is significant to groundwater capture zones and remediation design. One of the assumptions in well hydraulic theory is that the potentiometric surface is flat.

In Figure 9.6 the flat water table is indicated by a flat dashed line and the sloping surface is exaggerated for emphasis. A sloping water table yields an asymmetrical cone of depression. The down-gradient volume is extended in the up-gradient direction in a capture-zone scenario. As one can see, this is also an important component in pumping-test design.

9.3 Step-Drawdown Tests

In order for pumping tests to effectively stress the aquifer, a proper pumping rate must be established. Even though the wells are properly placed

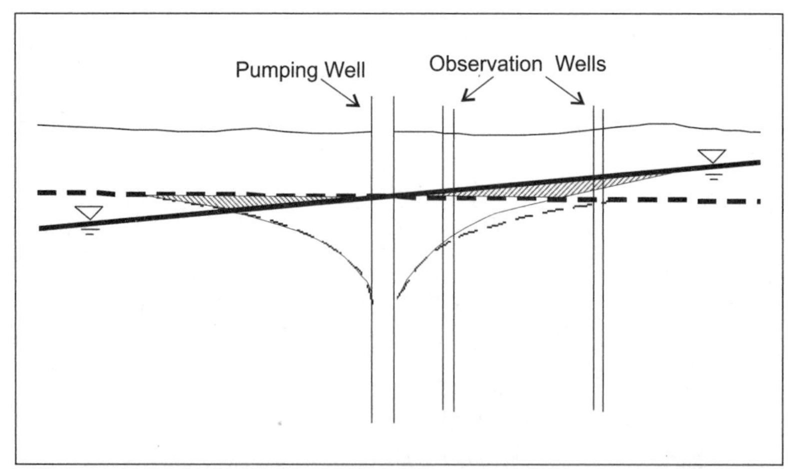

Figure 9.6 Cross section showing the difference between a cone of depression in a flat water table surface versus a sloping one in a capture zone scenario. (The hatched pattern represents an equal volume.)

and developed, if the pumping rate is too low, a small cone of depression will result and the drawdown in observation wells may not be detected. Conversely, if pumping rates are too high, then the test will not run very long because the pumping level will reach the pump, possibly causing cavitation. Another cause of cavitation occurs when the suction pipe is too small a diameter, producing turbulence and creating air bubbles. This is not a problem with submersible pumps unless the pumping level has reached the pump. Water-well pumps were designed to pump water, not air.

The procedure of changing pumping rates over a consistent time interval in a deliberate manner is known as a **step-drawdown test**. This is a single-well test where the initial pumping rate is lower than the maximum expected rate. After the drawdown stabilizes, then the rate is increased, or stepped up, to a higher rate for the same amount of time, usually 30 minutes to 2 hours (Kruseman and de Ridder 1990). The key is to run each step for the same amount of time and to run at least three steps. This procedure is also done to check well performance. Step tests were first devised by Jacob (1947) to check what would happen to the drawdown if the pumping rate varied during a pumping test.

Well-hydraulics theory is based on the concept of laminar flow for groundwater. In the near vicinity of a well bore, velocities may increase to the point that turbulent conditions result. If the conditions are laminar, then the drawdown is proportional to the pumping rate (Q). If turbulence occurs, there are other well losses that also contribute resulting in a nonlinear relationship. Collectively these can be represented by the following relationship presented in Equation 9.1 (Roscoe Moss 1990).

$$s = ds + ds' + ds'' + ds''' \qquad [9.1]$$

where:

s = total drawdown measured in the pumping well

ds = head loss in the aquifer (formation loss)

ds' = head loss in the damage zone (skin effect)

ds'' = head loss in filter pack

ds''' = well loss from water entering the screen

The total drawdown observed within a pumping well is the difference between the static water level and the pumping level, where the pumping level is the level inside the well casing. The actual water level that exists outside of the casing varies according to a group of additional head losses that occur from other affects. Included are head losses that occur as water moves through the aquifer (ds), head losses as the groundwater encounters a skin effect zone (ds'). The skin effect is a result of fine drill cuttings or films from fluids that remain on the borehole wall from drilling. Groundwater then passes through the filter pack material (ds''). Naturally developed wells may not have significant skin effects but will have the natural gravel packing materials. As groundwater enters the well screen and changes di-

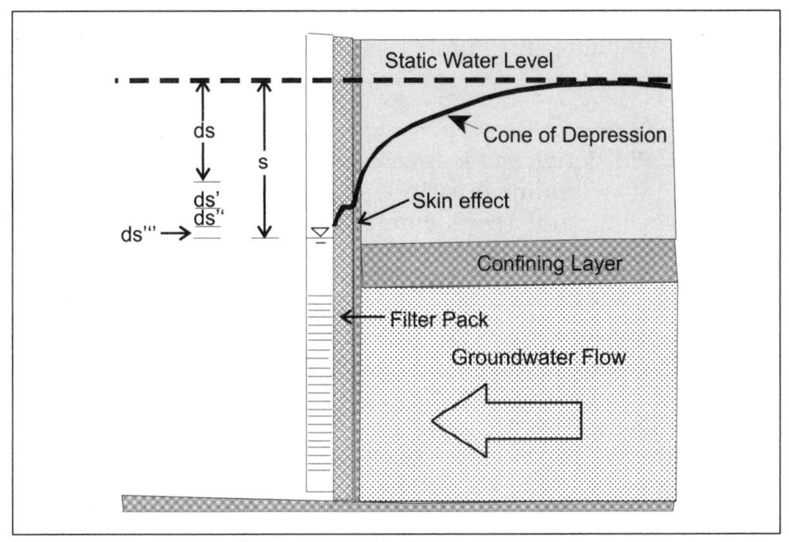

Figure 9.7 A schematic illustrating the individual contributing head losses to the total head loss observed at the pumping level. [Modified after JAWWA (1985).]

rection from horizontal to vertical flow towards the pump, additional wells losses take place (*ds'''*). A depiction of all these well losses is shown in Figure 9.7.

Laminar flow in a perfectly efficient well in a confined aquifer will yield the following relationship described as the modified nonequilibrium equation by Driscoll (1986):

$$B = \frac{264}{T} \log\left(\frac{0.3T\ t}{r^2\ S}\right) \qquad [9.2]$$

where:

Q	=	pumping rate (gal/min)
T	=	transmissivity (gal/day-ft)
t	=	time (day)
r	=	radial distance of observation point (ft)
S	=	storativity (dimensionless)

This is also okay for unconfined aquifers if the drawdown (s) is small relative to the saturated thickness (b). The next step is to let the laminar term be represented by (B).

$$s = \frac{264Q}{T} \log\left(\frac{0.3T\ t}{r^2\ S}\right) \qquad [9.3]$$

This yields the following simple relationship:

$$s = BQ \qquad [9.4]$$

If all of the additional well loss terms shown in Equation 9.1 are combined together into a nonlinear turbulence factor (C) and added to the laminar term, we have a crude representation of the total drawdown ([9.5).

$$s = BQ + CQ^2 \qquad [9.5]$$

A more rigorous analysis of these processes and how they can be solved is presented by Driscoll (1986) and Kruseman and de Ridder (1990). A method on how to interpret variable-rate pumping test is also described by Birsoy and Summers (1980). However, a more simplified approach was presented by Walton (1970), which is useful for field approximations of maximum pumping rates and for selecting pumps. Walton (1970) provides a procedure where the turbulence factor (C) can be estimated from a step-drawdown test. The process involves conducting a step-drawdown test with successive increases in Q and plotting the results on a time-drawdown graph (Cooper-Jacob plot, Chapter 10). Each step appears as a straight-line segment followed by a marked drop in drawdown between steps (Example 9.4b). Initially, during laminar flow, the slopes of the segments are approximately the same. As well efficiencies decrease and turbulence effects increase, the slopes of the straight-line segments will steepen. A rough estimate of well efficiency can be determined, and an appropriate maximum pumping rate can be estimated.

The turbulence factor can be approximated by:

$$C \approx \frac{\left(\frac{\Delta s_n}{\Delta Q_n}\right) - \left(\frac{\Delta s_{n-1}}{\Delta Q_{n-1}}\right)}{\Delta Q_n + \Delta Q_{n-1}} \qquad [9.6]$$

where:

C = turbulence factor, in $ft/(ft^3/sec)^2$

Δs_n = the change in drawdown from step n 1 to n, in ft

ΔQ_n= the change in pumping rate from step n 1 to n, in ft^3/sec

Values of C also provide a qualitative measure of well efficiency. Walton (1970) gives the following groupings in Table 9.1:

Table 9.1 Qualitative Values for Turbulence Factor C, in $ft/(ft^3/sec)^2$ [From Walton (1970)]

Turbulence Factors C	Comments
< 5	Great
5–10	Good
10–40	Fair to poor
> 40	Bad

The methodology will be illustrated with two examples: one from Walton (1970) with an efficient high-yield irrigation well followed by a lower pumping-rate test used in a monitoring-well field.

Example 9.4a

A 24-in. (0.61-m) diameter irrigation well was tested and the data are summarized in Table 9.2.

Table 9.2 Step-Drawdown Data from a 24-in. Irrigation well [From Walton (1970)]

Step	*Q* (gpm)	Δ *Q* (gpm)	Δ *Q* (ft³/sec)	Δ *s* (ft)
1	1,000	1,000	2.22	5.40
2	1,280	280	0.62	1.59
3	1,400	120	0.27	0.72

From Equation 9.6 the respective *C* values are determined.

$$C_{1,2} \approx \frac{\left(\dfrac{1.59\ \text{ft}}{0.62\ \text{ft}^3/\text{sec}}\right)-\left(\dfrac{5.40\ \text{ft}}{2.22\ \text{ft}^3/\text{sec}}\right)}{2.22\ \text{ft}^3/\text{sec}+0.62\ \text{ft}^3/\text{sec}} = 0.046\ \text{ft}\Big/(\text{ft}^3/\text{sec})^2$$

$$C_{2,3} \approx \frac{\left(\dfrac{0.72\ \text{ft}}{0.27\ \text{ft}^3/\text{sec}}\right)-\left(\dfrac{1.59\ \text{ft}}{0.62\ \text{ft}^3/\text{sec}}\right)}{0.62\ \text{ft}^3/\text{sec}+0.27\ \text{ft}^3/\text{sec}} = 0.115\ \text{ft}\Big/(\text{ft}^3/\text{sec})^2$$

By using Equation 9.5, one can determine the well losses from the turbulent factor at 1280 gpm (0.081 m^3/sec) and 1,400 gpm (0.088 m^3/sec), respectively.

s_{1280} = 0.046 ft/(ft^3/sec)^2 (2.85 (ft^3/sec)^2) = 0.37 ft

s_{1400} = 0.115 ft/(ft^3/sec)^2 (3.12 (ft^3/sec)^2) = 1.12 ft

The relative % losses can now be calculated:

% loss_{1280} = 0.37 ft/(5.4 ft +1.59 ft) × 100% = 5.3% very efficient!

% loss_{1400} = 1.12 ft/(5.4 ft + 1.59 ft + 0.72 ft) = 14.5% still efficient

At 1,400 gpm (0.088 m^3/sec), the well is still approximately 100% % $\text{loss}_{1400} \approx$ 85% efficient.

Example 9.4b

Prior to the 24-hour pumping test described in Example 9.2, a suitable pumping rate was not known, so a step-drawdown test was conducted (Figure 9.8). The results of the test are shown in Table 9.3 and the data are plotted in Figure 9.9. Note how the slope changes from step 1 to step 2. Step 3 was not pumped long enough to stabilize.

Figure 9.9 illustrates how the slope changes, given a particular pumping rate. The first pumping rate was not stressing the aquifer very much. By time-step 2, the slope had steepened significantly. Time-step 3 was not run long enough to evaluate a slope.

Figure 9.8 Students conducting a step-drawdown test.

Table 9.3 Results of a Montana Tech Hydrogeology Field Camp Step-Drawdown Test

Step	Q (gpm)	Δ Q (gpm)	Δ Q (cfs)	Δ s ft
1	8.1	8.1	0.18048	2.98
2	14.4	6.3	0.014037	3.73
3	18.5	4.1	0.009135	4.69

It is not always necessary to conduct a formal step-drawdown test prior to performing a pumping test. However, it is very helpful particularly if one is sizing a pump for production purposes and wishes to make a purchase. After one gains some experience, the appropriate pumping-discharge rates for a constant discharge pumping test can be estimated by observing the drawdown in the pumping well after several minutes. Control of the discharge can be made over a fairly good range by constricting the outflow. However, one must start with a pump size that is within the appropriate range.

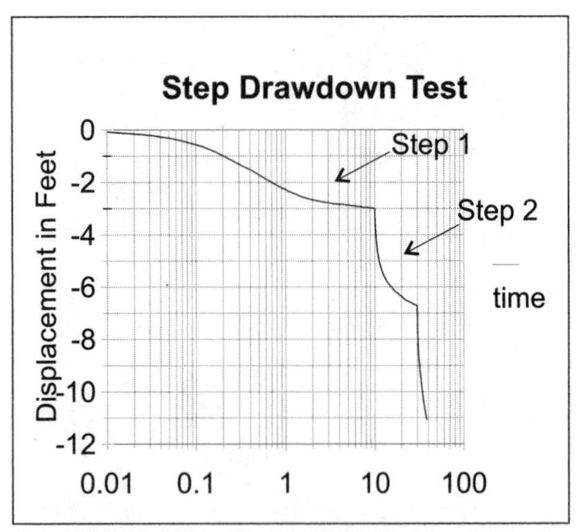

Figure 9.9 Step-drawdown test with 3 steps (pumping rates). (Note: Step 3 was not long enough to stabilize for calculations.)

9.4 Setting Up and Running a Pumping Test

Setting up a pumping test can be frustrating if something is left off the equipment list and one must retrieve something back at the office. This section discusses the typical equipment that one should bring to have a successful experience. Discussion of the pitfalls that may occur are also presented. A discussion of the information that should be recorded in field books is shown in Chapter 1.

The following is a list of equipment needed for a pumping test (items to bring are indicated in italics):

- *Power supply* (usually a generator, equipped with outlets that will allow the pump control box to be plugged in).
- *Extra fuel cans* for generator.
- *Control box and pump.*
- *Data logger*—for storing data and user's manual.
- *Transducers*—anticipate the drawdown expected for each well and have a transducer within range. Don't forget the *jumper cables* that connect the transducer to the data logger.
- *E-tapes*—for taking backup readings and to check conditions prior to setting up the test.

- *Field printer* with extra paper or *laptop computer* to view and plot the data *in the field,* with extra diskettes.
- *Discharge system*—this may consist of a *garden hose* for small discharge rates or *riser pipe, elbows,* and *connectors* to convey the discharge away from the site. A means of measuring discharge is needed, either a *flow meter,* or a *calibrated container* (buckets) and a *stopwatch.* An in-line flow constrictor, such as a *gate valve* (preferred) or ball valve to put back pressure on the pump and control the discharge. *Teflon tape* may be needed for the threads when connecting the discharge lines and riser pipe.
- *Pipe wrenches*—to construct the discharge line (aluminum ones are much lighter).
- *Duct tape* (the most versatile tool in the box).
- *Electrical tape.*
- *Miscellaneous tools* (screwdrivers, wrenches, electrical tester, etc.).
- *Field books,* *field forms,* and *pens or pencils* (with refills).
- *Well logs* and *field map.*
- *Semi-log and log-log paper* for each observation well. (Even if a laptop computer is available, data should always be plotted in the field.)
- *Kitchen timers* (for timing readings).
- *Keys* (for gate locks or well access).
- *Shovel* and *rope.*
- Miscellaneous field equipment (such as water-quality equipment to check changes in pH, temperature, or specific conductivity).
- Hat, sunscreen, bug spray, and personal items.
- Rain clothing, tarp, and/or tent.

Power Supply and Pumps

In considering the appropriate power supply for a pumping test, one must have a knowledge of the pump and electrical feed requirements. Because many pumping tests are remote, a generator is often used. Buying the cheapest generator may be a big mistake. To run a small sampling pump system, it is necessary to provide a steady feed of voltage. Cheaper units will often sputter and surge. Unless the power feed is steady, trying to use these sensitive sampling pump systems will be a problem. Good results for small pumps have been obtained from Honda generators (Figure 9.10).

Figure 9.10 Pumping test conducted with a steady feed Honda generator.

These generators run smooth and are relatively quiet. Some manufacturers overrate their products, so it is necessary to make sure there is sufficient capacity to start and run a pump. Other pumping systems may have their own recommended generator systems, so one should consult the user's manual.

In starting a generator, particularly if it has sat idle for several months, it may be necessary to activate a warmup switch for 15 to 30 seconds before switching the generator on. Changing the oil once a year is also a good idea, regardless of how little it may have been used. When the oil level gets low, many generators will shut down. It has been the author's experience, when the oil level has not been checked and is low, that the generator will make a couple of changes in sound (hiccups) from the normal purr of the engine approximately 1 hour before it suddenly sputters and stops. Newer generators have a safety feature that shuts the engine off when the oil level is low.

Generators may require diesel or unleaded gasoline. No particular preference is recommended. Another maintenance item is the battery. It may be that everything else is working fine, but the battery cables are corroded. Sometimes after cleaning, one can give the generator a jump start, and then during the pumping test the battery will become charged (Figure 9.11). It is helpful to test the equipment before taking it out into the field.

Generators are usually distinguished by the number of kilowatts they are capable of. Pumps require a boost of electrical power to get started. A general rule of thumb is: *A kilowatt is required for every horse power (hp).*

Figure 9.11 Jump-starting a generator battery.

This is not entirely true, but will get one in the ballpark. For example, a 20-kW generator might get a 25-hp submersible pump started, but would require 220 volts and 3-phase power. A 25-hp pump may need a 440-volt source. This is part of knowing what your electrical feed requirements are. Submersible pumps are wired to control boxes that are then plugged into the generators with 110-, 220-, or 440-volt outlets. These pumps also come with quite a range of hp and electrical requirements, such as how many phases and amperage.

Smaller pumps (up to 1 hp) are usually single phase and have simple wiring requirements (Figure 9.12). From 1.5-hp on, submersible pumps may require 3-phase configurations. *Unless you are very adept at electrical wiring, this should be done by a professional.* This can cause all sorts of problems, such as the pump impellers going the wrong way or causing damage to the pump. The higher the horsepower rating, the greater the voltage outlet needed. For example, in the 10- to 20-hp range, the electrical configurations may require a shift from 220 to 440 volts. This gets into the hundreds of gpm (thousands of m^3/day) discharge range. At this level, it may be advisable to use a turbine pump system (Figure 9.13).

Example 9.5

A new 1-hp pump was purchased for use during a field camp. The wiring was done on the tailgate of a pickup truck. The pumping test was set up, and no water would come out. The pump was retrieved and placed in a 55-gal drum to test and see if it could pump

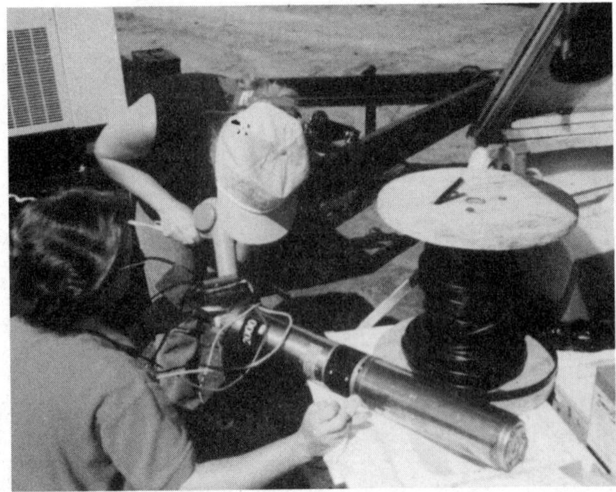

Figure 9.12 Using small pump in the field.

Figure 9.13 Turbine pump yielding 375 gpm.

water. It was determined that the wiring was done such that the impellers turned the wrong way. Once the wires were switched, the pump worked properly. Testing the pump in a drum is a quick way to see if the wiring is correct.

In another occasion, a sampling pump would not work. The impellers were seized up from sand (it is not a good idea to use a sampling pump for well development!). In this case, the pump impellers were cleaned and then put back in upside down. Again no water would come out and the pump got warm. After a careful inspection and comparison to diagrams in the user's manual, the impellers were placed properly and the pump worked once again (Figure 9.14).

Figure 9.14 Fixing the impellers on a small sampling pump.

The wire to the pump is usually spooled on a drum. The pump and wire can be lowered down the well if a broom handle or something like it is placed through the drum for unspooling. Pumps can be very heavy, and a security cable is recommended for retrieval in case it becomes hard to maneuver (Figure 9.15). For a sampling pump, the riser pipe or hose is already attached to the pump. In a submersible pump, lengths of riser pipe must be fitted into the top of the pump, extending upward. Each piece should be tightened with wrenches, with Teflon tape wrapped reverse style on the threading so it won't *bunch up*. Once the appropriate pumping depth is achieved, a 90° elbow fitting is needed to extend the discharge line away from the test site, preferably beyond the expected extent of the cone of depression or to a drain, if available (Figure 9.16).

Some pumps must be primed or have a chamber filled with water so they will produce water (Figure 9.17). This brings up a topic known as net positive suction head (NPSH). On the suction side of a pump, the NPSH is determined from subtracting the negative pressures (suction lift, friction loss, and vapor pressure) from the positive pressures (a function of atmospheric pressure) (Figure 9.18). The atmospheric pressure is a function of altitude (Table 9.4). If the distance between the intake and the pump chamber is too great, the pump will never produce water, no matter how large the motor is. The vapor pressure requirement is not needed for temperatures greater than 60°F (Jacuzzi 1992). The pump design varies by manufacturer.

Figure 9.15 Installing a 5-hp pump, with a security cable attached to prevent loss down the well.

Figure 9.16 Using a 90° elbow on a discharge line directed away from the well.

Figure 9.17 Sump pump requiring priming before successful function.

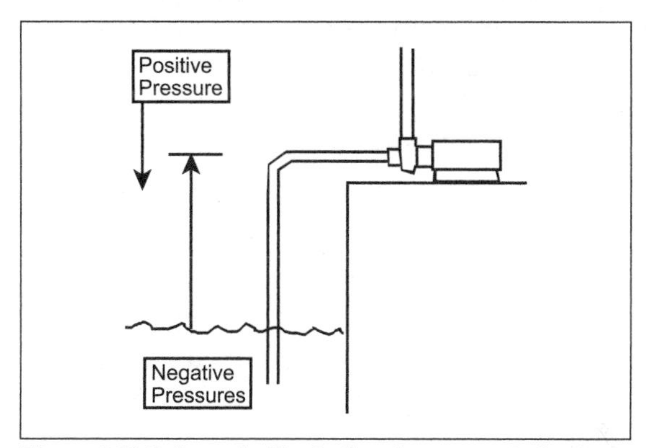

Figure 9.18 Schematic illustrating the concept of net positive suction head.

Table 9.4 Variation of Atmospheric Pressure as a Function of Altitude [From Jacuzzi (1992)]

Units	Sea Level	2,000 ft	4,000 ft	6,000 ft	8,000 ft	10,000 ft	12,000 ft
$lb/in.^2$	14.7	13.7	12.7	11.8	10.9	10.1	9.3
In. Hg	30.0	28.0	25.9	24.1	22.3	20.6	19.1
Ft Water	34.0	31.7	29.4	27.3	25.2	23.4	21.6

Data Loggers and Transducers

Data loggers are a vast improvement over manual methods, but can cause some field workers to become complacent and careless. It is always a good idea to collect manual backup measurements. One of the blessings of the data logger/transducer configuration is best illustrated by the effort required to collect single-well test data. In the "good old days," the trick was to finagle an E-tape down past the riser pipe and pump wiring in search of the pumping level. This would often result in hanging up the E-tape and missed data. The test would often have to be repeated several times with no guarantee of success.

Hydrogeologists should have access to data loggers and transducers to collect pumping-test data. Several companies have rental arrangements; however, if several pumping tests are expected to be run on a regular basis, it is wise to consider buying some equipment. The payback on the investment will take place in short order.

Transducers can come with a cable spool and extra cable lines greater than 350 ft (107 m) (Figure 9.19). The transducer should be lowered into the well within the depth range for the transducer and below the expected range of drawdown. The higher the transducer range reading, the less sensitive (lower precision) they become. It doesn't make much sense to place a 100 psia transducer in a well that may only detect 1 foot (0.3 m) of drawdown. An example of typical ranges are shown in Table 9.5. Drawdown in the hundreds of feet range occur in mine shafts and large production wells. The principle is to select a transducer based on water depth and the expected drawdown.

Figure 9.19 Transducers with cabled lines exceeding 300 ft (100 m).

Table 9.5 Transducer Ranges in psi Referenced from Atmospheric Pressure

Transducer Range in psia	Range in ft	Range in m
5	11.5	3.5
10	23.1	7.0
15	34.6	10.6
20	46.2	14.1
30	69.3	21.1
50	115.5	35.2
100	231.0	70.4
200	462.0	140.7
300	693.0	211.1
400	924.0	281.4

Once a transducer is lowered into a well, a loop (bite) greater than 1-in. should be made to make sure that the cable's air line is not kinked. The cable should then be wrapped around the well casing one or more times and then secured in place with duct tape (Figure 9.20). The cable then is run to the data logger and connected. It is a good idea to lower the transducer down the well early in the setup phase to allow for stretching and straightening of the cable. The transducer in the pumping well should be taped (electrician's or duct tape) to the riser pipe approximately 1 foot above the pump, with additional tape added every 5 ft (1.5 m) to the top of the well (Figure 9.21). If the cable must cross a road or a heavy traffic area, it is advisable to pass it under the road, if a culvert is available. Otherwise, two boards can be duct-taped to the road with the cable placed in a space between them, allowing for traffic to drive over the boards without impacting the cable directly.

Data loggers are usually capable of collecting data from multiple transducers. Each transducer needs to be registered and enabled. Transducer parameters, such as scale, offset, linearity, type, and ID number are required. The reference level is usually the last thing keyed in before starting the test. Extra care should be taken to ensure that the jumper cables are well connected to the data logger (please note that these are not the cables used to charge a dead vehicle battery). This can be checked by using the data logger to take a reading from the transducer. If "no signal" occurs or the reading is zero, check the connection and try again. The connections generally only work in one orientation. Look at the prongs of male and fe-

Figure 9.20 Transducer cable with greater than 1-in. bite for airflow.

Figure 9.21 Taping transducer cable to a riser pipe in a pumping well.

male connections to see if any wires may be bent. Many transducers have a strain gage near the probe tip that has a screw on protective cap.

It is a good idea to unscrew the cap and make sure this area is clean. The following suggestions may explain why a transducer may be giving an improper signal or no signal at all:

- Check all connections.
- Check for cable damage—an unsecured cable can fall out of a truck and become damaged on the highway.
- Check the strain gage end—strange field readings have been cleared up by unscrewing the end cap and washing the cap and strain gage. This can happen if there is some silty material in the bottom of the well that is not cleaned off.
- Make sure the transducer is above the bottom of the well and free hanging.
- Some transducers have breather lines in their cables to sense changes in air pressure—transducer cables should have a loop of at least 1-in. diameter, so that the cable does not become kinked. Kinked cables can result in strange readings. New transducers are vented to avoid this problem.
- The cable may be placed in water deeper than the range of the transducer—this will result in a reading of zero and may break the strain gage.
- Check for damage to the probe—it may need to be shipped off to the manufacturer for recalibration.

Once the transducers are connected to the data logger and all the transducer information has been entered, it is time to set the reference level. This can be an elevation or an arbitrary point. It is *not* a good idea to use 0.0 as the reference level, since negative readings may occur. This does not bode well when plotting on log or semi-log graphs (since negative log values are undefined). The user has the option of referencing all data from a positive downward or positive upward position. For example, top-of-casing (TOC) readings are referenced so that when water levels drop (like drawdown in a pumping test), the changes are recorded as positive downward. Surface referenced readings decrease downward. Readings can be set to log cycle or linear time scale. Usually after four log cycles have been recorded, a default linear scale is activated. Linear scales are useful for long-term monitoring or if readings for early-time details are not needed. Log-cycle scales allow one to obtain early-time data, which can capture the elastic response in unconfined aquifers (Chapter 11). Log cycles may have an interval sequence similar to those shown in Table 9.6.

In addition to pressure transducers, many companies provide additional probe types for collecting other kinds of data. For example, water-quality probes are available to collect pH, specific conductance, temperature, or specific ion data. This can be helpful, along with the drawdown data, to detect sources of recharge or for other interpretations. Some self-contained units have multiple capabilities of level and water-quality data all in one. These can be programmed directly by attaching a cable to a laptop computer. Access to the data during the pumping test is also possible (Figure 9.22).

Table 9.6 Typical Log Configuration of Data-logger Readings

Log Cycle	Time	Number of Readings
1	0–2 seconds	20
2	2–20 seconds	20
3	20 seconds–2 minutes	20
4	2–20 minutes	20
4	20–memory limit	linear every 20 minutes

Figure 9.22 Evaluation of real-time data on a laptop computer.

E-tapes

Electrical water-level meters or E-tapes have many uses at a pumping-test site. They are usually needed to take level measurements in all of the wells prior to lowering the transducers and the pump. They can be used to sound well depths for wells with missing well logs and even measure the distances between wells if a regular measuring tape is not available. One can determine the appropriate psia range for transducers from well logs and the depth to water. E-tapes are essential for backup readings. Backup readings are needed because they can "save" a test if the equipment malfunctions or the power supply unexpectedly shuts down. E-tape measurements also help keep the field person focused and occupied. Kitchen timers are useful to alert the field person when the next reading should be taken. A discussion of E-tapes and their use and design is presented in Chapter 5.

Discharge System

The riser pipe provides a pathway for pump discharge. Once it reaches the well head and is connected with an elbow to convey the water away from the pumping well, it becomes the discharge line. The discharge line should extend far enough to not affect the cone of depression. Sometimes steady-state conditions are reached because the discharge line is too close to the pumping well. In this case, the pumped water recharges the aquifer by infiltrating back from the surface.

For smaller sampling pumps, the discharge line is essentially a garden hose. Additional hose can be attached to extend the line outward. Control of discharge can be made by attaching a spigot to the end of the hose. The spigot acts as a mini gate valve to control discharge. Pumps are designed to deliver a certain amount of flow per head loss. It is a good idea to have a pumping rate that is less than the capabilities of the pump. Back pressure can be placed on the pump by partially closing a gate valve or ball valve while maintaining a steady discharge rate. As drawdown continues during the test, the lift required by the pump may increase. Discharge can be maintained by opening the valve slightly. *There is much more control of this process with gate valves than with ball valves.*

One of the assumptions required for well hydraulic theory is that the discharge (Q) is constant (Chapter 10). One of the requirements of a pumping test is to configure the discharge system so that a constant discharge rate can be maintained. This requires frequent checks of pumping rates throughout the test. The first 15 minutes or so of the test are especially critical, until the drawdown stabilizes and valve adjustments are less frequent. For discharge rates up to 100 gpm (545 m^3/day) or so, a container and a stopwatch can be used. This can get the field person very wet if the container is too small. For higher discharges, a U.S. 5-gal (18.9-L) bucket will be too small. Just because the bucket is supposed to be five U.S. gallons, that doesn't mean that it is. These have been observed to range from 4 to 5.5 gal (15.1 to 20.8 L). It is important to calibrate a bucket with a mark for every gallon, or some other desired volume, using a permanent marker. This way, there is some confidence in the readings. Time can be measured with a stopwatch or a watch with a second hand. The greater the discharge, the greater the precision needed for time. Table 9.7 may be helpful.

Table 9.7 Reference of Time to Discharge Rates in gpm and L/sec

Time to Fill 5 gal in sec	Discharge Rate in gpm	Discharge Rate in L/sec	Time to Fill 3 gal in sec	Discharge Rate in gpm	Discharge Rate in L/sec
2	150	9.47	1	180	11.36
3	100	6.31	2	90	5.68
4	75	4.73	3	60	3.79
5	60	3.79	4	45	2.84
6	50	3.16	5	36	2.27
7	42.9	2.71	6	30	1.89
8	37.5	2.37	7	25.7	1.62
9	33.3	2.10	8	22.5	1.42
10	30	1.89	9	20	1.26
12	25	1.58	10	18	1.14
14	21.4	1.35	12	15	0.95
16	18.8	1.19	14	12.9	0.81
18	16.7	1.05	16	11.3	0.71
20	15	0.95	18	10	0.63
22	13.6	0.86	20	8.2	0.52

Another method to measure discharge is with an in-line flow meter. These are usually installed in the discharge line within 10 ft (3 m) or so of the pumping well and towards the well from the control valve. The control valve creates a back pressure and fulfills the requirement of the pipe being full flowing for readings to be accurate. In-line flow meters may have impellers that rotate to calculate a velocity and are calibrated to read a Q in gpm. They are accurate within a small percentage of the flow and are convenient because all you have to do is read the gauge.

For larger flow volumes, such as those from a high-capacity submersible or turbine pump, a different arrangement is often used. A circular orifice weir is fitted at least 6 ft from the gate valve within the discharge line (Driscoll 1986). An access nipple is fitted to the discharge line at midpipe, 2 ft (0.6 m) towards the gate valve from the orifice plate. To the nipple a manometer tube is attached that extends upward to be measured on a scale (Figure 9.23). The manometer measures the pressure (head) in the pipe and should be made of clear tubing so that its height can be observed.

Figure 9.23 Monometer tube setup with circular orifice weir to measure flow.

Duct Tape

Duct tape is one of those essential items that is helpful in setting up a pumping test. Duct tape is used to attach transducers to well casing. It can be used to hold braces together to support discharge lines, patch things together, and do a host of other jobs. Any serious user of duct tape should take a look at *The Duct Tape Book* by Berg and Nyberg (1995). As an example pertinent to hydrogeology work, consider Figures 9.24 and 9.25.

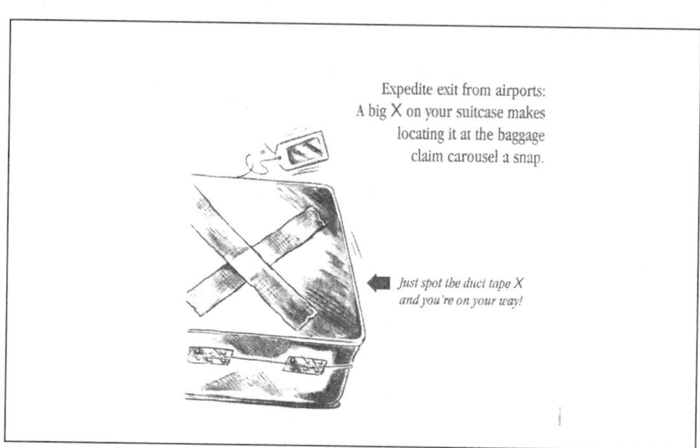

Figure 9.24 Hydrogeologists are always on the go. [After Berg and Nyberg. Used with permission from Pfeifer-Hamilton Publishers (1995).]

Figure 9.25 Duct tape holding a drill rig cab seat together.

Setup Procedure

Now that the essential equipment have been described, it may be useful to list the tasks generally associated with pumping tests. It is assumed at this point that all observation wells have been drilled, completed, and developed, and permission to be on the property has been obtained. One should have a lamp or source of light that can be plugged into the generator for lighting and flashlights (with extra batteries) for taking manual readings during the night. Having made the appropriate arrangements for food, supplies, and restroom facilities or designated areas will help make the time go smoother. A portable field table with chair and a tarp for inclement weather conditions are also recommended. Here is the list:

1. Gather together the equipment listed at the beginning of this section. Obtain extra fuel.

2. Check the weather forecast and bring any essential clothing for the anticipated conditions.

3. Measure and record the water levels in all wells to be used in the test. Monitoring of wells for a few days prior to the test is helpful to determine if any regional trends are occurring (general decreases or increases in water levels).

4. If there are any nearby surface streams, establish the stage with a physical marker, so that it can serve as a gauging point should changes in stage occur.

5. Offload the equipment and begin placing the transducers in the wells and securing them in place with duct tape.

6. Place the pump and riser pipe in the well along with the transducer designated to measure levels in the pumping well. This transducer is needed to determine how close the pumping level is to the pump intake.

7. Attach and support the discharge line so that the discharge water is a suitable distance from the pumping well (Figure 9.23). Configure the discharge line for the appropriate discharge volume (e.g, in-line flow meter or manometer and scale), or get the discharge measuring device (calibrated bucket) and place it near the discharge point if another measuring system is not being used.

8. Attach all transducers to the data logger with connecting cables. The data logger should be placed fairly close to the generator, so that they can both be activated simultaneously at start time.

9. Set up the test in the data logger by giving the test a name and enter all the transducer information. Make sure there is a good connection with each transducer by taking a reading of each one at the data logger. Do the readings make physical sense? If not, then check out why.

10. Run a step-drawdown test or test the pumping rate so that the gate valve or discharge control valve can be closed back to the appropriate discharge rate. Once there, leave it in the desired position.

11. Prepare fieldbooks for frequent readings (listed below) and synchronize all watches (Chapter 1, Section 1.8).

12. Prepare personnel with E-tapes for manual backup readings; one E-tape dedicated for each well (or two) is advised. If you are working alone, have an E-tape next to each well for ready use.

13. Make sure any safety issues are taken care of (listed below).

14. Prepare a signal so all parties can begin at the same time. The most critical is activating the data logger and the pump at the same time. A "1, 2, 3, go!" system works well.

15. Once the test has begun, frequent manual readings are necessary. If there are two of you, it is helpful to have one call out the readings and the other keep track of time and record the numbers. The pumping rate in the calibrated bucket should also be measured or the discharge read from the flow meter or manometer tube. If manual read-

ings coincide with data logger recordings splash effects may distort the data-logger value.

16. When the time spacing between readings gets longer (15 to 30 minutes), plot the data or graph it to see what is happening. One should never stop a test based on time alone. Look at the data and see whether boundary conditions or delayed yield or other effects are indicated in the data. When sufficient data have been collected, prepare for recovery mode.

17. If the data logger has a step function to go directly into recovery log mode, all that needs to be done is to synchronize the step-function button and deactivate the generator or pump. Some data loggers require that another test be defined and prepared beforehand to activate the recovery phase. By starting the "new" test, the "old" one is stopped automatically.

18. Once again, fieldbooks need to be prepared for frequent readings (listed below) and all watches synchronized.

19. The signal is given and the data logger and pump are shut off. Immediately recovery mode manual readings are taken and recorded.

20. Make sure to plot the data during the test when time allows. There is a tendency to stop the recovery phase too soon. It is an excellent additional source of information, so it should be collected. Plan accordingly and be patient.

Frequency of Manual Readings

The majority of drawdown takes place during the first hour or so of the test. Continued drawdown does occur, especially if a barrier condition is encountered; however, changes are usually not as dramatic. The same is true for recovery mode. The initial hour after pumping stops is when recovery occurs most rapidly. At these times, if extra personnel can be on hand to help with manual readings, it can make a big difference in reducing stress. The flurry of activity at the beginning of the pumping test and the beginning of recovery is similar to a WFD.

The frequency of readings depends on the response of drawdown in the observation wells. Wells farther away may not "see" a response for several minutes. This depends on the completion depth of the wells and the geology (Example 9.2). It also depends on whether confining or semi-confining conditions are possible. Low (confined) storativities correspond to a rapid expansion of the cone of depression (Chapter 10). The key is to watch the changes in the data. Usually, 0.03 to 0.05 ft (0.91 to 1.5 cm) is a fair indication that the drawdown is real. When drawdown changes between readings drop to approximately 0.02 ft (0.6 cm) per reading, one should advance to the next time interval (Table 9.8). As a general guide, the schedule shown in Table 9.8 is recommended, and then it should be adjusted according to what is observed.

Table 9.8 Recommended Time Intervals for Manual Readings in a Pumping Test

Time Since Pump is Turned on or off, in min	Length of Time Interval, in min	Comments
0–10	0.5	Readings can be staggered when observation wells are being measured.
10–15	1.0	
15–30	2.0	
30–60	5.0	All later readings should be within 1–15 seconds of the appropriate time interval.
60–90	10.0	
90–120	15.0	
120–360	30.0	
36–1,440	60.0	
1,440–end of test	240.0	

Safety Issues

As with any activity, there are always safety issues. Common sense will generally help one to be alert to most problems. A few specific items will be mentioned from experiences by the author.

- Make sure the fuel containers are placed away from the exhaust of the generator. During one test, it was observed that part of the plastic fuel container was beginning to melt.
- Make sure vehicle exhaust is away from where people are working. Notice the wind direction and park in a safe orientation.
- Make sure the generator and data logger are protected from rain and placed on a surface with good drainage so that puddles of water do not accumulate .
- Secure the cables from the transducers so tripping hazards are avoided.
- Secure the area, if possible, to keep children and animals away from the working area (Figure 9.26).
- Use a funnel when refueling the generator during a test. Splashing fuel can be a problem.

Figure 9.26 If possible, children and animals should be kept away from the work area.

- Have lighting at night and a first aid kit on site.
- Don't panic.
- Don't work alone. This way someone can go for help.

Once again, common sense will help one to avoid most problems. The buddy system is always advisable for keeping each other alert or making runs for supplies. It is terrible to have a problem with no one around to help.

9.5 Things That Affect Pumping Test Results

If none of the field conditions change from the beginning to the end of a pumping test, you are very fortunate. During the course of most pumping tests, some type of condition change may occur that affects the interpretation of the results. These changes should be recorded. If one is not paying attention to changes during the test and writing them down, they may not be accounted for during the interpretation phase (Chapter 10).

Weather and Barometric Changes

Consider the following scenario. At the beginning of a pumping test, it is sunny and clear. About 4 hours into the test, a storm front moves through. The sky is dark and a steady rain falls. During the night, the storm passes over and the sky clears again. A nearby stream increases in stage to 1 ft (0.3 m) higher than at the beginning of the test and subsequently drops 2 ft.

If an aquifer is confined to semi-confined, the water levels in wells are affected by barometric changes. Sunny and clear weather accompanied by high barometric pressure pushes down harder on the water surface in wells, thus decreasing the relative water elevation. When a storm comes into the area, the relative air pressure decreases. Thus the air is not pushing as hard on the water surface, allowing the water level to rise. Personally, the author has observed water-level rises as much as 1 ft when a storm has come in. During a pumping test, the water levels are supposed to be decreasing if they do not encounter any recharge sources. It may be that a recharge source is being interpreted when it is only a storm front affecting the data. This is where field notes come in handy.

Other Apparent Sources and Sinks

Other physical phenomenon occurring in the field may affect the interpretation of pumping-test results. Nearby wells that are pumping and shut down during a pumping test may affect the drawdown results. Since overlapping cones of depression are additive, then the drawdown rates are accelerated when nearby wells are pumping. The interpretation of the transmissivity will be lower than the actual value. When a nearby well stops pumping, the time/drawdown curve will flatten, and the transmissivity will appear to be greater or the curve will suggest that a source of recharge has been encountered. Conversely, if a well turns on that was previously not running during the pumping test, its cone of depression may overlap and it may appear that a barrier condition has occurred.

Example 9.6

A pumping test was being conducted to see whether a mining company supply well would impact other local residents. The geologic setting is depicted in Figure 9.27. The pumping well was completed within the granitic bedrock (granodiorite), and the residential wells were completed in both the granodiorite and in the surficial sediments. The sediments ranged in thickness from 30 to 55 ft (9.1 to 16.8 m) in the pumping-test area. In this instance, any resident within 2,000 ft (610 m) of the pumping well could have their wells monitored. Approximately seven residential wells were monitored by students. The instructions given to the residents were that no water use should take place during the test, which would begin at 8 a.m. on a Saturday morning. During the test, one of the observation wells seemed to be recovering the whole time. None of the wells completed within the sediments indicated a response except this well. After conferring with the well owner, it was discovered that someone used the bathroom earlier that morning and

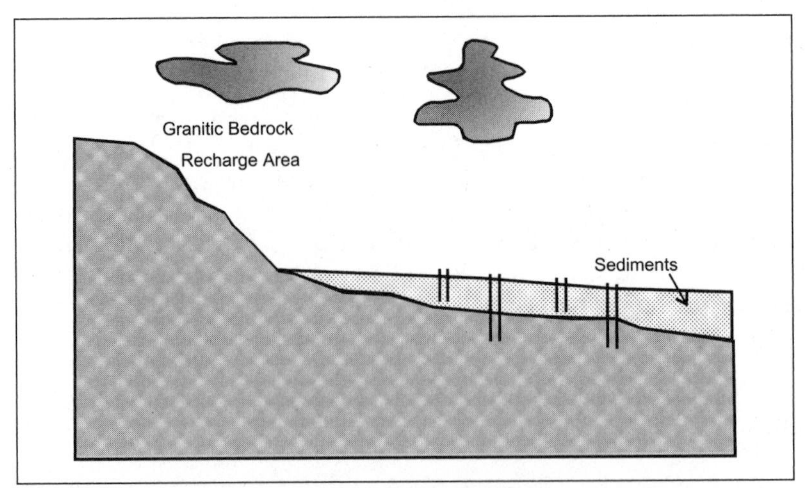

Figure 9.27 Geologic setting for Example 9.6. Residential wells are completed in both the granitic bedrock and the surficial sediments. (Not to scale.)

what was being observed was a recovery response from pumping. Wells completed within the granodiorite within 1,100 ft (335 m) did indicate a drawdown response.

Another factor that can affect the drawdown response is a change in stage of a local stream. At the beginning of the test, the stage of the stream should be marked so that its stage is known. If changes in stage occur during the test, they can be noted.

Did it rain during the test? This could be a potential source of recharge in a shallow unconfined aquifer. Is the discharge line too close to the pumping test site? The author has observed a test reach steady-state conditions merely because someone failed to conduct discharge beyond the cone of depression.

Another consideration is the proximity of railroads and heavy traffic areas to shallow confined systems. A documented case in Long Island, NY, showed a hydrograph response near a train station (Jacob 1940). The weight of the train compressing the aquifer resulted in water-level rise, and when the train departed, the water level would drop once again. Be alert to physical changes that may occur during a pumping test. Write them down and use your field notes during the interpretation phase. Additional discussion on the interpretation of pumping-test results is presented in Chapter 10.

9.6 Summary

Pumping tests are valuable field methods in estimating the hydraulic properties of aquifers. The time and effort required to perform a pumping test can provide information about the aquifer that cannot be obtained from slug tests or specific capacity tests. Pumping tests can be stressful because they require a fairly large list of equipment and many things can go wrong. A check list is useful to remind field personnel what needs to be brought and tasks that need to be done.

It is important to have an understanding of the geology so that observation wells can be oriented appropriately and completed at the proper depths. Appropriate pumping rates are needed to stress the aquifer so that drawdown can be observed. Transducers should be selected that will be compatible with the water depths and the expected drawdown ranges. Discharge systems should be designed to maintain a constant discharge rate and obtain accurate information. The discharge line should be configured with a constricting device, so that it can be opened to accommodate more flow should drawdown conditions cause the flow rates to decrease. Extra personnel are helpful to have around at the beginning of a pumping test to help set up equipment and collect the early-time manual readings. Changing weather and other field conditions that occur during the test should be recorded in the field books for use during the interpretation phase. The flurry of activity that can occur during a pumping test reminds the author of a WFD.

References

Berg, J., and Nyberg, T., 1995. *The Duct Tape Book.* Pfeifer-Hamilton, Duluth, MN, 64 pp.

Birsoy, Y. K., and Summers, W. K., 1980. Determination Of Aquifer Parameters From Step Tests And Intermittent Pumping Data. *Ground Water.* Vol. 18, pp. 137–146.

Donovan, J.J., 1983. *Hydrogeology and Geothermal Resources of the Little Bitterroot Valley, Northwestern, Montana.* Montan Bureau of Mines and Geology Memoir 58, 60 pp. and two sheets.

Donovan, J.J., 1985. *Hydrogeologic Test Data for the Lonepine Aquifer, Little Bitterroot Valley, Northwestern Montana.* Montana Bureau of Mines and Geology, Open-file Report 162, 9 pp.

Driscoll, F.G. 1986. *Groundwater and Wells.* Johnson Screens, St. Paul, MN, 1089 pp.

Fetter. C.W., 1993. *Contaminant Hydrogeology.* Prentice-Hall, Upper Saddle River, NJ, 500 pp.

Heath, R.C., and Trainer, F.W., 1968. *Introduction to Ground Water Hydrology.* Waterwell Journal Publishing, Worthington, OH, 285 pp.

Jacuzzi, 1992, *Jacuzzi Brothers Hydraulics Training Seminar Water Systems Workbook.* 40 pp.

Jacob, C.E., 1947. Drawdown Test To Determine Effective Radius of Artesian Well. *Trans American Society of Civil Engineers,* Vol. 112, Paper 2321, pp. 1047–1064.

Jacob, C.E., 1940. Correlation of Groundwater Levels and Precipitation in Long Island, NY, *Transcripts of American Geophysical Union,* Vol. 24, Pt.2, pp. 564-573.

Journal of American Water Works Association (JAWWA), Vol. 77, No. 9, September 1985.

Kruseman, G.P., and de Ridder, N.A., 1990. *Analysis and Evaluation of Pumping-Test Data. 2nd ed.* International Institute for Land Reclamation and Improvement, Wageningen, The Netherlands, Publication 47, 377 pp.

Roscoe Moss Company, 1990. *Handbook of Ground Water Development.* John Wiley & Sons, New York, 493 pp.

Walton, W. 1970. *Groundwater Resource Evaluation.* McGraw-Hill, New York, 664 pp.

Chapter 10

Aquifer Hydraulics

One of the key pieces of information hydrogeologists need to solve problems in the field of hydrogeology is an understanding of the hydraulic properties of an aquifer. Many questions, such as, how long will it take water from the recharge area to reach a production well? How long will it take a contaminant to move from point A to point B? If this production well is activated, how far will the cone of depression reach? And how many other wells will be affected? What distribution of hydraulic properties should be assigned to the layers within this groundwater-flow model? In order to answer such questions, it is necessary to perform pumping tests (Chapter 9) and slug tests (Chapter 11) to measure the aquifer stress response in wells over time. This chapter begins by discussing the traditional analytical methods followed by a discussion of applications when the data do not fit the ideal case. Unfortunately, this is the more common scenario found in the field. It is beyond the scope of this chapter to provide an in-depth discussion of aquifer hydraulics, but it hoped that the presentation here will be useful. There are excellent discussions in Hantush (1964), Lohman (1979), Driscoll (1986), and Kruseman and deRidder (1991). Slug-test analytical methods are all presented in Chapter 11.

The Theis (1935) method is used first to evaluate time-drawdown data to see what they look like. There is a danger in some individuals immediately launching into a more complicated analytical method without first attempting a Theisian (1935) fit. Any pumping-test data set should probably be evaluated with a Theis (1935) analysis. If the graphical image does not fit or make sense, alternative methods may be needed. Data are also easily evaluated using the Cooper-Jacob method (1946); however, more assumptions are required. Additional topics covered in this chapter include distance-drawdown relationships, image-well theory, and a discussion of the application of pumping-test data to aquifer properties. Explanations for deviations from the Theis (1935) curve are offered along with field examples. Dual porosity and fracture-flow data sets have their own characteristic response. These are also presented with field examples.

10.1 Wells

Wells provide a point of access to the water-bearing materials, and are one of the most important attributes of groundwater studies. When properly designed and constructed (Chapter 8), wells provide useful information about the characteristics of an aquifer. Characterizing the physical properties of aquifers is the study of **aquifer hydraulics**.

The water level in a well, when no pumping is occurring, is known as the **static water level** (SWL). The SWL reflects the total head at the midpoint of screened intervals of a well (Chapter 3). (It is recommended that the reader refer to Chapter 5 in the interpretation of water-level data). If the well is open-hole completion (open at the bottom with no screen) it is considered to be representative of the head at the bottom of the well. Typically, SWLs are measured from the top of the casing (TOC). Level surveys are then performed to determine the elevation of the TOC and the ground surface level (GSL) (Chapter 5). Once the reference level has been defined, the change in water level with time can be measured during a pumping or aquifer test to evaluate the hydraulic properties of an aquifer. The difference measured between the nonequilibrium water level at a particular time and the initial SWL is known as **drawdown** (Driscoll 1986)(Figure 10.1).

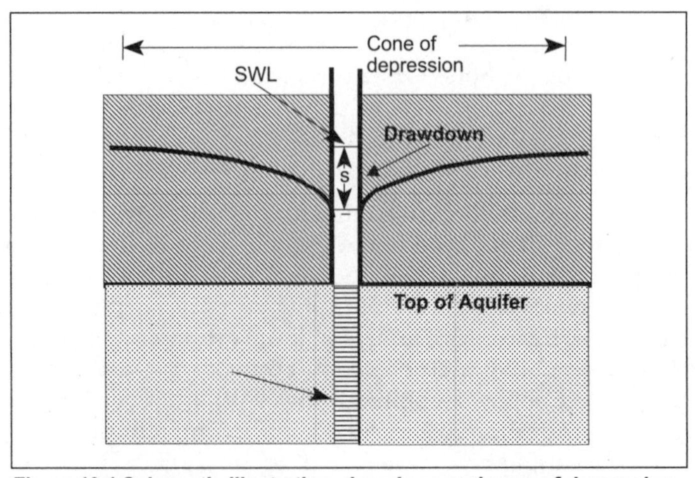

Figure 10.1 Schematic illustrating drawdown and cone of depression in a confined aquifer.

Cone of Depression

When pumping begins, the water level in the pumping well is lowered. This lowering (measured as drawdown) induces a gradient or slope all the way around the well bore known as the **cone of depression**. Flow proceeds radially towards the well, much as the spokes of a bicycle converge on the hub. The cone of depression extends outward until it can capture enough water to meet the demands of the pumping rate (Q). The shape of the cone of depression is estimated by observation wells that also show a drawdown response. If these are located in different azimuth orientations away from the pumping well (Chapter 9), one can evaluate anisotropy by plotting the drawdown and contouring it.

The cone of depression in a confined aquifer (Figure 10.1) occurs within the potentiometric surface above the top of the aquifer, whose position is determined by the SWL in cased wells. This means that the saturated thickness (b) is always maintained. Recall also that the potentiometric surface is an imaginary surface that reflects both the pressure head and elevation head (Chapter 5).

The cone of depression in an unconfined aquifer represents a physical draining of the porous materials near the pumping well, creating an actual depression in the water-table surface (Figure 10.2). It is indicated by the respective drawdowns observed in nearby observation wells. The shape and extent of the cone of depression depends on the pumping rate (Q) and the nature of the geologic materials. Finer-grained materials produce a steeper cone of depression because a steeper hydraulic gradient is required to

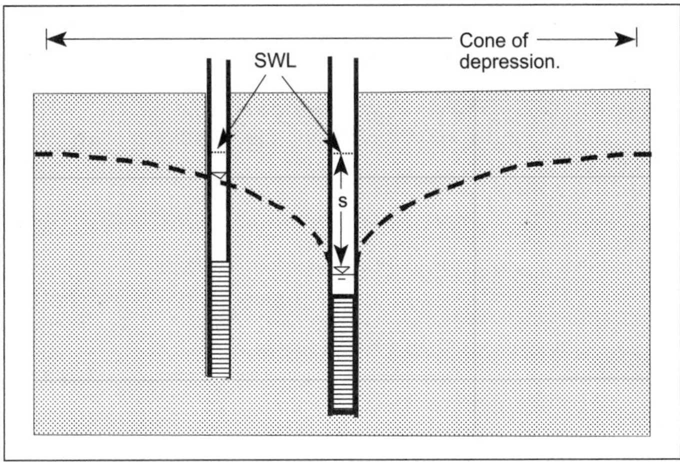

Figure 10.2 Schematic illustrating drawdown and cone of depression in an unconfined aquifer.

move groundwater toward the pumping well. Pumping within transmissive materials produces a flatter cone of depression (Figure 10.3) that may extend further out. The relationship between the steepness of the cone of depression and the transmissivity is a proportional relationship between the hydraulic conductivity and the hydraulic gradient expressed in Darcy's Law (Chapter 5). How this is applied can be illustrated in Example 10.1.

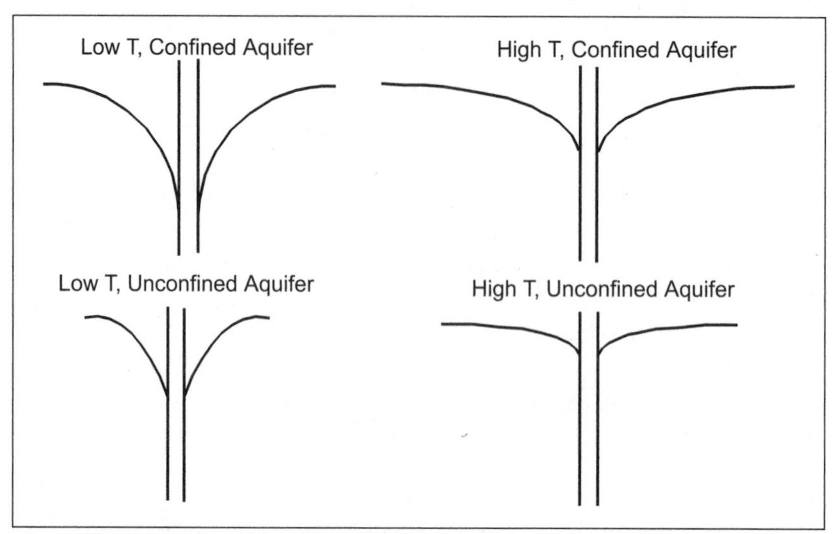

Figure 10.3 Schematic of the differences in the shapes of cones of depression based on aquifer properties.

Example 10.1

Within the Butte, Montana, area are a number of Superfund operable units associated with mining operations. During the remediation design and subsequent construction phase, mine tailings were to be excavated and removed to another location. It was required that dewatering of the mine tailings and the underlying alluvial materials take place to perform the excavations.

Adjacent to the excavation operation is another Superfund operable unit associated with "organics" contaminated groundwater resulting from chemicals used in the timber treating process. The chemicals were used in treating support timbers used as underground mine supports. In this operable unit, the shallow groundwater system became contaminated with a mixture of diesel and pentachlorophenol (PCP). The concern with dewatering the mine tailings operable unit was that if the hydraulic conductivity was high, then the cone of depression from dewatering would extend sufficiently to the southeast, potentially mobilizing the organic plume.

The crux of the problem was centered in the estimated value for the hydraulic conductivity of the alluvium. The hydraulic conductivity for the alluvial materials was believed to be approximately 80 ft/day (24.3 m/day). This was based upon groundwater models that could not reach calibration unless hydraulic conductivity numbers for the alluvium were this high. In the model, the alluvium was represented by a relatively thin package of sediments over impermeable bedrock. This layer could not allow sufficient water through it unless the hydraulic conductivity value was increased over the values found from pumping tests. Personnel from a local state agency (Montana Bureau of Mines and Geology) argued that the pumping tests suggest that the alluvial materials had a lower hydraulic conductivity. They pointed out that the required volume of water could move through the area using a lower hydraulic conductivity if an additional saturated thickness consisting of weathered granitic bedrock be used. The matter was settled by conducting a pumping test screened in the alluvial material. In the results, a hydraulic conductivity of 20 to 30 ft/day (6.1 to 9.1 m/day) was calculated for the alluvium, and a hydraulic conductivity for the weathered granite was estimated to be 1 to 5 ft/day (0.3 to 1.5 m/day). By adding an additional weathered bedrock layer to the model, calibration was reached with the lower hydraulic conductivity value more reflective of the local geology and pumping test results. This permitted the dewatering activities to take place (John Metesh personal communication, Butte, Montana, September 1995).

Figure 10.3 illustrates that a lower hydraulic conductivity results in a steeper, less extensive cone of depression, whereas a higher hydraulic conductivity results in a flatter, farther-reaching cone of depression. In example 10.1, the flatter cone from the higher estimated hydraulic conductivity would potentially reach the organic plume, mobilizing the contaminants. During the actual excavation and dewatering process, it was found that the lower hydraulic conductivity was correct.

Comparison of Confined and Unconfined Aquifers

The size of the cone of depression is also affected by the storage coefficient and aquifer transmissivity. Typical ranges of storage for a confined aquifer are between 10^{-3} and 10^{-6}. Unconfined aquifers have specific yields that range between 0.03 and 0.30 (Fetter 1993). Leaky-confined or semiconfined aquifers fall somewhere in between. The significance of the smaller storage value is that the cone of depression in a confined aquifer will extend faster and farther than in an unconfined aquifer. This is shown in Figure 10.3 and illustrated by the Example 10.2.

Example 10.2

While performing the technical analysis for a source-water protection plan in Ramsay, Montana, a pumping test was conducted on the local public water supply well (Figure 10.4). There are two production wells plumbed into a water tower pressurizing system. The wells were drilled in 1917, and the well logs are nonexistent. It was unclear whether the wells were drilled into younger layered sediments or deeper into the underlying Tertiary Lowland Creek Volcanics. The pump in each well can be activated manually or controlled by an automatic system. The well closest to the water tower (the south well) was used as the pumping well, and the well 510 ft (155 m) to the northwest was used as the observation well. During the test, the pumping well was activated manually after instrumenting the wells with drawdown measuring equipment. The pumping rate was 210 gpm (13.2 L/sec). The north observation well experienced 0.05 ft (1.5 cm) of drawdown within 30 seconds of pumping. This indicates that the cone of depression extended at a rate of at least 17 ft/sec (5.2 m/sec), which is at the rate of a very fast person running towards the well at full speed! A rapid expansion of the cone of depression implies the aquifer is confined!

Figure 10.4 Water tower for the Ramsay, Montana, public water supply.

Drawdown in a confined aquifer is from response to compression of the mineral skeleton and decompression of the water, and this occurs speedily. The cone of depression in an unconfined aquifer is a physical depression in the water table from drainage of the pores, is of more limited extent, and develops more slowly.

Example 10.3

To obtain a mental picture of what is occurring in a confined aquifer, imagine taking a household cleaning sponge (flat rectangular shapes). If this sponge is sandwiched between two pieces of plywood, then sealed with silicone around the sides, it could theoretically represent a confined aquifer. The sponge represents the aquifer materials, and the plywood represents the confining layers. A hole is drilled through the top plywood piece into the sponge, and a soda straw is inserted to represent a well (Figure 10.5). The sponge is saturated and the elevation of the water level in the straw is visible above the top plywood piece. Any slight compression of the "aquifer" results in water rising quickly up the straw. The amount of compression would be almost imperceptible, even though a dramatic expulsion of water takes place. Lifting your hand off of the upper plywood piece results is the water level immediately dropping back to static conditions. This represents the elastic-like response of aquifers.

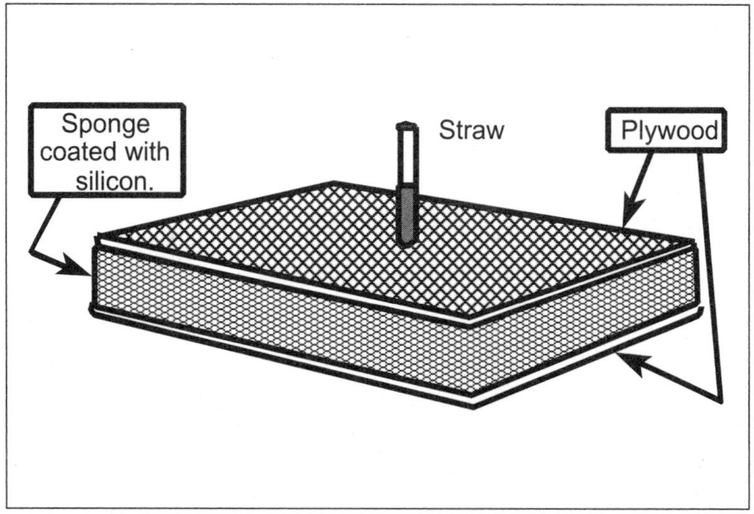

Figure 10.5 Model of a confined aquifer made with a sponge sandwiched between plywood and sealed with silicon calking. A hole is drilled and a straw inserted to represent a well.

The equations describing nonequilibrium drawdown with time were first attributed to C. V. Theis (1935). In analyzing pumping-test data sets, the usual first step is to perform a Theis analysis to see what the data look like.

10.2 Traditional Pumping-Test Analytical Methods

In the early 1930s, C. V. Theis noticed that there was a relationship between drawdown and well yield with time. He approached a friend named Lubin, a mathematician at the University of Cincinnati, and described his problem (Cleary personal communication, San Francisco, CA, 1993). Lubin stated that he had a solution that was well known from the heat-flow literature. The analog to a well in an aquifer was compared to a wire within a cube of steel, heated by a battery. Lubin said that the solution of the radial heat-flow equation being integrated was dependent upon the following assumptions:

- No edge effects (aquifer is of infinite areal extent).
- Uniform thickness (pumped well fully penetrates and receives water from the full thickness of the aquifer).

- Constant heat source (constant pumping rate).
- Homogeneous and isotropic (aquifer is uniform in character and the hydraulic conductivity is the same in all directions).
- No sources or sinks (water is discharged instantaneously from storage and not from external sources either adding or removing water).

Additional assumptions inherent in the Theis equations listed in Driscoll (1986) are:

- Pumping well is 100% efficient.
- Laminar flow (darcian) flow prevails throughout the well and aquifer.

The solution is known as the Theis equation:

$$s = \frac{Q}{4\pi T}\left[-05772 - \ln u + u - \frac{u^2}{2\times 2!} + \frac{u^3}{3\times 3!} - \frac{u^4}{4\times 4!} + - \ldots\right] \quad \textbf{[10.1]}$$

where:

Q = pumping rate (L^3/t)

T = transmissivity (L^2/t)

S = drawdown (L), and

[ ] = $W_{(u)}$, part in brackets is known as the well function

Note: The units need to be worked out correctly to get the appropriate drawdown in units of length!

The well function is a function of u, where:

$$u = \frac{r^2 S}{4Tt} \quad \textbf{[10.2]}$$

Note: u is dimensionless!

Where:

r = radial distance to the pumping well (L)

S = storage coefficient (unitless)

T = transmissivity (L^2/t)

t = time (in days or minutes)

The infinite series in the brackets of Equation 10.1 is known as the **well function ($W_{(u)}$)**. Interestingly enough, Theis (1935) published the solution of a partial differential equation (PDE) that did not exist until Jacob, a graduate student of his at the University of New Mexico, came up with it in 1941 (Cleary personal communication, San Francisco, CA, 1993).

Considering the assumptions given, flow towards a well can be thought of as moving towards the well in a polar coordinate system. In the case where the aquifer is isotropic and homogeneous, with sources or sinks, the radial flow equation can be expressed as:

$$\frac{\partial^2 h}{\partial r^2} + \frac{1}{r}\frac{\partial h}{\partial r} + \frac{W}{T} = \frac{S}{T}\frac{\partial h}{\partial t} \qquad [10.3]$$

where:

r = radial distance from the well (L)

t = time (t)

h = hydraulic head (L)

S = storativity (dimensionless)

T = transmissivity (L^2/t)

W = source or sink term (L/t)

Instead of coming up with Equation 10.3 and solving it numerically, Theis (1935) solved it graphically. The method involves superimposing two curves and picking a common point known as the **match point** (Figure 10.6). The first curve, the **normal type curve**, is a plot of the well function ($W_{(u)}$) (ordinate) versus u (abscissa)(see Equation 10.2) on log-log scale. The data curve represents the field observed drawdown (s) (ordinate) versus r^2/t (abscissa), also on log-log scale. (Please note that the values are being plotted on a log-log scale. No transformations of the data are needed.) Values for the well function and u are contained in Appendix D.

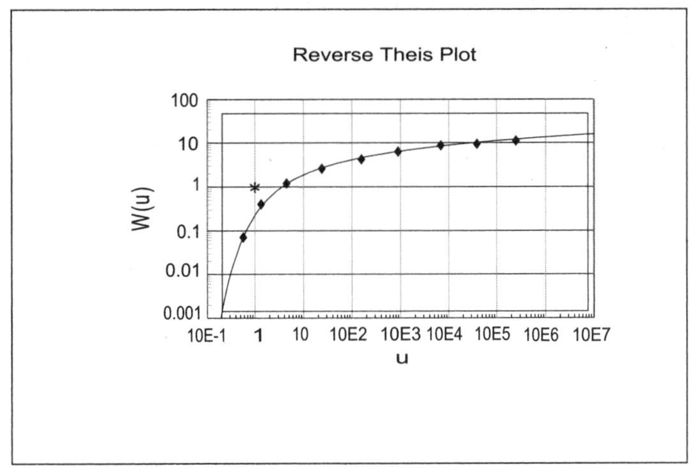

Figure 10.6 Theis method where data plot in drawdown versus time is superimposed on the reverse type curve. Match point is represented by the star where W(u) and u equal one.

A convenient match point is usually selected where $W_{(u)}$ and u both equal 1.0. The corresponding matching value for s and r^2/t can then be read from the match point location to solve for transmissivity from equation 10.1 and then for storage by rearranging Equation 10.2.

Most hydrogeologists today use a **reverse type curve**. Both the type curve and field observed data curve are "reversed" by taking the inverse of the abscissa information. The reverse type curve is a plot of $W_{(u)}$ versus $1/u$, both on log scale. The data curve is a plot of drawdown (s) versus t/r^2, also on log-log scale. The reason for using the reverse Theis curve is because any given observation well will have a constant radial distance (r). This means that the drawdown can be plotted directly with corresponding points in time (usually plotted in minutes and later *converted to days* in the calculations). Plots of more than one observation well on the same plot must use the calculated t/r^2 values for the respective distances (r). Software plots indicate transmissivity in L^2/min. The transmissivity is calculated by Equation 10.4.

$$T = \frac{Q}{4\pi s} W_u \qquad [10.4]$$

where T = transmissivity (L^2/t)

And the calculation for storativity:

$$S = \frac{u4Tt}{r_2} \qquad [10.5]$$

where S = storativity (dimensionless)

Remember to make sure all your units are consistent! Many errors occur in the calculations from not being careful with units! Calculations of transmissivity are usually in ft^2/day, m^2/day, or gal/day-foot. Not being careful with your calculations can lead to erroneous interpretations and poor design. More applications will be given in Section 10.4.

Example 10.4

One of the important requirements during a pumping test is that the pumping discharge rate be constant. This can affect the shape of the Theis curve. During a court case, a homeowner and a well driller were at odds over drilling a well too deep (deep-holing). The total depth of the well was 325 ft (99 m).

At 200 ft (61 m) the well driller told the well owner they had 4 to 5 gpm (13.1 to 16.4 L/min) but were going to "chase it a bit." They chased it to 325 ft while the owner was at work and declared the well a success making 8 gpm (30 L/min). An argument led to the disagreement. The well owner was advised that if another driller came in and pulled the casing, backfilled the hole with gravel, and retested the well, achieving the same production yield, it would indicate the additional footage was unnecessary. The well owner was cautioned that there was great risk in doing this because he may lose the well from materials caving into the well. The well owner was mad enough go ahead with it anyway. Another driller was hired, the casing was pulled, the well was backfilled with gravel to 212 ft, and then plugged with bentonite before reinstalling the well screen.

A pumping test was conducted by the other driller to see what kind of yield the well would make. The pump was set at 187 ft (57 m) and the pumping commenced at 10:00 am. The data collected are presented in Table 10.1.

Table 10.1 Pumping Test Data and Well Yield from Example 10.4

Time	Elapsed Time (min)	Water Level (depth)	Pumping Rate (gal/min)
10:00	0	18´10´´	7
10:25	25	61´5´´	7
10:30	30	65´10´´	7
10:35	35	69´8´´	7
10:40	40	73´9´´	5
10:45	45	79´0´´	7
10:50	50	85´0´´	7
10:55	55	89´10´´	7
11:00	60	93´9´´	7
11:05	65	98´7´´	7
11:15	75	105´8´´	6.5
11:25	85	115´	7
11:45	105	120´	5.5
11:55	115	140´	8.17
12:00	120	144´	7.5
12:12	132	164´	7.2
12:20	140	170´	7.2
12:35	155	174´	7.91
12:40	160	175´	8.3
12:45	165	175´	8.6
13:45	225	175	6

The expert witness representing the driller used the data from Table 10.1 to generate a time-drawdown Theis plot. The shape of the drawdown curve was disjointed, caused by variations in the pumping rate, and the departure from the “type curve” was explained using fracture-flow theory to an uneducated jury. The author thought it prudent to visit the site and conduct his own brief pumping test at 8 gpm (30.2 L/min). A very characteristic Theis curve resulted when the pumping rate was constant.

Cooper-Jacob Straight-Line Plot

Cooper and Jacob (1946) recognized that if u in the Theis equation was sufficiently small, the nonequilibrium equation could be modified to a logarithmic term instead a well function (Equation 10.9). Arguments arise as to when u is sufficiently small. Values range from 0.05 (Driscoll 1986; Fetter 1994) to 0.005 (Fletcher 1997). The software package AQTESOLV (Duffield 1991) indicates where u is 0.01. The smaller u the less the error is involved. We believe 0.02 is sufficiently small. With u sufficiently small, Equation 10.1 can be rewritten as:

$$s = \frac{Q}{4\pi T}\left[-0.5772 - \ln(u)\right] \qquad [10.6]$$

By substituting in u and finding a natural log value for 0.5772, we obtain:

$$s = \frac{Q}{4\pi T}\left[-\ln(1.78) - \ln\left(\frac{r^2 S}{4Tt}\right)\right] \qquad [10.7]$$

remembering that $\ln A = \ln(1/A)$ and $\ln B \times C = \ln B + \ln C$:

$$s = \frac{Q}{4\pi T}\left[+\ln\left(\frac{1}{1.78}\right) + \ln\left(\frac{4Tt}{r^2 S}\right)\right] \qquad [10.8]$$

Factoring out the natural log function and combining terms we obtain:

$$s = \frac{Q}{4\pi T}\ln\left[2.25\left(\frac{Tt}{r^2 S}\right)\right] \qquad [10.9]$$

Since the relationship between $\log_{10}X$ and $\ln_e X$ is $\log_{10}X = 0.43429 \ln_e X$, we obtain:

$$s = \frac{2.3Q}{4\pi T}\log\left(\frac{2.25Tt}{r^2 S}\right) \qquad [10.10]$$

where all units are assumed to be consistent

s = drawdown (L)

S = storage (dimensionless)

T = transmissivity (L^2/t)

t = time (t)

Q = pumping rate (L^3/t)

When one plots drawdown (arithmetic scale) versus time (log scale) a straight line results if the conditions of small u are met along with the other Theisian conditions (Figure 10.7). By evaluating Equation 10.2 one can see that u is small when r (the distance of the observation point) is small or when the time t is large. This eliminates having to work with the well function ($W_{(u)}$). It should be noted that for a given pumping rate and observation well, located at a constant distance (r), drawdown (s) and time (t) are the only variables. By separating the variable (t) from the constant terms:

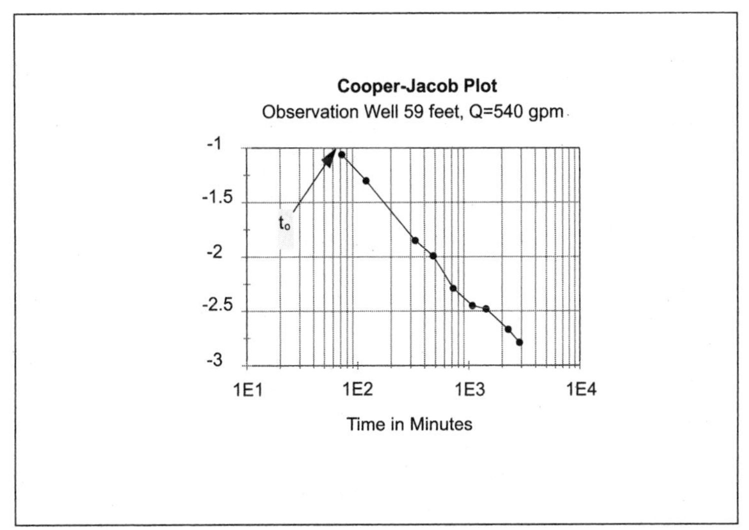

Figure 10.7 Cooper-Jacob plot showing straight-line behavior in drawdown versus time. Drawdown of zero occurs at time t_o= 62 min.

$$s = \frac{2.3Q}{4\pi T}\log\left(\frac{2.25T}{r^2S}\right) + \frac{2.3Q}{4\pi T}\log(t) \quad \textbf{[10.11]}$$

Cooper and Jacob (1946) noticed that this could be compared to the equation for a line:

$$s = C_1 + C_2\log(t), \text{ or } \quad y = b + mx \quad \textbf{[10.12]}$$

By selecting two times t_1, t_2, where: $t_2 > t_1$, then the drawdown at time t_1 is described by:

$$s_1 = \frac{2.3Q}{4\pi T}\log\left(\frac{2.25Tt_1}{r^2S}\right) \quad \textbf{[10.13]}$$

Drawdown at time t_2 is described by:

$$s_2 = \frac{2.3\,Q}{4\pi T}\log\left(\frac{2.25\,Tt_2}{r^2S}\right) \qquad [10.14]$$

The two drawdowns are expressed as:

$$s_1 - s_2 = \frac{2.3Q}{4\pi T}\log\left(\frac{2.25Tt_2}{r^2S}\right) - \frac{2.3Q}{4\pi T}\log\left(\frac{2.25Tt_1}{r^2S}\right) = \frac{2.3Q}{4\pi T}\log\left(\frac{t_2}{t_1}\right) \qquad [10.15]$$

By selecting time over one log cycle, the (t_2/t_1) term simplifies to 1.0. The transmissivity is calculated using consistent units by the familiar expression in Equation 10.16:

$$T = \frac{2.3\,Q}{4\pi\Delta T} \qquad [10.16]$$

where Δs is the drawdown over one log cycle.

The data, then, can be plotted in the field on semi-logarithmic paper (arithmetic scale) s (ordinate) versus time (in log scale, usually in minutes) or using a spreadsheet program, without worrying about $W_{(u)}$. How do we get the storativity? The straight-line plot is extended to where the zero drawdown axis occurs (Figure 10.7). The time where drawdown equals zero is known as t_o. Cooper-Jacob (1946) noticed that Equation 10.10 could be used to solve for the storativity. The first step is to move the constant to the other side:

$$\frac{s4\pi T}{2.3\,Q} = \log\left(\frac{2.25\,Tt_0}{r^2S}\right) \qquad [10.17]$$

By setting drawdown $(s) = 0.0$ and taking the exponent of both sides, we obtain:

$$10^0 = \frac{2.25\,Tt_0}{r^2S} \qquad [10.18]$$

Which simplifies to:

$$1.0 = \frac{2.25\,Tt_0}{r^2S} \qquad [10.19]$$

The storativity can be calculated by:

$$S = \frac{2.25\,Tt_0}{r^2} \qquad [10.20]$$

where:

T, r = as described previously

t_0 = straight-line projection of the time-drawdown curve up to where it intersects the zero-drawdown axis

When the results from the Theis curve are compared to the Cooper-Jacob method, the results are generally comparable; however, the Cooper-Jacob method is easier to plot in the field and does not require the curve matching. If the conditions are not met, the Theis method is still viable for interpretation.

Example 10.5

A pumping test was conducted in July 1979 with a pumping rate of 1,200 gpm (75.6 L/sec) The data are from an observation well located 850 ft away. The data are shown in Table 10.2. and plotted using the Theis (1935) and Cooper-Jacob (1946) methods, as shown in Figures 10.8a and 10.8b.

Table 10.2 Data from July 1979 Pumping Test

Time	Water Level
0700	40.0
0715	40.13
0730	40.5
0800	41.19
0830	41.7
0900	42.11
1000	42.78
1100	43.32
1300	11.01
1500	44.52
1700	44.89
1900	45.23
2100	45.55

Notice the characteristic Theisian fit in Figure 10.8a. The calculated values for both methods are comparable, however the Cooper-Jacob straight-line fit is a least-squares best-fit. The author advises that you hold the Cooper-Jacob plot away from you at a low angle to evaluate whether the data form a straight line or not. If you try this, you will note that the data after 180 min form a straight line. (If the plot is lying on a table, get your face close to the table to evaluate the straightness of the data). This is slightly steeper than

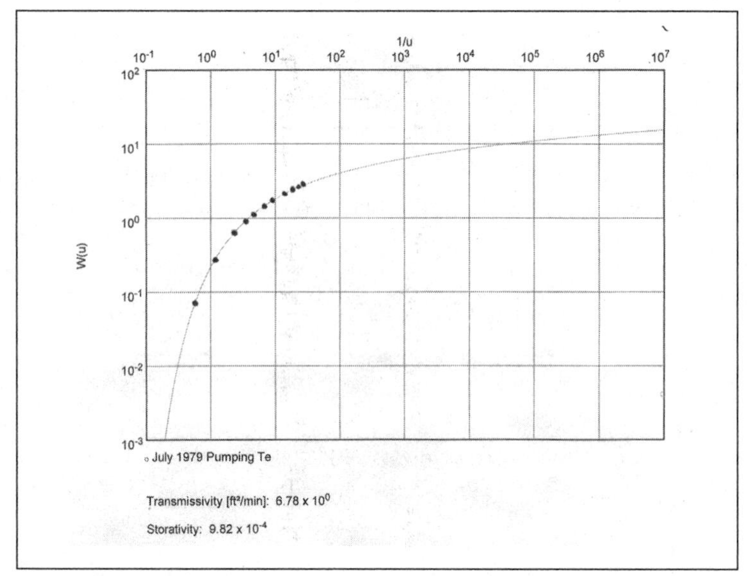

Figure 10.8a A typical Theis fit.

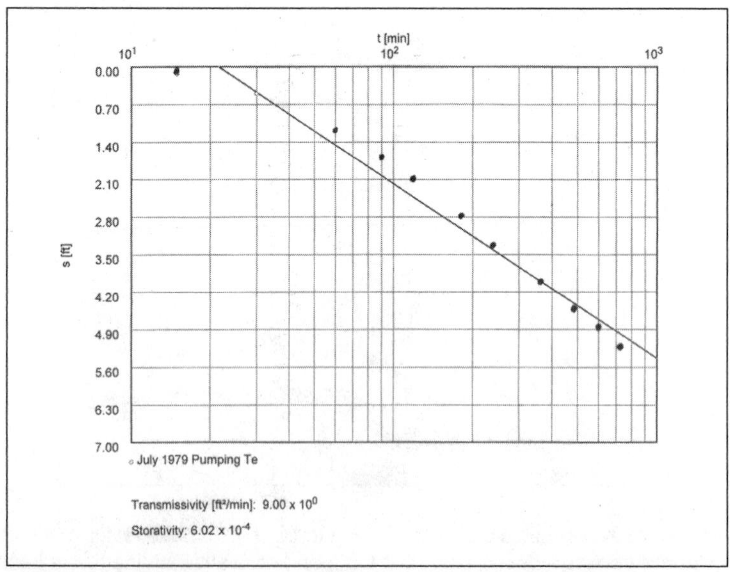

Figure 10.8b A least squares best-fit match. The data after 180 min form a straight line. (Straight line incorrectly fit by software.)

the best-fit line indicated on the plot and will result in the transmissivity value being slightly smaller than indicated but will better represent the later time.

Example 10.5 brings up a caution about software. These programs are wonderful for allowing you to plot the data, but tend to remove the thinking part of data analysis. If one accepts a best-fit line to the data using the Cooper-Jacob (1946) method, and there are multiple slopes resulting from boundaries and other conditions, the interpretation will be wrong. Spreadsheet programs are useful for plotting data using the Cooper-Jacob (1946) method; however, plots using the Theis method are more involved. Errors enter in when the data plot and the well function characteristic curve are not at the same scale and curve matching is attempted. They should both have the same log-log dimensions when overlaying and selecting a match point.

Leaky-Confined and Semiconfined Aquifers

Reporting that an aquifer is confined or unconfined when the storativity falls between confined and unconfined values suggests a lack of understanding of the aquifer system, miscalculations, or errors in interpretation. Storativities between 0.001 and 0.03 suggest that the aquifer could be leaky-confined or semiconfined. The basic difference between a leaky-confined aquifer and a semiconfined aquifer is resolved through extending the time of pumping and recording the drawdown response.

Example 10.6

The following personal story may help illustrate an example of improperly reporting storativity values. Back in the mid-1980s the author participated in writing up a report of aquifer test data results on the Eocene Wasatch Formation for a coal gasification study (Borgman et al. 1986). Dr. Shlomo P. Neuman of the University of Arizona was one of the reviewers, and his pen bled freely where I had incorrectly said that a particular unit was confined. In retrospect, the unit was semiconfined. A particular carbonaceous shale unit was confining a sandy aquifer unit where it existed; however, there were places where the shale unit pinched out and thickened once again, allowing the aquifer unit to become connected to another unconfined sandy unit above. Had the author been more experienced or aware, the data would have properly been interpreted.

Both types initially follow the Theis type A curve and then flatten, indicating a recharge condition. Physically, there is induced recharge through a leaky **aquitard** (Figure 10.9). The rate of **leakance** is determined by the hydraulic conductivity of the aquitard unit and head differential. The hydraulic conductivity would normally be a couple of orders of magnitude lower than the aquifer unit. In this case, the stress caused by the pumping rate induces a change in head sufficient to cause groundwater to flow from above or below the aquifer unit through the aquitard into the aquifer being pumped. Since all sources of recharge are moving through a semiconfining unit that is not releasing any storage, the flatness of the recharge effect on a Theis curve will continue in a horizontal manner (Lohman 1979).

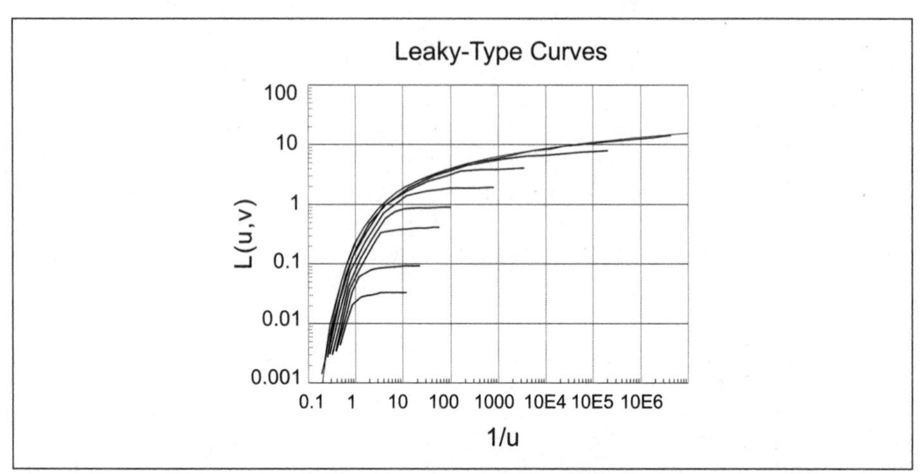

Figure 10.9 Leaky-type curves with the flat response being values of v. From top to bottom, v values are 2.0, 1.5, 1.0, 0.5, 0.2, and 0.1. The smaller the v value, the closer the curve is to the reverse Theis type.

When the recharge effect (the drawdown data fall below the Theis curve) on a log-log plot is curved rather than becoming flat, like Figure 10.9, it is possible that the semiconfining layer is releasing water from storage. This occurs because the semiconfining layer is saturated and has a head higher than the head in the aquifer being pumped (Figure 10.10). This head differential may cause the aquitard to release water from storage, in an attempt to reach equilibrium. Type curves for this approach were created by Hantush (1960). The approach is to fit the field (time-drawdown) data to a Beta (β) curve and select a convenient match point to obtain values for $H(u,\beta)$, $1/u$, time, and drawdown (s).

In a semiconfining scenario, units may pinch out laterally to encounter another part of the aquifer that is actually unconfined. The flattening of the data on the Theis curve reflects a time delay of water from the aquifer yielding to the well as an apparent source of recharge or "seeing" the continued saturated thickness. Another source of recharge may come from the semiconfining layers themselves or through induced recharge from a layer above or beneath the aquifer being stressed. Most confining layers are leaky to one extent or the other. As in the semiconfining scenario, the initial drawdown tends to follow the confined Theis curve. As the aquifer becomes stressed and the head lowers in the pumping well, the head change stimu-

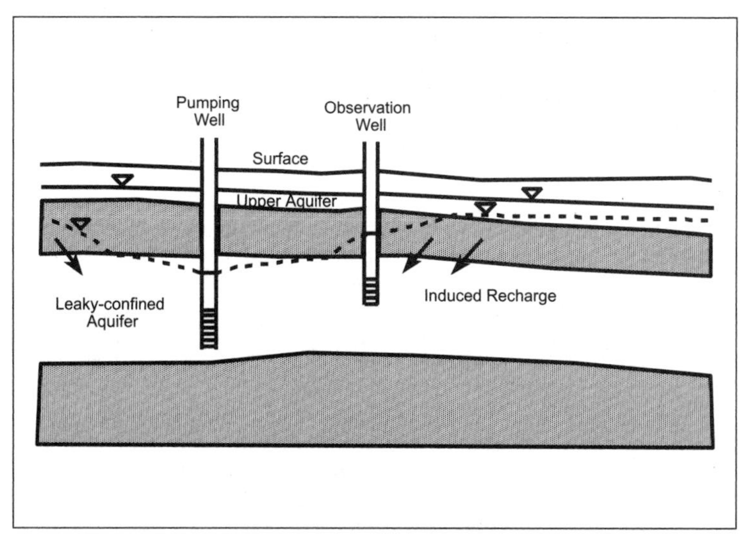

Figure 10.10 Pumping test conducted in a leaky-confined aquifer, with induced recharge through a silty layer, separating the upper and lower aquifer. The cone of depression is indicated by a dashed line.

lates leakage from above or below to act as recharge to the system. The result is a flattening of the curve rather than following the Theis curve. Hantush (1956) developed a whole suite of these curves. They are also given in Lohman (1979). If the semiconfining layer pinches out, the new saturated thickness becomes stressed again resulting in the data following the Theis nonequilibrium (Type B) curve once again. The "climbing" of the drawdown data back onto the Type B curve at a later time is known as **delayed yield** (Neuman 1972, 1975, 1979, 1987; Boulton 1973). Fitting the later-time data results in roughly a three-order of magnitude shift (increase) in the calculation of storativity. This puts the results closer to the unconfined range. If one only fits the early-time data, ignoring the later delayed-yield response, a lower storage coefficient results.

The question becomes, is this a leaky-confined aquifer or really an unconfined aquifer with delayed yield? This question can only be answered by running the pumping test sufficiently long (2 to 3 days, depending on the geologic setting). If the semiconfining layer extends far enough away, then the recharge effect will remain flat, while one that pinches out within the reach of the cone of depression will show delayed yield if pumped long enough.

Understanding the significance of storativity values is important in evaluating aquifer conditions and the proper application of problem evaluation, such as the impacts of pumping wells on neighboring wells. The radial distance from the pumping well to zero drawdown in the cone of depression is known as the **range of influence**. This happens to be a sensitive parameter according to the storativity value (Example 10.7).

Example 10.7

A pumping test was conducted near Miles Crossing (Chapter 6, Figures 6.4 and 6.5) in a fluvial setting impacted by mine tailings from flooding. The pumping test was conducted during the Montana Tech hydrogeology field camp. Well MT97-2 is the pumping well, and the other three are observation wells. The pumping rate was approximately 4.5 gpm (17 L/min). The results of each log-log plot at the same scale are shown in Figures 10.11a, b, c, and d, from top to bottom, respectively. The direction of groundwater flow is generally from east-northeast to west-southwest.

One notes the immediate confined-like response followed by a recharge event (flattening of the curves). It appears that the static water level is up in the finer-grained silty materials and the initial response is confined, followed by a draining of the silty materials. Near the end of the test, it is apparent in the pumping well that delayed yield is beginning indicated by the upward trend (Figure 10.11b). It is interesting to see the same phenomenon with a different perspective in a Cooper-Jacob plot (Figure 10.12). Figure 10.12 represents the same time and drawdown for observation well MT97-3, but in semilog form. Notice the shift in t_0 from the early time to the position at late time where delayed yield is occurring. This results in a significant change in the estimate of storativity. Delayed yield did not occur until approximately 800 min. Had the pumping test been terminated at 600 min, this phenomenon would not have been observed. Notice also if a computer software package had selected a "best-fit" line through all the data, a very different interpretation would have resulted.

Distance-Drawdown Relationships

Back in 1906, G. Thiem developed a method of estimating transmisssivity and storage after steady-state conditions have been achieved (i.e., the cone of depression had stabilized). Once steady state occurs, time becomes inconsequential, and the hydraulic properties of an aquifer estimated from drawdown becomes a function of distance.

The Cooper-Jacob (1946) time-drawdown approach is modified to plot drawdowns at various radial distances from the pumping well. These plots are known as **distance-drawdown plots** (Jacob 1950). By graphing drawdown (arithmetic scale) versus distance (on a log scale) a straight-line fit results (Jacob 1950). Estimates of the hydraulic properties can also be made, if the assumption of u being sufficiently small (0.02, previous section) can be made. The intercept where the straight-line fit crosses the zero drawdown represents the range of influence of the cone of depression (Figures 10.1 and 10.2).

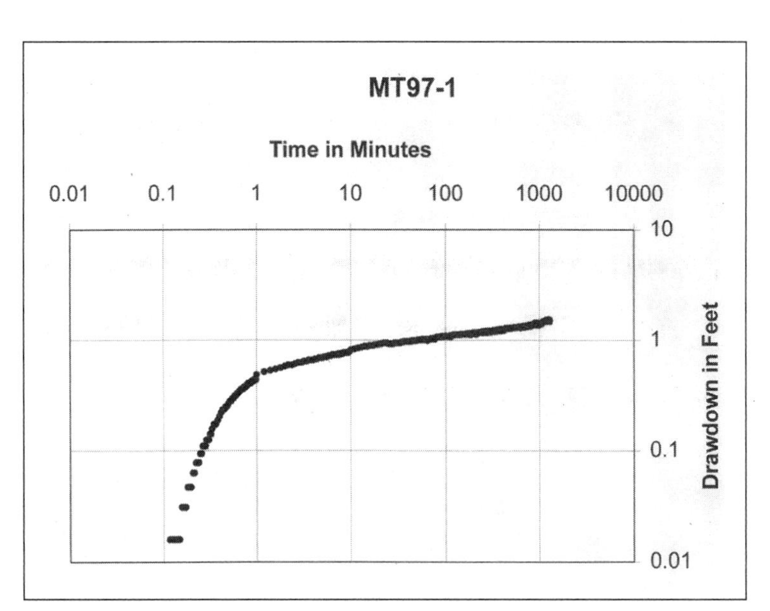

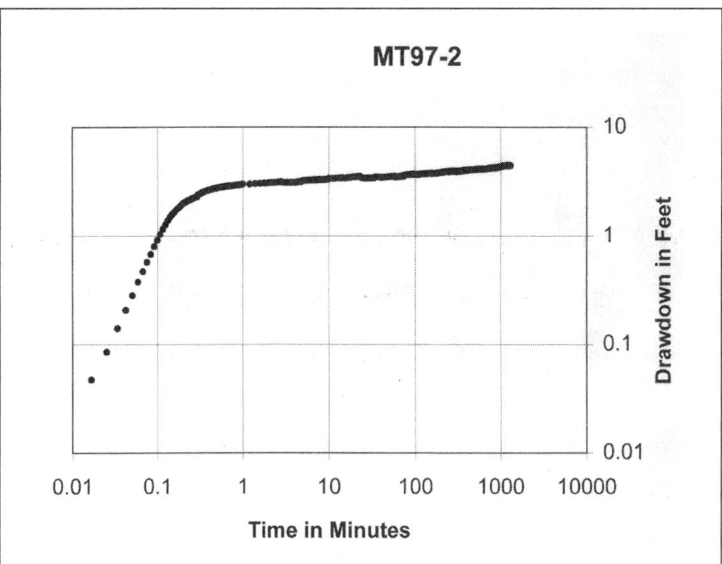

Figure 10.11a and b Theis plots for the pumping well MT97-2 and an observation wells, from top to bottom, respectively. Each indicate the beginnings of a delayed yield effect at time > 800 min.

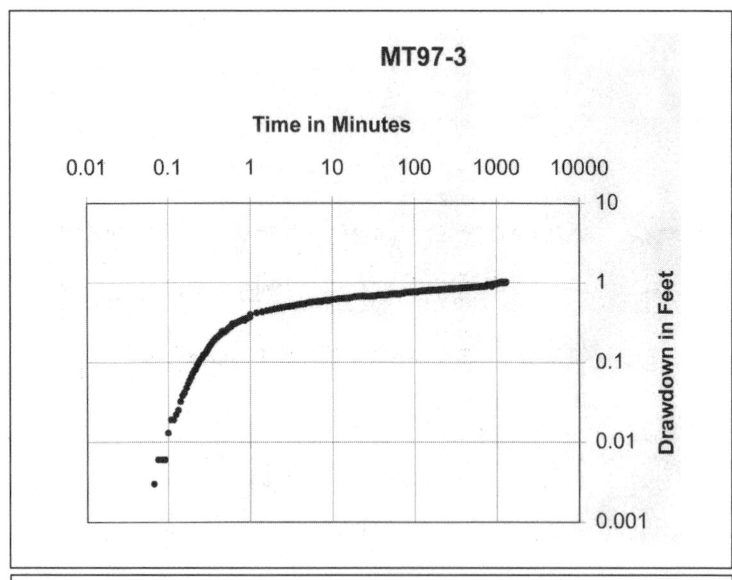

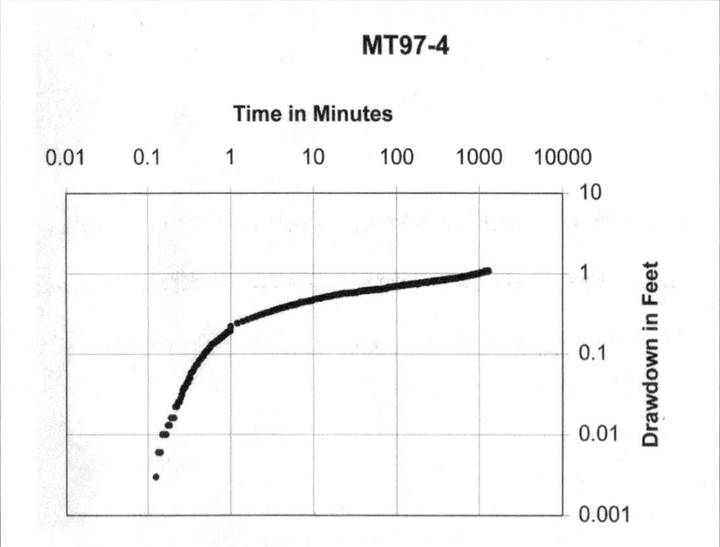

Figure 10.11c and d Theis plots for two observation wells from top to bottom, respectively. Each indicate the beginnings of a delayed yield effect at time > 800 min.

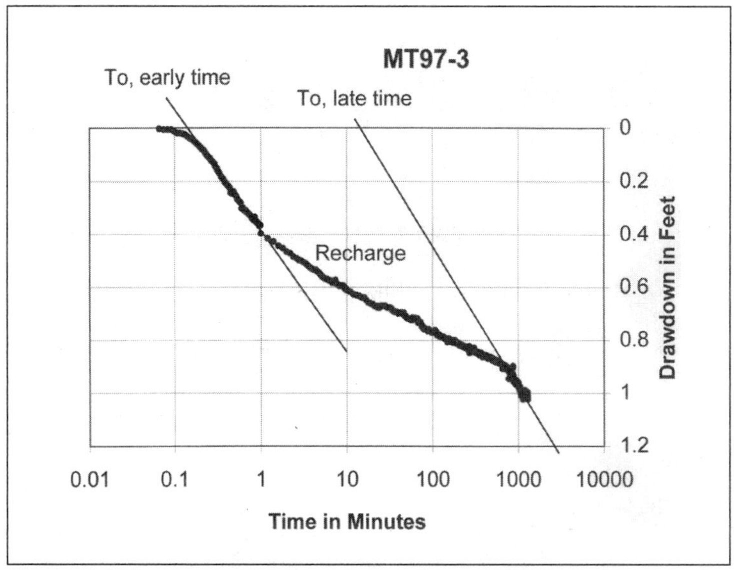

Figure 10.12 Copper-Jacob plot of observation well MT97-3 in Miles Crossing pumping test (compare with Figure 10.11c).

There is a powerful relationship between the time-drawdown plot and a distance-drawdown plot. The slope of the straight-line fit of the latter is exactly twice that of the time-drawdown plot. The reason for this can be seen by evaluating Equations 10.10 and 10.22. Within the log term notice that time (*t*) is to the first power and radial distance (*r*) is to the second power. The double slope results from the following relationship:

$$\log(r^2) = 2 \times \log(r) \qquad \textbf{[10.21]}$$

This ratio is a fixed relationship. The approach is to choose a particular point in time. Assume that the cone of depression is held stationary. This allows an estimation of the distance the cone of depression extends, at any point in time with only one observation well. Clearly, more observation wells in different azimuth orientations and varying distances from the pumping well would improve the understanding of the "true" distance-drawdown effect. A comparative plot of drawdown versus distance can be made from several different observation wells during a pumping test to see if a straight-line plot results. If this is true, it indicates that the cone of depression is symmetrical around the pumping well. If this is not true, there is likely some anisotropy to the shape of the cone. The equations used to calculate transmissivity and storage analogous to Equations 10.16 and 10.11 using consistent units are (note difference in denominator):

$$T = \frac{2.3Q}{2\pi\Delta s} \quad [10.22]$$

$$S = \frac{2.25Tt}{r_0^2} \quad [10.23]$$

where:

T,t = as described previously

r_0 = range of influence, straight-line projection of the distance-drawdown curve up to where it intersects the zero-drawdown axis

---

Example 10.8

A mining operation producing railroad ballast required water to provide control dust for their operations (Figure 10.13). They spent two years preparing for the operation and had ignored the water issue until the end. A wildcat well drilled over 570-ft (173-m) deep near their operations failed to produce water. Approximately 0.8 mile (1 to 1.5 km) to the south is a small community of approximately 60 persons, including a recreational vehicle park. Each home or business has its own well and septic tank. The geology consists of approximately 50 ft of sediments overlying granitic rocks. The production zones for wells are in fractures in granite that parallel a creek flowing down from the mountains to the east and from the upper sandy sediment units. The mining company made a deal with a local resident to drill a well on his property that could pump water up the hill to the north. When the drill rig was set up in the yard, the residents started asking questions.

Figure 10.13 Mining operation producing ballast.

The author was asked to represent the mining company in answering a "few questions" at a public meeting. Unaware of the sensitive nature of the problems to be discussed, he agreed to be present. Before the meeting, the author obtained lithologic logs for each of the wells in the area. By evaluating the production testing performed on each well recorded on the well log, an estimate of the transmissivity and storativity were made. By manipulating Equation 10.23, an estimate of the range of influence for the proposed pumping rate could be performed. The transmissivity was estimated using a specific capacity equation for fracture flow (Section 10.3), and the storativity of the granite layer was estimated to be 0.001. A time of 100 days continuous pumping was used as a worst-case scenario.

$$r_0 = \sqrt{\frac{2.25(13\ \text{ft}^2/\text{day})100\ \text{days}}{0.001}} = 1{,}710\ \text{ft (520 m)}$$

At the meeting, the room was filled with angry residents. By the time the author was given the floor to speak, everyone was at peek irritation, similar to a hissing group of geese. (Ironically, the lady who was most upset did have geese that attacked us every time we took a water level in her well). This distance-drawdown analysis was used to tell all residents who lived more than 2,000 ft from the proposed well that they could go home. During the meeting, the mining company committed to a pumping test and groundwater flow model to evaluate pumping and recovery scenarios. This solution appeased the residents and ultimately led to a practical solution.

Predictions of Distance-Drawdown from Time-Drawdown

As was mentioned, a powerful predictive tool results from the slope relationships between time-drawdown and distance-drawdown graphs. If a short-duration pumping test (less than 24 hours) is conducted and a straight-line relationship develops, the straight line of the time-drawdown graph can be extended to longer periods of time (days, weeks). Once the drawdown at the longer period of time has been determined, a distance-drawdown graph (using twice the slope of the time-drawdown graph) can be used to estimate the range of influence of the cone of depression. This is useful when evaluating interference effects on other wells in the area.

If you have no pumping test data because you are trying to get a permit to drill a well, you can use Equation 10.23. Notice that if you estimate a value for T and S, you can obtain r_o for a particular time (Example 10.8).

Example 10.9

Suppose there is a need to irrigate 30 acres. You need to find out from your local state agency if any wells exist near the proposed well site. The well logs may give an idea of how deep it is to water and what lithologies the wells are completed in. They also may indicate how much water they make. Since $T = K \times b$; if an estimate for hydraulic conductivity can be obtained, and the saturated thickness (b) is known, a value for T can be estimated (Your estimate of T should be conservative). The storativity can be estimated from the lithology (Chapter 3) and whether the aquifer is confined (10^{-3} to 10^{-6}) or unconfined (0.03 to 0.30). In this example, a range for T and times should be used to evaluate impacts to other wells Table 10.3. By estimating the drawdown in the pumping well and the range of influence (distance of zero drawdown) one can perform a distance-drawdown plot to evaluate the impacts to neighboring wells. To produce Table 10.3, it was necessary to first use Equation 10.2 to calculate u. It was assumed for the pumping well, that the effective radius was 1 ft (0.3 m). A corresponding value for W_u was looked up in Appendix B. The drawdown was calculated using Equation 10.1. As a sample calculation, suppose the transmissivity is estimated to be 10,000 ft^2/day and the storativity is estimated to be 0.15, the range of influence after 30 days from Equation 10.23 is:

$$r_0 = \sqrt{\frac{2.25(10{,}000\ \text{ft}^2/\text{day})30\ \text{days}}{0.15}} = 1200\ \text{ft}\ (640\ \text{m})$$

Table 10.3 Comparison of Drawdowns, Transmissivity, and Time

Trans-miss-ivity	1/3 Day		30 days		60 Days		180 Days	
ft²/ day	Draw down @1 ft	Dist. to cone edge	Draw down @1 ft	Dist. to cone edge	Draw down @1 ft	Dist. to cone edge	Draw down @1 ft	Dist. to cone edge
1000	16.3	71	24.9	671	26.2	949	28.3	1643
3000	6.1	122	9.0	1162	9.4	1643	10.1	2846
5000	3.9	158	5.6	1500	5.9	2121	6.3	3674
10000	2.1	224	2.9	2121	3.1	3000	3.3	5196

From Example 10.9, notice that the pumping rate Q was not used to evaluate r_0. The range of influence is dependent on T, S, and time (t). Since, the pumping rate does *not* affect the range of influence (see Equation 10.23), what is observed then is how the pumping rate affects the slope of the distance-drawdown plot. For a particular time t, the r_0 is fixed, but the slope of the distance-drawdown line drawn back to the vicinity of the pumping well depends on the pumping rate (Figure 10.14). Notice that the drawdown in Figure 10.14 at approximately 160 ft is expected to be 1.0 ft (0.3 m) for the pumping rate of 125 gpm (425 L/min). For most practical problems, the distance (r) at the pumping well is the (**effective radius, r_e**), or the distance out to where the formation was disturbed during drilling (damage zone).

The Cooper-Jacob (1946) method of time-drawdown and Jacob (1950) method of distance-drawdown are a modification of the Theis equation and assume that u is small. The Theis method is always a valid first approach to evaluating pumping-test data, and the Cooper-Jacob (1946) and Jacob (1950) methods are valid when the time (t) is sufficiently large or radial distance (r) is sufficiently small. It is useful to evaluate data using as many

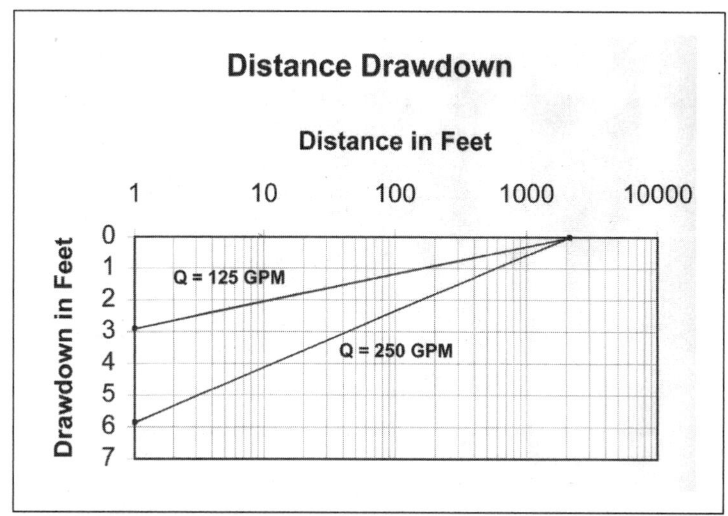

***Figure 10.14** Distance-drawdown relationship assuming T= 10,000 ft^2/day, S= 0.15, and time = 30 days.*

methods as possible. Each may help provide a different perspective and aid in a better interpretation.

Recovery Plots

If you are running a pumping test, it makes a lot of sense to also collect the **recovery data** once the pump is turned off. There are a number of errors made in evaluating recovery data. There is a tendency to plot the recovery versus time in a similar manner as in a Cooper-Jacob (1946) plot. This does not work out. Instead the **residual drawdown** versus time should be plotted. Once the pump and generator are shut off, the data logger should be activated to collect recovery data. If this is being done manually, a similar frequency of data collection should be taking place as when the pumping test started (Chapter 9). As one might expect the early-time recovery data changes quickly at first and then slows down over time. One must be careful to make sure there is a check valve in the pump *or* water from the discharge line and riser pipe will drain, turning the impellers backwards and giving a "false" recharge.

When performing a residual drawdown versus time plot the following information is needed: the time since pumping started (t), the time since pumping stopped (t'), and the residual drawdown (s'). The actual plot is the residual drawdown versus the *ratio of* t/t', where residual drawdown is in arithmetic scale and the ratio of t/t' is in log scale. This produces a relationship where the early time data are indicated by large numbers and the late-time data become smaller as the ratio of t/t' get closer to 1. If a straight-line relationship develops, then projecting the straight line to where the residual drawdown is zero represents the time at which full recovery will take place. Theoretically, this should occur at the projection to where the ratio of t/t' is 2. In a homogenous isotropic aquifer with no recharge sources, the time it took to create the drawdown is the same time it will take to recover. Ratios less than 2 indicate slower recovery, and ratios greater than 2 indicate more rapid recovery. Once the pump is shut off, an imaginary injection well with a Q representative of the weighted average Q used during the pumping phase will cause the aquifer to recover. Data should be collected from all of the observation wells to provide a suite of recovery values for interpretation.

Example 10.10

During the recovery phase for the pumping test described in Example 10.8, recovery data were collected at 11, 20, and 36 hours. The duration of the pumping test was 12 hours. A plot of these data is shown in Figure 10.15. Projected full recovery is at approximately t/t' = 1.25, or near 48 hours. Time (t) keeps increasing after the pump shuts off, so the following relationship occurs: $1.25 = (12 + x)/x$, or $1.25x = 12 + x$, therefore, $0.25x = 12$ and $x = 48$ hours. A value of 1.25 indicates that the recovery is much greater than the drawdown part.

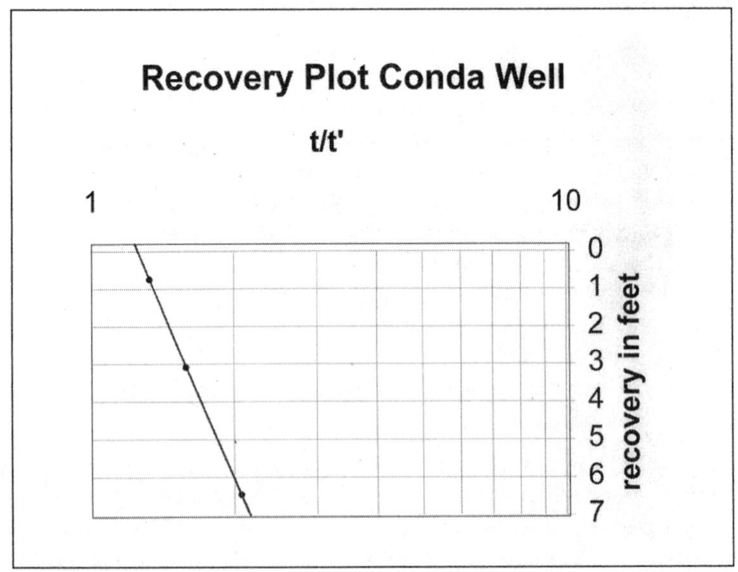

Figure 10.15 Recovery plot of well in fractured granite, with a pumping duration of 12 hr.

This becomes extremely important if you are evaluating pumping and recovery scenario impacts. Suppose, for example, that your distance-drawdown plot does show impact to other wells. It may be that by pumping and then letting the well recover before pumping again, it would provide a reasonable compromise to everyone's water use needs. The author's general experience is that many unconsolidated aquifers take *1 ½ times* the pumping rate time to fully recover and that fractured bedrock aquifers may take up to *4 times* the pumping rate time to fully recover (Example 10.10). This is significant because a simple 12 hours pumping–12 hours recovery scenario may have a detrimental affect over time.

10.3 Specific Capacity

One "first cut" approach to estimating transmissivity from drill logs is to use the **specific capacity** (S_c) relationship. Specific capacity is defined as:

$$S_c = \frac{Q}{s} \quad [10.24]$$

where:

Q = gpm

s = drawdown in feet after 24 hours pumping

From the specific capacity equation, a rough rule of thumb derived from distance-drawdown relationships yields the following relationship for transmissivity (Driscoll 1986).

For confined aquifers:

$$T = S_c \times 2{,}000 \quad [10.25]$$

For unconfined aquifers:

$$T = S_c \times 1{,}500 \quad [10.26]$$

where T is in gal/day-ft.

A specific capacity approach applied to fractured bedrock aquifers was developed experimentally by Huntley and Steffey (1992), yielding the following relationship for bedrock aquifers:

$$T = 38.9(S_c)^{1.18} \quad [10.27]$$

where T is in ft^2/day.

The usefulness of the specific capacity can be summarized as follows:

- Gives a rough estimate of T (even with lousy data)
- Can be applied to confined, unconfined, or fractured bedrock aquifers
- Gives a rough idea of well completion (should the well yield more water than it is getting?)

If the specific capacity value is grossly smaller than expected, the following reasons could explain the results:

- Poorly completed, not screened in the right zone, or not placed at all.
- Well screen is clogged, either poor well development, precipitation of minerals in the screen, or biofouling.
- Contractor fraud? It may be that if there is a pattern to poorly completed wells in an area where specific capacity values are noticeably higher in adjacent wells. If the poor wells were all completed by the same contractor, you could be suspicious of fraud.

Equation 10.27 was used in Example 10.8 to perform the distance-drawdown relationship needed for the public meeting.

10.4 Well Interference and Boundary Conditions

When the drawdown cones of multiple wells overlap at a particular point, the net affect is the *sum of the drawdowns* from all of the wells. This can be determined either graphically, using distance-drawdown methods, or mathematically, using Equations 10.1 or 10.10. Equation 10.1 can be used anytime, whereas Equation 10.10 can only be used if u is small (i.e., less than 0.2). The reason the effect of neighboring wells is additive is because the LaPlace equation is linear.

$$\frac{\partial^2 h}{\partial r^2} + \frac{\partial^2 h}{\partial r^2} = 0 \qquad [10.28]$$

Drawdown can be positive (downward, s) or negative (mounding, $-s$). When there are injection or recharge wells ($-Q$), instead of creating a cone of depression, there is a **cone of impression**. A cone of impression can be thought of as an upside-down cone of depression, creating a mound in the potentiometric surface at the location of the injection well (Figure 10.16). The net drawdown affect of pumping and injection wells can be evaluated independently and added at a particular point of interest (Example 10.11).

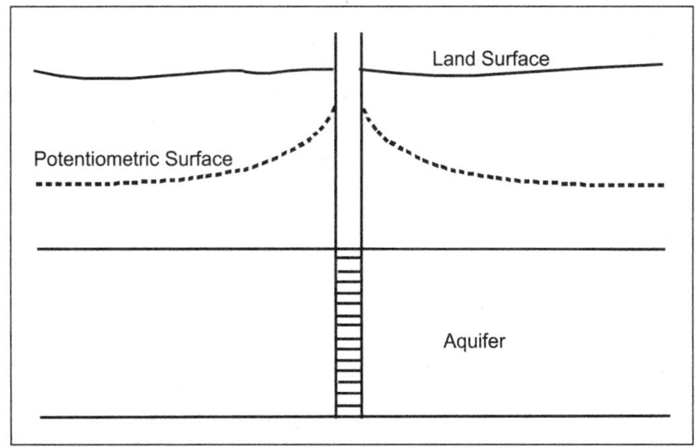

Figure 10.16 Schematic illustrating a cone of impression (mounding of the potentiometric surface).

Example 10.11

Suppose we wish to evaluate the net drawdown at point A from the effect of three wells. Two of these wells are pumping wells, and one of the wells is an injection well (Figure 10.17). The pumping rates (500 and 600 gpm) and injection rate (450 gpm) in Figure 10.17 are shown, and the duration of pumping is one year (365 days). The transmissivity is 5,000 ft^2/day, and the storage coefficient is 0.20. The steps are to calculate evaluate u and then look up a corresponding value for $W_{(u)}$ in Appendix D and apply Equation 10.1. If u is small, Equation 10.10 can be used, but requires more effort. It is usually easiest to look up a $W_{(u)}$ for a calculated value of u and apply Equation 10.1. The results are shown in Table 10.4.

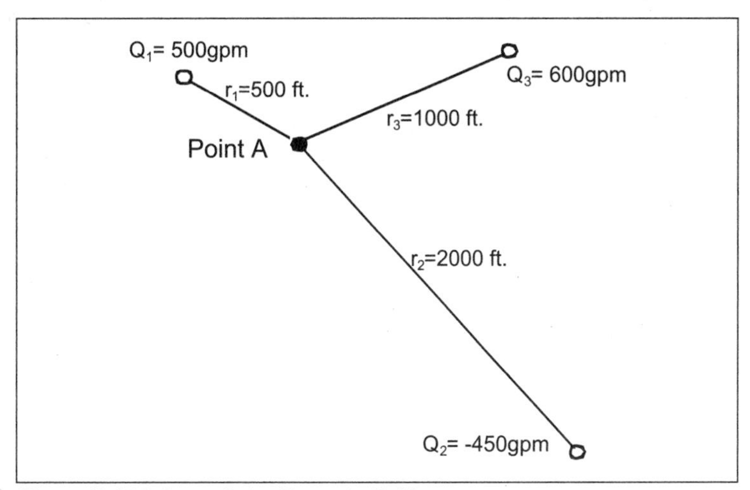

Figure 10.17 Schematic for Example 10.11.

Table 10.4 Parameters Associated with Drawdown Calculations

Well	*r* (ft)	*u*	$W_{(u)}$	*Q* gpm	Drawdown (ft)
1	500	0.00685	4.413	500	6.76
2	2,000	0.1096	1.737	-450	-2.39
3	1,000	0.0274	3.044	600	5.60
				Net	9.97 (10 ft)

The net value is 10.0 ft of drawdown. The values for W_u are given in Appendix D. A sample calculation follows.

$$s = \frac{Q}{4\pi T} \times W_u = \frac{500\,\text{gal/min}}{4\pi(5{,}000\,\text{ft}^2/\text{day})} \times \left(\frac{1\,\text{ft}^3}{7.48\,\text{gal}} \times \frac{1{,}440\,\text{min}}{1\,\text{day}}\right) 4.413 = 6.76\,\text{ft}$$

Graphically, a distance-drawdown graph should be created separately for each pumping or injection well in Example 10.11. The injection well would be treated the same way as the pumping well except that the drawdown would be negative. The drawdown (plus or minus) at the respective distance from the point of interest (point A) would be picked off of the distance-drawdown graph and then summed up. The results should be pretty much the same. This brings up a point of the power of a groundwater model because the drawdown is determined not only at point A, but every cell on a grid. If the aquifer is not of infinite aerial extent, boundary conditions must also be considered.

Aquifer Boundary Conditions

Up to this point, we have been considering the aquifer to have infinite areal extent. This means that the cone of depression continues outward forever. The cone of depression will stop if it

- Becomes large enough to capture enough water demanded by the pumping rate from recharge and leakage; thus reaching steady-state conditions or
- Reaches a hydrogeological recharge boundary.

In reality, most hydrogeologic settings rarely have aquifers that are of infinite areal extent; rather they change laterally in grain size, shape, or lithology. These changes affect the shape of a time-drawdown curve in characteristic ways. Some boundaries that are encountered are less permeable than the aquifer unit being pumped. These are known as barrier boundaries. Examples of **barrier boundaries** include:

- Decrease in aquifer thickness from thinning or erosion

- Decrease in aquifer permeability from decreasing grain size (e.g., facies change)
- Encountering a fault plane, bedrock contact, or some other barrier
- Startup of a nearby pumping well, or effect of another nearby pumping well

Some cases occur where the boundaries encountered are more permeable than the aquifer unit being pumped. Examples of **recharge boundaries** include:

- Increase in aquifer thickness (increase *T*)
- Increase in aquifer permeability from increasing grain size (e.g., facies change)
- Encountering a recharge source, such as a lake, stream, or gravel channel
- Shutdown of a nearby pumping well, or the effect of another nearby pumping well shutting off
- Leakance from adjacent aquifers

The effects of boundaries are usually seen when the drawdown is plotted, either by the Theis (1946) or Cooper-Jacob methods (1946) (Figures 10.18 and 10.19). Notice how in a barrier boundary the drawdown increases with time, and in a recharge boundary the drawdown decreases with time. The way boundaries are accounted for in drawdown calculations is by applying well interference to **image well theory**.

Image Well Theory

When a boundary is encountered, its effects are simulated by an image well. Perhaps it would be better to explain with an example. Suppose a pumping well begins pumping at time (*t*). After 100 min, the cone of depression hits a bedrock barrier boundary. The effect on a time-drawdown plot is a steepening of the slope; thus the drawdown is now increasing more rapidly with time. To simulate this effect, an *imaginary well* is placed at an equal distance on the other side of the boundary the same distance the pumping well is to the boundary and at the same pumping rate. (This is like a mirror image). Since the boundary is a barrier boundary, the imaginary well is also a pumping well (barrier boundaries result in the pumping rates of the image wells having the same sign). The drawdown effects at the boundary, or anywhere within the radius of influence of the imaginary well, are added together with the effects of the real pumping well according to the principles explained in the section on well interference.

If the boundary is a recharge boundary, the image wells are located in the same way except that the "sign" of pumping for the imaginary well is opposite that of the real well. The drawdown from pumping becomes negated by the recharge effect from the imaginary well with the opposite sign. These effects are seen in Figures 10.18 and 10.19. If two boundaries are encountered, the resulting effect would be an image well for each boundary. You would have to have more than one observation well to "see" two boundaries.

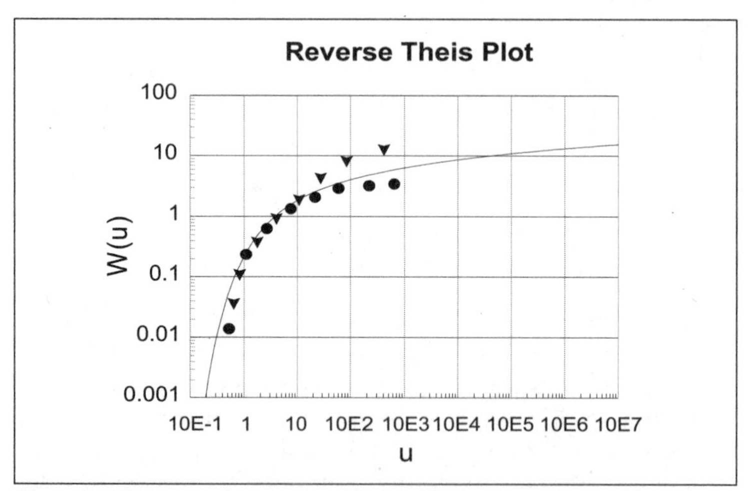

Figure 10.18 Reverse-type curve with data illustrating recharge (dots) and barrier (triangle) conditions.

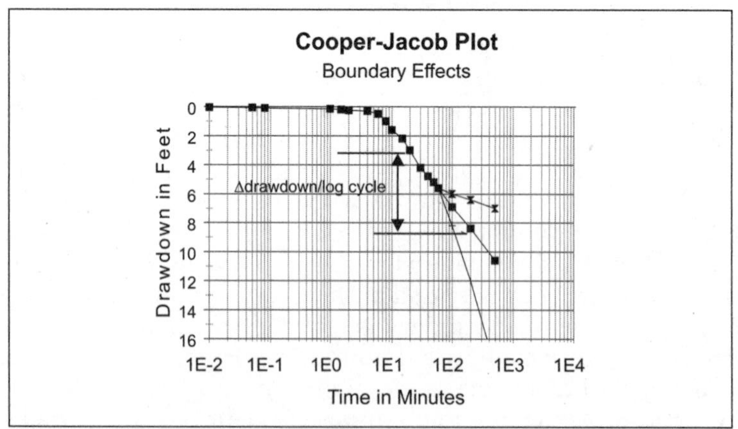

Figure 10.19 Cooper-Jacob plot of three wells: Hourglass pattern indicates recharge, boxes represent straight-line pattern (no boundary effects), and pluses represent a barrier condition. Deviations take place at 60 min.

10.5 Partial Penetration of Wells and Estimates of Saturated Thickness

Wells that fully penetrate the aquifer and are fully screened have radial (horizontal) flow towards the well bore. If the well only partially penetrates the aquifer, the flow paths have a vertical component to them (Figure 10.20). The flow paths are, therefore, longer and converge on a shorter well screen, resulting in an increase in head loss and decrease in well efficiency (Driscoll 1986).

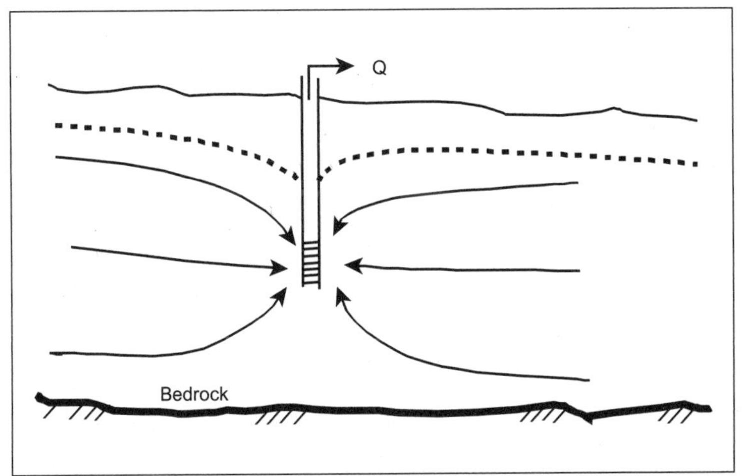

Figure 10.20 Schematic of partial penetration effects illustrating horizontal and vertical flow components.

Many times it is not possible or desirable to design a well that fully penetrates an aquifer. This is particularly true in an unconfined aquifer (Figure 10.2.0) that may be hundreds of meters thick; hence these wells are partially penetrating. In the Helena, Montana, Valley, many domestic wells only have an "open-hole" completion or have a few circular saw slots in PVC casing or torch cut slots in steel casing, while partially penetrating hundreds of feet (meters) of aquifer. These limited openings provide the only access to the aquifer to retrieve water. In an unconfined aquifer, it is practical to place screen in the lower one-third of the aquifer to allow maximum drawdown in the production zone. (It is much easier for water to move downward than upward.) Observation wells for pumping tests should be placed far enough away from the pumping well to avoid partial penetration effects.

When is an observation well in a pumping test far enough away from the pumping well to avoid the effects of partial penetration? Hantush (1964) indicates that if the observation well fully penetrates the aquifer or if it is partially penetrating and more than $1.5b\ (K_h/K_v)^{0.5}$ away from the pumped well, the effects are negligible. In the distance equation, b is the saturated thickness, K_h and K_v are the horizontal and vertical hydraulic conductivities. The vertical hydraulic conductivity is approximately an order of magnitude less that the horizontal hydraulic conductivity. Driscoll (1986) suggests that at a distance of $2b$, horizontal flow prevails.

Estimates of the Saturated Thickness in Unconfined Aquifers

In the preceding discussion, there is a big question about the saturated thickness b. In a confined aquifer that is partially penetrating, estimates of saturated thickness can be challenging. The estimation process is harder for unconfined aquifers. Significant errors may result when assuming the saturated thickness is the length of the screen. This usually results in an over estimation of the hydraulic conductivity K; however, estimates for K can be underestimated if one ignores the geology. Based upon our experience from conducting many aquifer pumping tests and from many geologic settings, we offer a couple of rules of thumb for estimating the saturated thickness. These estimates should be followed based upon the following two general principles:

Time—When a pumping test is being conducted in relatively homogenous sediments and time is short (several hours), then the contributing thickness b' is approximately 1.3 times the length of the screen. When time is longer (greater than 24 hours) the reference thickness in a unconfined aquifer becomes the distance from the water table down to the bottom of the screen (L). The estimated saturated thickness b'' becomes L times 1.3. This is illustrated in Figure 10.21.

Geology—When the lithologic units are layered or interbedded with significant changes in grain size or physical properties in the vertical direction, then the transmissivity assigned to the saturated thickness must be evaluated according to the contributing hydrogeologic units. If the hydraulic conductivity of a given hydrogeologic unit is one order of magnitude or greater than other units, the majority of water will be produced from the higher K unit. It will be easier for water from the lower hydraulic conductivity units to find a pathway up or down to the higher hydraulic conductivity unit than to take a long pathway to the well screen. This may result in the water being produced from an aquifer thickness less than the screen length in shorter duration pumping tests (Figure 10.22).

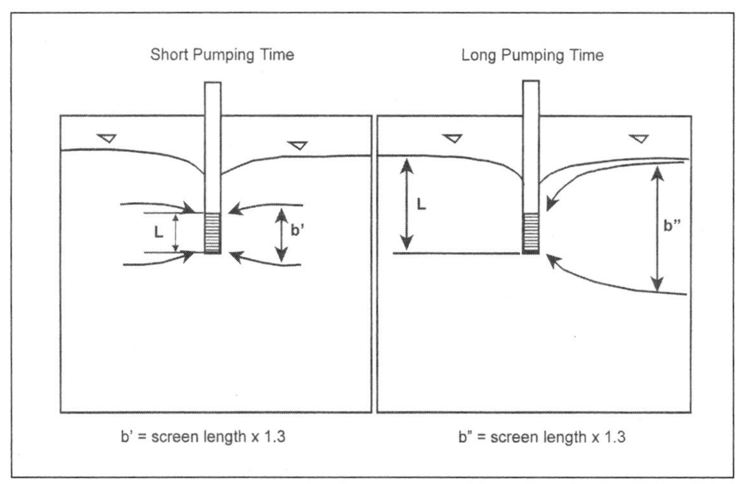

Figure 10.21 Contributing thickness of an aquifer depending on time for a relatively homogeneous hydrogeologic setting.

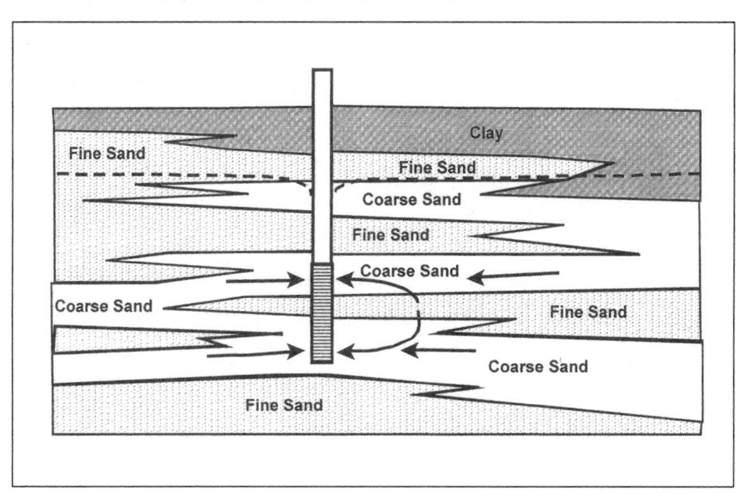

Figure 10.22 Pumping test in variable hydrogeologic units. In short duration tests, the actual contributing thickness may be less than the screen length.

The issue concerning the saturated thickness b is important and one that is not given enough attention. It has been the author's experience that many hydrogeologists working for consulting companies are mainly concerned with obtaining a transmissivity value during the pumping test and divide by the screen length to obtain a hydraulic conductivity value without regard whether the K value matches the lithology. Worse yet is the case where hydraulic conductivity values are used in calculations, along with reported transmissivity values from pumping tests, and the calculated saturated thicknesses make no sense whatsoever. It is as though the pumping tests were performed because they were necessary to estimate the hydraulic properties of an aquifer, and then in the analysis the hydraulic conductivity didn't come out as expected so the field results were ignored.

Example 10.12

While performing a technical analysis of a consultant's report of a flow and transport model for a landfill, the author noticed that a very careful analysis was performed during the site characterization phase. A significant amount of expense, time, and effort was made to establish a conceptual model based upon drill hole information and constructed cross sections. Additionally, pumping tests were performed with both transmissivity and hydraulic conductivity values reported that were used in the numerical model. The hydraulic conductivity values used in the groundwater-flow model were inconsistent with the lithologic units and reported transmissivities. (Also, the length of pumping was on the order of a few minutes to a few hours). The respective saturated thicknesses were calculated at 50 to 60 ft (15.2 to 18.3 m) thick, when the screen length was only 10 ft (3 m) and pumping-test duration was short. It was as though a list of field exercises were performed to fulfill certain requirements, with little effort made to be consistent when it came time for the numerical model. Calibrating the model to yield a particular flow scenario seemed to be more important than reality.

10.6 Fracture Flow Analysis

The previous section has been primarily concerned with pumping tests in porous media. Pumping tests in fractured porous media have a very different time-drawdown behavioral response. The physical setting is one where high conductivity fractures drain quickly, while smaller fractures and block media drain at later time. Diagrammatically there is a different flow-path length for groundwater in fractures compared to porous media. The flow-path length compared to lateral distance is known as the **tortuosity** factor. In fractured media, the tortuosity factor can be significantly less (Figure 10.23).

The distances over which groundwater moves are more direct and therefore can result in significant velocities. As the velocity of groundwater flow increases, it begins to gain kinetic energy and become turbulent. Turbulent flow in fractured aquifers is a common phenomenon (Carriou 1993). Once turbulent flow dominates, then Darcy's law is no longer valid and a different equation for hydraulic gradient is used (Bear 1979).

$$\frac{dh}{ds} = A_L q + B_T q^n \qquad [10.29]$$

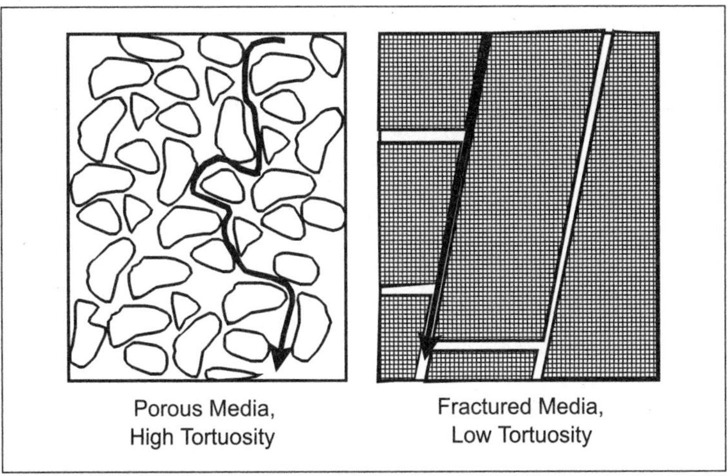

Figure 10. 23 Differences in tortuosity (flow-path length) in porous versus fractured media.

where:

dh/ds = hydraulic gradient (L/L)

q = specific discharge $(L/t) = K \times dh/ds$

A_L = laminar flow constant (t/L)

B_T = turbulent flow constant (t/L)

n = turbulent flow exponent

K = hydraulic conductivity (L/t)

Turbulence increases the drawdown within the pumping well by a factor of $B_T q^n$, where n increases with the degree of turbulence, which has experimentally been found to range from 1.6 to 2.0 (Bear, 1979).

The dynamics of groundwater flow to a pumping well was first considered by Barenblatt et al. (1960). They gave the following equation for describing flow in a fractured aquifer to a pumping well:

$$\Delta s_f = \frac{Q}{4\pi T} \int J_0(xr) \left[1 - \exp \frac{ktx^2}{1 + x^2 \eta_f} \right] \frac{\partial x}{x} \qquad \textbf{[10.30]}$$

where:

s_f = drawdown in the fractured aquifer (L)

Q = pumping rate (L^3/T)

T_f = transmissivity of fractured aquifer (L^2/t)

J_o = Bessel function of the first kind, or zero order

r = radial distance from the pumping well (L)

t = time since pumping began

κ = ratio of fractured aquifer transmissivity to aquifer storativity, or T_f / S_p (L^2/t)

η = fissure characteristic (dimensionless)

S_p = porous block storativity (dimensionless)

d_x = differential × operator (radial distance from center of well to the limit of integration (L)

The part after the integral sign in Equation 10.30 is the well function associated with a discretely fractured aquifer. The type curve was plotted by Carriou (1993) and is shown in Figure 10.24. Notice the linear early-time portion of the curve. This represents water yield from discrete fractures crossing the well bore coming from fracture storage (Carriou 1993). As more water is withdrawn, the hydraulic gradient within the bedrock induces water from the bedrock and smaller fractures to move to the larger fractures, thus transitioning to a more Theisian-like response at later time. Additional type curves are shown in Streltsova (1976). Field studies in petroleum settings indicate that fractures greater than 0.05 mm are large enough to contribute significantly to fluid extraction volume while contributing very little to overall aquifer block storativity (Gringarten et al. 1974; Streltsova 1976).

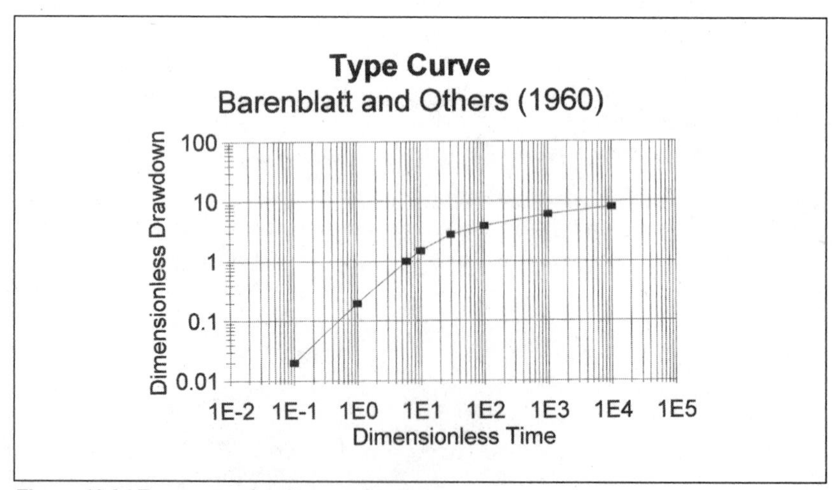

Figure 10.24 Type curve for discrete fractures.

The analytical approach for pumping tests in discrete fractures is to plot field values of drawdown (*s*) versus time (*t*) in log-log scale. This plot overlain on top of the type curve for dimensionless fracture drawdown (s_{df}) versus dimensionless time, also in log-log scale (Figure 10.24). The axises are kept square similar to the Theis curve-matching approach (Section 10.2), except that both the linear pattern (representing fracture dewatering) and the late time are both included in the Barenblatt et al. (1960) solution. This means that there is no need to emphasize early- or late-time in the "fit." The transmissivity and storativity are solved using Equations 10.31 and 10.32 from Barenblatt et al. (1960).

$$T_f = \frac{Qs_{df}}{4\pi\Delta s} \qquad \textbf{[10.31]}$$

$$S_p = \frac{4T_f t}{t_d r^2} \qquad \textbf{[10.32]}$$

where:

Q = withdrawal rate ($ft^3/min.$)

S_{dr} = dimensionless fracture drawdown from match point

T_f = fracture transmissivity (ft^2/min)

S_p = aquifer storativity (dimensionless)

t = time from match point (minutes)

t_d = dimensionless time from match point

r = radial distance from pumping well to observation well (ft)

The pumping test described in Example 10.8 resulted in some of the observation wells indicating a fracture response (Figure 10.25). The test was only run for 12 hours, so only the linear (fracture dewatering) response is observed.

Additional fracture responses are possible from fractured media depending on the physical properties of the block materials. For example, if there is a fractured sandstone being stressed by pumping, both the fractures and the blocks will yield water in a dual porosity model. The dual porosity model is similar to a delayed yield response in porous media (Neuman 1979), where the fractures behave linearly; once the blocks begin to yield water, a recharge effect (flattening of the drawdown curve) occurs. Later on when the blocks become stressed, another curvilinear response occurs, indicated by increased drawdown once again (Figure 10.25).

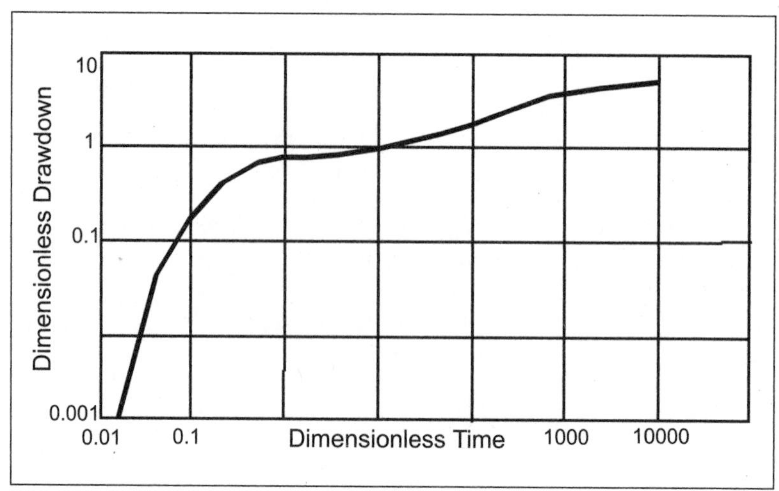

Figure 10.25 Dual porosity type curve indicating a delayed yield-like response.

Pumping Tests in Flowing Wells

Pumping tests can also be conducted in flowing (artesian) wells if the "pumping well" can be controlled with a shutoff valve. At start time, the pumping well is opened up, allowing free flow of water. A flowing observation well is plumbed with a transducer to measure the reduction in head over time (Figure 10.26). If the flow does not decrease more than 10% or so, a reasonable fit can be made to the pumping-test data. "Pumping" times should be similar to conventional pumping tests.

Figure 10.26 Plumbing a transducer in a flowing well used as an observation well.

Example 10.13

In Petroleum County, in central Montana, are several flowing wells from differing Cretaceous hydrogeologic units. In a study performed by Brayton (1998), pumping tests were conducted in the basal member of the Cretaceous Eagle Formation. A flowing pumping well was shut down, while the flowing observation well was plumbed with a pressure transducer. The pumping test was conducted by opening the valve on the "pumping well" and allowing the data logger to record the decrease in pressure over time. Decreased pressure was converted to drawdown in feet over time. An example plot of the data is shown in Figure 10.27, indicating a dual porosity fracture-flow response.

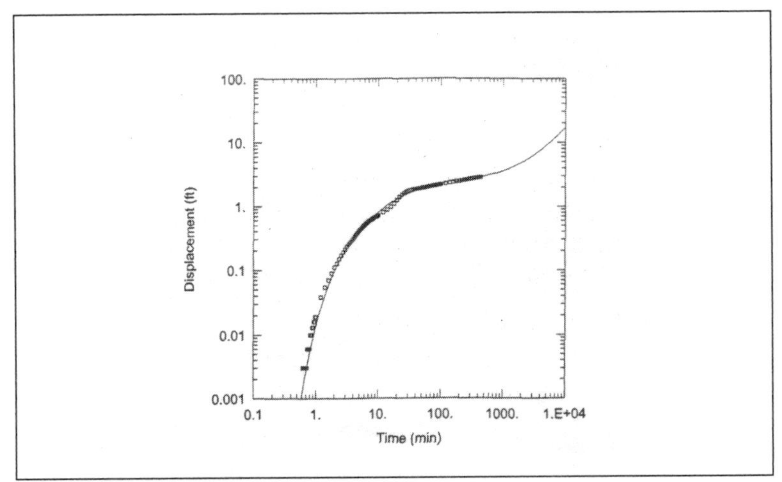

Figure 10.27 Dual porosity fracture-flow response from a pumping test in the Cretaceous Eagle Formation, central Montana. [After Brayton (1998).]

10.7 Summary

One of methods of evaluating the hydraulic properties of aquifers is through conducting pumping tests. How to conduct a pumping test is described in Chapter 9, while helpful suggestions on what to record in your fieldbook are found in Chapter 1. This chapter addressed several of the analytical methods used to analyze pumping-test data. The most common approach is to perform a Theis analysis or do a Cooper-Jacob plot *in the field during the test* to see how the data behave. Boundary conditions and other effects are observed directly in the plot. Care must be taken to ensure that the pumping rate was kept constant, or strange pattens my occur that are an artifact of varying pumping rates.

One must also make sure the pumping test was conducted long enough when a recharge response was observed early on. Does the recharge response reflect a leaky confined setting, or is delayed yield occurring? This can only be resolved through recording additional time-drawdown responses.

Direct relationships exist between time-drawdown and distance-drawdown plots. These can be used to evaluate the impacts of a longer duration pumping test on neighboring wells.

The analysis of pumping-test data is closely connected with having an understanding of the geologic setting and a knowledge of the well-completion details. Pumping tests yield an estimate for transmissivity (*T*) and storativity (*S*). These are two-dimensional parameters, in which the vertical dimension is averaged. When estimating the hydraulic conductivity (*K*) from the transmissivity value, it is important to consider test duration and geology, particularly in a partial penetration scenario.

Fracture-flow responses have their own characteristic behavior. Type curves for discrete-fracture or dual-porosity models are very different. A common aspect of both is the early-time linear response on a log-log plot of drawdown versus time (Figure 10.28). Care must be taken to make sure the analysis method, field observations, and geologic setting all make sense.

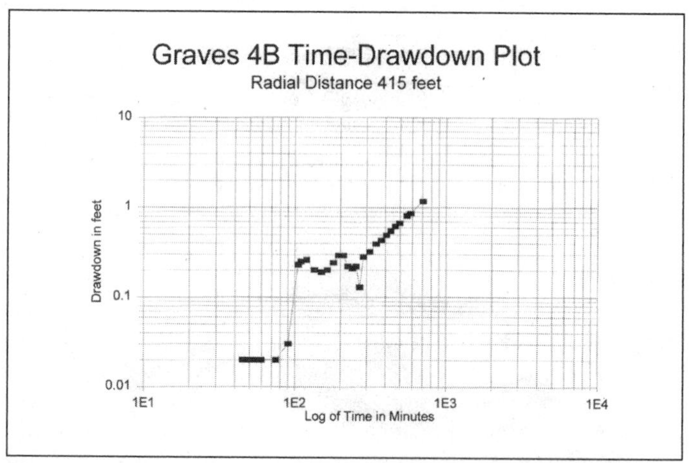

Figure 10.28 Observation well drawdown response from Example 10.8.

References

Barenblatt, G.I., Zheltov, I.U.P., and Kochina, I.F., 1960. Basic Concepts in the Theory of Seepage of Homogenous Liquids in Fissured Rocks. *Journal of Applied Mathematical Mechanics*, Vol. 24, No. 5, pp. 1286–1303.

Bear, J. 1979. *Hydraulics of Groundwater.* McGraw-Hill, New York, 569 pp.

Borgman, L.E., Weight, W.D., and Quimby, W.F., 1986. *Geological and Hydrological Characteristics of the Hoe Creek Underground Coal Gasification Site.* Technical Report 86-2. Prepared for the United States Environmental Protection Agency EMS Laboratory, Las Vegas, NV, 71 pp.

Boulton, N.S., 1973. The Influence of The Delayed Drainage on Data from Pumping Tests in Unconfined Aquifers. *Journal of Hydrology,* Vol 19, pp. 157–169.

Brayton, M., 1998. *Recovery Response From Conservation Methods in Wells Found in the Basal Eagle Sandstone, Petroleum County, Montana.* Master's Thesis, Montana Tech of the University of Montana, Butte, MT, 68 pp.

Carriou, T.M., 1993. *An Investigation into Recovering Water Levels in the Butte, Montana, Bedrock Aquifer.* Master's Thesis, Montana Tech of the University of Montana, Butte, MT. 160 pp.

Cooper, H.H., Jr., and Jacob, C.E., 1946. A Generalized Graphical Method for Evaluating Formation Constants and Summarizing Well-Field History. *Transactions of the American Geophysical Union,* Vol. 27, pp. 526–534.

Driscoll, F.G., 1986. *Groundwater and Wells.* Johnson Screens, St. Paul MN, 1108 pp.

Duffield, G.M., and Rumbaugh, J.O., III, 1991. *AQTESOLV, Aquifer Test Solver Software. Version 1.00,* Geraghty and Miller Modeling Group, Reston, VA., 90 pp.

Fetter, C.W., 1993. *Applied Hydrogeology, 3rd Edition.* Macmillan, New York, 691 pp.

Fletcher, F.W., 1997. *Basic Hydrogeology Methods: A Field and Laboratory Manual with Microcomputer Applications.* Technomic, Lancaster, PA, 310 pp.

Gringarten, et al., 1974. Unsteady-State Pressure Distributions Created by a Single Infinite-Conductivity Vertical Fracture. *Society of Petroleum Engineers Journal,* Vol. 14, No. 4, pp. 347.

Hantush, M.S., 1956. Analysis of Data From Pumping Tests in Leaky Aquifers. *Transactions of the American Geophysical Union,* Vol. 37, pp. 702–714.

Hantush, M.S., 1960. Modification of the Theory Of Leaky Aquifers. *Journal of Geophysical Research,* Vol. 65, pp. 3713–25.

Hantush, M.S., 1964. Hydraulics of Wells. In *Advances in Hydroscience, Vol. 1,* V.T. Chow, (ed.) Academic Press, New York, pp. 281–432.

Huntley, D.R., and Steffey, D., 1992. The Use of Specific Capacity to Assess Transmissivity in Fractured-Rock Aquifers. *Ground Water,* Vol. 30, No. 3, pp. 396–402.

Jacob, C.E., 1950. Flow of Ground-water. In *Engineering Hydraulics,* H. Rouse (ed.), John Wiley, New York, pp. 321–386.

Kruseman, G.P., and deRidder, N.A., 1991. *Analysis and Evaluation of Pumping Test Data, 2nd Edition.* International Association for Land Reclamation and Improvement, Publication 47, Wageningen, The Netherlands, 377 pp.

Lohman, S.W., 1979. *Ground-Water Hydraulics.* U.S. Geological Professional Paper 708, 70 pp.

Neuman, S.P., 1972. Theory of Flow in Unconfined Aquifers Considering Delayed Response to the Water Table. *Water Resources Research,* Vol. 8, pp. 1031–1045.

Neuman, S.P., 1975. Analysis of Pumping Test Data from Anisotropic Unconfined Aquifers Considering Delayed Gravity Response. *Water Resources Research,* Vol. 11, pp. 329–342.

Neuman, S.P., 1979. Perspective on "Delayed Yield". Water *Resources Research,* Vol 15, No.2, 99899–99908.

Neuman, S.P., 1987. On Methods of Determining Specific Yield. *Ground Water* , Vol. 25, No. 6, pp. 679–684.

Streltsova, T.D., 1976. Hydrodynamics of Groundwater Flow in a Fractured Formation. *Water Resources Research,* Vol. 12, No. 3, pp. 405 –414.

Theis, C.V., 1935. The Lowering of the Piezometer Surface and the Rate and Discharge of a Well Using Ground-Water Storage. *Transactions of the American Geophysical Union,* Vol. 16, pp. 519–524.

Thiem, G., 1906. *Hydrologische Methoden.* Gebhardt, Leipzig, 56 pp.

Chapter 11

Slug Testing

Slug testing has been used over the years to obtain a cost-effective quick estimate of the hydraulic properties of aquifers. Slug testing has gained in popularity since the 1980s since it can be used to obtain hydraulic property estimates in contaminated aquifers where treating pumped waters is not desirable. Slug-testing analysis methods were first developed during the 1950s (Hvorslev 1951; Ferris and Knowles 1954). The simple method devised by Hvorslev (1951) led to its wide use for both confined and unconfined aquifers as an estimate of the hydraulic conductivity within the screened interval. Improvements of the methods by Ferris and Knowles (1954) were made by Cooper et al. (1967) and Papadopulos et al. (1973) for confined aquifers. Later, Bouwer and Rice (1976) and Bouwer (1989) developed a method for analysis in confined, semi-confined, or unconfined aquifers that takes into account aquifer geometry and partial penetration effects. Campbell et al. (1990) summarizes the various methods used and their relative merits with an extensive list of references. An article by Butler et al. (1996) describes the protocol that should be used in the field to reduce errors of estimation. A recent book by Butler (1998) presents a fairly comprehensive summary on slug-test methodology and analysis. If this chapter does not satisfy the reader on the topic of slug testing, the book by Butler (1998) is highly recommended.

Constant-duration pumping tests are a better method of analyzing the physical properties of aquifers (Chapter 9), since their influence goes beyond the immediate vicinity of the borehole, although in high-transmissivity aquifers the influence can be detected several hundred feet away (Bredehoeft et al. 1966). Slug tests are especially useful in locations where handling the discharge of contaminated pumping waters is prohibited, and yet a rough estimate of the hydraulic properties of the aquifer is needed. Another great thing about slug tests is that they can be conducted relatively quickly, so that several point estimates of hydraulic conductivity can be collected within a day's work. The chances of a hydrogeologist needing to perform slug tests are pretty high. The intent of this chapter is to discuss the methodology of performing slug tests in the field followed by methods used to analyze the data. This is done for both the overdamped and underdamped cases. Discussion of the problems that can occur in the field will also be presented along with the typical mistakes that are made when performing the analysis.

11.1 Field Methodology

Slug testing involves lowering a "slug" (a cylinder of known volume, usually constructed of PVC pipe or some other suitable material filled with sand or gravel that is capped at both ends) tied to a rope or cable, down a well bore to displace the static water with an equivalent volume of fluid (Figure 11.1). A pressure transducer, previously placed below the slug level, senses the water-level responses, and the data are recorded in a data logger (Figure 11.2). The slug remains in place until the water level equilibrates. Once at equilibrium, the data logger is activated and the slug is briskly retrieved above the static water level (Figure 11.3). The displaced water recovers to the original static level. The time required for static conditions to occur once again is proportional to the hydraulic conductivity of the aquifer materials. Graphical methods are used to obtain parameters that are used to calculate the hydraulic conductivity values.

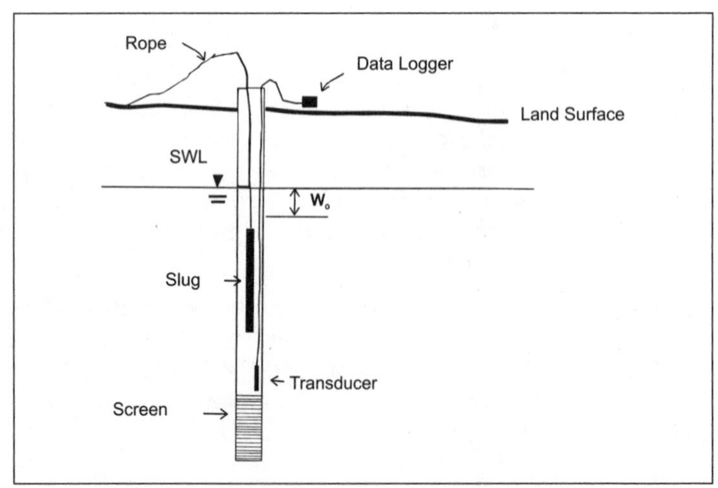

Figure 11.1 Slug testing schematic. [Weight and Wittman (1999). Used with permission of Groundwater Publishing Co.]

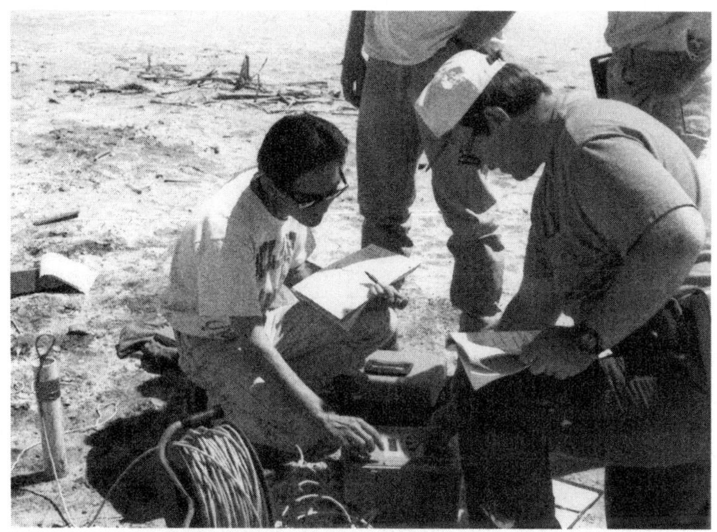

Figure 11.2 Using the data logger in the field to gather slug-test data.

Figure 11.3 Lowering a slug down a well.

Traditionally, water-level changes during slug tests were measured using steel tapes or electrical tapes (E-tapes). This only allowed slug tests to be conducted on relatively low hydraulic conductivity materials. Therefore, this was applicable when water levels slowly recovered over a matter of minutes to several hours. With the current wide use of high-speed data loggers and pressure transducers, slug-test analysis may be conducted on materials over a wide range of hydraulic conductivities.

How to Make a Slug

It is important that the size of the slug be appropriate to the amount of water desired to be displaced. For example, 1 1/4-in. PVC works well for 2-in. wells, and 3-in. PVC works well in 6-in. wells (larger-diameter slugs can be used, but become difficult to handle). The lengths should be designed so that at least a foot (0.3 m) or two (0.61 m) of displacement can occur. One can calculate the approximate volume of the slug before going into the field. Typical slug lengths are 3 ft (1 m), 6 ft (2 m), or 10 ft (3 m). The general procedure for constructing slugs is described next:

- Take a blank piece of PVC casing and cut it to the desired length, allowing for the top and bottom caps to contribute to the total length.
- The bottom of the casing is capped, and either glued with PVC glue or screwed together, or both. Preferably, this is reinforced with duct tape.
- Sufficient sand or gravel is added, so that when the slug is capped at the top, it will sink.
- A hole is made in the top cap, large enough for the rope or cable to thread through. A large knot, like a figure eight, can be tied to keep the rope or cable from pulling back through.
- The top cap is then secured into place. Once again, this should have duct tape wrapped *around the cap and rope* or cable for additional security.
- Make a variety of sizes and have additional materials with you in the vehicle to make additional slugs or if adjustments are needed.

Example 11.1

While performing several slug tests in the Beaverhead Valley, near Dillon, Montana (Figure 11.4), in 1996, a supposedly secure top cap separated from the slug and we helplessly watched the rest of the slug drop out of sight. It had been held on with PVC glue. Reinforcing each slug, top and bottom, with duct tape became standard procedure ever since and even after several tens of more tests, this security step proved itself again and again.

In another setting, a heavy pump was to be lowered down a well, via a braided stainless-steel cable. A loop through the top of the pump was used

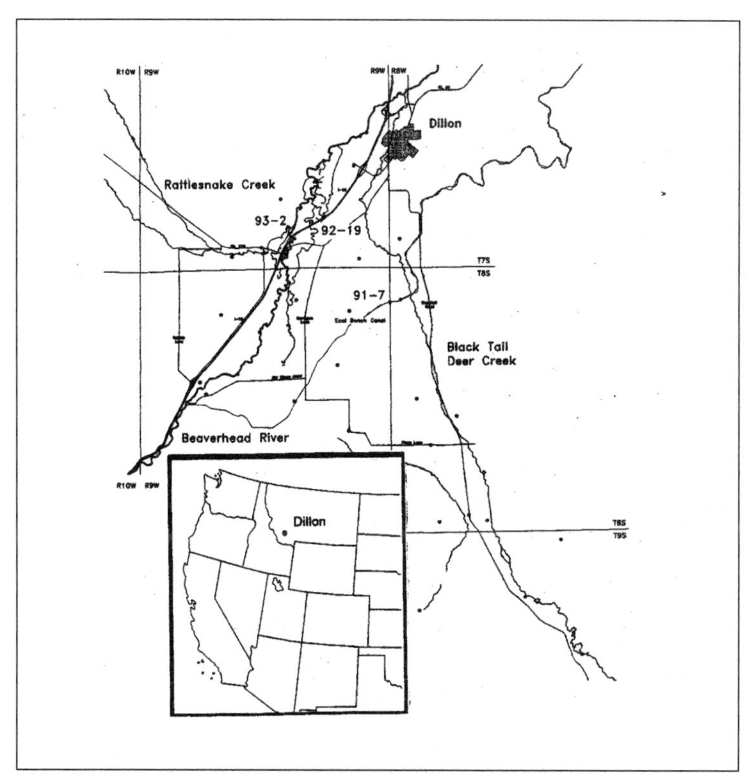

Figure 11.4 Slug test area, Beaverhead Valley, Dillon, Montana.

to link the cable, and a purchased clamp that was wrenched tight was supposed to hold the loop. Fortunately, after raising the pump, the faulty clamp slipped, revealing its inability to hold the cable before it was positioned over the well. A new approach of tying a knot and duct taping the knot, in addition to the clamp, worked beautifully. It is much easier to cut away duct tape than "fish out" lost equipment.

Performing a Slug Test

Performing a slug test is fairly straightforward and should take less than a half-hour per well, unless the hydraulic conductivity is very low. The following steps are presented to help the field person remember what to do:

- Obtain general information about the well, such as well completion information, total depth, location of screen, diameter, height of water column above the screen. This is important to select an appropriately sized slug, and to know the transducer range that will be necessary (Chapter 9).
- Remove the well cap and take a static water-level measurement. This should be checked a few times before starting the test to see if changes or trends are occurring.
- Select the appropriate transducer and lower it *below* the location where the bottom of the slug will be placed. Note that if the transducer is rated at 10 psi if it lowered into more than 23 ft of water, it may become damaged and not take a reading (Chapter 5). Allow the transducer to stretch and hang for a few minutes, so that vertical movement is not occurring during the test.
- Create a loop (bite) of at least 1-in. diameter in the transducer cable and secure it with duct tape or electrical tape. The loop rests on the top of the well casing, and an additional wrap of the transducer cable is made around the well casing. This is all secured into place with duct tape so that the transducer doesn't move during slug retrieval (Figure 11.3).
- Connect the transducer to the data logger, establish the transducer parameters, and prepare for the starting of the test. Before lowering the slug and setting the reference level, take a reading from the data logger to see if the water depth looks reasonable. Reasons for strange readings on transducers are discussed in Chapter 9.
- Lower the slug with a rope or cable below the water level and tie it off until the water level equilibrates. Note that if the slug floats, it needs to be retrieved and filled with more sand and gravel to make sure it sinks.
- Once the water level has equilibrated, the test is ready. Recheck the data logger and establish the reference level to be some value greater than the maximum expected displacement (a value of 10 or 100 is usually what the author uses).
- While holding the line, carefully untie the slug rope and be ready for the signal to retrieve the slug. Start the data logger in log mode, and with only a second or two delay, briskly retrieve the slug. One way to think of it is to count to four and on three start the data logger.
- Check the data logger to see whether the water level has equilibrated and stop the test. This process should be repeated three times to make sure the data behave similarly.
- Decontaminate the slug and make sure well caps are replaced and the site is left the way you found it.

What was just described is a **rising-head test**. Theoretically, one could get the same results by suspending the slug above the static water level and lowering the slug into the water. This would cause the water level to be displaced upward and gently return down to the original level (a **falling-head test**). Another way to perform a falling-head test is to pour a known volume of water down a well and record the "fall" in water level to equilibrium. This assumes that adequate well development has occurred. The authors prefer the rising-head test because the water that is displaced must come from the aquifer, rather than pushed out into the aquifer as is done in a falling-head test.

The reason for turning the data logger on just a few seconds before the slug is "instantaneously" retrieved is to make sure that the slug clears the water. If the slug hasn't cleared the water surface, "splash effects" and an erroneously large reading will occur (Table 11.1). One must look for the maximum displacement (h_o) in the data. This is presumed to occur at time zero. All other data (h) are adjusted and referenced from the maximum displacement point. In Table 11.1, the measurement at time 0.0133 min represents where the slug did not break the water surface yet, thus creating an erroneously large displacement value. The subsequent oscillation downward represents the splash effect. Time zero has been adjusted from 0.0266 min. Displacement versus time plots are used to estimate the hydraulic properties.

11.2 Analyzing Slug Tests—The Damped Case

Slug-testing methods and other topics in well hydraulics presume that Darcy's Law is valid. Specifically, this means that viscous forces or frictional effects during groundwater flow predominate over inertial forces, which are usually considered to be negligible. The viscous forces allow sufficient friction to keep the water column from oscillating. This is known as the damped case. When the water column is displaced to the maximum (h_o), recovery occurs quickly at first and then tapers off. This is viewed in a data plot as a flattening in the slope. In high-transmissivity aquifers, water-level responses recover quickly and may have sufficient momentum to overcome the viscous forces and oscillate as an underdamped spring. What to do in this situation is discussed in Section 11.3.

This section will present how to analyze slug tests using the most common methods of slug test analysis, the Hvorslev (1951) and Bouwer and Rice (1976) methods. Frequently these are used for analysis in unconfined aquifers, but they have been used in analyzing partially penetrating and confined aquifers if the screen is well below the confining layer (Fetter

1994). Both methods provide an estimate of the hydraulic conductivity, but have no means for estimating storativity. Papadopulos and others (1973) provide a set of type curves that data can be matched to calculate transmissivity and storativity for confined aquifers. The data are plotted on semi-log paper as the ratio of h/h_o (arithmetic) versus the log of time. A description of this method is summarized in Fetter (1994). Additional discussion and methodologies are presented by Butler (1998) in his book.

Table 11.1 Slug Test Illustrating a Slug not Clearing the Water Surface with Initial Reference of 100 ft (3-ft slug in 2-in. well)

Time (min)	Displacement (ft)	Time (min)	Displacement (ft)
0.0	99.984	0.12	101.218
0.0033	99.984	0.14	101.155
0.0066	99.984	0.16	101.91
0.01	100.095	0.18	101.033
0.0133	107.954	0.20	100.947
0.0166	101.392	0.30	100.767
0.02	101.423	0.40	100.623
0.0233	101.55	0.50	100.509
0.0266	101.566	0.60	100.442
0.03	101.534	0.70	100.378
0.0333	101.518	0.80	100.316
0.0366	101.502	0.90	100.269
0.04	101.487	1.00	100.236
0.05	101.455	1.20	100.175
0.06	101.408	1.40	100.148
0.07	101.376	1.60	100.116
0.08	101.329	1.80	100.092
0.09	101.297	2.00	100.079
0.10	101.265	2.50	100.042

Hvorslev Method

One of the most popular methods of analyzing slug-test data is known as the **Hvorslev** (1951) method (pronounced horse-loff). It is probably the quickest and simplest method. It is restricted to estimates of hydraulic conductivity within the screened zone of a well and can be applied to confined and unconfined aquifers. The maximum displacement value (h_o) is identified in the data. This is presumed to occur at time zero. The ratio of all other displacement data (h) to h_o or h/h_o are plotted on the ordinand (y-axis) in log scale, and time is plotted arithmetically along the abscissa (x-axis). There are no type curves with which to match the data. The plot is needed to identify the parameter (T_o) or the time it takes the water level to rise or fall to 37% of the initial maximum, h_o. This 37% comes from the inverse of the natural log e. The value e is 2.7182818..., and the inverse of this number is 0.368 or roughly 0.37, hence the 37%. (It will be shown after the Bouwer and Rice (1976) section how to analyze these data without using a ratio approach to obtain T_o. There is an easier way). Hvorslev (1951) developed several equations to meet certain conditions, but the most commonly used equation is:

$$K = \frac{r^2 \ln(L/R)}{2LT_o} \qquad [11.1]$$

where:

K	=	hydraulic conductivity
R	=	radius of well casing
R	=	radius of screen plus packing (i.e, effective radius)
L	=	length of well screen plus filter packing if it extends above the top of the screen
T_0	=	time to reach 37% of H_o

Equation 11.1 is good for any case where $L/R > 8$. This is easily satisfied unless the well screen and packing interval are very short. The other parameters are derived from the well completion information (Figure 11.5). The radius of the well casing r is straightforward unless the displacement water level drops within the range of the screened interval. In monitoring wells that are completed to monitor LNAPLs such as gasoline, the screen typically extends above the static water level (Figure 11.6). In this case, the casing radius becomes an effective casing radius. If the gravel-packing material also extends up above the top of the screen, the porosity of the gravel-packing material is taken into account to obtain an adjusted casing radius (r_A) by using Equation 11.2 (Bouwer 1989). The parameter η represents the porosity of the packing materials.

$$r_A = [(1.0 - \eta)r^2 + \eta R^2]^{1/2} \qquad [11.2]$$

Example 11.2

As an example to this adjustment for r_A, suppose the radius of the well screen is 0.167 ft (5.1cm) and there is a gravel-packing material of 0.25 ft (7.6 cm) with a porosity of approximately 25% all around the well screen. The radius of the borehole is 0.42 ft (12.7 cm). The adjusted radius of the casing using Equation 11.2 would be:

$$r_A = [(1.0 - 0.25)(0.17\text{ ft})^2 + 0.25(0.4167\text{ ft})^2]^{1/2} = 0.253\text{ ft}$$

If the water level is affected within the packing materials, then the height or magnitude of displacement is proportionally less than would be

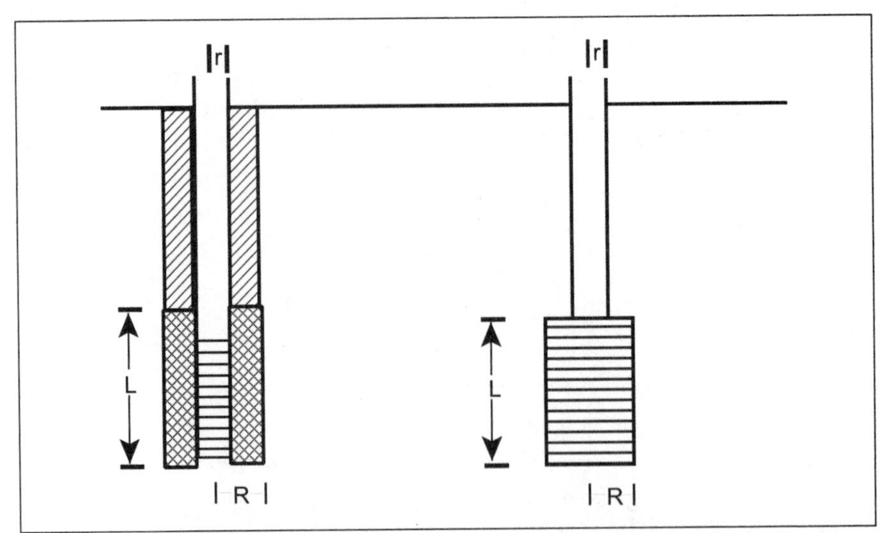

Figure 11.5 Well completion information used to obtain parameters used in the Hvorslev (1951) method in Equation 11.1. [Modified from Fetter (1994).]

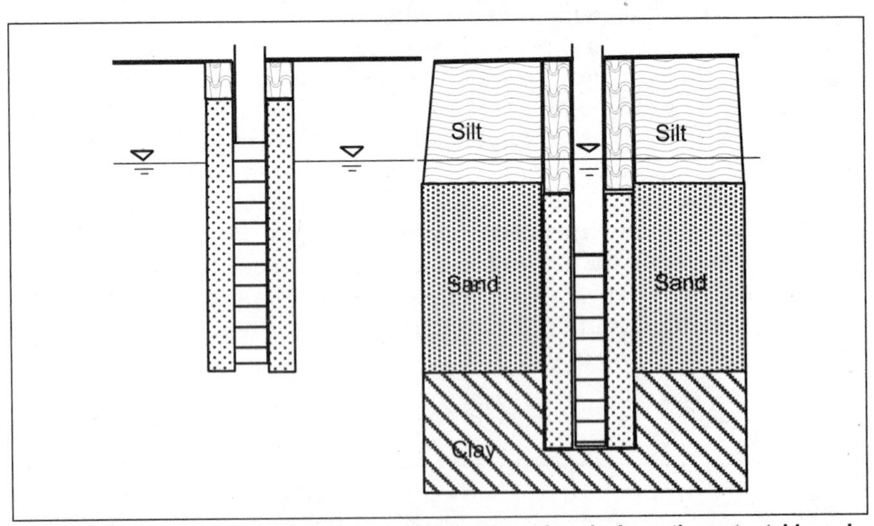

Figure 11.6 Examples of well completion showing gravel pack above the water table and the screen and gravel pack partially completed in clay.

seen if all displacement changes occurred within the casing itself. The parameter R is the effective radius of the borehole. The diameter of the borehole may vary, according to problems encountered during drilling. If the borehole is damaged or coated with drilling fluids, then a "skin effect" will separate the borehole from the aquifer and the hydraulic conductivity may reflect the "skin," not the aquifer. The parameter L *includes* the gravel-packing materials. It is a common error to use only the screen length. The full gravel-pack length is capable of allowing water from the aquifer to enter the packing material and thus the screen. One must properly apply the well completion information to perform the calculations. For example, Figure 11.6 shows an example of a well completed with screen completed above the static water level and an example of a well partially screened within a clay unit. It doesn't make much sense to extend the parameter L above the water table or into the clay unit if they are not contributing water. Example 11.2 shows how to adjust for the radius of the casing when displacement levels drop within the screen interval.

The author is aware of cases where the driller wishes to complete a monitoring well with packing materials at a lower hydraulic conductivity than the aquifer. If this is true and the well is supposed to be used for estimating hydraulic properties, then the driller should be "slapped up the side of the head." The hydraulic parameters that will be estimated will only reflect the gravel pack and not the aquifer. It is the hydrogeologist's responsibility to make sure that well completion reflects the purposes of the well.

Example 11.3

The data plotted in Figure 11.7 comes from a well completed in a streamside mine tailings area in southwestern Montana. Monitoring wells were installed to learn about the flow characteristics in the impacted area. The raw data are found in Table 11.1. The adjusted time zero is at 0.0266 min. All data previous to this time are presumed to be attributed to splash effects and are not used. The reference of 100 ft (30.5 m) is subtracted from all the displacement data before plotting.

If one looks at the data from a low angle (now get your face near the table surface), roughly three straight-line segments can be identified. The first break is at approximately 0.25 min (15 sec) and the second break is after 1.0 min. Approximately 50% of the total recovery took place during the first straight-line segment. The second and third segments are collectively referred to by the authors as a "tailing effect" as a tapering recovery takes place. This is described in the paper by Bouwer (1989). From the diagram, a T_o of approximately 0.36 min is identified and used in Equation 11.1. The diameter of the casing is 2 in. (5.1 cm), $r = 0.083$ ft, and the diameter of the borehole is approximately 8 in. (22.8 cm), radius of $R = 0.33$ ft. The length of the screen (L) is 5 ft (1.5 m) and the packing materials are 7 ft (2.1 m).

$$K = \frac{r^2 \ln(L/R)}{2LT_o} \qquad = \frac{(0.083\text{ ft})2 \times \ln(7\text{ ft}/0.33\text{ ft})}{2 \times 7\text{ ft} \times 0.36\text{ min}}$$

$$= 4.2 \times 10^{-3}\text{ ft/min} \times 1440\text{ min/day} = 6\text{ ft/day}$$

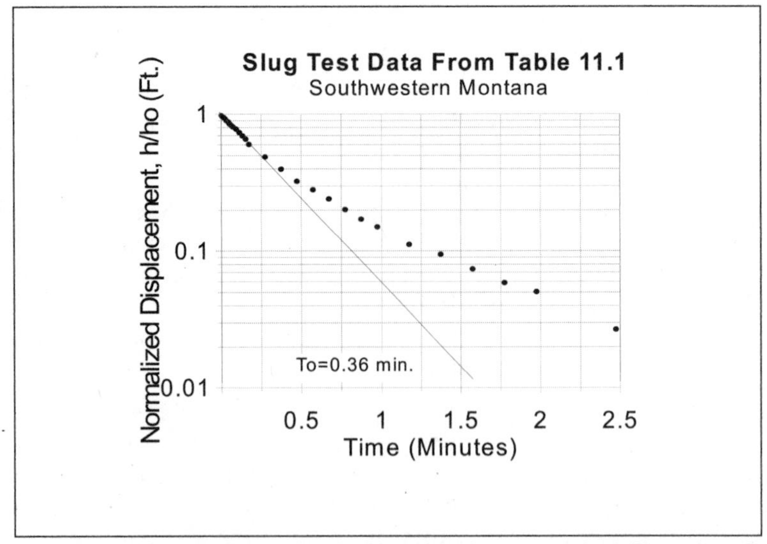

Figure 11.7 Slug test illustrating the Hvorslev slug test method. Flattening of the curve shows the tailing effect.

Bouwer and Rice Slug Test

Another commonly used slug-test method was developed by Bouwer and Rice (1976) that would estimate the hydraulic conductivity of aquifer materials in a single well. It was designed to account for the geometry of partially penetrating or fully penetrating wells in unconfined aquifers (Bouwer 1989). Many software packages provide the Hvorslev (1951) and Bouwer and Rice (1976) methods as the only choices for slug test analysis. Having more than one method to calculate the hydraulic conductivity is helpful. It is especially useful for cross-checking values for comparison purposes. If the results of one method differ significantly from the other, then this is helpful in detecting calculation errors.

The slug-test data are collected, as before, except that Bouwer and Rice (1976) use a slightly different notation. They prefer to call the maximum displacement y_o instead of h_o and refer to the other displacement values as y_t instead of h. There is also no ratio of the data from y_o. Instead, y_o is chosen from the plot at time zero, unless there is a double-line effect (discussed later on in this section). The geometry of the variables used in the analysis is shown in Figure 11.8.

The variables are defined as:

h = depth from bedrock to water table

L_w = length from bottom of screen to the water table

L_e = length of well screen plus any packing materials

R_w = effective radius of well bore

y = displacement interval (shown as H in Figure 11.8)

r_c = radius of the casing where the rise of the water level is measured

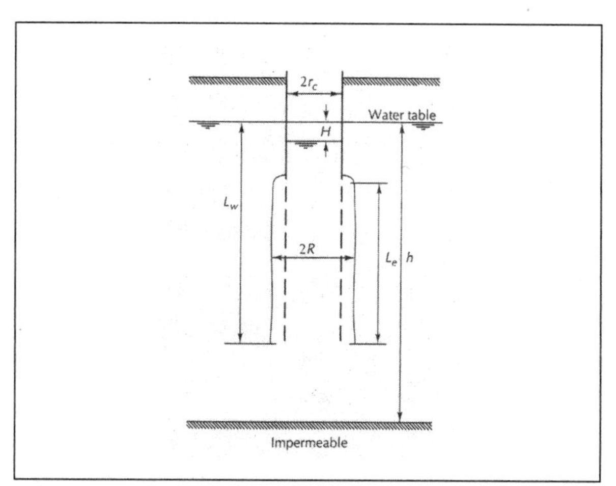

Figure 11.8 Geometry and parameters for a partially penetrating well screen in an unconfined aquifer. [Bouwer (1989). Used with permission of Ground Water Publishing]

As was mentioned earlier, Bouwer (1989) suggested that if the water level rises within the open or screened portion of the well, an adjusted r_c value needs to be calculated that accounts for the porosity of the packing material. This adjustment is not necessary if the rise in water level back to SWL is confined within the well casing. The hydraulic conductivity is calculated from:

$$K = \frac{r_c^2 \ln(R_e / R_w)}{2L_e} \frac{1}{t} \ln \frac{y_o}{y_t} \qquad [11.3]$$

where:

y_o = the maximum displacement at time zero

y_t = displacement at time t

$\ln(R_e/R_w)$ = dimensionless ratio that is evaluated from analog curves (R_e is the effective distance over which the head displacement dissipates)

The analog data in Bouwer and Rice (1976) is used to fit one of two equations, one for the case that $L_w < h$ (i.e., partial penetration), and one for $L_w = h$ (i.e., fully penetrating). The partial penetration case is:

$$\ln\frac{R_e}{R_w}\left[\frac{1.1}{\ln(L_w / R_w)}+\frac{A + B * \ln[(h - L_w)R_w}{L_e / R_w}\right]^{-1} \quad \textbf{[11.4]}$$

and the fully penetrating case is:

$$\ln\frac{R_e}{R_w}\left[\frac{1.1}{\ln(L_w / R_w)}+\frac{C}{L_e / R_w}\right]^{-1} \quad \textbf{[11.5]}$$

Where A, B, and C are dimensionless numbers plotted as a function of L_e/R_w.

The plots of A, B, and C are shown in Figure 11.9. One possible benefit of the Bouwer and Rice (1976) method is that the aquifer geometry is taken into account. A problem arises in using this method if the depth to bedrock is not known or cannot be estimated.

Another phenomenon that may be observed in pumping tests is what is known as the double straight-line effect (Bouwer 1989). During initial displacement, the water in the gravel-packing material drains quickly into the well, thus showing a steep slope in the displacement data (Figure 11.10). When the water level in the packing materials equals the water level in the well, then the water level slows to reflect the contribution of the aquifer materials, thus forming a second straight line. Any deviations after the second straight line may be attributed to the "tailing effect" described earlier in the Hvorslev (1951) method. The steep first-line portion of the curve usually takes place over a matter of a few seconds. In Example 11.3, one may argue that a double-line effect is observed; however, the first straight line takes place over 15 sec, and the calculated value is consistent with the lithology (fine to medium sand). It is the author's experience that steep first-line segments reflecting drainage from the gravel pack occur over a few seconds or a very small proportion of the total time of recovery.

Since the first (very steep) straight line is representative of the gravel-packing materials, then the second straight line is used in performing the calculations. Notice that y_o is projected back to time zero. The straight line for the aquifer response is projected below the tailing effect. Any time t can be chosen along the aquifer response line for the calculations. In Equation 11.3, the time t has a corresponding displacement value y_t that is used when calculating the hydraulic conductivity. Since there are no requirements concerning 37% of the recovery, as in the Hvorslev method, it is convenient to select a time and displacement that are most easily read from the plot.

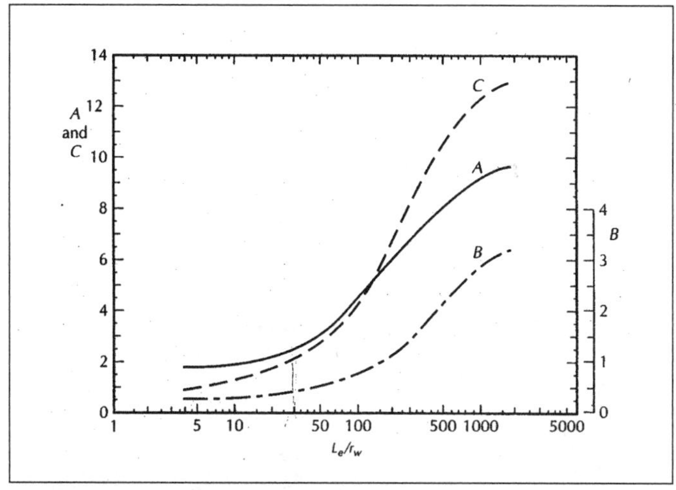

Figure 11.9 Parameters A, B, and C (dimensionless) plotted as a function of L_e/R_w used in Equations 11.4 and 11.5. [Bouwer (1989). Used with permission of Ground Water Publishing.]

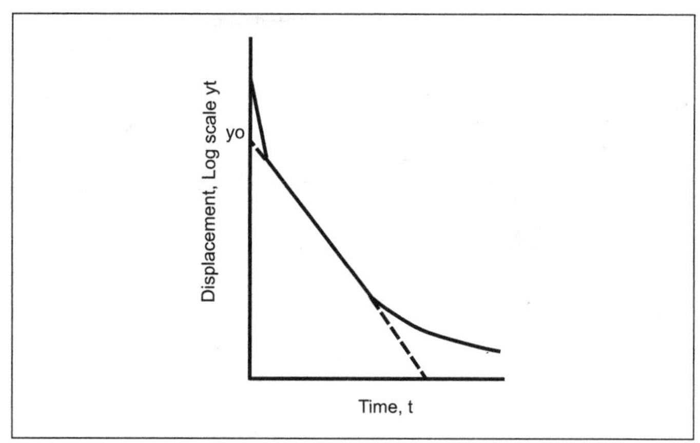

Figure 11.10 Displacement of head in a well versus time, showing the double straight-line effect. The first segment represents drainage of the gravel pack, the second segment the aquifer response, and the curved the tailing effect. [Modified from Brouwer (1989).]

Example 11.4

Three students were sent out to a tailings pond site to conduct slug tests. Figure 11.11 shows a cross section of the setting. Each student selected a slug that would displace the same amount of water (in feet) as the well number. All students evaluated the data using the Bouwer and Rice (1989) method. Notes on the lithologic materials were misplaced. Implications concerning the nature of the tailings were to be evaluated from the slug test results. The data are shown in Table 11.2, and the plots and calculations are shown below.

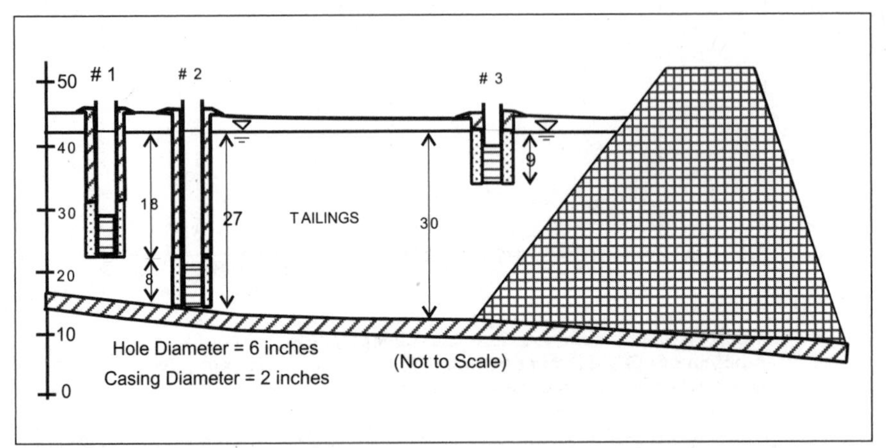

Figure 11.11 Cross section view of the tailings pond with monitoring well completion information for Example 11.4.

Well 1

Since well 1 is partially penetrating, Equation 11.4. is used, where:

$$L_e = 9 \text{ ft},\ L_w = 18 \text{ ft},\ r = 0.083 \text{ ft},\ R_w = 0.25 \text{ ft},\ h = 26 \text{ ft},$$

$$A = 2.6,\ B = 0.45,\ t = 9 \text{ sec},\ y_t = 0.1 \text{ ft},\ y_o = 1.0 \text{ ft}$$

$$\ln\frac{R_e}{R_w}\left[\frac{1.1}{\ln(18\ \text{ft}/0.25\ \text{ft})}+\frac{2.6+0.45\times\ln[(26\ \text{ft}-18\ \text{ft})/0.25\ \text{ft}]}{9\ \text{ft}/0.25\ \text{ft}}\right]^{-1}=2.683$$

$$K=\frac{(0.083\ \text{ft})^2(2.683)}{2(9\ \text{ft})}\frac{1}{9\,\text{sec}}\ln\frac{(1.0\,\text{ft})}{(0.1\text{ft})}=2.63\times10^{-4}\,\text{ft}/\text{sec}=23\ \text{ft}/\text{day}$$

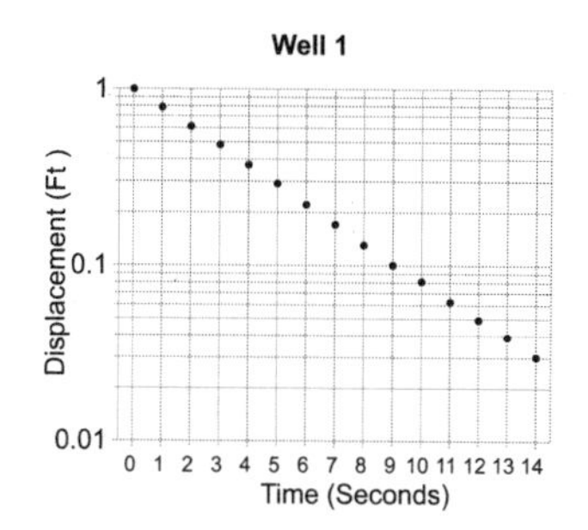

Well 2

Since well 2 is fully penetrating, Equation 11.5. is used, where:

L_e = 8 ft, L_w = 27 ft, r = 0.083 ft, R_w = 0.25 ft, h = 27 ft, C = 2.2
t = 5 sec, y_t = 1.0 ft, y_o = 2.0 ft

$$\ln\frac{R_e}{R_w}\left[\frac{1.1}{\ln(27\text{ ft}/0.25\text{ ft})}+\frac{2.2}{8\text{ ft}/0.25\text{ ft}}\right]^{-1}=3.29$$

$$K=\frac{(0.083\text{ ft})^2(3.293)}{2(8\text{ ft})}\frac{1}{5\text{ sec}}\ln\frac{(2.0\text{ ft})}{(1.0)\text{ ft}}=1.98\times10^{-4}\text{ ft/sec}=17\text{ ft/day}$$

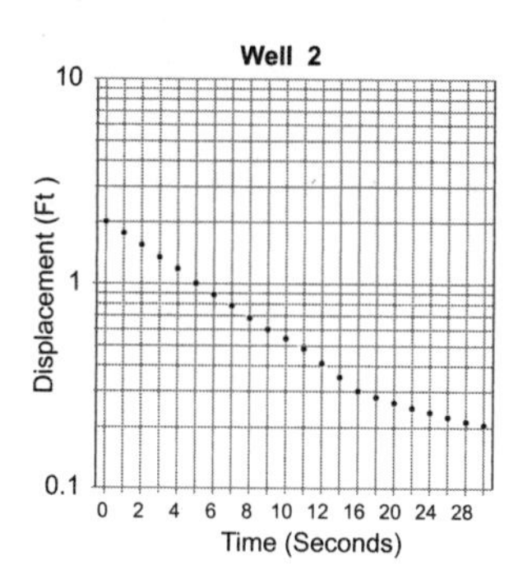

Well 3

Since well 3 is partially penetrating, Equation 11.4. is used, where:

L_e = 9 ft, L_w = 9 ft, r = 0.083 ft, R_w = 0.25 ft, h = 30 ft,

$A = 2.6$, $B = 0.45$, $t = 9$ sec, $y_t = 0.9$ ft, $y_o = 1.12$ ft

$$\ln\frac{R_e}{R_w}\left[\frac{1.1}{\ln(9\,\text{ft}/0.25\,\text{ft})}+\frac{2.6+0.45\times\ln[(30\,\text{ft}-9\,\text{ft})/0.25\,\text{ft}]}{9\,\text{ft}/0.25\,\text{ft}}\right]^{-1}=2.3$$

$$K=\frac{(0.083\,\text{ft})^2(2.30)}{2(9\,\text{ft})}\frac{1}{9\,\text{sec}}\ln\frac{1.12\,\text{ft}}{0.9\,\text{ft}}=2.14\times10^{-5}\,\text{ft/sec}=2\,\text{ft/day}$$

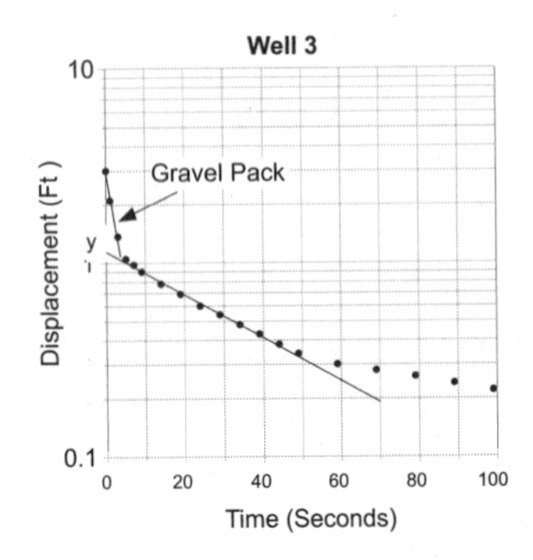

Even though the lithologic logs were lost, one can make some inferences concerning the hydraulic and lithologic properties of the tailings sediments from the location of the wells and the calculations. It is evident that the hydraulic conductivity decreases towards the dam. This is something that might be expected. As the tailings are slurried towards the dam, the coarsest fraction would likely settle out first. The finer materials would continue towards the dam. It appears that the coarsest fraction lies near the bottom with a fining upward sequence to the sediments. This suggests that higher hydraulic conductivities may be found nearer the bedrock.

Common Errors Made in Analyzing Slug Test Data

There are some common ways that individuals make errors in performing calculations of slug tests. The most routine ones are presented below.

- Using the diameter of a casing or borehole instead of the radius. All of the methods require the radius instead of the diameter.
- Forgetting to include the gravel-packing materials above and below the screen as the contributing length *L*. Many have only used the screen interval, which leads to inaccurate results.
- Be careful not to average the data to make a single "best fit" line. If the data are roughly in a straight line, then this is no problem, but including the tailing effect portion along with the aquifer response data will result in underestimating the hydraulic conductivity. Get your face close to the table to evaluate the plot to detect the appropriate straight-line section from which to draw the line.

Table 11.2 Slug Test Data from a Tailings Pond Site

Well 1		Well 2		Well 3	
Time (sec)	Displacement (ft)	Time (sec)	Displacement (ft)	Time (sec)	Displacement (ft)
0	1.00	0	2.00	0	3.00
1	0.78	1	1.76	1	2.10
2	0.61	2	1.54	3	1.37
3	0.48	3	1.34	5	1.05
4	0.37	4	1.18	7	0.98
5	0.29	5	1.00	9	0.90
6	0.22	6	0.88	14	0.78
7	0.17	7	0.78	19	0.69
8	0.13	8	0.68	24	0.60
9	0.10	9	0.60	29	0.54
10	0.081	10	0.54	34	0.48
11	0.062	11	0.48	39	0.43
12	0.049	12	0.41	44	0.38
13	0.039	14	0.35	49	0.34
14	0.03	16	0.30	59	0.30
		18	0.28	69	0.28
		20	0.263	79	0.26
		22	0.248	89	0.24
		24	0.236	99	0.22
		26	0.224		
		28	0.213		
		30	0.205		

- Don't forget to square the casing radius value. In both methods described above, the radius of the casing is squared in the calculations. In the Bouwer and Rice (1976) method, it is important to take the inverse of the bracketed quantity in Equations 11.4 and 11.5.

- In the Bouwer and Rice (1976) method, in the partial penetration case, make sure that the constants *A* and *B* are picked from the appropriate scales (the *B* scale is on the opposite side of the graph as the *A* scale.

How to Analyze Slug Tests for Both Damped Methods from a Single Plot

Now that the Hvorslev and Bouwer and Rice methods have been presented, it is useful to see how to perform the calculations for both methods from a single plot. The simplest way of dealing with slug-test data is to plot displacement versus time. Recall that in the Hvorslev (1951) method the data were normalized with h_o so that at time zero the value would be near 1.0. From this plot, the time at 37% was selected as T_o to be used in the calculations. Any displacement versus time plot can be used for the Hvorslev (1951) method. One only needs to determine what h_o or y_o is and then take 37% of this value for the corresponding T_o value.

For example, in well 3 in Example 11.4, if the projected h_o in the tailings pond is 1.12 ft, then 37% of this is 0.41 ft. T_o would be approximately 40 sec. This would be used in Equation 11.1 to obtain:

$$K = \frac{(0.083\ \text{ft})^2 \ln(9\text{ft} / 0.25\text{ft})}{2(9\text{ft})40\text{sec}} = 3.4 \times 10^{-5}\ \text{ft} / \text{sec} = 3\text{ft} / \text{day}$$

In comparing the calculated values between the two damped case methods, the Hvorslev (1951) method yields a larger number, although both values are smaller than those of the other two wells. The reader is invited to make a comparison of well 1 and well 2 using the Hvorslev (1951) method (taking 37% of y_o to determine T_o) to recalculate the hydraulic conductivity. Similar trends to the Bouwer and Rice (1976) method should be observed; however, the numbers are not exactly the same. This is typical of aquifer hydraulics; in some cases, one method may yield a higher or lower value than another method, but should yield similar trends.

The author noticed that when the well was full penetrating (the case of well 2) the numbers from the two methods were almost the same. The Hvorslev (1951) method produces a higher hydraulic conductivity value for wells 1 and 3. In both of these cases, the wells are partially penetrating. The Bouwer and Rice (1976) method takes into account partial penetration effects by including the aquifer geometry and may more accurately reflect the conditions. This is a subject for additional study and debate.

11.3 Analyzing Slug Tests—Underdamped Case

Unlike the previous sections where the methodologies for the damped case have been presented, there are occasions in high-transmissivity aquifers where the water-level response during slug tests behaves like an under-damped spring. The water level literally oscillates back and forth above and below the static equilibrium level in a methodical motion. This behavior is unexpected by many field personnel, since slug tests were primarily developed for low-transmissivity aquifers (Lohman 1972). Although different methods for analyzing these data sets exist, it is the author's experience that many individuals who encounter oscillatory water-level data during slug tests deem the data either "confusing or unsolvable" or simply assign a "high" hydraulic conductivity. Analytical methods for oscillatory data in the literature are mathematically complex and challenging to use (van der Kamp 1976; Uffink 1984; Kipp 1985). This can be intimidating to the practitioner and field hydrogeologist and may have contributed to its limited use as an analytical tool or method of choice among software companies that produce well hydraulics packages.

Simplifying methods for practitioners have been developed for spreadsheets for the van der Kamp (1976) method (Wylie and Magnuson 1995) and for the Kipp (1985) method (Weight and Wittman 1999). One potential drawback is that both methods assume there is no skin effect, although the equations in the Kipp (1985) method do allow for skin effects. It is known that both positive and negative skins can occur (Yang and Gates 1997; Butler 1998); however, the impact of skin effects is often evaluated with lower hydraulic conductivity formations and can be reduced with proper well development. The authors believe that a skin effect is a "fudge factor" associated with poor well completion and is difficult to quantify.

A relatively small group of researchers and practitioners have published information on water-level responses that indicate inertial effects resulting in oscillations. van der Kamp (1976) presented the first significant paper leading to the analysis of oscillatory data using a sinusoidal approximation method. In 1985, Kipp expanded the theory presented by Bredehoeft et al. (1966) to produce a series of type curves. There are other underdamped methodologies that have been developed by Uffink (1984), Springer and Gelhar (1991), and McElwee et al. (1992). Some of these are also similar to the method by Kipp (1985) in that they also estimate a damping parameter (ζ) and the effective water-column height (L_e) and have been tested in the field.

Nature of Underdamped Behavior

The first significant explanations of underdamped oscillatory behavior were offered by Bredehoeft and others (1966). They were intrigued by a couple of examples that occurred in Florida and Georgia. In Florida, a 12-inch well completed in a cavernous limestone aquifer (transmissivity approximately 120,000 ft^2/day) responded in an underdamped oscillatory fashion when a float was periodically raised and lowered. This behavior was picked up by a pressure transducer within the well. In Georgia, a recorder located 200 ft from a city supply well showed an oscillatory response every time the city well's pump kicked on (Bredehoeft et al. 1966). Additional geologic field settings are described by Butler (1998). According to Bredehoeft and others (1966), systems are said to be:

- Overdamped or damped, where no oscillations occur following initial disturbance, indicating that viscous forces or Darcian conditions dominate the system,
- Underdamped, where oscillations occur as a damped sine wave, indicating that inertial forces are significant, and
- Critically damped, where the system is in transition between the two.

Bredehoeft et al. (1966) first studied inertial effects by simulating an oscillatory system using an electric analog, but did not develop a method of slug-test analysis. They present a relationship between transmissivity and effective column height for a given transmissivity and storativity (Figure 11.12). The fit line in Figure 11.12 represents where the critically damped transition takes place. A sinusoidal approximation to the underdamped oscillatory slug test response was presented by van der Kamp in 1976. Another sinusoidal approximation, using a pneumatic procedure, is described by Uffink (1984) in Kruseman and deRidder (1990). Kipp (1985) was later able to develop a method of slug-test analysis with type curves. The van der Kamp (1976) approximation method is easily performed using a spreadsheet solution developed by Wylie and Magnuson (1995). The type curves of the Kipp (1985) method along with a spreadsheet solution were developed by Weight and Wittman (1999).

Example 11.5

During the 1988 Hydrogeology Field Camp of Montana Tech of the University of Montana, underdamped oscillatory responses from slug tests were observed in wells west of Butte, Montana (Manchester 1990). Matching oscillations were noticed from two wells located 40 ft apart, although the observation well was significantly dampened. Follow-up pneumatic slug tests were performed, including one in a well "completed" in the swimming pool on the Montana Tech campus to simulate "infinite" transmissivity. A spectacular oscillatory response was observed (Manchester 1990).

Another "infinite" transmissivity example was observed in a 12-in. production well after a blasting charge was set off. The 12-in. well was supposed to be drilled into a mine adit that was encountered by two adjacent wells A and B (Figure 11.13). The adit was believed to be missed by only 1 to 3 ft. A blasting charge was set off to breach the connection (Figure 11.14). A pressure transducer in well B was destroyed during the process so no data

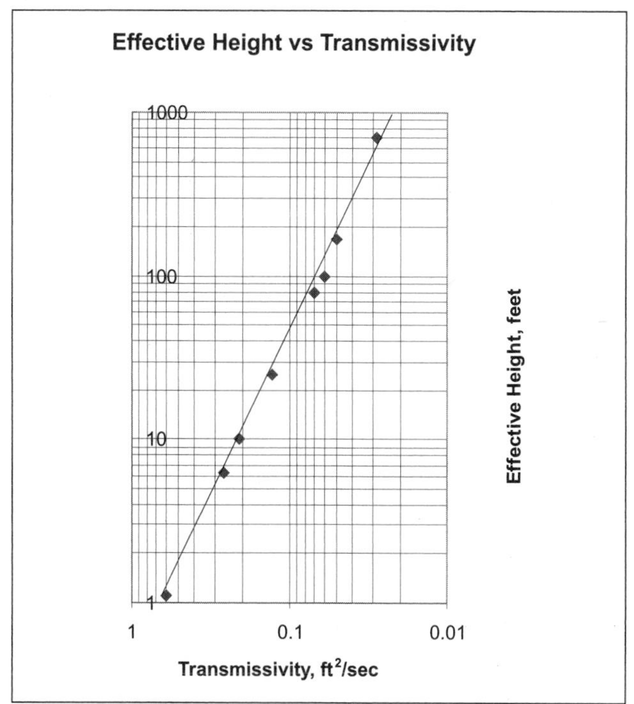

Figure 11.12 Relationship of transmissivity to effective water column height. Area above the plotted points represent the underdamped case, area below represents the damped case, and modified fitted line represents the critically damped transition. [Modified after Bredehoeft et al. (1966.)]

were collected; however, a reflected beam of light from a mirror revealed a clear oscillatory response for over seven minutes. The production well was subsequently tested at 200 gpm with 0.6 ft of drawdown observed, indicating that an excellent production well was achieved.

Weight and Wittman (1999) describe oscillatory water-level responses while collecting slug-test data from wells scattered within the Beaverhead Groundwater Project Area (BGPA) in southwestern Montana (Figure 11.3). Point estimates of hydraulic conductivity values within the BGPA were needed to provide additional hydraulic control within a large area that was to be modeled. These data would augment other hydraulic parameter estimates from pumping tests and previous slug tests. Of the 30-plus wells that were tested, over half of the wells showed an underdamped oscillatory response. It became necessary to analyze these data or have significantly less hydraulic control within the model area.

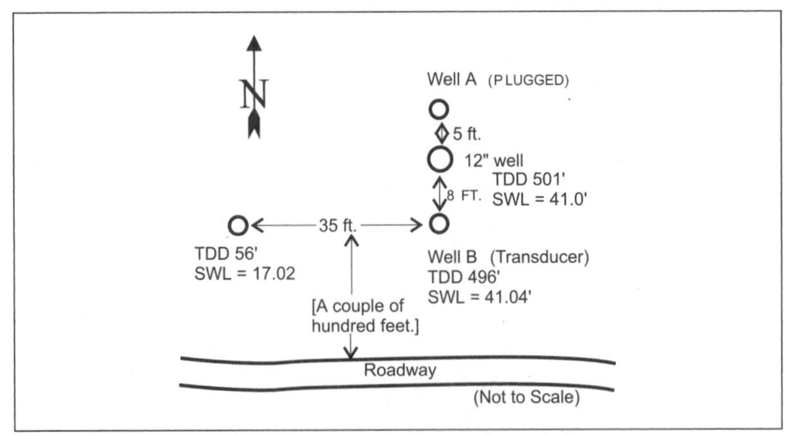

Figure 11.13 A 12-in. production well missed a mine adit encountered by wells A and B. A blasting charge successfully breached the adit as oscillatory behavior occurred in the water level for over 7 min.

Figure 11.14 Constructing explosives pack.

Example 11.6

Perhaps one of the more spectacular oscillatory data sets collected by the authors occurred west of Butte, Montana, in July 1996, during the Hydrogeology Field Camp at Montana Tech of the University of Montana. The slug test was conducted prior to a constant duration rate test (Figure 11.15). The normalized oscillatory response data are shown in Table 11.3 and illustrated in Figure 11.16.

Table 11.3 Normalized Oscillatory Response Data, Sand Creek Well, West of Butte, Montana

Time (sec)	Displacement (ft)	Time (sec)	Displacement (ft)
4.0	-0.019	25.0	1.471
4.5	0.529	26.0	1.377
5.0	1.011	27.0	1.112
5.5	1.458	28.0	0.718
6.0	1.846	29.0	0.252
6.5	2.158	30.0	-0.223
7.0	2.384	31.0	-0.642
7.5	2.529	32.0	-0.954
8.0	2.583	33.0	-1.121
8.5	2.548	34.0	-1.134
9.0	2.438	35.0	-0.995
9.5	2.249	36.0	-0.734
10.0	1.997	37.0	-0.390
10.5	1.682	38.0	-0.012
11.0	1.320	39.0	0.344
11.5	0.923	40.0	0.636
12.0	0.510	41.0	0.832
12.5	0.095	42.0	0.907
13.0	-0.315	43.0	0.863
13.5	-0.696	44.0	0.712
14.0	-1.039	45.0	0.476
14.5	-1.335	46.0	0.189
15.0	-1.571	47.0	-0.107
15.5	-1.741	48.0	-0.375
16.0	-1.845	49.0	-0.576
16.5	-1.880	50.0	-0.693
17.0	-1.849	51.0	-0.715
17.5	-1.754	52.0	-0.639
18.0	-1.603	53.0	-0.485
18.5	-1.398	54.0	-0.277
19.0	-1.152	55.0	-0.044
19.5	-0.872	56.0	0.186
20.0	-0.573	57.0	0.375
21.0	0.044	58.0	0.510
22.0	0.621	59.0	0.570
23.0	1.081	60.0	0.551
24.0	1.373		

Figure 11.15 Constant rate pumping test, Hydrogeology Field Camp, July 1996.

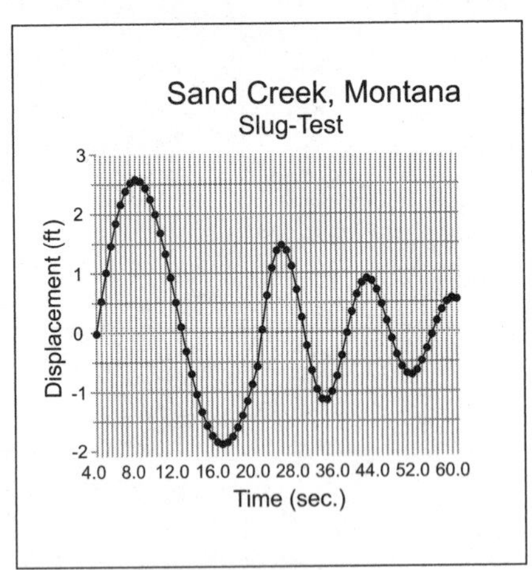

Figure 11.16 Slug test illustrating oscillatory behavior (data from Table 11.3).

Effective Stress and Elastic Behavior

Physically, the oscillatory behavior phenomenon can be partially explained by the concept of effective stress and inertia, resulting in an elastic wave propagation. The effective stress is the difference between the total stress and the fluid pressure. Within an aquifer, the total load from the water and geologic materials is "felt" above a given plane (Fetter 1994). The load carried by the mineral skeleton is somewhat offset by the fluid pressure, resulting in an effective stress.

$$\sigma_t = \sigma_e + P_f \qquad [11.6]$$

where:

σ_t = total stress

σ_e = effective stress

P_f = fluid pressure

In confined aquifers, changes in pressure can occur with very minute changes in actual thickness (Fetter 1994). In this condition, there is very little change in the total stress; however, the effective stress increases as the fluid pressure decreases and vice versa. The exchange back and forth in stresses produces an elastic-like response in a confined aquifer from being under pressure. The inertial effects of the displaced water column eventually balance out with the frictional forces of the geologic media and well completion materials (van der Kamp 1976).

The phenomenon of elastic behavior and does not appear to be restricted to confined aquifers only. It is the author's observation that *any* aquifer that is being stressed during the first few seconds to a minute or so produces water from compression of the mineral skeleton or expansion of water (Chapter 10), even in unconfined aquifers (Prickett 1965; Neuman 1979; and Moench 1993). This essentially describes the specific storage (S_s) or elastic storage viewed on a type A Theis (1935) curve (the early-time data, Chapter 10). Higher-transmissivity aquifers tend to propagate perturbations within an aquifer faster, indicating significant inertial effects over lower-transmissivity aquifers. An example of the quick transition of confined to unconfined conditions is presented in Example 11.7.

Example 11.7

Near Kalispell, in northwestern Montana, is a glacio-fluvial setting where a shallow coarse gravel aquifer (less than 30 ft, 9.1 m) overlies lacustrine clays. A constant-rate pumping test at 350 gpm. (1,908 m^3/day) was conducted in a 10-in. well to evaluate the hydraulic properties of the aquifer (King 1988). A plot of the data from an observation well located 37 ft away from the pumping well is shown in Figure 11.17. No transducer was on hand to collect the earliest time data (less than 1 min); however, delayed yield is evident within a couple of minutes. Delayed yield indicates the aquifer is unconfined and that the elastic response of the aquifer must have occurred within a minute or less, the approximate time of a slug test in a high-transmissivity aquifer. The transmissivity calculated was very high for a porous media aquifer, 70,000 ft^2/day.

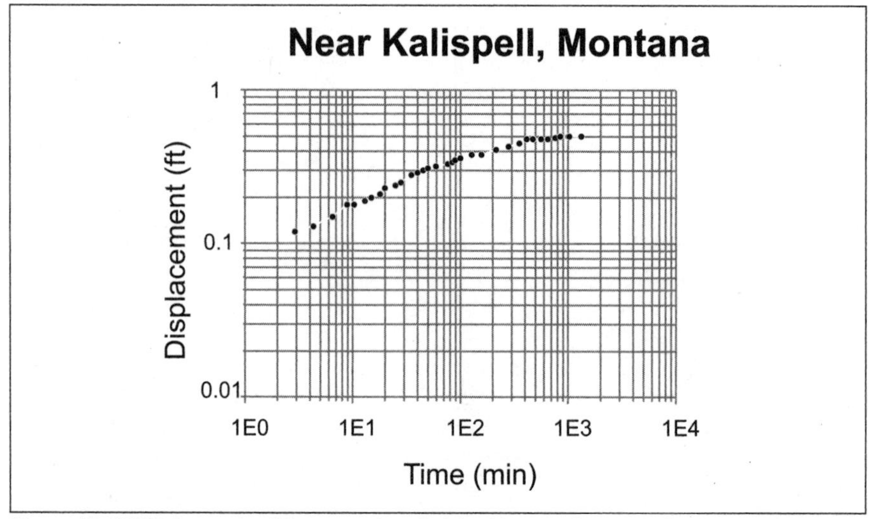

***Figure 11.17** Displacement (ft) versus time (min) with pumping rate of 350 gpm, in a pumping test near Kalispell, Montana.*

During an instantaneous removal of a slug, the fluid pressure suddenly decreases, causing an increase in the effective stress. The mineral skeleton has to take more of the load, and this squeezes the aquifer slightly. Water from the aquifer then pushes into the well bore past the original "preslug" equilibrium position resulting in an upward oscillation. The inertial effects, i.e., rate of change of momentum of the water column, pulled downward by gravity, gushes back out into the aquifer, increasing the fluid pressure once again during the downward oscillation. This can occur because the water column height is sufficiently high or the transmissivity is sufficiently large to overcome the critically damped transition zone Bredehoeft et al. (1966). This process repeats itself until the aquifer materials and well completion materials finally slow the process down. It can be thought of as an elastic response created by a pressure wave propagating radially from the well. The higher the transmissivity, the easier it is for the pressure wave to propagate outward. Wylie and Magnuson (1995) draw the analogy of a mechanical system of a spring in a viscous medium, where the water column is the mass and the aquifer is the spring. In the "infinite" aquifer scenario of a cavernous limestone, swimming pool, or mine adit, eventually the well completion materials and boundary effects of the cavity cause damping to finally occur. In the above discussion, it is assumed that there were no skin effects.

Two methods will be presented on how to analyze data that behave in an underdamped fashion, the van der Kamp (1976) and Kipp (1985) methods. Each of these will incorporate spreadsheet algorithms developed by Wylie and Magnuson (1995) and Weight and Wittman (1999) to facilitate the process.

van der Kamp Method

Garth van der Kamp (1976) developed a sinusoidal approximation method to underdamped slug-test data. A brief theoretical development is presented, followed by the use of a spreadsheet analog using an example. The geometry needed to discuss the analysis of the well response is shown in Figure 11.18.

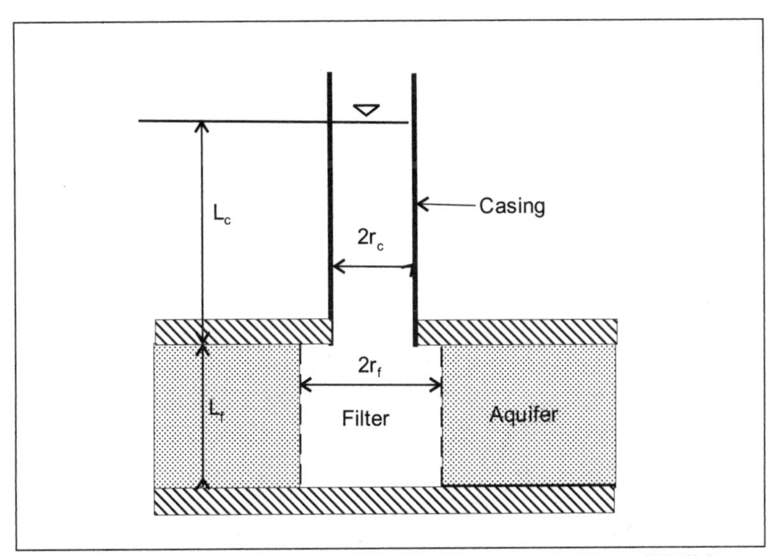

Figure 11.18 Geometry of slug test. [Modified from van der Kamp (1976).]

The effective length of the water column (L), as defined by Cooper et al. (1967) is:

$$L = L_c + 3/8L_f \qquad [11.7]$$

The equation for the balance of forces within the well filter, including inertial effects, is given in Equation 11.8.

$$\frac{d^2w}{dt^2} + \frac{g}{L}w = \frac{g}{L}h_f \qquad [11.8]$$

where:

w = transient water level in the well (L)

g = acceleration due to gravity

h_f = hydraulic head of water in the filter (L)

L = length of the effective water column

If the transient part of the hydraulic head can be assumed to be governed by the standard equation for radial flow and that h_f is the same everywhere inside of the filter pack an approximate solution can be used to simulate an exponentially damped cyclic fluctuation (Wylie and Magnuson 1995):

$$\omega(t) = \omega_o e^{-\gamma t} \cos(\omega t) \qquad [11.9]$$

where:

γ = damping constant (gamma)(T^{-1})

ω = angular frequency of the oscillation (omega)(T^{-1})

ω_o= initial water-level displacement from slug removal (L)

T = time from the start of the test (T).

The field data are fitted to the approximation equation to determine values for γ and ω. It is presumed that Storativity (S) and the filter radius (r_f) are known or can be estimated. In the spreadsheet program created by Wylie and Magnuson (1995) the transmissivity values are estimated by fitting the field data to the approximation algorithm. Access to the spreadsheet program via the Internet was through the Idaho National Engineering and Environmental Laboratory (INEEL) home page at http://www.ineel.gov/groundwater. Essentially, once the spreadsheet is loaded, one must follow the format of setting up the field data and then manually make adjustments to γ and until there is a good match to the approximation algorithm. These will be graphed as two different plot series whose patterns overlay once the appropriate γ and ω are chosen. The following suggestions will help one to quickly evaluate the data.

- Retrieve the spreadsheet program from the Internet created by Wylie and Magnuson (1995).
- Copy this spreadsheet to another worksheet so your field data can be evaluated, using the first spreadsheet as a reference.
- On the reference sheet notice that the *maximum* displacement value is used as the time zero position. If maximum the displacement value is not used at time zero, the graph of the algorithm approximation will not make sense.

- Increasing the gamma (γ) values tightens up the frequency of the field data, and decreasing them will cause them to spread out more.
- Increasing the omega (ω) values tends to decrease the amplitude height of the field data.
- Decreasing the storativity values tends to increase the transmissivity significantly.

Example 11.8

An example of a sequential fit to the data from Table 11.3 is shown in Figure 11.19. As in the Kipp (1985) method, the storativity values are somewhat sensitive to the transmissivity outcomes. A discussion about this follows the presentation of the Kipp (1985) method.

Kipp Method

Another method that can be used to estimate transmissivity values for aquifers was developed by Kenneth Kipp (1985). He was the first to expand the earlier theory developed by Bredehoeft et al. (1966) that takes into account both well-bore storage and inertial effects and produce a series of type curves that allow for oscillatory responses to be analyzed (Figure 11.20).

A summary of the basic assumptions given by Kipp (1985) was summarized by Manchester (1990) with additional discussion by Weight and Wittman (1999).

- Confined aquifer, bounded on the top and bottom
- Uniform aquifer thickness
- Well fully penetrates the aquifer
- Aquifer of infinite areal extent
- Homogeneous porosity (η) and matrix compressibilities
- Flat potentiometric surface
- Delayed yield not considered
- Constant water density in the well bore and a constant compressibility in the aquifer

Kipp (1985) was able generate dimensionless variables and parameters which were plotted into a series of type curves (Figure 11.20). For example, there is a dimensionless inertial parameter (β) that can be held constant

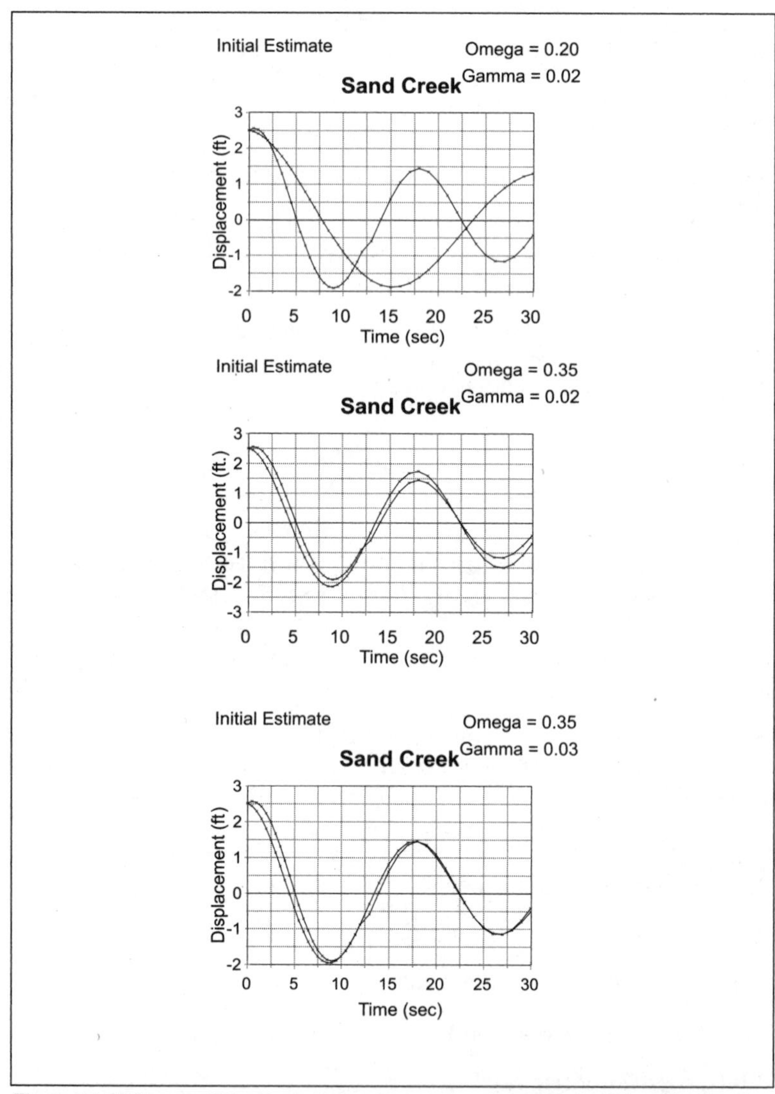

Figure 11.19 Sequential adjustments of omega and gamma in fitting the Sand Creek field data (Table 11.3) using the sinusoidal approximation spreadsheet algorithm by Wylie and Magnuson (1995).

while different values of the storage parameter (α) are used to produce the dimensionless damping parameter (ζ). The tabled values from the Kipp (1985) paper were entered these into a spreadsheet (Weight and Wittman 1999). Field data can appropriately be scaled to fit the type curves to yield a value for transmissivity. A copy of this file is available via the Internet from the home page of the Department of Geological Engineering at Montana Tech (www.mtech.edu).

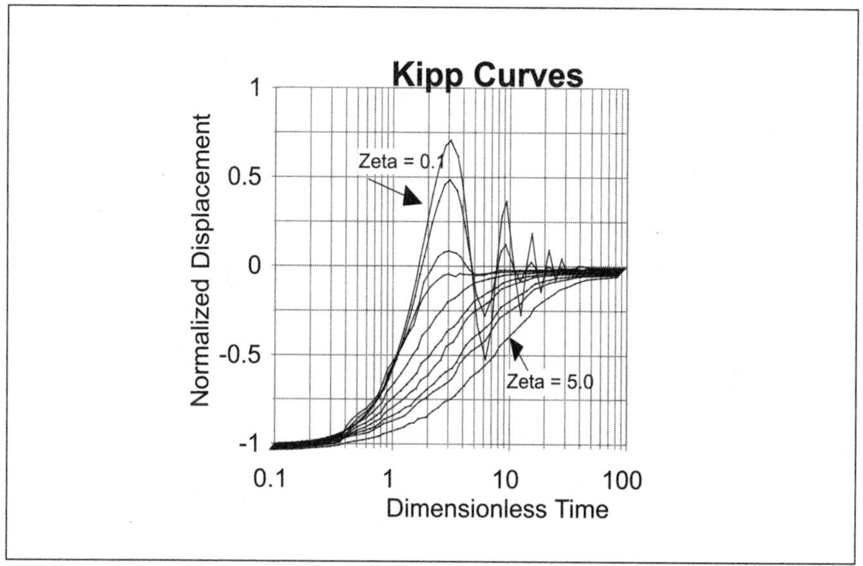

***Figure 11.20** Type curves for normalized oscillatory slug-test data sets. Damping parameters Zeta are 0.1, 0.2, 0.5, 0.7, 1.0, 1.5, 2.0, 3.0, 4.0, and 5.0, with the smallest value representing the most oscillatory responses.*

Methodology

The steps of the Kipp methodology will first be described from the paper by Weight and Wittman (1999) along with examples. To be able to follow along, it is assumed that the data collected in the data logger (displacement versus log time) have been uploaded to a PC or notebook computer into a spreadsheet program.

1. The uploaded data should first be converted into time units of seconds (x-axis) and displacement values in meters (y-axis) to match the type curves. Please read step 2 before this step is completed.

2. Evaluate the raw data and determine the reference point that the oscillations eventually dampen out to. This should compare closely to the original SWL or reference point. For example, if the reference point in the raw data is 10.0 ft (3 m) and the data eventually dampen out to 10.019 ft, then 10.019 ft becomes the reference point. This also becomes the zero displacement reference level (w). All other displaced

data be referenced to this final damped value so that differences will oscillate above and below zero.

3. Determine the maximum displacement (w_o). Note: sometimes there is some confusion here if the data logger is turned on *before* the slug has been pulled. The authors have seen cases where the data logger records data as the slug is being pulled from the well bore and has not broken the water surface yet. This results in an abnormally large maximum displacement value (Table 11.1). If this occurs, use the next largest displacement value as w_o. This is also the reason for activating the data logger a few seconds before pulling the slug. Furthermore, it is important to also rescale the time value back to when the slug has just been pulled (time zero).

4. Calculate $-w/w_o$ values for the displaced data. The minus sign is there because the type curves are viewed from a negative displacement perspective (i.e., rising-head test). Graph time in seconds (x-axis in log scale) versus $-w/w_o$ (y-axis in arithmetic scale).

5. Match the type curves with the graphed data. As you do this you will determine three values: the damping parameter (ζ)), the first peak $\hat{t}$, and the subsequent inflexion point t. Highly oscillatory systems, with parameters smaller than 0.5, should be matched to the first largest inflexion point or peak to select time $\hat{t}$. Time t is picked at the second inflexion point (where the displaced data first changes from concave downward to concave upward, usually near where it crosses the displacement axis; see the examples that follow). For oscillatory systems with a parameter greater than 0.7, use the lower inflexion point (where concave upward just begins, again consult the example).

6. Calculate the effective column length (L_e) using $\hat{t}$ and t using Equation 11.9 (Kipp 1985):

$$\hat{t} = \frac{t}{(L_e / g)^{1/2}} \qquad \textbf{[11.10]}$$

where:

L_e = effective static water length (m)

g = acceleration of gravity (m/s^2)

$\hat{t}$ = dimensionless time

t = time

7. Calculate the dimensionless storage parameter (α). This takes into account well-bore storage by considering the radius of the casing. It is presumed that storativity S is known or can be estimated (Kipp 1985).

$$\alpha = \frac{r^2_c}{2 \times S \times r^2_s} \qquad [11.11]$$

where:

α =	dimensionless storage parameter
r^2_c =	radius of the casing (m)
S =	storage
r^2_s =	radius of the screen (m)

8. Solve for the dimensionless inertial parameter (β), (Equation 28 in Kipp 1985). The critical damping parameter (ζ) is selected or estimated from the fit of the data to the type curves. This will also include additional discussion following step 9.

$$\zeta = \frac{\alpha(\sigma + 1/4 \times \ln(\beta))}{2 \times \beta^{1/2}} \qquad [11.12]$$

where:

ζ = dimensionless damping factor

α = dimensionless storage parameter

β = dimensionless inertial parameter

σ = dimensionless skin factor

9. Calculate the transmissivity (T) *in* m^2/sec, using the previously derived parameters (Kipp 1985).

$$\beta = \frac{L_e}{g}\left(\frac{T}{S \times r^2_s}\right)^2 \qquad [11.13]$$

Equation 11.12, used to solve for β , is most easily derived by setting up an iterative solver on a spreadsheet. The solver can be set up from the following logic also presented by Weight and Wittman (1999). Assuming the dimensionless skin factor (σ) is zero, Equation 11.12 can be rewritten in the following form:

$$\zeta = \frac{\alpha \times \ln(\beta)}{8 \times \beta^{1/2}} \qquad [11.14]$$

Equation 11.14 can be rearranged to:

$$\beta \times C^2 = (\ln \beta)^2 = 2 \times \ln \beta \qquad [11.15]$$

where:

$$\zeta \times 8 / \alpha = C \qquad [11.16]$$

and C = a constant

$$\frac{\beta \times C^2}{2} = \ln \beta \qquad [11.17]$$

Steps 6 through 9 are performed on the spreadsheet developed by Weight and Wittman (1999), including an iterative solver for Equation 11.14. The solver uses a fixed cell for the constant (C) and another cell is assigned to β. Three columns with formulas are set up to compare order of magnitude changes in β with the two sides of the equation expressed in Equation 11.14. When the values in the two columns match closely, the appropriate value for β can be identified. The following examples will illustrate the spreadsheet approach developed by Weight and Wittman (1999).

Example 11.9

This example describes the step-by-step process to evaluate the data in Table 11.3 as a comparison to Example 11.8.

Key in the data from Table 11.3 onto a spreadsheet. The displacement values in feet need to be converted to meters (multiply by 0.3048 m/ft), the time data in seconds are already in the appropriate format.

Evaluate to reference point where the data dampen to. In this case, assume this is at 0.0 ft since the time/displacement data stop before full dampening takes place.

The maximum observable displacement w_o is 0.787 m (2.583 ft). The data previous to 4 sec are clouded by splash effects, this makes it difficult to determine where to rescale time zero. The author subtracted 6 sec from the time data to rescale the data back to time zero.

A column on the spreadsheet is set up for the $-w/w_o$ values. A plot of time versus $-w/w_o$ values is shown in Figure 11.21.

From the plot, the values for $\hat{t}$ and t are estimated to be 10.1 and 13.8 sec, respectively. The zeta value appears to be less than 0.1 (Figure 11.21), so a value of 0.09 is estimated.

The radius of the screen is 0.0254 m, and the radius of the casing is 0.0762 m, so alpha is calculated to be 4500.

The value for beta from the spreadsheet matches well at 1.66×10^9, yielding a transmissivity of 0.01924 m^2/sec (17,900 ft^2/day). This compares with an estimated

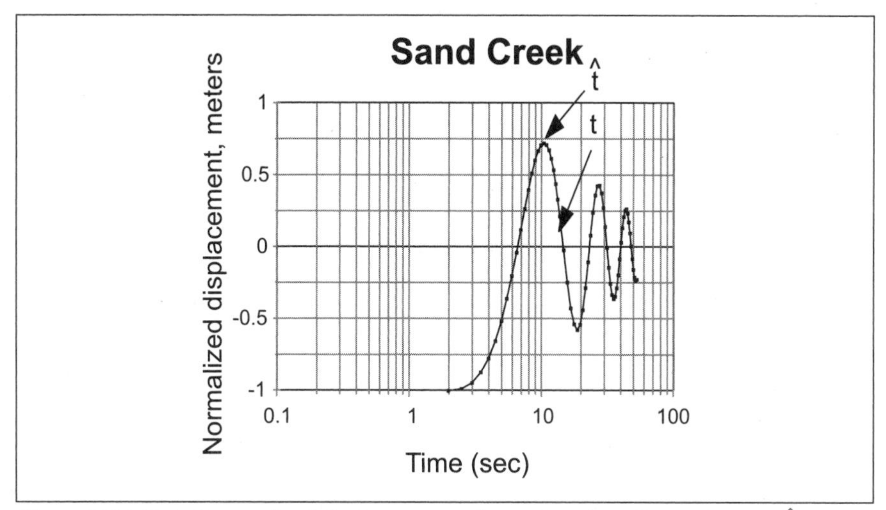

Figure 11.21 Plot of normalized displacement $-w/w_0$ versus time, illustrating how $\hat{t}$ and t are estimated in the Kipp Method (1985) for the Sand Creek data (Table 11.3).

value of 9600 ft^2/day using the van der Kamp method (1976) developed on spreadsheet by Wylie and Magnuson. Both suggest a high transmissivity.

Example 11.10

To illustrate an example of a larger damping parameter zeta, here is a data set from Well 93-2 south of Dillon, Montana, along Blacktail Deer Creek (Figure 11.3). The data are shown in Table 11.4.

Key in the data from Table 11.4 onto a spreadsheet. The displacement values in feet need to be converted to meters (multiply by 0.3048 m/ft); the time data in seconds are already in the appropriate format.

Evaluate to reference point where the data dampen to. In this case, dampening takes place near 0.0 ft (0.003 ft).

The maximum observable displacement w_o is 0.417 m (1.367 ft). The data previous to 1.8 sec are clouded by splash effects. The author subtracted 1.8 sec from the time data to rescale the data back to time zero.

A column on the spreadsheet is set up for the $-w/w_o$ values. A plot of time versus $-w/w_o$ values is shown in Figure 11.22.

From the plot, the values for $\hat{t}$ and t are estimated to be 2.2 and 8.5 sec, respectively. The zeta value appears to be about 4.0 (Figure 11.22).

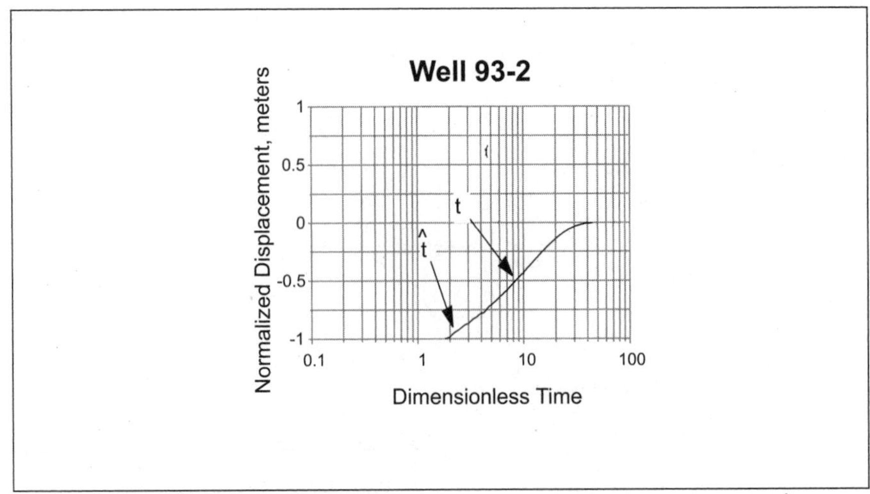

Figure 11.22 Plot of normalized displacement -w/w$_0$ versus time, illustrating how $\hat{t}$ and t are estimated in the Kipp Method (1985) for Well 93-2 in the Beaverhead, Montana, area.

Table 11.4 Data Set from Well 93-2 South of Dillon, Montana

Time (sec)	Displacement (ft)	Time (sec)	Displacement (ft)	Time (sec)	Displacement (ft)
0.20	.213	4.8	0.980	20.0	0.194
0.4	2.261	5.2	0.948	21.0	0.171
0.6	0.546	5.6	0.914	22.0	0.149
0.8	0.371	6.0	0.875	23.0	0.133
1.0	0.476	6.4	0.844	24.0	0.114
1.2	1.082	6.8	0.812	25.0	0.102
1.4	1.183	7.2	0.780	26.0	0.089
1.6	1.316	7.6	0.749	27.0	0.076
1.8	1.367	8.0	0.72	28.0	0.067
2.0	1.348	8.6	0.679	29.0	0.054
2.2	1.297	9.2	0.638	30.0	0.048
2.4	1.266	9.8	0.600	31.0	0.041
2.6	1.234	10.4	0.565	32.0	0.035
2.8	1.202	11.0	0.533	33.0	0.029
3.0	1.186	12.0	0.479	34.0	0.026
3.2	1.158	13.0	0.431	35.0	0.019
3.4	1.129	14.0	0.387	36.0	0.016
3.6	1.113	15.0	0.346	37.0	0.013
3.8	1.085	16.0	0.311	38.0	0.007
4.0	1.066	17.0	0.276	40.0	0.007
4.4	1.037	18.0	0.247	42.0	0.003
		19.0	0.219	44.0	0.003

The radius of the screen is 0.0254 m and the radius of the casing is 0.0762 m, so alpha is calculated to be 45000.

The value for beta from the spreadsheet matches well at 7.15E+07, yielding a transmissivity of 0.0192 m^2/sec (130 ft^2/day). This compares with an estimated value of 0.0095 m^2/sec (64 ft^2/day) using the van der Kamp method (1976) using an omega value of 0.05 and a gamma value of 0.08 from the spreadsheet developed by Wylie and Magnuson (1995). Both suggest a relatively low transmissivity. Data like this would more likely be analyzed with the Hvorslev (1951) or Bouwer and Rice (1976) method, but are presented here for comparison purposes.

Comparison of Two Methods and Discussion of Storativity

Weight and Wittman (1999) provide a table illustrating a comparison of 16 slug tests using the two spreadsheet algorithms presented above. In most cases the results are comparable (within a factor of 2 or 3). The author's experience is that one method does not necessarily yield a higher or lower value on a consistent basis. Sometimes the van der Kamp (1976) method yields a higher value than the Kipp (1985) method, with the reverse also being true. The benefit of having two methods is to get a range of conductivity values that may be representative of the aquifer.

Generally, the more oscillatory the data are, the smaller the damping parameter (ζ) and the higher the transmissivity. Wells with larger water columns also tend to have higher transmissivities (Figure 11.12). Butler (1998) points out that the water-column length is usually measured from the midpoint of the screen to the static water level. Wells with estimated damping parameters greater than 1.0 can probably be analyzed using the Hvorslev (1951) or Bouwer and Rice (1976) methods (Weight and Wittman 1999).

Both methods require an estimate of storativity. If the aquifer is confined, values between 10^{-3} and 10^{-5} are typical. However, as was discussed earlier with Example 11.7, both spreadsheet algorithms can be applied to leaky-confined or even unconfined aquifers (Weight and Wittman 1999). In all cases, it is necessary to use a confined storativity value or the results will not match the lithologies represented. Wells that are more shallowly completed and have less confining materials should begin evaluation using storativities of 0.001, and wells completed deeper or have semi-confining materials present should be evaluated using a storativity of 0.0001 (Weight and Wittman 1999). Storativity values can be adjusted a half-order of magnitude to see where both underdamped methods agree best. Decreasing the storativity value results in an increase in the calculation of β and subsequently a larger transmissivity value. It is the author's experience that the Kipp (1985) method appears to be more sensitive to changes in storativity than the van der Kamp (1976) method. One should use the same storativity values for both methods for a particular data set.

11.4 Other Observations

It is the author's experience that the results from slug testing tend to underestimate hydraulic conductivity values compared with constant duration pumping tests. This is based upon our field experience where we have observed data from wells that have been tested by both pumping tests and slug tests and from discussions with other colleagues. The amount of underestimation may range from approximately 30 to 100% to over an order of magnitude. This may be a function of well development and, therefore, skin effects and the limited effective radius of influence of slug testing.

Within contaminated plume areas, where monitoring wells are installed, wells are not designed to produce much water or tend to be well developed. Monitoring wells completed in high-transmissivity aquifers are installed with the intention of providing hydraulic head values and estimates of hydraulic properties. It is in high transmissivity areas where monitoring wells that are slug tested may also show oscillatory behavior. The methods described in the this chapter can be used to evaluate slug-test data.

References

Bouwer, H., and Rice, R. C., 1976. A Slug Test for Determining hydraulic Conductivity of Unconfined Aquifers With Completely or Partially Penetrating Wells. *Water Resources Research*, Vol. 12, pp. 423–428.

Bouwer, H., 1989. The Bouwer and Rice Slug Test—An Update. *Ground Water*, Vol. 27, No. 3, pp. 304–309.

Bredehoeft, J. D., Cooper, H. H., Jr., and Papadopulos, I. S., 1966. Inertial and Storage Effects in Well-Aquifer Systems: An Analog Investigation. *Water Resources Research*, Vol. 2, No.4, pp. 697–707.

Butler, J. J. Jr., McElwee, C. D., and Liu, W., 1996. Improving the Quality of Parameter Estimates Obtained From Slug Tests. *Ground Water*, Vol. 34, No. 3, pp. 480–490.

Butler, J.J., Jr., 1998. *The Design, Performance, and Analysis of Slug Tests.* Lewis Publishers CRC Press, Boca Raton, FL.

Campbell, M.D., Starrett, M.S., Fowler, J.D., and Klein, J.J., 1990. Slug Tests and Hydralulic Conductivity. *Proceedings of Petroleum Hydro-*

carbons and Organic Chemicals in Groundwater, October 31–November 2, Houston, TX, NWWA, Dublin, OH, pp. 85–99.

Cooper, H. H. Jr., Bredehoeft, J. D., Papadopulos, I. S., 1967. Response of a Finite-Diameter Well to an Instantaneous Charge of Water. *Water Resources Research*, Vol. 3, pp. 263–269.

Crump, K. S., 1976. Numerical Inversion of Laplace Transforms using a Fourier Series Approximation. *Journal of ACM*, Vol. 28, No. 1, pp. 89–96.

Driscoll, F.G., 1986. *Groundwater and Wells*, 2nd Edition, Johnson Division. St. Paul, MN, 891 pp.

Ferris, J. G., and D. B. Knowles, 1954. *Slug Test for Estimating Transmissibility*. U. S. Geological Survey, Note 26, 7 pp.

Fetter, C. W. 1994. *Applied Hydrogeology, 3rd Edition*. Macmillan, New York, 691 pp.

Hvorslev, M. J., 1951. *Time Lag and Soil Permeability in Ground Water Observations*. U. S. Army Corps of Engineers Waterway Experimentation Station, Bulletin 36.

King, J. B., 1988. *Hydrogeologic Analysis of Septic-System Nutrient-Attenuation Efficiencies in the Evergreen Area, Montana*. Master's Thesis, Montana College of Mineral Science and Technology, Butte, MT, 91 pp.

Kipp, K. L. Jr., 1985. Type Curve Analysis of Inertial Effects in the Response of a Well to a Slug Test. *Water Resources Research*, Vol. 21, No. 9, pp. 1397–1408.

Kruseman, G.P., and de Ridder, N.A., 1990. *Analysis and Evaluation of Pumping-Test Data*. International Institute for Land Reclamation and Improvement, Publication 47, 377 p.

Lohman, S.W., 1972. *Well Hydraulics*. U.S. Geological Survey Professional Paper 708.

Manchester, K. R. 1990., *Oscillatory Responses Due to Inertial Effects and Observation Well Water Level Fluctuations Induced by Pneumatic and Vacuum Slug Test Methods*. Master's Thesis, Montana College of Mineral Science and Technology, Butte, MT, 85 pp.

McElwee, C. D., Butler, J.J., Jr. and Bohling, G.C., 1992. *Nonlinear Analysis of Slug tests in Highly Permeable Aquifers Using a Hvorslev-Type Approach*, Kansas Geological Survey, Open-File Report 92-39.

Moench, A.F., 1993. Computation of Type Curves for Flow to Partially Penetrating Wells in Water-Table Aquifers. *Ground Water*, Vol. 31, No. 6, pp. 966–971.

Neuman, S. P., 1979. Perspective on "Delayed Yield". *Water Resources Research*, Vol. 15, pp. 899–908.

Papadopulos, I. S., Bredehoeft, J. D., and Cooper, H. H., Jr., 1973. On the Analysis of "Slug-Test" Data. *Water Resources Research*, Vol. 9, pp. 1087—1089.

Prickett, T.A., 1965. Type-Curve Solution to Aquifer Tests under Water-Table Conditions. *Ground Water*, Vol. 3, No. 3., pp 5–14.

Springer, R.K. and Gelhar, L.W., 1991. *Characterization of Large-Scale Aquifer Heterogeneity in Glacial Outwash by Analysis of Slug Tests with Oscillatory Responses, Cape Cod, Massachusetts*, U.S. Geological Survey Water Resource Investigation Report 91-4034.

Theis, C.V., 1935. The Relation Between the Lowering of the Piezometric Surface and the Rate and Duration of Discharge of a Well Using Ground-water Storage. *Transactions of the American Geophysical Union*, Vol. 16, pp. *519–524*.

Uffink, G.J.M., 1984. *Theory of the Oscillating Slug Test.* National Institute for Public Health and Environmental Hygiene, Bilthoven. Unpublished research report, 18 pp.

van der Kamp, G., 1976. Determining Aquifer Transmissivity by Means of Well Response Tests: The Underdamped Case. *Water Resources Research, American Geophysical Union*, Vol. 12, No.1, pp. 71–77.

Weight, W.D. and Wittman, G.P., 1999, Oscillatory Slug-Test Data Set: A Comparison of Two Methods. *Ground Water*, Vol. 37, No. 6 pp. 827–835.

Wittman, G.P., 1997. *Computer Simulated Flow Model of the Groundwater Resources of the Beaverhead Valley in the Dillon Area, Beaverhead County, Montana.* Master's Thesis, Montana Tech of the University of Montana, Butte, MT, 101 pp.

Wylie, A., and Magnuson, S., 1995. Spreadsheet Modeling of Slug Tests Using the van der Kamp Method. *Groundwater*, Vol. 33, No. 2, pp. 326–329.

Yang, Y.J., and Gates T.M., 1997. Wellbore Skin Effect in Slug-Test Data Analysis for Low-Permeability Geologic Materials. *Ground Water*. Vol. 35, No. 6, pp. 931–937.

Chapter 12

Vadose Zone

The vadose zone is composed of the materials from the land surface down to the water table, including the capillary fringe or zone (Figure 12.1). Typically, this includes a significant soil zone, partially disintegrated bedrock, and bedrock, where the land surface is not underlain by alluvial or glacial deposits. Generally, the soil zone consists of the finest particles, constitutes the limiting factor for recharge, and contains much of the stored soil water used by plants; hence, this zone has received the majority of the attention in past studies (not to mention that it is the most accessible horizon and the easiest to install instruments into). Understanding the dis-

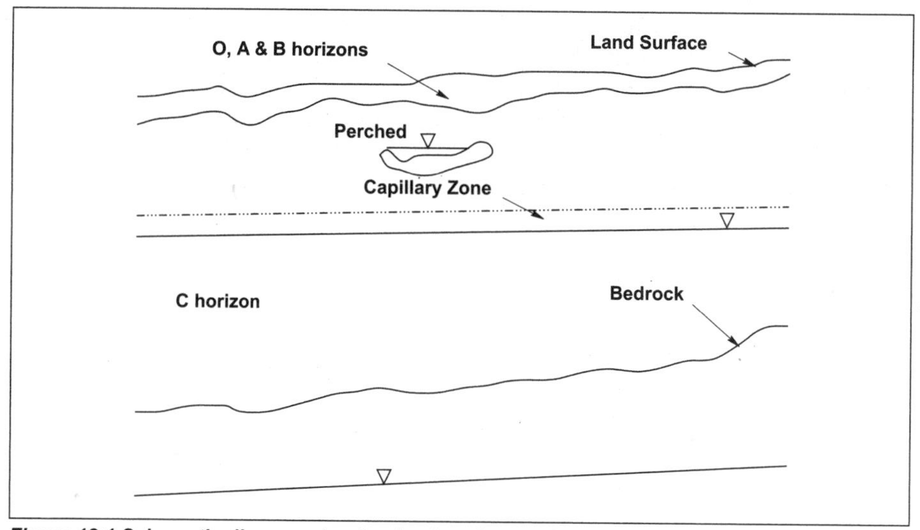

Figure 12.1 Schematic diagram showing land surface, soil horizons, a small perched water area, shallow water table capillary zone, and deeper water table. In some areas, the water table positions could represent seasonal range of the water table position.

tinction between the vadose and saturated zones, and the strengths and weaknesses of vadose-zone monitoring and vadose-zone flux calculations or estimates are your goals for this chapter.

Water budgets (Chapter 1) generally have fairly large (percentage-wise) error estimates for evapotranspiration and recharge. Rainfall reaching the land surface initially wets the surficial soil layer. Soil scientists will refer to the depth to which a "wetting front" has advanced (from the surface downward) when water reaches the land surface. Ponding of water on the land surface may be natural (simply the accumulation of runoff in topographic lows) or result from an artificial structure to evaluate infiltration rates. Unsaturated soils have mainly air in the larger void spaces; however, there will be a coating of water on the grains. If some of these voids are "small" (submillimeter in size), they can draw in water much as a paper towel does because of stronger capillary forces augmenting the gravity head. This means that the initial rate of water uptake by a soil is more rapid than at a later time when the wetting front has advanced deeper into the soil profile.

The evapotranspiration (ET) factor in a water balance may be estimated based upon the vegetation coverage (the type of vegetation is far less significant than the amount of the surface containing vegetation). But if runoff is measured from a "representative" area, and the amount of water that passes below the root and capillary rise depths (let's call this "true recharge") is also measured, the ET value can be calculated as precipitation minus the runoff and true recharge.

Thus measurements of natural plot runoff and natural infiltration can provide a better ET value as well as give fairly accurate information on recharge to the surficial aquifer at the study site, to get a better handle on the water budget. However, soil and vegetation variability will make the extension of the results beyond the study site more tenuous as the area assigned the study site values increases.

Another important aspect of soils is their ability to attenuate the concentration of contaminants before they reach the water table. Organic matter in the soil acts much as activated charcoal to adsorb organic molecules. Because organic matter is concentrated at the top of the soil profile, this attenuation occurs near the land surface, and you have to worry about small children eating this dirt.

When contamination, such as a spill, does occur, the desired response is to "get" the contaminant while it is still in the vadose zone. This is because the recovery cost is roughly 1/10 of what it will cost once the contaminant has reached the water table and started to disperse as a groundwater plume requiring interception ditches and skimmers (for floating product) or pump-back wells and separators, and a destruction technology (air stripping, etc). Calculations are commonly made to estimate how much time is available for the recovery, but these calculations are commonly made using bulk matrix characteristics, which may be too optimistic as indicated below.

Last, a caution about what follows is needed. Unless it is explicitly stated otherwise, models and approaches discussed refer to bulk soil or sediment characteristics. The study is of what would be called the matrix characteristics in a fracture-flow model. Thus, macropores and joints or fractures in the soil are ignored unless encountered by our core holes. The reader is strongly urged to peruse "Field Testing Some Hydrogeologic Assumptions," by Wayne Pettyjohn (1997) for some striking examples of assumption failure, such as a fourfold increase in dissolved nitrate two days after a rainfall event (macropore recharge, grass fertilized shortly before rainfall).

12.1 Summary of Vadose Zone Terms

Typically the soils literature papers use volumetric water content (percent of water by volume) as do most of the hydrogeologic studies. However, most geotechnical studies employ gravimetric water content (percent of water by weight). Laboratory results are nearly always expressed in weight loss upon heating, and when the results give this as a percentage of the original sample weight, this also is the gravimetric water content. Consequently, the first thing that has to be done in using existing data for a site is to determine what type of water content was used in the paper.

The equations for calculating these contents are:

$$\%H_2O_w = \left(\frac{W_w - W_d}{W_d}\right)100 \qquad [12.1]$$

$$\%H_2O_v = \left(\frac{V_w}{V_s}\right)100 \qquad [12.2]$$

where:

$\%H_2O_w$ = water percent by weight or gravimetric water content

$\%H_2O_v$ = water percent by volume

W_w = wet weight of soil

W_d = dried weight of soil

$\%H_2O$ = volumetric water percent

V_w = volume of water

V_s = volume of soil (undisturbed, e.g., volume cored)

Similarly, in order to discuss water movement, we need to define a few terms; these should help you realize that if you are using a published method to analyze your data, you need to know to what "type" of porosity the author is referring.

Total Porosity—The total void volume (whether or not the voids are interconnected) of a soil or rock divided by the total volume of the material. This is typically represented as:

$$V_V = V_T - V_S \qquad [12.3]$$

where:

V_V = volume of the total void space

V_T = total volume of soil and voids

V_S = volume of the solid material

Fillable Porosity—This definition is a mixed field and lab term. Originally this term was defined as the amount of water required to cause a unit length rise in the water table of an unconfined aquifer (Bouwer 1978) or soil. In the laboratory, it is the volumetric fraction of water added to saturate a dried sample kept at "field volume." The problem is that it is difficult to maintain the original volume in the lab. Fillable porosity is commonly less than drainable porosity because of air entrapment and hysteresis effects.

Drainable Porosity—The fraction of the total volume that is water-released by gravity drainage. This assumes that the soil or rock was completely saturated at the start. The length of time that the sample is allowed to drain will vary depending upon the protocol used. The amount of water collected as a function of time varies with the grain size of the material, with finer materials releasing water much more slowly. Typical drainage times range from 1 to 10 days, and the result in finer soils is often a function of the length of time the test is run. In hydrogeologic terms, this is roughly equivalent to the effective porosity—the pore volume that actually transmits water (Chapter 5).

It may be beneficial to look at the soil grains to understand some additional terms: hysteresis, capillary rise, and other effects that you will encounter.

Water has a strong affinity for rock and soil particles. Consequently, there will be a thin coating of water surrounding a particle. This affinity is normally demonstrated in grade-school science classes with capillary tubes, showing that water rises in the tube (glass tubes are similar to quartz) because of surface attraction between the water and the tube and the surface tension between the water and the air, and that the rise is approximately proportional to the radius of the tube. Actually, the rise is also affected by the surface of the solid phase and the purity of the water (dissolved inorganic species raise the surface tension while dissolved organics lower the surface tension of the water). The capillary rise equation may be expressed as:

$$hc = \frac{2\sigma \cos \alpha}{\rho g r_t} \quad [12.4]$$

where:

hc = height of the capillary rise (L)

σ = surface tension between the air and the liquid (water) (M/T^2)

α = contact angle (<900 if liquid rises and >900 if liquid is depressed)

ρ = density of the liquid (M/L^3)

g = force of gravity (L/T^2)

r_t = radius of the tube (L)

Translating this relationship from glass tubes to geologic materials leads to some difficulties. Just what is the equivalent of the capillary tube radius in a silty sand or a sandy loam? One can assume that the d_{10} radius (10% finer on a sieve test plot) represents the typical pore throat diameter, and thus, the presumed pore throat radius is ½ the d_{10} diameter. This is at best a crude first approximation. Soil scientists prefer to derive a statistical model of pore-size distributions from moisture-retention curves; however, this approximation is sufficient for most hydrogeology field courses.

The capillary rise is a function of the throat size. In clean gravels, we really can't detect much of a capillary zone. As the matrix of the sediment or soil becomes finer, so do the pore throat diameters, and some water is held above the level of the water table (you do remember that the level of the water table is determined by the level of water in a well that just barely penetrates the water-releasing zone of an unconfined aquifer, don't you? In fact, the definition is based upon the pressure of the pore water being equal to the atmospheric pressure; see Fetter 1994).

Example 12.1

As an example, let's calculate the theoretical capillary rise above the water table for a loess deposit with a d_{50} and d_{10} of 0.004 mm (loess is very well sorted by the wind, has a very loose packing with little or no layering, and little shear strength when wetted, so it is susceptible to collapse when saturated). We will need to assume an α value. For silicates, α is approximately 0° and cos α is approximately 1°. σ is about 72.8 gm/sec^2 (or 72.8 erg/cm^2 in most chemical tables) at 20°C. Thus, for pure water at 20°C we calculate:

$$h_c = \frac{(2)(72.8\ gm/sec^2)}{(0.988\ gm/cm^3)(980\ cm/sec^2)(0.0002\ cm)}$$

$$= \frac{(145.6\ gm/sec^2)}{(0.1956\ gm/cm)(sec^2)}$$

$$= 744.35\ cm = 7.44\ m(24.4\ ft)$$

The surface tension of water does vary with temperature (about 75.6 gm/cm^2 at 0°C) and decreases to zero at the critical temperature, (374°C). However, it should be noted that the calculated capillary height for the loess example is not matched by field observations. Observed capillary rises are up to about 1 meter (3 ft) in these deposits in the state of Washington where the net flux of soil water is downward (Hayden Ferguson, Montana State University, retired, personal communication, March 2000).

Above the water table (i.e., in the vadose zone) the pressure of the fluid phase is less than atmospheric. How can this be? Why is the pressure of the water in the soil above the water table less than atmospheric pressure? Let us revisit the concept of potential.

The Concept of Potential

All calculations of pressure or head have to be based upon some reference elevation datum. It is common to use the water table (where relatively flat), the land surface, or an arbitrary elevation for head calculations. If land surface is used, all numbers are normally negative. Most hydrogeologists seem to prefer using the water table as the calculation datum when possible, and most soil scientists use the land surface as the reference datum. When using data from others, be sure to determine what datum was used in the calculations.

The negative (relative to atmospheric) pressure results from the adsorption of water in a thin layer on the grain surface and the suction of the pore throats, which cause capillary rise. It is common to ignore adsorbed water (which only becomes important in very dry soils) and the osmotic potential (typically caused by the migration of a liquid solvent, e.g., water, across a semipermeable membrane from a more dilute solution into a more concentrated solution) in most studies. Thus the matric potential is commonly employed as being the force necessary to balance the capillary head rise. It must have an opposite sign for the forces to balance, hence the matric potential is always a negative quantity. The basic equation (using the nomenclature from Stephens (1996) but shifting from volume nomenclature (using) to pressure nomenclature (using Ψ) to determine the soil-water potential is:

$$\Psi_{sw} = \Psi_{\text{matric}} + \Psi_{\text{pressure}} + \Psi_{\text{osmotic}} \qquad [12.5]$$

where the soil-water potential is the sum of the matric, pressure (actually, the soil air pressure or hydraulic head if fully saturated, but typically $\cong$ 0 when not saturated), and osmotic potential (only significant when you have soil or geologic materials with membrane properties and chemical potential gradients caused by different concentrations of dissolved materials, typically not important). Thus, the total potential (which governs movement) may then be written as:

$$\Psi_T = \Psi_{SW} + \Psi_G \qquad [12.6]$$

where:

Ψ_T = total potential (hydraulic head)

Ψ_G = elevation potential

The total potential Ψ_T is zero at the water table and greater than zero below the water table *and,* under perched water conditions within the vadose zone, while in the rest of the vadose zone, the total potential is less than zero. Potential is being described on a unit weight basis so that it can be measured in terms of length, as you are used to doing with hydraulic head.

The total potential is what is actually measured by a tensiometer (Yeh and Guzman-Guzman 1995). This means that the tensiometry data from nested sets of tensiometers installed at increasing depth increments can be used to determine if the movement of moisture is upwards or downwards. In order to do this, the tops of the tensiometers should be at the same elevation with the same amount of stick up. This also permits the assumption that the length of the column of water (Ψ_G) is approximately equal to the length of the tensiometer, if you want to go back and calculate the matric potential from the total potential at each depth.

Infiltration

Having looked at capillary effects, one can better understand the nature of infiltration. If rain falls on a somewhat dry soil, the matric potential that causes the capillary rise causes the soil to suck water downward from the land surface (i.e., capillary forces working with gravity instead of against it to draw water down into the soil). Classic cumulative infiltration curves for a uniform soil look like what is shown in Figure 12.2. Depending upon how dry the soil was initially (antecedent moisture content), infiltration will start somewhere on the steep limb of the curve and decrease as the wetting front advances downward as shown by Figure 12.3, which is a plot of the same data showing the rate of infiltration, $i(T)$ [$i(T) = dI/dT$]. This figure is based upon maintaining a constant head of free water (ponded) at the top of the soil profile. The water-supply rate exceeds the soil-uptake rate; this is commonly referred to as a flood-plot analysis. To establish a flood plot, a relatively flat land surface area is chosen. A rectangular box, typically made with 10- or 12-in. (25- or 30-cm) planks is laid out and channels 1- to 2-in. (2.5 to 5 cm) deep are cut into the soil to aid when setting the planks on their side. About 1/3 to 1/2 in. (1 cm) of powdered bentonite is put in the channel to reduce leakage under the planks. The planks are positioned and stakes driven (on the outside) to support the planks. Then the joints at the corners of the box are caulked. The box should have a fairly large area (normally about 60 ft^2 or 5.6 m^2; typically, the planks are the length that just fit in the pickup bed with the tailgate closed) as we hope that the box will overlie typical soil macropore features such as soil joints and major root tubes. The water needed to maintain ponding to a depth (typically averaging about 6 inches [15 cm] at the deepest point) is measured and recorded along with the time, as it is added to maintain the head within the box.

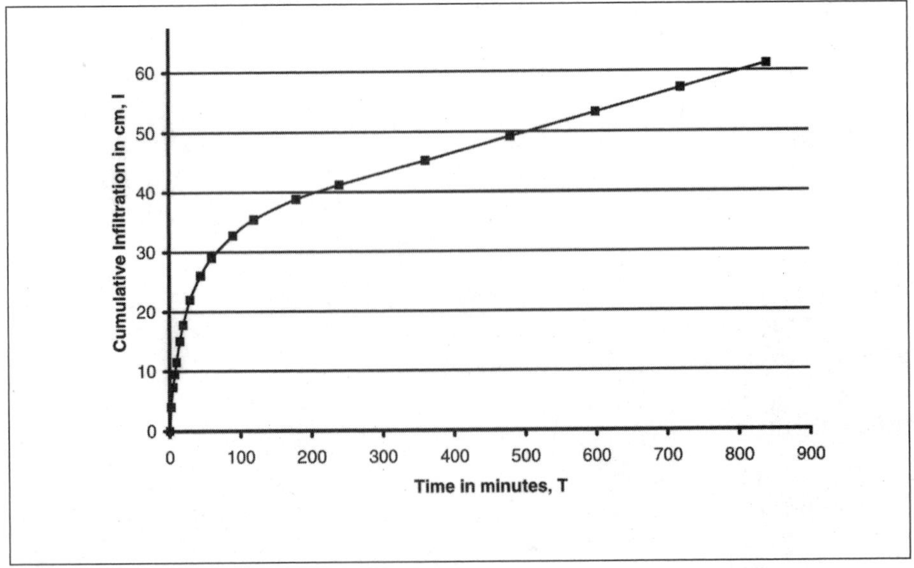

Figure 12.2 Plot of cumulative infiltration (I) versus time (T) for fine-grained soil.

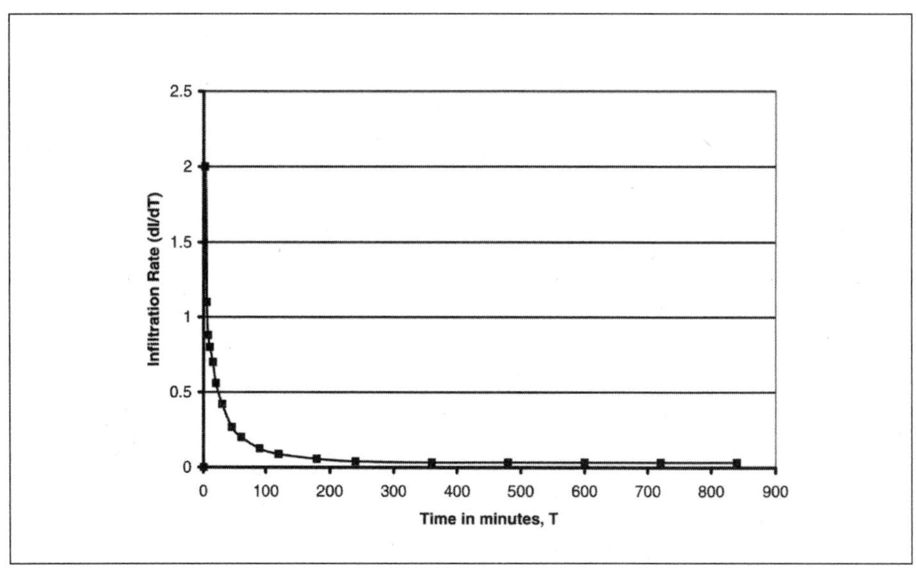

Figure 12.3 Replot of Figure 12.2 data depicting the infiltration rate (dI/dT) versus time (T).

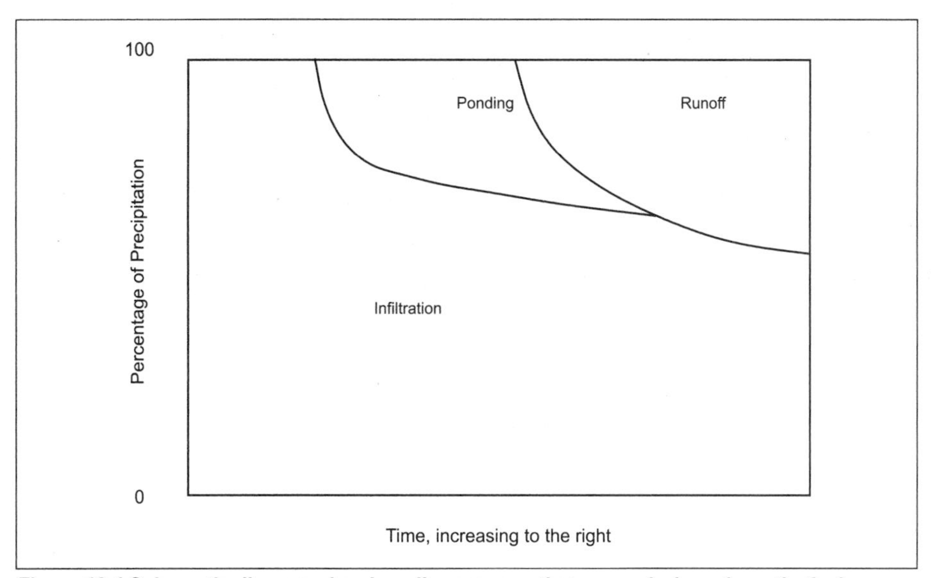

Figure 12.4 Schematic diagram showing all processes that occur during a hypothetical constant-rate precipitation event with rainfall greater than the saturated vertical permeability of the soil.

Flood-plot data are useful for establishing soil hydraulic characteristics, but are not usually typical of the soil response to a rainfall event. Figure 12.4 depicts what occurs in a hypothetical rainfall event of constant rainfall intensity. During the early stage, all rainfall is absorbed by the soil and the wetting front becomes deeper. Initially, the gradient is infinite (the head differential is the sum of the elevation potential and the matric potential at the wetting front, the distance over which the head is expended is equal to the depth to the wetting front) as the distance is zero. The hydraulic head maintains a constant value of 1.0 (saturated gravity drainage), so the gradient is:

$$\Delta H = \frac{\Psi_T + L + h}{L} \quad [12.7]$$

where:

L = depth to the wetting front

h = average height of ponded water

Initially the matric potential (Ψ_{matric}) at depth L (the value at the land surface is zero by definition) is more significant than the pressure potential ($\Psi_{pressure}$) or gravity potential (Ψ_G), and for small values of L, the gradient is very large, permitting all rainfall to be absorbed. You should note that the pressure and gravity potentials cancel each other out in the absence of air entrapment and $\Psi_T = \Psi_{matric}$. L/L is the standard unit head for saturated material.

Once the wetting front deepens, the gradient decreases towards a value of 1.0, and the vertical hydraulic conductivity of the soil limits the water uptake. This is indicated by the flattening of the curved-line portion of Figure 12.3. Note that once the soil cannot take up all of the rainfall, some ponding normally occurs on the surface and then runoff starts to occur (Figure 12.4). This is probably a good time to shift gears and examine methods of measuring infiltration in the field.

12.2 Direct Measurement of Infiltration Rates

Several approaches may be used, and when dealing with siting septic drainfields these are called percolation tests. The flood plot described previously normally provides data as depicted in Figures 12.2 and 12.3. However, to allow for retained gas phase (the infiltrating water has to displace the air in the pore spaces and it takes time to dissolve the last of the air trapped in smaller pore spaces and constrictions), the saturated vertical hydraulic conductivity is typically calculated as twice the apparent steady state shown by the approximately horizontal data at the end of the test for short-term tests (Everett et al. 1984). We have found that this usually works for tests of 4 hours or less. For longer duration tests, assuming that the final rate is the saturated vertical hydraulic conductivity usually works better.

An excellent, brief summary of methods for determining vadose-zone hydraulic conductivities is presented by Bouwer (1978), while more detailed information on the double-tube method is presented in Bouwer (1962, 1964).

A more recent version of the single-tube, constant-head method is employed in the Guelph Permeameter® (Figure 12.5), which has become quite popular because you can easily do sequentially deeper determinations with the tool by augering through the wetted zone of the previous test and doing another test at a deeper level. The intent is to avoid the early data while there is significant lateral spreading of the infiltrated water, using the later stage where water movement is essentially downward with two different head levels; the equations used permit calculation of saturated vertical hydraulic conductivity, soil sorptivity, and soil matric flux potential. Having

Figure 12.5 Guelph Permeameter in the field.

calculated negative values for numbers that can only be positive, you should be forewarned that variations in soil characteristics with depth can alter the second head-level rates and yield erroneous results. Consequently, we recommend coring through the wetted zone at the end of the test and recording the soil description. A data set collected in the early afternoon of March 25, 2000, on fill material that is part of a streamside restoration project is presented in Appendix E as an explanation of how to do the Guelph calculations.

Percolation tests for septic systems are usually conducted with augered or hand-dug holes. Depths and methodology required will vary from state to state, but in general they require an initial determination of the soil type (sand and silt versus silt and clay based upon initial infiltration rates). For the more permeable materials, 6 to 12 in. (15.2 to 30.5 cm) of water are added and the time per inch (2.5 cm) of decline is recorded. As long as the time isn't too long (roughly 15 min or longer) or too short (roughly less than 3 min), the measurements are continued with periodic replenishment of water is continued and the rate recorded. The scenario with too rapid an infiltration rate means that an alternate site must be found (insufficient time for the oxidation of organics before reaching the water table). The "too slow" scenario often requires filling the hole with water one or more times and retesting the infiltration rate after clays close to the hole have had time to hydrate. In Montana, infiltration rates of less than 1 in./hr (2.5 cm/hr) are not accepted, and an alternate site must be found.

12.3 Soil Moisture Measurement and Sampling Tools

Just because precipitation is taken up by the soil, doesn't mean that it will recharge the groundwater system. Water-balance equations can be written in a number of ways, but basically they all boil down to inputs (precipitation, stream inflow, groundwater inflow, imported water) minus outputs (evaporation, transpiration, sublimation, stream and groundwater outflows, exported water). Most soil studies typically focus upon the vertical components within a small study area. A number of tools can be used to provide either direct or indirect measurement of water in the soil profile. The accuracy of these measurements is a function of the accuracy and precision of the instruments and a function of the representativeness of the instrumented ground relative to the site as a whole.

Should you be installing access tubes or fixed equipment (lysimeters, tensiometers, etc.), always perform soil coring so that you can generate soil-coring logs that can be used to develop the equivalent of fence diagrams to show what can be deduced about the variability of the soil layers at the study site(s). This is particularly critical if surface geophysical methods are being considered to help fill in the gaps; conversely, you might try to use such methods to delineate boundaries if soils show significant variation.

Neutron Probes (Indirect Measuring Devices)

Neutron probes are tools that estimate the water content of a soil based upon the slowing and back scattering of neutrons from a radioactive source by hydrogen atoms in water and hydrogen and other light atomic weight elements in the access casing and soil. The carbon and attached hydrogen atoms in the organic matter in the uppermost soil zone usually ensures a slightly "too high" water content. A typical probe unit consists of the probe (usually having an americium-beryllium source and a boron trifluoride gas detector, Figure 12.6), cable, a housing (containing paraffin or some other

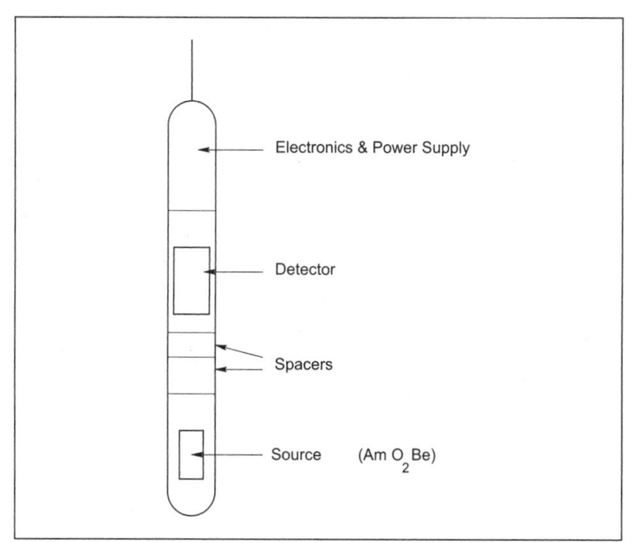

Figure 12.6 Schematic depiction of a neutron probe. [After Keys and MacCary (1971).]

hydrogen-rich neutron velocity reducer) and a display unit giving the instrument readout value. If the neutrons have been slowed sufficiently (about 1/500 of the initial velocity, requiring about 500 collisions with hydrogen atoms) to be seen by the detector, they are counted. The housing is placed on top of the uncapped access tube, the probe is lowered down the access tube, and readings are taken starting about a foot below the land surface. The access tubes are normally fabricated using aluminum, but thin-walled plastic tubing is sometimes used. Their inside diameter should be only slightly greater than the outside diameter of the probe.

Figure 12.7 depicts the result of soil moisture determinations compared with neutron probe results using the factory calibration. If the factory calibration was correct, the data should plot from the lower left-hand corner to the upper right-hand corner; instead all of the data plot below this, indicating that the factory calibration overestimates the soil-water content. The fitted curve is Jim McCord's best fit of his data from the sand and loam sites with 2-in. (5.1-cm) aluminum access tubes in tight holes. Since plastics are made from petroleum, both its hydrogen and carbon affect the neutron velocity attenuation measured by the probe, and correction of the factory calibration curve is normally required. The chapter in Greacen (1981) on calibration is good reading for anyone who really needs to modify factory settings. The zone measured by the detector is approximately spherical, with a typical radius of 8 to 35 centimeters depending upon whether the soil is saturated or quite dry, respectively (Stephens 1996).

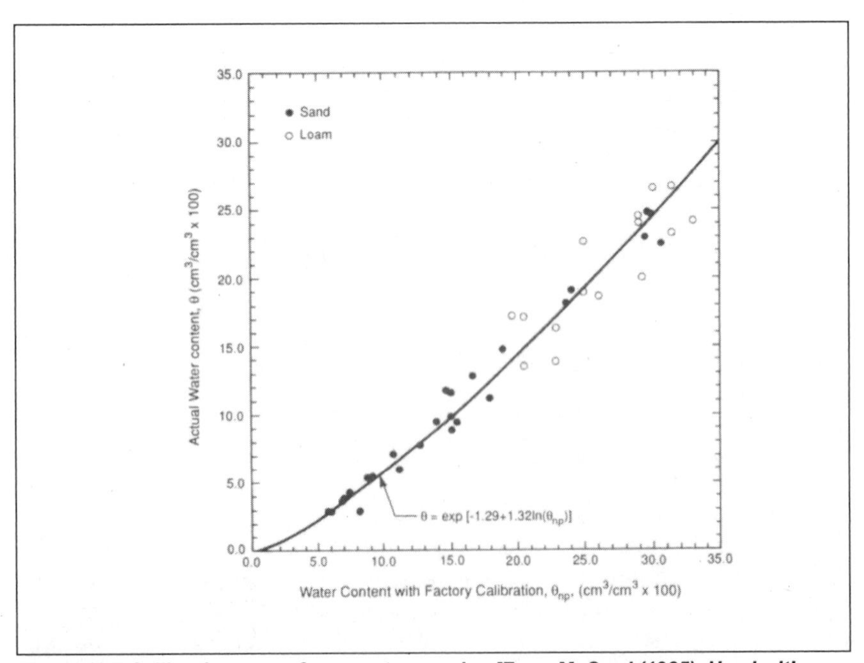

Figure 12.7 Calibration curve for a neutron probe. [From McCord (1985). Used with permission.]

The access tube is simply a casing for the probe (typically 1/8-in. larger inside diameter than the probe outside diameter) that is sealed at the bottom to prevent groundwater from entering the tube if the water table rises above the bottom of the access tube. These tubing seals do not always survive installation unscathed. As part of an integrated saline seep investigation near Glascow, Montana, in December 1974, I watched a soils grad student studying soil moisture "fry" (the pseudoscientific term for short-circuiting the probe electronics) a probe in an access tube that had water in the bottom, so you can be assured that not all tubes and probes have watertight seals. This problem can be avoided by using an E-tape or chalked steel tape to check the access tube for water or other liquids (vandalism is sometimes a problem) and total depth prior to inserting the probe in the access tube.

Use of the neutron probe is typically accomplished by getting a series of measurements at fixed depth increments such as 1-ft (30.5-cm) increments starting 1 ft (30.5 cm) below the land surface.

Resistance Methods (Indirect Measuring Devices)

These techniques are grouped, relative to the way most books present them. Basically there are surface measurements with portable tools, or subsurface measurements with dedicated tools.

The land-surface methods use common geophysical tools (Chapter 4) such as DC resistivity (conventional resistivity using Wenner or Schlumberger electrode arrays), induced polarization (sometimes used to help map contaminant plumes), or electromagnetic induction. The most common resistivity approach employs the Wenner array at a number of locations with the electrode spacings increased in increments to generate depth profiles at these locations. In Wenner arrays, the electrodes have a fixed spacing with all separations of equal distance. If some resistivity array locations are placed close to tensiometer or neutron probe sites, rough values for soil moisture relative to resistance can be generated and used at nearby locations. However, this geophysical method is "blind" below the first highly conductive (i.e., quite moist or wet fine-grained) layer as it effectively short-circuits the current flow. The method often can help delineate the water table (actually it picks the top of the capillary fringe) in areas with relatively shallow water table conditions.

Subsurface methods require that the measuring device be installed in the soil. In practice, this is normally limited to the base of the cultivation depth to avoid extreme differences in the soil characteristics. Agriculturists may choose to use gypsum blocks, which have a field life expectancy of 3 to 5 years under irrigated conditions (Soilmoisture 1999). Basically, the porous gypsum "block" (which is actually a cylinder about 1-inch in diameter and height) is buried to the desired depth after cultivation or using coring tools with the cable lead coming to the land surface where it can be connected to a portable meter when being measured. The block is a essentially a conductivity cell measuring the conductance of the water sorbed onto the surface of the porous gypsum material. The conductance of the liquid is, for practical purposes, constant because of the gypsum dissolution, so the conductance of the cell is proportional to the "soil-water content" in the cell.

Gypsum blocks may be installed to monitor for leaks under or adjacent to impoundment or treatment facilities, but should be regarded as semipermanent tools because even without irrigation, they have a limited life span.

Tensiometers (Direct Measuring Devices)

Tensiometers let us use the material previously discussed under potential. By definition, the water-table surface is in equilibrium with atmospheric pressure. Below the water table, the pressure is greater than atmospheric, caused by the force of gravity upon the column of water extending from the point of interest up to the water table. Above the water table, water in the vadose zone is at less than atmospheric pressure. This negative pressure (always calculated relative to the current atmospheric pressure) is commonly referred to as the matric potential, matric pressure, or soil or matric suction. A tensiometer is simply a device for measuring the total potential, and the magnitude of this matric pressure is calculated by subtracting the elevation potential. Tensiometers work with relatively moist soils (matric suction ranging from 0 to about 0.8 atmosphere [Everett et. al 1984]). The Hanks and Ashcroft text, *Applied Soil Physics*, while designed for agricultural science students and with agricultural application emphasis, has an excellent introduction to the necessary concepts and has proven to be a valuable hands-on reference. Additionally, Stephens (1996) is recommended for students who wish to delve deeper into this material, and a more detailed discussion of tensiometery is presented by Yeh and Guzman-Guzman (1995) in the Wilson, Dorrance et al. handbook (1995).

A basic tensiometer consists of a tube body made up in various lengths, a short porous ceramic cup at the bottom (usually threaded to the body with an O-ring seal), and a top that seals the tube (typically a rubber stopper) or a needle valve to a storage reservoir. The minimum needed for a functional tensiometer is shown schematically in Figure 12.8, and a commercial version with a water reservoir and gauge is shown in Figure 12.9. The stopper may have a tube penetrating it and a gauge on top, or a gauge

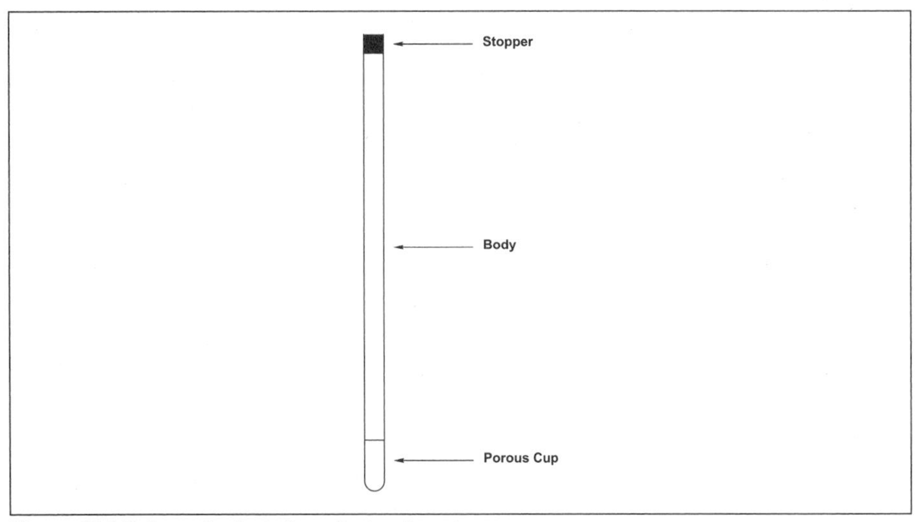

Figure 12.8 Schematic depiction of a tensiometer.

Figure 12.9 Soilmoisture Jet Fill tensiometer with water reservoir (1-quart thermos for scale).

with a needle that may be pushed through the stopper to sense the pressure. The latter approach allows a single, high-quality measuring device to be used on all of the tensiometers and is the preferred approach. Measured values are attributed to the depth of the center of the porous cup.

The hole for installation of the tensiometer is typically created by driving a soil coring probe (tube) with a slide hammer device to the total depth in increments no greater than 75 to 80% of the core tube capacity. This vertical hole should be only slightly larger in diameter than the tensiometer; typically a 7/8-in. coring probe is used. The soil is commonly described or bagged for later description for each increment before driving another increment to prevent mixing up the samples. We have had the best luck with the shorter basic probes (they have a cutout portion extending about 2/3 of the capacity length) with threaded, replaceable cutting tips. The replaceable tips are absolutely necessary in ground with pebbles and cobbles as you will quickly ruin probes without the replaceable tips (Figure 12.10).

***Figure 12.10** Slide hammer, two coring probes, and an extension rod. All items are upside down from use position. (Note the damage to the tip of the right-hand coring probe, which is not the replaceable tip type.)*

A soil slurry made from the cored material from the bottom of the hole is added to the hole before installing the tensiometer. The tensiometer is then installed in the hole, filled with water, and sealed. The tensiometer will need some time to come to equilibrium with the soil moisture, and readings for the first day or two are normally discarded for coarser soils; it may take up to a week in clay-rich soils until readings stabilize. Normally a series of tensiometers are installed at varying depths in a cluster to permit the determination of the direction of water movement. How these values are used will be discussed in later examples.

Psychrometers (Direct Measuring Devices)

A psychrometer is a tool for measuring the relative humidity in dry soils where the soil suction is more negative than 2 atmospheres down to about 75 bars. This limits their use to very dry, relatively fine-grained soils, and they are used when tensiometers fail because of large matric suction values. It should be noted that many plants will wilt and die at matric suctions more negative than 15 atmospheres. The authors have not dealt with these with these conditions in the field and do not have any practical guidelines. Instead we suggest that papers by people working in desert environments be investigated; Stephens (1996) has a good, brief summary on these devices.

Lysimeters (Direct Sampling Devices)

While other methods exist, the most common method used to sample the soil water from the unsaturated zone uses a tube with a porous ceramic cup (Figure 12.11) that is evacuated to create a partial vacuum. The porous cup is often constructed out of ceramic, but also may be of Teflon or sintered stainless steel, depending upon the application needs. You can visualize this as a larger-diameter tensiometer that, instead of being filled with water, contains two small-diameter, hard-plastic tubes inserted through a rubber stopper at the top of the lysimeter body, as shown in Figure 12.12. The tubes are used to establish a pressure gradient and to collect the sample. The sampling tube extends to the bottom of the porous cup and is typically cut at a 45^{0} angle to avoid tube "sealing" on the bottom of the cup. The second tube only goes a short distance beyond the stopper; different colors for these two tubes are preferred to easily distinguish which tube is which. Above the stopper, short sections of softer, compressible tubing (typically Tygon© or Teflon tubing) fit tightly over the rigid tubing and tubing clamps are installed. Figure 12.13 shows a homemade lysimeter being installed; the hand on the left is adding bentonite above the silica slurry, which was put in the hole before inserting the lysimeter.

Figure 12.11 Porous ceramic cups on a lysimeter and a tensiometer.

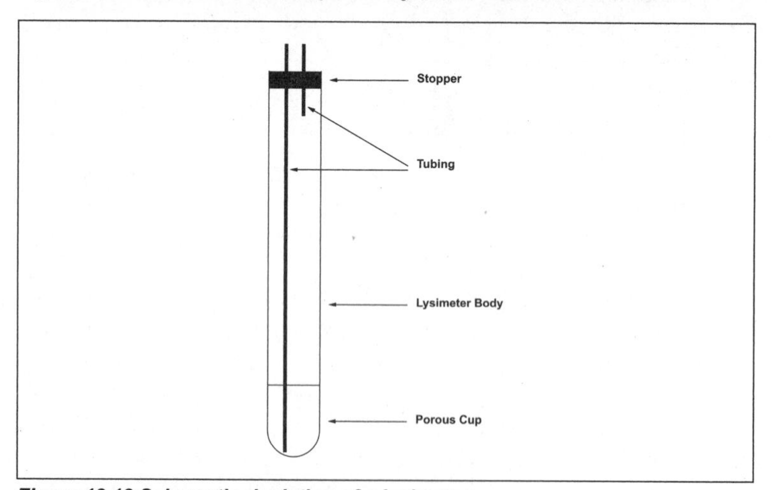

Figure 12.12 Schematic depiction of a lysimeter.

A cover or security device is required to protect the tubing from weather and animals. We have had tubes chewed up by wintering elk and rodents. Homemade lysimeters or lysimeters that do not have the body extending to the land surface can be protected using a plastic dishpan with several bricks or heavy rocks to hold the cover in place. Commercial lysimeters

Figure 12.13 Installing a homemade lysimeter.

usually offer screw-on caps to protect the tubing that can be used if the body extends to the land surface. When using these caps, it is important that the tubing not be too long to fit inside of the cap.

Installation of the lysimeter into shallow core depths is normally accomplished using a hand-cored hole. For deeper installations, a truck-mounted auger rig using the smallest diameter flights available (typically 4- or 6-in. diameter) is preferred. We have successfully installed and sampled lysimeters to depths of 55 ft in glacial sands (Reiten 1991), and others have installed them to greater depths. Note that for depths below about 25 ft, suction lift of the sample in the tubing will not support the water column rising to the surface; high-pressure versions, which utilize a check valve and an upper chamber that can be pressurized, will be needed.

Before installing the lysimeter, enough silica-flour paste is added to the hole to positively cover the porous cup of the lysimeter when the lysimeter is installed. The silica-flour paste provides a fine-grained "conduit" to provide transmission of liquid from the sides of the access hole to the porous cup. Fine-grained material is needed to support the vacuum (typically 0.75 atmosphere) that will be applied to the lysimeter when collecting the sample. Depending upon the depth of installation, cuttings and bentonite plugs can be slurried in above the silica paste. For shallow installations, with the lysimeter body extending to the surface, you may have the silica paste flow out at the land surface. In this case, scrape away the paste and dig a 6-in. deep trench around the lysimeter and install a mounded bentonite seal (actually a mixture of fine-grained cuttings with 5 to 10% bentonite works better as it is less likely to have severe dessication cracking).

Once lysimeter installation is completed, clamp shut the softer tubing to the sampling line (longer) tube. Apply a vacuum to the short line using a hand pump with a vacuum gauge, and clamp the second soft tubing line once approximately 0.75 atmosphere of vacuum has been achieved. The first few samples collected should be thrown out as they reflect moisture from the installation process (silica-flour paste).

Sample collection can be done two ways. The standard approach is to unclamp both lines and apply a vacuum to the sampling line with a stoppered Erlenmeyer flask in-line to catch the sample. Alternatively, you can gently apply increasing pressure to the short tubing until fluid starts flowing out of the sampling tube and direct the flow into the sample collection container. In either case, you want to remove all of the fluid, so a vacuum on the sampling line may be required once the sample has been collected. Try applying pressure until you get no flow, then shift to vacuum and see if you can actually get any extra fluid out. In theory you shouldn't get any extra fluid, but in practice you often can as the water "plug" is less stable under pressure. This may be because of the pressure surges. For deep samples (using high-pressure lysimeters), the vacuum approach is used to move the sample into the upper chamber and then a gas pressure application is used to bring the sample to the surface.

Some other considerations when using lysimeters include:

1. Conditioning or cleaning the porous cup with one or more acid washes to remove/dissolve residual ceramic powder is frequently advised. If you do this, use a diluted acid. Fully submerge the porous cup, extract a volume equivalent to the porous cup capacity, disassemble the lysimeter, rinse the inside and the tubing as well as the outside of the unit at least three times with distilled water, reassemble the unit, put the unit into another container with distilled water, and finally extract samples until the pH is above 5.5.

You will find that if the rinsing was not done carefully, this last operation can be very time consuming. The acid used is normally the same one that will be used to preserve water-quality samples (typically HNO_3); a 1% volumetric solution will give an initial pH of less than 2 and should be sufficient to dissolve powders and exchange with cations bound on the porous cup material. The more concentrated the acid, the harder it is to bring the extracted pH back up to the desired range with distilled water.

2. Check bubbling pressure. We have had the best luck with the low-pressure or "high-flow" ceramic cups. With all shallow installation lysimeters (purchased or homemade) you may wish to check the bubbling pressures by inserting the porous cup into water and pressurizing the unit until gas bubbles start escaping; this is also a good check for construction flaws. If you then put a .5-atmosphere partial vacuum on the unit, you should be collecting a sample. If you don't get water into the lysimeter under this condition, you probably will never collect any in the field, because it is difficult to get much more than 0.75 to 0.8 atmosphere partial vacuum under typical field conditions, and the partial vacuum will decrease as fluid enters the lysimeter.

3. Estimate the soil-water pressure before installation. While we have used lysimeters without tensiometers or neutron probe access holes, it is best to have some idea of the matric pressure so you don't assume that your lysimeters have failed when, in fact, they cannot overcome the matric potential. Should you be dealing with a lysimeter-only site and think you may have plugged lysimeters, try making a small ponding box (or use a piece of large-diameter casing) and install it around your shallowest lysimeter, add several gallons of water to infiltrate, and see if your problem isn't matric potential before digging up the lysimeters.

It is common in semi-arid regions for lysimeters to produce soil-water samples in the late spring and early summer, and then not produce samples because the plant roots have reduced the soil-water content to less than 0.8 atmosphere. With weekly sampling, this can be indicated by the failure of progressively deeper lysimeters to obtain a sample.

12.4 Data Collection and Use

Neutron Probe Data

Neutron probe data start out as detector counts, which are converted to percent moisture content. These are typically plotted as vertical profiles, as shown in Figure 12.14, which also shows some seasonal variation in a cropped field. One of the simplest methods of using these data, assuming that you get a profile similar to Figure 12.14, is to use the zero flux plane concept (Figure 12.15) with data from shortly after the period of greatest precipitation during the nongrowing or low-growth season (you want the maximum soil-water content period for this method). Basically, the volume of moisture below the maximum moisture content on a plot of total hydraulic head versus depth is calculated to drain to the water table (minus the

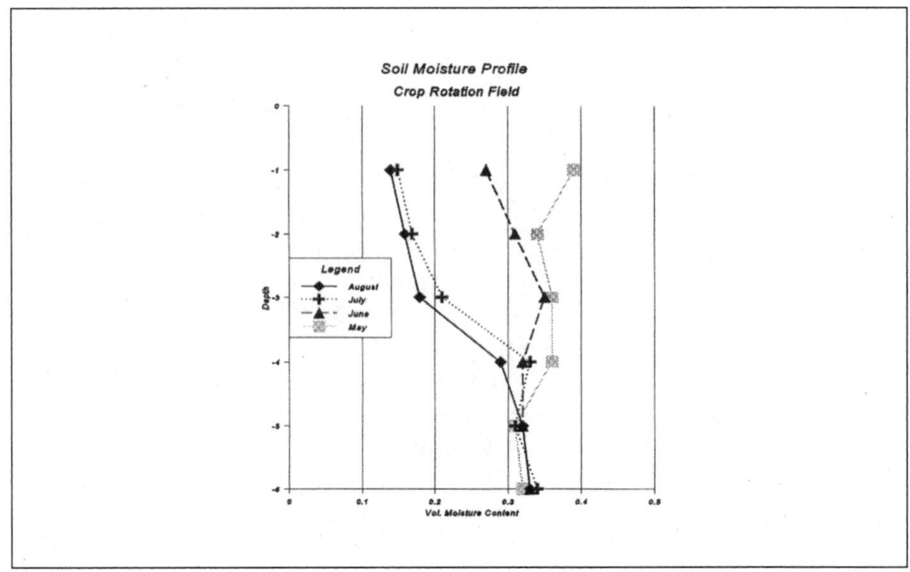

Figure 12.14 Soil moisture content at a crop rotation plot determined by a Troxler 3300 neutron probe at the Western Triangle Agricultural Research Center near Conrad, Montana. [From Freshman (1996). Used with permission.]

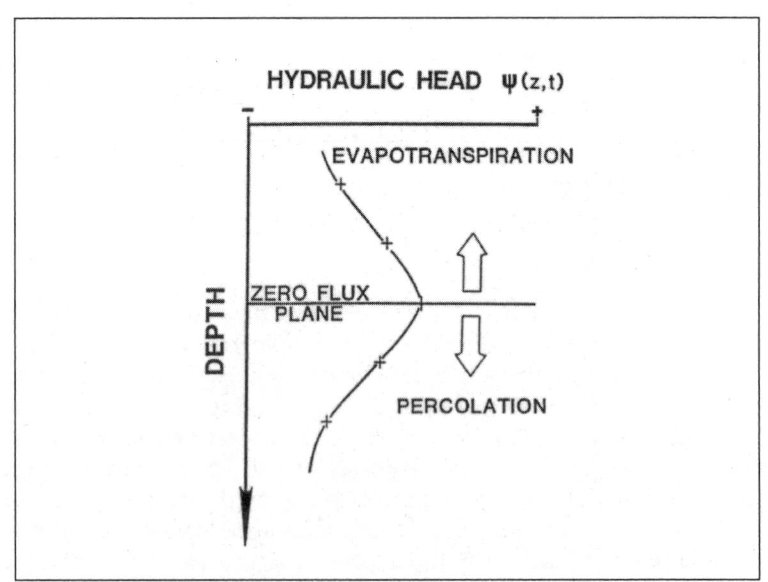

Figure 12.15 Schematic diagram of the zero flux plane concept. [From Dreiss and Anderson (1985). Used with permission.]

"background" or nondraining moisture content), while the surplus moisture above the maximum is assumed to be lost to evapotranspiration.

For a study site in central coastal California, Dreiss and Anderson (1985) found that this approach yielded reasonably accurate results compared to more rigorous approaches. As a cross check on that conclusion, let us try this approach on data from Socorro, New Mexico (Stephens et al. 1986), shown in Figure 12.16. Note that in a uniform sandy soil, the hydraulic head is equivalent to the moisture content. Using the moisture profile for January 24, 1985, and the 1.5- to 2.5-m depth increment which, averages about 7.5% moisture and has a background water content of about 4.5%, yields an infiltration of 100 cm × (0.075 to 0.045) or 3 cm (30 mm). This yields a value at the upper end of estimates for annual recharge, which was initially 7 to 37 mm (Stephens and Knowlton 1986) but was later revised downward to 4 to 9 mm (Stephens et al. 1991). This check suggests that the method may overestimate recharge in a semi-arid area by up to a factor of 4. Conversely, in humid areas (such as most of Alabama) where recharge commonly occurs during the growing season, it probably is in the ball park or underestimates by a bit the amount of recharge. However, as a first cut approach, the zero flux plane method doesn't do badly and is probably sufficient for most purposes.

Tensiometer Data—North-Central Montana Example

The potentials measured by your tensiometers provide an indication of water movement because the readings are of the total potential, as discussed earlier with the theory. Figure 12.17 shows the tensiometer data for approximately the same time points that were shown in Figure 12.14 (moisture content), plotted in a similar manner such that increasing (less negative) total potential is depicted similarly to increasing water content. These data are not internally consistent; note that the 1-ft data point for July "bends" the wrong way (probably the result of a rainfall event shortly before the measurement) and that the May data (first recorded data collection) don't match well with the moisture data (zero centibars implies saturation at the 6-ft depth versus the lowest moisture content shown in Figure 12.14). The problem with the May data may be that the tensiometer values may not yet have stabilized, especially at the deepest intervals. This hypothesis was partially justified, when the 5-28-94 data (taken 4 days later) were added to the plot as half-tone data points and lines. The 5- and 6-ft data shifted as predicted if it was a reequilibration after installation problem. This suggests that the tensiometer plot and the neutron probe plot, while adjacent to each other, may have minor differences in the soil characteristics or initial conditions. Also, if the 3- to 4-ft zone is a bit more

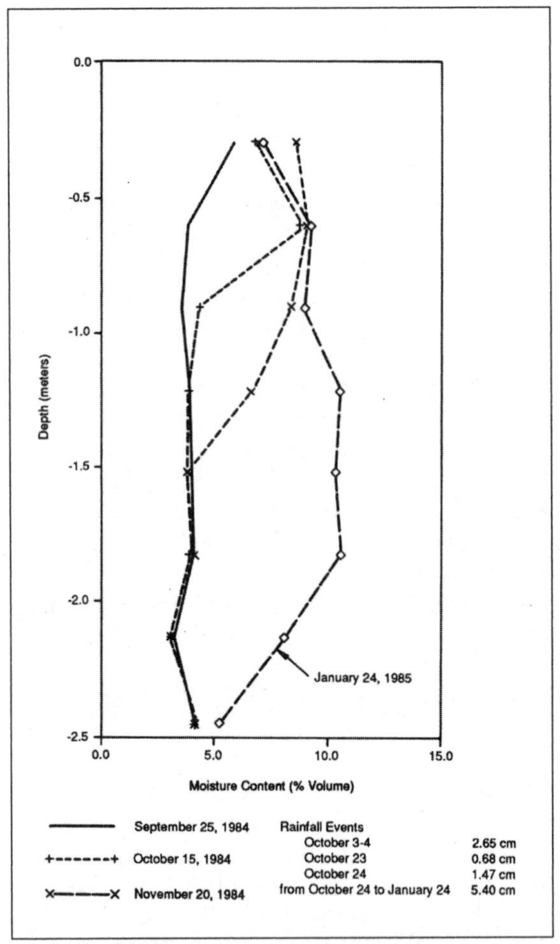

Figure 12.16 Water content profiles following precipitation at a sandy site north of Socorro, New Mexico, with the water-table depth at roughly 5 m. [From Stephens (1996). Used with permission.]

clay-rich, the differences between the moisture content and total potential can be explained.

Such discrepancies are not uncommon in dealing with soil studies. With that caveat, examination of the plotted data shows that as the growing season gets going and the plants establish deeper roots, the upper soil zone (down to 2 ft) is being depleted and water is moving up from deeper depths, as shown by the late June data.

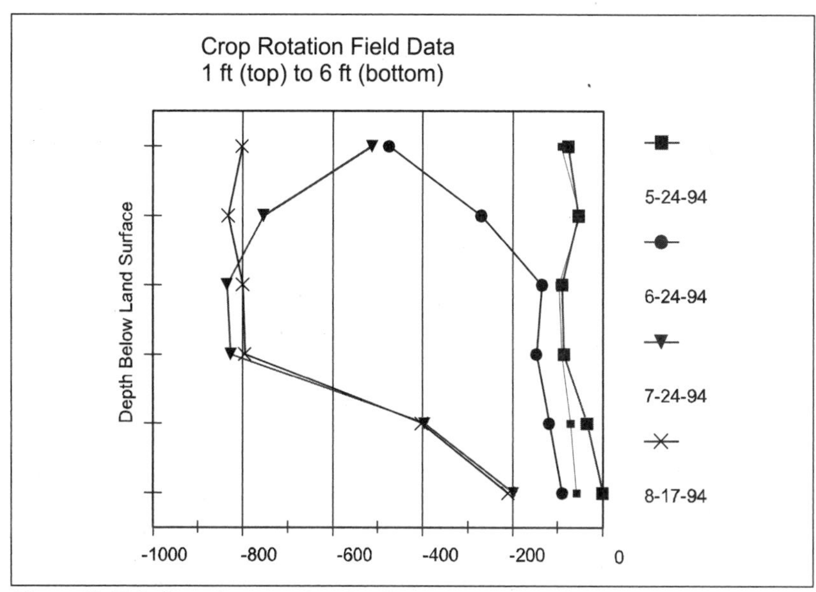

***Figure 12.17** Soil tensiometry data for the same crop rotation plot presented so that increasing soil moisture (and less negative total potential) is to the right. (Half-tone data are from May 28, showing that May 24 data probably did not provide equilibrated values at the 5- and 6-ft depth. [From Freshman (1996).]*

The 1-ft depth value from July 24th is truly suspect, as by that time of year shallow soil moisture is pretty well depleted. Two days earlier, the value for that depth was measured as 1000 millibars, and two days later it was measured as 781 millibars, so the plotted value (513 millibars) may be a meter-reading or transcription error. This leads us to the variability of tensiometer data. Just how variable is it?

The data used from this study (Freshman 1996) were collected carefully on land with a known history at Montana State University's Western Triangle Agricultural Research Center. The student involved was meticulous about methods and procedures, and these particular values were late enough into the project that there should have been no problems with familiarity with equipment or data recording. The data, in millibars, are presented in Table 12.1.

These data suggest that even with a readout that registers in millibars, variability of the soil system and equipment limits significant differences to ±2 centibars (20 millibars). Additionally, the 1000 values should be suspect. This was probably the upper end of the calibration range being used and the instrument couldn't show a more negative value. The variability of the two uppermost tensiometers is far more extreme than what is seen in the four lower tensiometers.

Table 12.1 Study Data in Millibars

Elevation	7/22/94	7/24/94	7/26/94
-1	-1000	-513	-781
-2	-1000	-754	-857
-3	-823	-835	-836
-4	-852	-826	-837
-5	-386	-396	-412
-6	-209	-199	-212

This leads to the question, "Did it rain in a time frame that would explain this?" A call to the Western Triangle Agricultural Research Center yielded the following rainfall data: July 15, 0.02 in.; July 20, 0.18 in.; and July 24, 0.02 in. It is assumed that the July 20. precipitation event did provide the increase in shallow soil moisture at the 1-ft depth noted on July 24. However, complete plots depicting precipitation, soil moisture, and total potential for each of the six depths at the study site would have made interpretation of the data collected for this part of the project far easier to accomplish and might have aided the researcher in evaluating macropore effects (one of the research goals).

This data set was used to demonstrate both the application and presentation methods of tensiometry data. It shows the importance of considering the components of your project goals, how to select which data to gather, and how the presentation of that data affects your research report.

Mass Flux Calculations

Calculations of mass flux are used with groundwater flow systems to calculate contaminant loadings, and the same sort of calculation can also be used in the vadose zone to calculate water flux and dissolved constituent loadings. To evaluate constituent loadings, the water flux determination is needed. However, anything beyond the zero flux plane approach is well beyond the scope of this text, and we will restrict ourselves to that approach. Let's see if it is possible to extend the concept to a non-bell-shaped curve using the north-central Montana data.

The change in storage of the soil moisture will give us an approximate flux number. Figure 12.18 shows the portion of the zero flux plane curve below the plane that is integrated to calculate the annual recharge flux. By using the schematic figure, we presume that the baseline soil moisture from the end of the growing season (typically late August or September) is known or can be estimated. As shown from the north-central Montana data set depicted in Figure 12.14, the water content at depth may not be vary greatly.

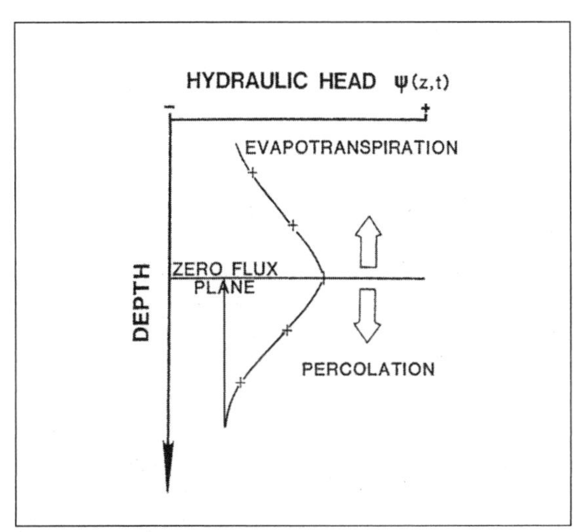

Figure 12.18 Schematic diagram showing the area to be integrated below the zero flux plane to calculate recharge. [Modified from Dreiss and Anderson (1985).]

Despite the lack of the proper bell-shaped form, we can attempt to estimate the recharge from that site as shown in Figure 12.19. By making an arbitrary decision to use the average moisture content at the 6-ft depth (approximately 32.8%) as the baseline water content for downward migration, and extending a line up to where it intersects the June data line, we can estimate a fairly conservative amount of recharge as the surplus moisture from 6 ft to that point of intersection (3.7 ft). Note that a larger amount of recharge would be calculated if just the soil-moisture losses from May to June from 3 to 5 ft were all assigned to downward movement. This approach yields three triangles and a rectangle to be integrated for the moisture content. The box and the uppermost triangle are to the right of the vertical line and constitute surplus moisture, while the lower two triangles represent a moisture deficit and must be subtracted from the recharge flux as indicated in Table 12.2.

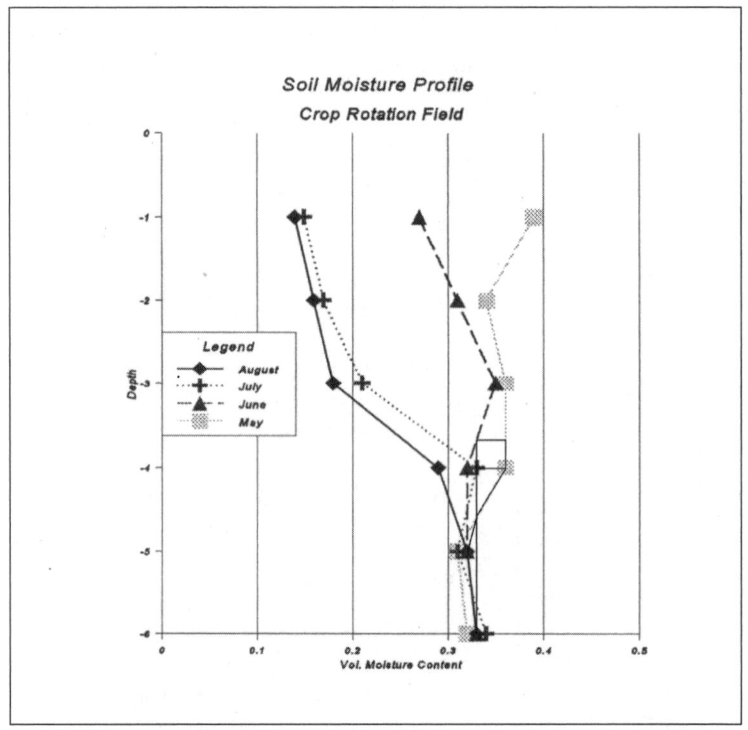

Figure 12.19 Revised from Figure 12.12 with lines needed to calculate the net recharge flux at the WTARC plot. [Modified from Freshman (1996).]

Table 12.2 Recharge Estimate

Item	Height	Width	Area
Box	0.339 ft	0.0306 vol%	0.01037
Triangle	0.546 ft	0.0306 vol%	0.00835
2 Triangles	1.42 ft	0.0102 vol%	0.00724
Sum			0.01148

Note that the net annual recharge calculated is in feet of water, which translates to 0.138 in. (0.35 cm) per year. The error associated with this crude calculation is probably on the order of ± 50% of the value. This is not a huge amount of recharge. If a typical household uses 1,000 gal of water per day, what would be the minimum lot size to avoid mining groundwater (i.e., neglect building and paving effects)? The answer is at the end of the references. Incidentally, the answer that we arrived at is on the low end of the range determined by tritium analysis in the study, and suggests that these lower recharge rates are not significantly impacted by macropores.

Lysimeter Water-Quality Samples—Sources of Error

Lysimeter water-quality samples are subject to certain limitations. Because the samples are drawn into the lysimeter under a partial vacuum, fractions of the gas phases are lost. Additionally, because of the residual air content and the holding time, oxidation reactions will occur if the water contains reduced iron, and the like. Hence iron, and to a lesser extent manganese, may be lost from the water sample. If the soil water is in equilibrium with calcite, the suction of the water into the lysimeter will cause some CO_2 degassing, causing a rise in pH leading to calcite precipitation, some of which will occur in the porous cup, decreasing its permeability and possibly leading to plugging. Obviously, the calcium content of the sample will be depleted under these conditions.

Another problem includes a typically limited sample volume from the sampling source, reducing the available sizes of sample volumes, prohibiting field determinations of pH, Eh, alkalinity, and specific conductance, and limiting or prohibiting sample rinsing of containers and filtration equipment, especially in fine-grained soils. This problem can be minimized by working with your laboratory personnel. See if the lab can provide special in-lab sample handling to reduce the sample volumes needed and if it will provide fresh, precleaned filtration equipment and containers for each sample. Some of volume deficiency may be overcome by resampling the lysimeter a day or two later for field parameters and by recognizing that these values have been impacted by the vacuum effects.

Finally, the porous cup material may react with the sample. Ceramic cups tend to release aluminum while adsorbing metals, bacteria, and viruses; conversely, Teflon tends to adsorb organic compounds (Wilson, Dorrance, et al. 1995). This is another reason why the initial samples are discarded and early samples collected from a lysimeter are given less weight when evaluating the data.

12.5 Concluding Thought

Soil systems are rarely homogeneous, and data collected may be affected by soil variability, locations of study plots, seasonal factors, equipment installation, and equipment choices. This does not mean that calculations cannot be performed and conclusions drawn from the data, but the results should be approached with scepticism and with an open mind to subtle contributing factors.

References

Bouwer, H., 1962. Field Determination of Hydraulic Conductivity Above a Water Table with the Double-Tube Method. *Soil Science Society Proceedings*, pp. 330–335.

Bouwer, H., 1964. Measuring Horizontal and Vertical Hydraulic Conductivity of Soil with the Double-tube Method. *Soil Science Society Proceedings*, pp. 19–22 and 134.

Bouwer, H., 1978. *Groundwater Hydrology*. McGraw-Hill, New York, 480 pp.

Dreiss, S.J., and Anderson, L.D., 1985. Estimating Vertical Soil Moisture Flux at a Land Treatment Site. *Ground Water*, Vol. 23, No., 4, pp. 503–511.

Everett, L.G., Wilson, L.G., and Hoylman, E.W., 1984. *Vadose Zone Monitoring for Hazardous Waste Sites*. Noyes Data Corporation, Park Ridge, NJ, 358 pp.

Fetter, C. W., 1994, *Applied Hydrogeology, 3rd ed.* Macmillan, New York, 691 pp.

Freshman, C. D., 1996. *Hydrogeologic Investigation of 68 Sections of Land in North-central Montana, with Emphasis on Groundwater Recharge*. Master's Thesis, Montana Tech of the University of Montana, 137 pp.

Greacen, E. L., 1981. *Soil Water Assessment by the Neutron Method*. CSIRO, East Melbourne, Australia, 140 pp.

Hanks, R. J. and Ashcroft, G. L., 1980. *Applied Soil Physics*. Springer Verlag, Berlin, 159 pp.

Keys, W. S., and MacCary, L.M., 1971. Application of Borehole Geophysics to Water Resources Investigations. *Technical Techniques of Water-Resources Investigations of the U.S. Geological Survey, Book 2*, 126 pp.

McCord, J.T., 1985. Topographic Controls on Ground-water Recharge at a Sandy, Arid Site. Manuscript of independent study, New Mexico Institute of Mining and Technology, socorro, NM.

Morrison, Robert D., 1983. *Ground Water Monitoring Technology, Procedures, Equipment And Applications*. Timco Manufacturing, Prairie Du Sac, WI, 111 pp.

Pettyjohn, Wayne A., 1997. Field Testing Some Hydrogeologic Assumptions. *The Professional Geologist*, July, 1997, pp. 4–12.

Reiten, J. C., 1991. *Impacts of Oil Field Wastes on Soil And Ground Water in Richland County, Montana: Part II, Contaminant Movement Below Oil Field Drilling Mud Disposal Pits Near Fairview, Montana.* Montana Bureau of Mines and Geology Open-File Report 237-B, 115 pp.

Sanders, L. L., 1998. *A Manual of Field Hydrogeology.* Prentice-Hall, Upper Saddle River, NJ, 381 pp.

Soilmoisture, 1990. Product data obtained from company web site at http://www.soilmoisture.com.

Stephens, D. B., 1996, *Vadose Zone Hydrology.* Lewis Publishers, Boca Raton, FL, 347 pp.

Stephens, D. B. and Knowlton, R. G. Jr., 1986. Soil Water Movement and Recharge Through a Sand at a Semi-arid Site in New Mexico. *Water Resources Research,* Vol. 22, pp. 881.

Stephens, D. B., Knowlton, R. G. Jr., McCord, J., and Cox, W., 1986. Field Study of Natural Groundwater Recharge in a Semi-arid Lowland. *New Mexico Water Resources Research Institute WRRI Report No. 177,* New Mexico State University, Las Cruces, N.M.

Stephens, D. B., Hicks, E., and Stein, T., 1991. Field Analysis on the Role of Three-Dimensional Moisture Flow in Groundwater Recharge and Evapotranspiration. *New Mexico Water Resources Research Institute, Technical Completion Report No. 260,* New Mexico State University, Las Cruces, N.M.

Wilson, L. G., Dorrance, D. W., Bond, W. R., Everett, L. G., and Cullen, S. J., 1995. In Situ Pore-Liquid Sampling in the Vadose Zone. In Wilson, L. G., Everett. L. G., and Cullen, S. J., (eds.), *Handbook Of Vadose Zone Characterization & Monitoring,* pp. 477–521.

Wilson, L. G., Everett, L.G., and Cullen, S.J., 1995. *Handbook of Vadose Zone Characterization & Monitoring.* Lewis Publishers, Boca Raton, FL, 730 pp.

Yeh, T.C. Jim, and Guzman-Guzman, Amado, 1995. Tensiometery. In Wilson, L. G., Everett. L. G., and Cullen, S. J., (eds.), *Handbook Of Vadose Zone Characterization & Monitoring,* pp. 319–328.

Answer (pp. 520, 521): Lot Area = 4,252,934.4 ft^2 = 97.63 acres (39.51 hectares)

Chapter 13

Tracer Test

Marek Zaluski, Ph.D.

A tracer test is a field method used in hydrogeology to quantify selected hydraulic or hydrochemical parameters. In its most common form, a tracer test is conducted by emplacing a defined quantity of traceable substance or a heat source within the aquifer and tracking it down hydraulic-gradient, where at certain points of space and time, its quantity is measured. This form of a tracer test can be viewed as a positive-displacement slug test or an injection test in which stress to the aquifer is delivered in the form of chemical substance or a heat source, rather than by injecting a known volume of water. Dissipation of such a stress with respect to time and the aquifer space provides a data set from which many hydrogeologic conditions may be concluded. Quantification of the groundwater flow direction and rate, aquifer porosity, anisotropy, dispersivity, retardation factor, and other physicochemical characteristics are the most common results of the tracer-test interpretation.

Other types of tracer tests deal with quantifying the dissipation of elements that naturally occur in groundwater and/or injection of certain substances that react with hydrocarbon contaminants, allowing determination of the contaminant concentration. These two types of tracer tests, as well as a large family of surface-water tracer tests (Chapter 6), are not discussed in this chapter.

Tracer tests, their objectives, theory, performance, and interpretation have been addressed in relatively broad literature. Information provided in this chapter is based on these sources as well as actual field experience of the author; it does not, however, aspire to be a handbook for all or any selected method or application. It summarizes practical guidelines for planning and performing tracer tests for common applications. For details on specific application and a comprehensive review of the subject, the reader is referred to the references included at the end of this chapter.

13.1 Tracer Test Objectives

In many cases, a tracer test is part of site-characterization efforts. The appropriate tracer test procedure is selected depending on questions remaining after all other information has been collected. For example, if the dispersivity value is all that is needed, then a single-well tracer test may be performed. However, if the isotropic conditions of the aquifer are questioned, a more time-intensive tracer test, which may involve numerous wells, would need to be considered. Therefore, the kind of the tracer test that needs to be conducted depends upon its objectives as specified in Table 13.1.

For all methods, the retardation coefficient may be obtained if adsorptive and conservative tracers are injected simultaneously and the results compared. The only exception is a recirculating test, with the pumped water recirculated to the point-source.

13.2 Tracer Material

There are quite a variety of substances that may be used as tracers for groundwater investigations. Detailed information regarding these tracers and the application methods can be found in Davis (1985) or other publications. Table 13.2 summarizes information on various tracers.

13.3 Design and Completion of Tracer Test

There are several components of a tracer-test procedure that, if considered, planned, and carefully implemented, will usually lead to a successful completion of the project. These are:

- Conceptual design
- Selection of the initial mass of the tracer or its concentration
- Point-source infrastructure
- Observation wells
- Sampling schedule
- Monitoring
- Equipment used

Table 13.1 Most Common Tracer Tests for Groundwater Investigations

Test	Subcategory	Source Duration	Information Obtained	Hydraulic Stress	Comments
Single well	Borehole dilution	Instantaneous	Flow direction, seepage velocity	Natural flow conditions	Special instrumentation needs to be installed in the well, e.g., thermistors
	Injection/ pumping	Instantaneous	Seepage velocity dispersion coefficient	Injection period precedes pumping period	
Point-source/ one sampling well	Natural flow	Instantaneous or continuous	Seepage velocity dispersion coefficient	Natural flow conditions	Wells must be located at the same flow line
	Diverging test	Instantaneous	Porosity, dispersion coefficient	Injection in the point-source; sampling well is not pumped	
	Coverging test	Instantaneous	Porosity, dispersion coefficient	Pumping from the sampling well; point-source well is not stressed	Pumping at discrete intervals enables vertical differentiation
	Recirculating test	Continuous	Porosity, dispersion coefficient	Both wells are stressed. Injecting well receives pumped water or “clean” water from other source	Tests a considerable larger portion of the aquifer, than a converging test
Point-source/ two sampling wells	Recirculating test	Continuous	Porosity, dispersion coefficient, anisotropy	All wells are stressed. Injecting well receives pumped water or “clean” water from another source	
Point-source/ multiple sampling wells		Instantaneous	Flow direction, seepage velocity, anisotropy*	Natural flow conditions	

*Only if contours of hydraulic head are known

Table 13.2 Most Common Tracers Used for Groundwater

Tracer	Sub-category	Permeable Medium	Detection		Advantages	Disadvantages	Comments
			Field Method	Limit			
Ions						At high concentrations tend to sink	Anions are more conservative than cations
	Br^-	Any kind	Bromide electrode, hack kit	1 ppm	(1) Low [<1 ppm] background concentration in ground-water (2) Very low toxicity (3) No MCL*	At low pH may be adsorbed by clay	1. May be the most common tracer used 2. Common source: NaBr
	Cl^-	Any kind	Chloride electrode, hack kit	1 ppm	Availability	High background concentration drinking water standard** = 250 ppm	Very conservative
	I^-	Any kind	Iodine electrode, hack kit	1 ppm	Very low [<0.01 ppm] background concentration in ground-water	May be affected by micro-biological activity	Radioactive isotope of iodine was used as a radioactive tracer

Table 13.2 Most Common Tracers Used for Groundwater (continued)

Tracer	Sub-category	Permeable Medium	Detection		Advantages	Disadvantages	Comments
			Field Method	Limit			(1) Maybe the oldest tracer used (2) Less conservative than anions
Dyes		Perform best in fracture rock and karst					
	Fluorescein	Karst rock, clean sand sandstone basalt	Fluor-imeter	0.3 ppb over back-ground	(1) Less expensive than other dyes (2) Visually detected if >40 ppm	(1) High natural background fluorescence (2) Adsorbs strongly to organic matter (3) Loses color at pH<6	Green opalescent color
	Rhodamine WT	Fracture and karst rocks; works better in porous medium than fluorescein	Fluor-imeter	0.013 ppb over back-ground	The least adsorbed dye	Biologically not as safe as fluorescein	Orange color
Radio-nuclides	^{131}I, ^{82}Br,	Any kind	Radia-tion meters	Extrem-ly low	(1) Short half life [8 & 1.5 days] in comparison to other radionuclides (2) Very mobile	Despite the demonstrated safety, public perception, and regulations prohibit the field usage of these tracers	Superior tracers

Table 13.2 Most Common Tracers Used for Groundwater (continued)

Tracer	Sub-category	Permeable Medium	Detection		Advantages	Disadvantages	Comments
			Field Method	Limit			
Heat		Any kind	Thermistors	0.05°C	Direct measure of groundwater flow direction		(1) Used in a single well (2) Feasible for multiwell tracer test, if a point or line sink of warm water is active for long time
Bacteria	Variety	High permeable media	None, Requires laboratory procedure			(1) Regulatory concerns (2) Strain used must be mobile	Bacteria must be nonpathogenic
Virus	Variety	High permeable media	None, Requires laboratory procedure			(1) Regulatory concerns (2) May move faster than advection (preferential path due to large size)	Virus must be nonpathogenic

* MCL = Maximum contaminant level

** National secondary drinking water regulations

Conceptual Design

This preliminary phase includes

- Review of site hydrogeologic information available
- Setting the objectives for the tracer test
- Selection of the kind of the tracer test to be conducted
- Selection of the tracer material and its cost evaluation
- Preliminary costing of a necessary infrastructure
- Assessment of operational costs

If the predicted expenses are within the available budget, the tracer concentration may be assessed in detail. Otherwise, it may be necessary to revise (limit) the objectives of the tracer test and choose a less expensive method as presented in Example 13.1.

Example 13.1

Initial Conceptual Design

Hydrogeologic information

A contaminated site in question is located between two perennial creeks. An unconfined aquifer beneath the site is formed in alluvial deposits. A 2-in. diameter monitoring well is located in the center of the reported incidental gasoline spill and two 2-in. diameter monitoring wells are located adjacent to each creek. There is no information regarding the site location with respect to the groundwater divide.

Objectives of the tracer test:

Determination of the flow direction at the spill location
Defining seepage velocity
Defining anisotropy ratio

Tracer test initially selected:

Point source (an existing central well) and multiple (six) sampling wells located in hexagonal pattern around the point source well.

Tracer material selected:

One kilogram of bromide provided as NaBr
NaBr is available in 50 Lb bags @ \$60 a bag — \$ 60

Necessary infrastructure:

Six 2-in. diameter observation wells @ \$500 — \$ 3,000

Operational costs:

Technician 12 hr @ \$50 — \$ 600

Measuring equipment (bromide electrode and a meter) — \$ 1,000

Miscellaneous expenses (containers, glassware, & incidentals)	$ 500
Total expenses	$5,160
Available budget	$3,050

The deficit of $2,160 could be offset by decreasing the number of observation wells from six to two. Such a reduction, however, would jeopardize the accomplishment of the three objectives for the tracer test. However, dropping the objective of anisotropy-ratio determination and changing the multiwell tracer test to a borehole dilution test would allow for achieving two remaining objectives within the budgetary requirements. Moreover, it would be possible to determine these parameters for three separate locations as demonstrated below.

Revised Conceptual Design

Objectives of the tracer test:

Determination of the flow direction at the spill location
Defining seepage velocity

Kind of tracer test:

Single well borehole dilution test

Tracer material selected:

Heat stress delivered by a down-hole instrument with a central heat electrode and six thermistors

Rental fee: @ $300 a field-day (shipment cost and time covered a vendor)	$ 300

Necessary infrastructure:

Three 4-in. diameter test wells @ $700	$ 2,100

Operational costs:

Technician 9 hr (3 hr per well) @ $50	$ 450
Measuring equipment, included in the rental fee	
Miscellaneous expenses (incidental)	$ 200
Total expenses	$ 3,050

Selection of the Initial Mass of the Tracer or Its Concentration

As stated in Table 13.1, a tracer can be introduced into the hydrologic system either in an instantaneous (slug) or a continuous manner. Depending on the type of the tracer test and the method of interpretation, either an initial mass of the tracer or its initial concentration needs to be considered to secure appreciable concentration of the tracer at the receptor point. For a continuous injection of a tracer, its initial concentration is to be evaluated. For an instantaneous placement, depending on the tracer test method, either the tracer mass or its concentration needs to be considered. An inherent contradiction of the successful tracer test is that it is necessary to assume the results of the tracer test before it is actually conducted. To do so, the effect of the tracer dilution, dispersion, adsorption, and reactivity need to be preevaluated to predict the expected peak concentration at the receptor and to compare it with the detection limit and operational range of available instruments.

The peak concentration at the receptor point can be estimated by using an appropriate closed-form equation, which can be found, for instance, in Javandel et al. (1984), Sauty (1980), (Bear) 1979, Domenico and Schwartz (1990), and Freeze and Cherry (1979), or using a public domain program SOLUTE (a proprietary version compiled by Paul K.M. van der Heijde is distributed by International Ground Water Modeling Center) or any other solute transport model. In this process, an assumed mass or an initial concentration is plugged into an appropriate method, and the tracer concentration at the receptor can be calculated for the given time and at its maximum.

At this point, however, it is necessary to emphasize that spending extensive time for premodeling of a tracer test is usually not a good time investment. This is because there are too many unknown variables at that stage of the hydrogeologic investigation. Had those variables been known, there would not have been any need to conduct the tracer test. Nevertheless, a good match of the hydrogeologic conditions and the kind of the tracer test with the stipulations set for applicability of the given formula is of prime importance. For example, for a simple case of a two-well tracer test conducted at natural flow regime with a instantaneous release of the tracer, the following formulas may be used.

For a three-dimensional flow regime (Freeze and Cherry 1979):

$$C_{(x,y,z,t)} = \frac{M}{8(\pi t)^{3/2}\sqrt{D_x D_y D_z}} \exp\left(-\frac{X^2}{4D_x t} - \frac{Y^2}{4D_y t} - \frac{Z^2}{4D_z t}\right) \quad [13.1]$$

For a two-dimensional flow regime (Bear 1979):

$$C_{(x,y,t)} = \frac{M}{4\pi t\sqrt{D_x D_y}} \exp\left(-\frac{X^2}{4D_x t} - \frac{Y^2}{4D_y t}\right) \quad [13.2]$$

For a one-dimensional flow regime (Bear 1979):

$$C_{(x,t)} = \frac{M}{2\sqrt{\pi D_x t}} \exp\left(-\frac{X^2}{4D_x t}\right) \quad [13.3]$$

In Equations 13.1, 13.2, and 13.3:

C = Contaminant concentration at location x, y, z at time t (M/L^3, M/L^2, M/L for three, two and one dimensional flow, respectively, where L is the linear dimension)

x, y, z = Coordinates (L) of the receptor with respect to the point source

M = Mass (M) of contaminant introduced at the point source—It can be expressed as the product of initial concentration C_0 and initial volume V_0.

t = Time elapsed (t) from the instantaneous injection of the tracer.

$D_x D_y D_z$ = Dispersion coefficient (L^2/t) in longitudinal, transversal, and vertical (transversal in the vertical plane) direction, respectively

X, Y, Z = Distance (L) to the receptor from the center of gravity of the contaminant mass

For the specific case where the point source and the receptor are at the same flow line, the position of the center of gravity of the contaminant mass at time t will lie along the flow path in the x direction at coordinates x_t, y_t, z_t (L). Consequently $y_t = z_t = 0$, but $x_t = vt$ where v (L/t) is the average velocity (seepage velocity) and n (dimensionless) is the effective porosity. Thus, in Equation 13.1, 13.2, and 13.3, $X = x\ vt$ and Y, Z = zero

Formulas 13.1, 13.2, and 13.3 are valid for calculation of the tracer concentration until it reaches its maximum (the rising limb of the bell-shaped curve). Thus, they may be used for prediction of the time for first arrival of the tracer at the receptor point. Certain modification of the definition for the time term (t) is needed to have these formulas valid for times longer than this, which correspond to the maximum concentration.

For the prediction of the maximum concentration (C_{max}) at the receptor point, i.e., when $x = vt$ and consequently $X = 0$, these formulas simplify to contain only their first member. Such formulas reflect a condition that for a nonreactive transport the maximum concentration of tracer coincides with the time t, which equals an advective travel time from the injection source to the receptor point. For a two-dimensional flow, the maximum concentration at the receptor point is expressed as:

$$C_{(max)} = \frac{M}{4\pi t\sqrt{D_x D_y}} \qquad [13.4]$$

An example of a spreadsheet for prediction of the maximum concentration of a tracer at the receptor point is presented in Example 13.2. For the tracer-test configuration as defined in the spreadsheet, the maximum concentration of 116.7 mg/L will be recorded after 1.93 days in the monitoring well located 12 ft from the point source. Using the same spreadsheet, one can calculate that the maximum concentration in a well located twice as far, i.e., at the distance of 24 ft, would be only 29 mg/L and would be recorded after 3.86 days.

Example 13.2

Spreadsheet software can be used to estimate maximum solute concentration at the receptor (2-dimensional, advective-dispersive transport, and instantaneously released tracer at natural hydraulic gradient).The following information is entered in the spreadsheet cells, represented by the table shown in this example:

hydraulic conductivity [ft/day]

*d*20 [mm] if used to estimate hydraulic conductivity (shaded cell)

effective porosity [dimensionless fraction]

distance [ft] between the injection and receptor points

hydraulic gradient [dimensionless fraction]

hydraulic head (*H*) difference [ft] at the injection and receptor locations if used for hydraulic gradient calculation (shaded cell)

tracer mass [mg]

tracer initial concentration [mg/L] and volume of water injected with tracer [L] if used for calculating mass of the tracer (shaded cells)

Assumptions:

Injection and receptor points are on the same flow line.

Lateral extent of the aquifer may not impact transversal spread of the tracer.

Longitudinal dispersivity is 0.1 of the distance between the injection and receptor points.

Transversal dispersivity is 0.01 of the distance between the injection and receptor points.

Hydraulic conductivity (*K*) may be estimated based on sieve analysis using the U.S. Department of Agriculture formula:

$$K=0.36*d20^{2.3} \text{ cm/s} \quad [13.5]$$

where:

*d*20 = Particle diameter (mm) that together with smaller particles constitutes 20% of weight of the sample

Hydraulic conductivity value	91.23 ft/d
*d*20	0.35 mm
K	0.03 cm/s 91.23 ft/day
Effective porosity	0.25
Distance from injection point	12 ft
Hydraulic gradient	0.02
Del *H*	0.20 ft
Calc. hydraulic gradient	0.02
Tracer mass, *M*	10,000 mg
Co	5,000 mg/L
Vo	2.00 L
M	10,000 mg

Formula:

*C*max = *M* / [4 PI**t**(*Dx***Dy*)^0.5*9.29]

[13.6]

where:

*C*max	=	maximum concentration at the receptor point [mg/dec^2, an equivalent of mg/L for 3-D flow]
Co	=	tracer initial concentration [mg/L]
Vo	=	volume of injected water with the tracer [L]
t	=	travel time for advective flow from the injection point to the receptor point [*d*]
Dx	=	coefficient of hydrodynamic dispersion in direction *X* [ft^2/day]
Dy	=	coefficient of hydrodynamic dispersion in direction *Y* [ft^2/day]
9.29	=	conversion factor from ft^2 to dec^2

Calculation

v	=	6.20 ft/day
t	=	1.93 days
Dx	=	1.20 ft^2/day
Dy	=	0.12 ft^2/day
*C*max	=	117 mg/dec^2

After the initial mass of the tracer or its concentration is calculated, it needs to be reevaluated with respect to uncertainty of the values used for calculation. It is common to use a safety factor and increase the calculated initial concentration or its mass one order of magnitude for tracer tests that will involve hydraulic stress at the point source or at the observation well. Depending on hydrogeologic conditions, a safety factor of 1,000 may be used for tracer tests that are conducted under the natural hydraulic gradient of groundwater. Conversely, a recirculating tracer test is the least vulnerable to miscalculations of the initial concentration.

There are two main reasons for using a safety factor:

- Intensive and costly efforts may be undertaken for tracer test preparation and its monitoring. Thus, it is hardly possible to imagine a more embarrassing scenario than the one with a no-show of the tracer at the receptor point.
- It is practically impossible to repeat the tracer test after an apparent failure of an earlier attempt and still have confidence regarding the origin of detected concentration of the tracer at the receptor point. Either a long time (relatively to the seepage velocity) has to separate two attempts or the second tracer needs to be different.

Point-Source Infrastructure

Injection wells

In most cases, a tracer is introduced to the hydrogeologic system using an injection well. Construction requirements for such a well depend on the source duration.

For a continuous injection of the tracer a well should have some characteristics similar to a pumping well, that is:

- The well needs to have an appropriate transmitting capacity to accommodate the designed injection rate without excessive well loss.
- Its screen should not straddle the water table, otherwise water will be injected directly to the unsaturated zone.

Conversely, if a tracer is to be injected between two packers that are placed within the screened interval, the well must not be gravel packed and it should be poorly developed to disallow a shortcut of water through the gravel pack back to the well.

One also needs to realize that for an unconfined aquifer, even if the screen does not straddle the water table, mounding of groundwater will occur. Therefore, the injection rate should be slow enough to accommodate the mass injected. Groundwater mounding might affect the dynamics of the tracer test, especially at the beginning when the transient phase groundwater "invades" the vadose zone whose hydraulic conductivity is always lower until it reaches full saturation (Chapter 12). This is one of the reasons why, for a recirculating test with a continuous tracer injection, the tracer should not be introduced to the system until a steady-state flow is approached.

For most tracer tests with an instantaneous release of the tracer, the injection well should be designed so that:

- Its screened interval and its diameter are large enough to allow for placement of the tracer solution as one slug. Otherwise, the requirements of the instantaneous release will not be satisfied.
- Its screened interval and its diameter are small enough to disallow or minimize the effect of dilution of the tracer in the well before it enters the aquifer.
- Its construction allows for placement of the tracer with no or minimal changes in the natural flow regime in the aquifer. Otherwise the requirement of point-source release will not be satisfied, and the calculated values will include additional errors. For instance, values of transversal dispersivity in horizontal and vertical planes will be exaggerated. Needless to say, this requirement is not important for a single well–injection/extraction test or the diverging test for which a radial flow pattern is a designed feature.
- Its screened interval coincides with the section of the aquifer to be tested by the tracer test.
- The screen has high transmitting capacity, and the well is well developed (Chapter 8).

A design that satisfies the above requirements would allow for placing in the well a tightly fitting and closed container with the entire volume of the tracer. After the natural flow regime has recovered, that is, the water table in the well has returned to its initial level, the bottom valve is opened and the container removed from the well exposing the tracer to a screened portion of the well. This method of releasing a tracer to a hydrogeologic system is especially recommended for a test that is conducted at natural flow conditions.

For such a case, a common recommendation of "pushing" the tracer by injecting a three-well-volumes slug of clean water is in error. If a well is screened over the tested interval, the tracer does not need to be "pushed" out of the well, because it will leave the well itself being "pushed" by the flowing groundwater that actually converges in the well. The convergence is caused by the principle of streamlines refraction (tangent law) at the interface of the aquifer material and a circular object, the well, of an infinite hydraulic conductivity. Specifically, twice as much of the groundwater will flow horizontally through a circular cross section of the well than through an equally wide section of the aquifer.

Sometimes, however, in a real field situation, some of the above recommendations are difficult to implement and the tracer test is conducted with several simplifications, as described in Example 13.3. While analyzing the results of such a test, it is necessary to limit the conclusions to only those whose correctness was not jeopardized by the deficiencies of the field procedure.

Example 13.3

A tracer test was conducted as an experiment to see whether a forced gradient setting (the hydraulic stress) could mobilize a NaBr tracer approximately 15 ft from an injection point towards a pumping well. The pumping well was pumped at a constant rate of 5 gpm (18.9 L/min) for 2.5 hr prior to the introduction of a NaBr solution. The solution was mixed in a 13-gal (49-L) container and pumped via a peristaltic pump over a period of 1 hr and 45 min (Figure 13.1). The solution was released at the screen depth and was supposed to be observed at the pumping well at some time in the future.

Figure 13.1 Conducting a tracer test.

The concentration of the solution at the pumping well was desired to be approximately 100 mg/L. An estimate of the starting concentration was devised through a simplified version of a pie approach (Domenico and Robbins 1985). The starting concentration at the pumping well was assumed to be 0 mg/L, and the starting concentration at the injection point was the unknown. The pumping rate (Q) was 5 gpm, and the proportion of the contributing Q for the pie shaped area near the injection point was estimated by assuming that it would take one hour to pump in the 13-gal solution (49-L). The mixing equation was used to estimate the unknown initial concentration to be used in the field test. Schematically this is illustrated in Figure 13.2.

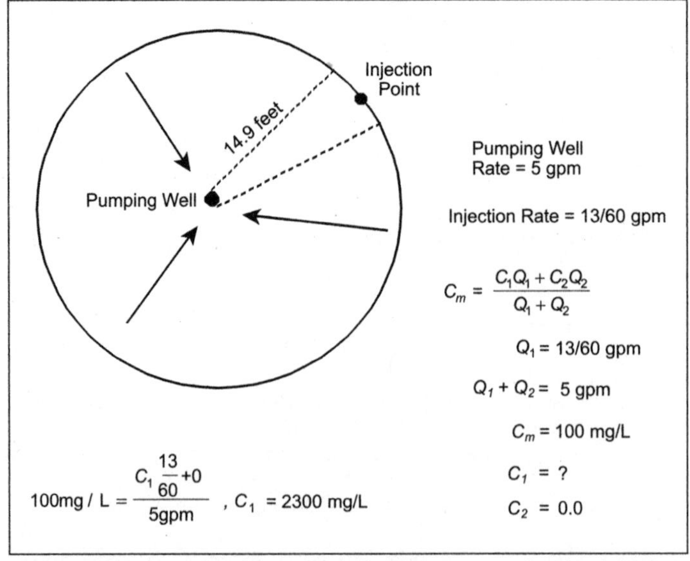

Figure 13.2 Schematic of forced gradient tracer test, showing initial concentration calculations.

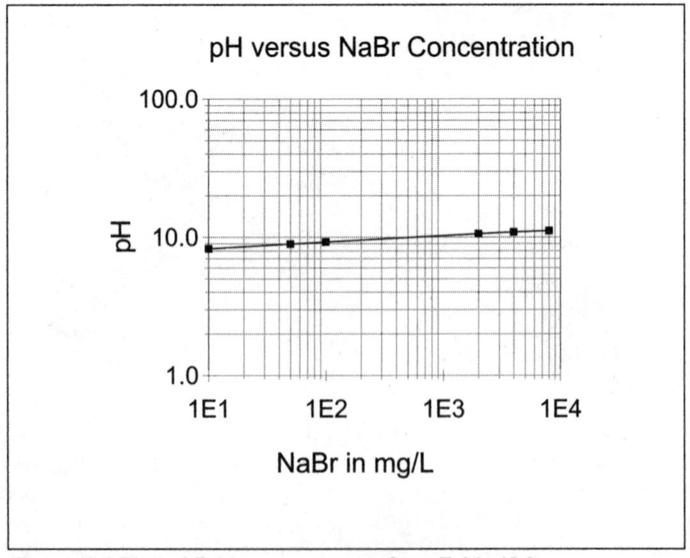

Figure 13.3 Plotted field measurements from Table 13.3

Because of unavailability of a bromide electrode in the field, the pH electrode was used instead. Standards were made up in the lab in mg/L and brought out into the field (Figure 13.3). Table 13.3 lists the values measured in the field.

Table 13.3 Field Measurements for Example 13.3

Concentration NaBr in mg/L	pH units (10^{-n})
Distilled water	7.09
10	8.27
50	8.97
100	9.26
Solution	10.64
4000	10.93
8000	11.18

The injection solution was estimated to be near 2300 mg/L. The effluent discharge remained at pH 7.05 for 2 ½ hours, and it occurred to the participants that it could take many hours before the plume moved 14.9 ft. Thus a piezometer was hand-drilled along the pathway at a distance of 6.9 ft from the injection point, and a reading of pH = 8.36 was obtained at 3 hours. The gradient was 4/15, the effective porosity was assumed to be 0.15, and the movement 7 ft in 1/8 of a day yielded a seepage velocity of 56 ft/day. The hydraulic conductivity was calculated to be 31 ft/day. This correlated well with estimates from constant discharge pumping tests, yielding an estimate between 25 and 40 ft/day.

Nonwell injections

In special cases like investigating surface water/groundwater relationship or groundwater recharge conditions, it is necessary to introduce a tracer at the location where the recharge is assumed to take place (Chapter 6). In many cases, especially in areas with well-developed morphology, it is a body of surface water that recharges groundwater system. An example of such a case may be a stream, flowing through a karst area, that may be losing a portion of its water through a sinkhole located in the stream bottom. Another case might be an artificial pond created either for landscaping or industrial purposes. In such situations, there is hardly a standard method for emplacement of the tracer, leaving investigators to their own inventiveness. This is illustrated in Example 13.4.

Example 13.4

The photograph in Figure 13.4 depicts the emplacement of the bromide tracer in a flow-through settling pond located on the top of a waste rock pile accumulated in front of an adit of the abandoned mine. The pond contained relatively good quality water. The water balance of that pond indicated that approximately 10% or 2 gpm of water flowed through the pond leaks, through the pond bottom, and presumably reappeared at the toe of the rock pile heavily contaminated (enriched) with heavy metals. To confirm and quantify this stipulation, a tracer test was conducted using bromide as the tracer. The following procedure was used to emplace the bromide.

Figure 13.4 Placing a bromide tracer in a flow-through settling pond.

The required initial concentration of bromide (Br) in the pond was determined as 2,000 mg/L. Depth of the pond was surveyed, and volume of water in the pond was determined to be 46,550 L (12,300 gal).Thus, 93 kg (205 lb) of bromide was needed for the tracer test. This amount of bromide was provided in 120 kg (264 lb) of dry sodium bromide (NaBr). Mine water usually flowing into the pond was diverted, and the outflow was dammed to create a standing body of water in the pond.

Dry NaBr, 120 kg (264 lb), was dissolved in a plastic container to make 465 L (123 gal) of solution that contained 258 g/L of NaBr, an equivalent of 200 g/L of Br. This solution had a specific gravity of 1.193. A battery-powered, boat-trolling motor was used to mix dry NaBr in the container with water.

A 50-ft length of 1.5-in diameter hose was attached to the bottom of the NaBr solution container. The other end of the hose was extended to a small pontoon floating on the pond. A trolling motor was attached to the pontoon to be used for mixing NaBr solution in the pond water. Ropes were attached to the pontoon, so that two people standing on the shore could pull the pontoon across and along the pond.

Another person boarded the pontoon and directed the outlet of the hose with gravity-flowing bromide solution to the propeller of the trolling motor, while the pontoon was dragged back and forth across the pond. This procedure provided good 1:100 mixing of the bromide solution with water in the pond, preventing stratification of the heavy bromide solution in the bottom. Upon mixing, samples of pond water were collected, and the concentration of bromide was checked using a bromide electrode.

An approximately 2,000 mg/L bromide solution was left in the pond to leak through its bottom. After four days, approximately 90% of the pond water leaked through its bottom, and the natural flow condition through the pond was restored to maintain constant hydraulic head over the moving tracer.

Observation Wells

For a recirculating tracer test, a sampling well should have some characteristics similar to a pumping well to accommodate required discharge. For all other tracer tests, the diameter of a sampling well should be small to minimize the effect of the dead water residing in the well above the sampling interval. More important, however, is to screen the well over the entire investigated interval and have no blank casing (sump) below the screen (Chapter 8).

Sampling Schedule

The sampling schedule is usually a by-product of the preliminary calculation of the tracer's initial concentration. Theoretically, first sampling should be scheduled for the calculated time of arrival of the front portion of the dispersed tracer. Then, the sampling and measurements of the tracer concentration need to be continued until the tracer test is terminated. For a continuously injected tracer, the test may be terminated when there is no change of the tracer concentration as measured in the real time at the receptor point. For an instantaneously injected tracer, which should yield a bell-shaped curve of the concentration versus time, it is best to terminate the test when the measurements are again close to the detection limit.

Prediction of the time for an earliest sampling may be done using Equations 13.1, 13.2, 13.3, or other formulas whose applicability match a configuration of the given test. A spreadsheet for determination of the tracer concentration versus time for a two-dimensional flow at natural hydraulic condition is given in Example 13.5. In this case, because of an assumption that the point source and the receptor point are placed at the same flow line, Equation 13.2 simplifies to:

$$C_{(x,y,t)} = \frac{M}{4\pi t\sqrt{D_x D_y}} \exp\left(-\frac{X^2}{4D_x t}\right) \qquad [13.7]$$

This simplification is possible because $Y = y = 0$, as explained earlier in this chapter, and there is no movement of the tracer with respect to the y coordinate, except for the transversal dispersion. Example 13.5 uses the same tracer-test configuration assumed for Example 13.2. The final result of the calculation shown in Example 13.5 is a predicted time for the first measurable record of the tracer at the receptor point if the instrument detection limit is 0.005 mg/L. This time of arrival is 0.21 day. Maximum concentration of 116.7 mg/L will be recorded at 1.93 days.

Example 13.5

Spreadsheet software can be used to predict the tracer concentration versus time (2-dimensional, advective-dispersive transport, and instantaneously released tracer at natural hydraulic gradient). The following information is entered in the spreadsheet cells, represented by the table shown in this example:

Limitation: Values are calculated for a valid time (t) not greater than the time ($t$max) corresponding to the maximum concentration.

Time (t) [day] for the calculated concentration of the tracer

Hydraulic conductivity [ft/day]

d20 [mm] if used to estimate hydraulic conductivity (shaded cell)

Effective porosity [dimensionless fraction]

Distance (x) [ft] between the injection and receptor points

Hydraulic gradient [dimensionless fraction]

Hydraulic head (H) difference [ft] at the injection and receptor locations if used for hydraulic gradient calculation (shaded cell)

Tracer mass [mg]

Tracer initial concentration [mg/L] and volume of water injected with tracer [L] if used for calculating mass of the tracer (shaded cells)

Assumptions:

Injection and receptor points are on the same flow line.

Lateral extent of the aquifer may not impact transversal spread of the tracer.

Longitudinal dispersivity is 0.1 of the distance between the injection and receptor points.

Transversal dispersivity is 0.01 of the distance between the injection and receptor points.

Instrument detection limit is 0.005 mg/L.

Hydraulic conductivity (K) may be estimated based on sieve analysis using the U.S. Department of Agriculture formula:

$$K=0.36*d20^{2.3} \text{ cm/s}$$

where:

*d*20 = particle diameter (mm) that together with smaller particles constitutes 20% of weight of the sample

time (*t*)	0.21 day
Hydraulic conductivity value	91.23 ft/day
*d*20	0.35 mm
K	0.03 cm/s 91.23 ft/day
Effective porosity	0.25
Distance from injection point	12 ft
Hydraulic gradient	0.02
Del *H*	0.20 ft
Calc. hydraulic gradient	0.02
Tracer mass, *M*	10,000 mg
Co	5,000 mg/L
Vo	2.00 L
M	10,000 mg

Formula:

Ct = *M* / [4*PI t (*Dx***Dy*)^0.5 × 9.29]exp((-*X*^2/(4*Dx** t*9.29))

where:

Ct = concentration of the tracer at the receptor point at time (*t*) [mg/dec^2, an equivalent of mg/L for 3D flow]

Co = tracer initial concentration [mg/L]

Vo = volume of injected water with the tracer [L]

t = time for calculated concentration of the tracer at the receptor point [*d*]

Dx = coefficient of hydrodynamic dispersion in direction *X* [ft^2/day]

Dy = coefficient of hydrodynamic dispersion in direction *Y* [ft^2/day]

9.29 = conversion factor from ft^2 to dec^2

X = *x vt*

Calculation			
Seepage velocity	v	=	6.20 ft/day
	X	=	10.70 ft
Max. concentration time	$tmax$	=	1.93 days
	Dx	=	1.20 ft^2/day
	Dy	=	0.12 ft^2/day
	Ct	=	116 mg/dec2, for t = 0.21 day

The time of arrival calculated in Example 13.5, however, needs to be adjusted (shortened) proportionally to the uncertainty of dispersivity values used for calculations. The frequency of measurements may also be determined through preliminary calculation to ensure that a reasonably smooth curve can be obtained.

Monitoring

Results of the tracer test must be monitored at real time as soon as a sample is collected. The monitoring equipment used depends on the kind of tracer. For instance, for dye tracers, a fluorimeter is required; for a radioactive tracer (currently very seldom used), an appropriate radiation meter is used. If bacteria are used as a tracer, it is necessary to provide an appropriate analytical field kit. Unfortunately, field kits for bacteria are notorious for unreliable performance.

For most popular tracers like bromide or iodine, a single ion electrode is commonly used. Such electrodes have certain operational range, usually starting from one part per million (ppm), and a very specific calibration procedure. If several samples a day are collected, an electrode must be calibrated at least twice a day, otherwise an electrode drift may remain unnoticed. It is a common mistake to save little time for a field calibration of the measurement instrument, only to end up spending excessive time later during the data-reduction process for removing a guesstimate drift portion of the field measured values.

Maintaining detailed notes of the recorded noise (i.e., measurements below the detection limit) is also strongly recommended especially for tracer tests with multiple observation wells and large vertical component of hydraulic gradient. It happened once to the author that the flow direction was traced only by recording consistent noise at one of many observation wells.

Equipment

Equipment used for a tracer test depends of the procedure used. The least equipment-intensive tracer tests are those conducted at natural flow conditions. Conversely, recirculating tracer tests usually include a lot of equipment that amplify proportionally to the rate of recirculated fluid. Figure 13.5 depicts a setup for a recirculating tracer test conducted with two sampling/extraction wells and an injection well operated at 7-gpm injection rate over a 4-day period. Two 1,000-gal tanks, electrical mixers, a battery of flowmeters, submersible and centrifugal pumps, and the like, were used for this project, which included needs for anisotropy ratio determination. The list of equipment may include the following.

Figure 13.5 Setting up equipment for a recirculating tracer test.

General Equipment

- First-aid kit
- Safety gloves
- Safety goggles
- Emergency telephone
- Weather radio
- Good supply of paper, notebooks, and pencils
- Paper towels

- String, duct tape, and insulation tape
- Measuring tape
- A kit with pH, specific conductance and temperature meters and probes
- Shelter for rainy weather (e.g., a pickup with a topper)

Tracer-Test Implementation Equipment

- Appropriate amount of a tracer
- Material safety data sheet (MSDS) of the tracer used
- Tarp or waterproof boxes to store supply of a tracer (if continuous injection)
- Weighing scale to weigh bulk tracer while preparing the tracer solution
- Containers for preparing and storing the tracer solution
- Stir or mechanical mixer for tracer solution preparation
- Calibrated bucket for discharge measurements, if applicable
- Flowmeters (for a recirculating tracer test)
- Submersible pumps (for a recirculating tracer test)
- Centrifugal pumps for transfer tracer solution (for a recirculating tracer test)
- Hoses and valves
- Generator, if applicable
- Data logger with pressure transducers (for recirculating tracer test)
- Water-level indicator
- Stopwatch
- Alarm clock
- Good source of light for night sampling
- A tool box with tools appropriate for the infrastructure used

Tracer Test Monitoring Equipment

- Measuring instrument(s) for monitoring of tracer concentration
- Instruments' instruction manuals

- Weighing scale for preparation of a set of calibration solutions.
- Calibrated cylinders and pipettes for preparing calibration solutions
- Tissues for blotting electrodes, etc.
- Deionized water
- An icebox with ice for cooling calibration solutions if temperature-sensitive
- A propane/butane stove for heating water to warm calibration solutions if temperature-sensitive
- Bottles (containers) to store calibration solutions
- Bottles (containers) to store and keep electrodes hydrated between measurements
- Boxes with lids to store measuring equipment and calibration solutions between measurements
- Forms for keeping records of instruments calibrating
- Forms for recording measurements
- A laptop computer with a spreadsheet allowing real-time plotting of the concentration curve.

13.4 Common Errors

A tracer test justifies the expression "those who do nothing make no errors." A variety of tracer-test procedures, tracer-test materials, and diversity of hydrogeologic conditions make each tracer test a unique task for which no standard operating procedure exists. It is thus inevitable that mistakes will be made despite all efforts undertaken to avoid them. The key issue is to minimize the number of mistakes. Some common field errors are listed below.

- The receptor was located too far from the injection point.
- Monitoring of the tracer test was terminated prematurely because:

  Assumed effective porosity was too low.
  Assumed hydraulic conductivity was two high.

- Monitoring of the tracer started too late because:

  Assumed effective porosity was too high.
  Assumed hydraulic conductivity was too low.

- Measured concentrations were too low to be used for quantitative interpretation because:

  Dispersivity values were underestimated.
  Tracer was less conservative than expected.
  Ratio of anisotropy and/or principal directions of anisotropy were "misevaluated."
  Location of the receptor was incompatible with the actual flow pattern.
  Vertical component of hydraulic gradient was neglected or "misevaluated."
  Instrument detection level was too high.

- A continuous-injection tracer test was initiated before steady-state conditions for flow had been established.
- Emplacement of the tracer significantly altered natural flow conditions while the conditions supposed to be undisturbed.
- Calibration check of the instruments used for measuring tracer concentrations at the receptor point were not frequent enough.

Errors committed during the test design or its result interpretation are usually related to inappropriate match of the formula or the method used with the hydrogeologic setting or the tracer-test conducted. An example of such an error would be using a formula developed for a three-dimensional flow to design a tracer test in a shallow flow system with a long distance between injection and receptor points.

References

Bear, J., 1979. *Hydraulics of Groundwater.* McGraw-Hill, New York.

Davis, S.N., Cambell, D.J., Bentley, T.J, Flynn T.J., 1985. *Ground Water Tracers.* National Water Well Association.

Domenico, P.A., and Robbins, G.A., 1985. A New Method of Contaminant Plume Analysis, *Ground Water,* Vol. 23, No. 4, pp. 476–485.

Domenico, P.A., and Schwartz, F.W., 1990. *Physical and Chemical Hydrogeology.* John Wiley & Sons, New York.

Freeze, R.A., and Cherry, J.A., 1979. *Groundwater.* Prentice-Hall, Upper Saddle River, NJ.

Javandel, I., Dughty, C., Tsang, C.F., 1984. *Groundwater Transport: Handbook of Mathematical Models.* American Geophysical Union. Water Resources Monograph Series 10. American Geophysical Union, Washington, DC.

Sauty, J. P., 1980. An Analysis of Hydrodispersive Transport in Aquifers. *Water Resources Research,* Vol. 16, No. 1, pp. 145–158.

van der Heijde, P.K.M., 1997. *Solute, Analytical Models for Solute Transport in Ground Water.* IGWMC-BAS 15, International Ground Water Modeling Center, Colorado School of Mines, Golden, CO.

Appendix A

Unit Conversions Tables

Table A.1 Unit Conversions for Length

Unit	Inch	Feet	Yard	Mile	mm	cm	mr	km
1 in.	1	0.08333	0.02777	1.5783e-05	25.40	2.54	0.0254	2.54e-05
1 ft	12	1	3	5280	304.8	30.48	0.3048	3.048e-04
1 yd	36	3	1	5.682e-04	914.4	91.44	0.9144	9.144e-04
1 mile	63,360	5280	1760	1	1.609e06	1.609e05	1609.3	1.609
1 mm	0.03937	3.2808e-03	1.0936e-03	6.2137e-07	1	0.1	0.001	1e-06
1 cm	0.3937	0.03281	1.0936e-02	6.2137e-06	10	1	0.01	1e-05
1 m	39.37	3.2808	1.0936	6.214e-04	1000	100	1	0.001
1 km	39.370	3280.8	1093.6	0.621	1e06	100,000	1000	1

Table A.2 Unit Conversions for Area

Unit	Feet²	Mile²	Acre	Meter²	km²	Hectare
1 feet²	1	3.587e-08	2.296e-05	0.09291	9.291e-08	9.291e-06
1 mile²	2.788e06	1	640	2.59e06	2.59	259
1 acre	43,560	1.5625e-03	1	4047	4.047e-03	0.4047
1 meter²	10.764	3.861e-07	2.471e-04	1	1e-06	0.0001
1 km²	1.0764e07	0.3861	247.1	1e06	1	100

Table A.3 Unit Conversions for Volume

Unit	mL	Liters	Meter^3	Feet^3	Gallons	Acre-ft	Million gal.
1 ml	1	0.001	1e-06	3.531e-05	2.6414e-04	8.107e-10	2.6414e-10
1 liter	1000	1	0.001	0.03531	0.2641	8.107e-07	2.641e-07
1 meter^3	1e06	1000	1	35.31	264.14	8.107e-04	2.641e-04
1 feet^3	28,317	28.317	2.8317e-02	1	7.481	2.2957e-05	7.481e-06
1 gallon	3785.9	3.7859	3.786e-03	0.1337	1	3.0691e-06	1e-06
1 acre-ft	1.2335 e09	1.2335 e05	1233.5	43560	3.2583 e05	1	0.32583
1e06 gal	3.7859 e09	3.7859 e06	3786	1.3367 e05	1e06	3.0691	1

Table A.4 Unit Conversions for Time

Unit	Seconds	Minutes	Hours	Days	Years
1 second	1	0.01667	2.778e-04	1.1574e-05	3.171e-08
1 minute	60	1	0.01667	6.944e-04	1.9026e-06
1 hour	3600	60	1	0.04167	1.1416e-04
1 day	86,400	1440	24	1	2.74e-03
1 year	3.1536e07	5.256e05	8760	365	1

Table A.5 Unit Conversions for Hydraulic Conductivity

Multiply	By	To Obtain
1 gal/day/ft^2	0.1337	ft/day
1 cm/sec	2835	ft/day
1 m/day	3.2808	ft/day
1 gal/day/ft^2	4.72×10^{-5}	cm/sec
1 ft/day	3.53×10^{-4}	cm/sec
1 m/day	1.16×10^{-3}	cm/sec
1 gal/day/ft^2	4.07×10^{-2}	m/day
1 cm/sec	864	m/day
1 ft/day	0.3048	m/day
1 cm/sec	21203	gal/day/ft^2
1 ft/day	7.48	gal/day/ft^2
1 m/day	24.54	gal/day/ft^2

Table A.6 Unit Conversions for Flow

Unit	m^3/sec	m^3/day	L/sec	ft^3/sec	acre-ft/yr	gal/min	million gal/day
1 m^3/sec	1	86,400	1000	35.31	6.255e-07	15,849	22.822
1 m^3/day	1.1574 e-05	1	0.01157	4.087e-04	0.2959	0.1834	2.641e-04
1 L/sec	0.001	86.4	1	0.03531	25.57	15.849	0.02282
1 ft^3/sec	0.02832	2446.6	28.32	1	724.2	448.8	0.6463
1 acre-ft/yr	1.5987e-08	3.3795	0.03911	1.381e-03	1	0.612	8.927e-04
1 gal/min	6.31e-05	5.4526	0.0631	2.228e-03	1.613	1	1.44e-03
1 million gal/day	0.04382	3786	43.816	1.5473	1120.2	694.4	1

Table A.7 Conversion of Pressure

Pressure mm Hg	Feet	Meters	Calibration
760	0	0	100
752	278	85	99
745	558	170	98
737	841	256	97
730	1126	343	96
722	1413	431	95
714	1703	519	94
707	1995	608	93
699	2290	698	92
692	2587	789	91
684	2887	880	90
676	3190	972	89
669	3496	1066	88
661	3804	1160	87
654	4115	1254	86
646	4430	1350	85
638	4747	1447	84
631	5067	1544	83
623	5391	1643	82
616	5717	1743	81
608	6047	1843	80
600	6381	1945	79
593	6717	2047	78
585	7058	2151	77
578	7401	2256	76
570	7749	2362	75
562	8100	2469	74
555	8455	2577	73
547	8815	2687	72
540	9178	2797	71
532	9545	2909	70
524	9917	3023	69
517	10293	3137	68
509	10673	3253	67
502	11058	3371	66
495	11455	3492	65

Appendix B

Relationship of Water Density and Viscosity to Temperature

Temp (°C)	Density (g/cm^3)	Viscosity (g/sec-cm)
0	0.99984	0.01782
2	0.99994	0.01673
4	0.99997	0.01567
6	0.99994	0.01473
8	0.99985	0.01386
10	0.99970	0.01308
12	0.99950	0.01236
14	0.99924	0.01171
16	0.99894	0.01111
18	0.99860	0.01056
20	0.99820	0.01005
22	0.99777	0.00958
24	0.99730	0.00914
25	0.99704	0.00894
26	0.99678	0.00874
28	0.99623	0.00836
30	0.99565	0.00801
35	0.99403	0.00723
40	0.99221	0.00656
45	0.99021	0.00599
50	0.98805	0.00549

Source: Handbook of Chemistry and Physics, CRC Press, Boca Raton, FL, 1990.

Appendix C

Chemistry Example

PHREEQC Calculations—Berkeley Pit

```
{Reading data base}
    SOLUTION_MASTER_SPECIES
    SOLUTION_SPECIES
    PHASES
    EXCHANGE_MASTER_SPECIES
    EXCHANGE_SPECIES
    SURFACE_MASTER_SPECIES
    SURFACE_SPECIES
    RATES
    END
{Reading input data for simulation 1}
```

SOLUTION

temp	13.5	Al	193.
pH	3.15	Cd	1.87
pe	8.228	Cu	203.
units	mg/L	Fe	1020.
density	1.01	Pb	0.522
Ca	482.	Mn	162.
Mg	280.	Zn	497.
Na	70.8	Cl	22.1
K	18.7	S	6760
S	6760		

END

```
{Beginning of initial solution calculations}
{Initial solution 1}
```

Solution Composition

Elements	Molality	Moles
Al	7.151e-03	7.151e-03
Ca	1.202e-02	1.202e-02
Cd	1.663e-05	1.663e-05
Cl	6.232e-04	6.232e-04
Cu	3.194e-03	3.194e-03
Fe	1.826e-02	1.826e-02
K	4.781e-04	4.781e-04
Mg	1.151e-02	1.151e-02
Mn	2.948e-03	2.948e-03
Na	3.079e-03	3.079e-03
Pb	2.519e-06	2.519e-06
S	7.035e-02	7.035e-02
Zn	7.601e-03	7.601e-03

Solution description

pH	3.150
pe	8.228
Activity of water	0.998
Ionic strength	1.484e-01
Mass of water (kg)	1.000e+00
Total alkalinity (eq/kg)	-1.786e-03
Total carbon (mol/kg)	0.000e+00
Total CO2 (mol/kg)	0.000e+00
Temperature (deg C)	13.500
Electrical balance (eq)	-3.420e-03
Percent error, 100*(Cat-|An|)/(Cat+|An|)	-2.23
Iterations	10
Total H	1.110142e+02
Total	O5.578762e+01

Distribution of species					
Species	Molality	Activity	Log Molality	Log Activity	Log Gamma
H+	8.717e-04	7.079e-04	-3.060	-3.150	-0.090
OH-	7.615e-12	5.596e-12	-11.118	-11.252	-0.134
H2O	5.551e+01	9.982e-01	-0.001	-0.001	0.000
Al	7.151e-03				
AlSO4+	4.721e-03	3.608e-03	-2.326	-2.443	-0.117
Al(SO4)2-	1.713e-03	1.309e-03	-2.766	-2.883	-0.117
Al+3	7.141e-04	1.098e-04	-3.146	-3.959	-0.813
AlOH+2	2.095e-06	7.143e-07	-5.679	-6.146	-0.467
AlHSO4+2	6.109e-07	2.083e-07	-6.214	-6.681	-0.467
Al(OH)2+	3.521e-09	2.691e-09	-8.453	-8.570	-0.117
Al(OH)3	2.162e-13	2.237e-13	-12.665	-12.650	0.015
Al(OH)4-	6.922e-16	5.290e-16	-15.160	-15.277	-0.117
Ca	1.202e-02				
Ca+2	6.891e-03	2.453e-03	-2.162	-2.610	-0.449
CaSO4	5.132e-03	5.310e-03	-2.290	-2.275	0.015
CaOH+	7.512e-13	5.740e-13	-12.124	-12.241	-0.117
Cd	1.663e-05				
CdSO4	6.975e-06	7.217e-06	-5.156	-5.142	0.015
Cd+2	6.507e-06	2.219e-06	-5.187	-5.654	-0.467
Cd(SO4)2-2	3.029e-06	1.033e-06	-5.519	-5.986	-0.467
CdCl+	1.214e-07	9.279e-08	-6.916	-7.032	-0.117
CdCl2	1.630e-10	1.687e-10	-9.788	-9.773	0.015
CdOH+	1.403e-13	1.072e-13	-12.853	-12.970	-0.117
CdCl3-	5.301e-14	4.051e-14	-13.276	-13.392	-0.117
Cd(OH)2	1.905e-20	1.971e-20	-19.720	-19.705	0.015
Cd(OH)3-	4.080e-30	3.118e-30	-29.389	-29.506	-0.117
Cd(OH)4-2	1.149e-40	0.000e+00	-39.940	-40.407	-0.467
Cl	6.232e-04				
Cl-	6.169e-04	4.557e-04	-3.210	-3.341	-0.132
FeCl+	3.388e-06	2.589e-06	-5.470	-5.587	-0.117
MnCl+	1.624e-06	1.241e-06	-5.790	-5.906	-0.117
ZnCl+	1.142e-06	8.729e-07	-5.942	-6.059	-0.117
CdCl+	1.214e-07	9.279e-08	-6.916	-7.032	-0.117

Distribution of species					
Species	Molality	Activity	Log Molality	Log Activity	Log Gamma
PbCl+	4.109e-09	3.140e-09	-8.386	-8.503	-0.117
FeCl+2	9.527e-10	3.249e-10	-9.021	-9.488	-0.467
ZnCl2	3.837e-10	3.970e-10	-9.416	-9.401	0.015
MnCl2	2.385e-10	2.468e-10	-9.622	-9.608	0.015
CdCl2	1.630e-10	1.687e-10	-9.788	-9.773	0.015
PbCl2	2.741e-12	2.836e-12	-11.562	-11.547	0.015
FeCl2+	1.265e-12	9.663e-13	-11.898	-12.015	-0.117
ZnCl3-	2.472e-13	1.889e-13	-12.607	-12.724	-0.117
CdCl3-	5.301e-14	4.051e-14	-13.276	-13.392	-0.117
MnCl3-	4.054e-14	3.098e-14	-13.392	-13.509	-0.117
PbCl3-	1.248e-15	9.535e-16	-14.904	-15.021	-0.117
ZnCl4-2	1.151e-16	3.925e-17	-15.939	-16.406	-0.467
FeCl3	4.256e-17	4.404e-17	-16.371	-16.356	0.015
PbCl4-2	5.562e-19	1.897e-19	-18.255	-18.722	-0.467
Cu(1)	2.509e-09				
Cu+	2.509e-09	1.790e-09	-8.601	-8.747	-0.147
Cu(2)	3.194e-03				
Cu+2	1.772e-03	6.448e-04	-2.751	-3.191	-0.439
CuSO4	1.421e-03	1.471e-03	-2.847	-2.833	0.015
CuOH+	1.221e-08	9.092e-09	-7.913	-8.041	-0.128
Cu(OH)2	2.588e-11	2.678e-11	-10.587	-10.572	0.015
Cu(OH)3-	2.978e-21	2.276e-21	-20.526	-20.643	-0.117
Cu(OH)4-2	1.877e-30	6.402e-31	-29.726	-30.194	-0.467
Fe(2)	1.825e-02				
Fe+2	1.131e-02	4.115e-03	-1.946	-2.386	-0.439
FeSO4	6.895e-03	7.134e-03	-2.161	-2.147	0.015
FeHSO4+	4.259e-05	3.255e-05	-4.371	-4.487	-0.117
FeCl+	3.388e-06	2.589e-06	-5.470	-5.587	-0.117
FeOH+	9.823e-10	7.507e-10	-9.008	-9.125	-0.117
Fe(HS)2	0.000e+00	0.000e+00	-114.787	-114.773	0.015
Fe(HS)3-	0.000e+00	0.000e+00	-173.287	-173.404	-0.117
Fe(3)	6.536e-06				
FeSO4+	4.608e-06	3.521e-06	-5.336	-5.453	-0.117

Distribution of species					
Species	Molality	Activity	Log Molality	Log Activity	Log Gamma
Fe(SO4)2-	1.167e-06	8.920e-07	-5.933	-6.050	-0.117
FeOH+2	4.554e-07	1.553e-07	-6.342	-6.809	-0.467
Fe+3	2.243e-07	3.449e-08	-6.649	-7.462	-0.813
Fe(OH)2+	6.028e-08	4.606e-08	-7.220	-7.337	-0.117
FeHSO4+2	2.009e-08	6.853e-09	-7.697	-8.164	-0.467
FeCl+2	9.527e-10	3.249e-10	-9.021	-9.488	-0.467
Fe2(OH)2+4	7.868e-11	1.064e-12	-10.104	-11.973	-1.869
Fe(OH)3	4.801e-12	4.968e-12	-11.319	-11.304	0.015
FeCl2+	1.265e-12	9.663e-13	-11.898	-12.015	-0.117
Fe3(OH)4+5	2.569e-14	3.088e-17	-13.590	-16.510	-2.920
FeCl3	4.256e-17	4.404e-17	-16.371	-16.356	0.015
Fe(OH)4-	5.169e-18	3.950e-18	-17.287	-17.403	-0.117
H(0)	2.469e-26				
H2	1.235e-26	1.278e-26	-25.908	-25.894	0.015
K	4.781e-04				
K+	4.463e-04	3.297e-04	-3.350	-3.482	-0.132
KSO4-	3.182e-05	2.431e-05	-4.497	-4.614	-0.117
KOH	1.558e-15	1.612e-15	-14.808	-14.793	0.015
Mg	1.151e-02				
Mg+2	6.550e-03	2.457e-03	-2.184	-2.610	-0.426
MgSO4	4.963e-03	5.136e-03	-2.304	-2.289	0.015
MgOH+	5.590e-12	4.272e-12	-11.253	-11.369	-0.117
Mn(2)	2.948e-03				
Mn+2	1.837e-03	6.684e-04	-2.736	-3.175	-0.439
MnSO4	1.109e-03	1.148e-03	-2.955	-2.940	0.015
MnCl+	1.624e-06	1.241e-06	-5.790	-5.906	-0.117
MnCl2	2.385e-10	2.468e-10	-9.622	-9.608	0.015
MnOH+	1.196e-11	9.137e-12	-10.922	-11.039	-0.117
MnCl3-	4.054e-14	3.098e-14	-13.392	-13.509	-0.117
Mn(3)	6.848e-21				
Mn+3	6.848e-21	6.086e-22	-20.164	-21.216	-1.051
Na	3.079e-03				
Na+	2.915e-03	2.226e-03	-2.535	-2.652	-0.117

Distribution of species					
Species	Molality	Activity	Log Molality	Log Activity	Log Gamma
NaSO4-	1.642e-04	1.255e-04	-3.785	-3.901	-0.117
NaOH	2.004e-14	2.074e-14	-13.698	-13.683	0.015
O(0)	0.000e+00				
O2	0.000e+00	0.000e+00	-44.548	-44.533	0.015
Pb	2.519e-06				
PbSO4	1.535e-06	1.589e-06	-5.814	-5.799	0.015
Pb+2	6.828e-07	2.328e-07	-6.166	-6.633	-0.467
Pb(SO4)2-2	2.966e-07	1.011e-07	-6.528	-6.995	-0.467
PbCl+	4.109e-09	3.140e-09	-8.386	-8.503	-0.117
PbOH+	8.377e-12	6.402e-12	-11.077	-11.194	-0.117
PbCl2	2.741e-12	2.836e-12	-11.562	-11.547	0.015
PbCl3-	1.248e-15	9.535e-16	-14.904	-15.021	-0.117
Pb2OH+3	3.755e-16	3.337e-17	-15.425	-16.477	-1.051
Pb(OH)2	3.394e-18	3.512e-18	-17.469	-17.454	0.015
PbCl4-2	5.562e-19	1.897e-19	-18.255	-18.722	-0.467
Pb(OH)3-	7.439e-26	5.685e-26	-25.128	-25.245	-0.117
Pb(OH)4-2	5.385e-34	1.836e-34	-33.269	-33.736	-0.467
S(-2)	0.000e+00				
H2S	0.000e+00	0.000e+00	-56.683	-56.669	0.015
HS-	0.000e+00	0.000e+00	-60.535	-60.668	-0.134
S-2	0.000e+00	0.000e+00	-70.319	-70.792	-0.473
Fe(HS)2	0.000e+00	0.000e+00	-114.787	-114.773	0.015
Fe(HS)3-	0.000e+00	0.000e+00	-173.287	-173.404	-0.117
S(6)	7.035e-02				
SO4-2	3.655e-02	1.213e-02	-1.437	-1.916	-0.479
FeSO4	6.895e-03	7.134e-03	-2.161	-2.147	0.015
CaSO4	5.132e-03	5.310e-03	-2.290	-2.275	0.015
MgSO4	4.963e-03	5.136e-03	-2.304	-2.289	0.015
AlSO4+	4.721e-03	3.608e-03	-2.326	-2.443	-0.117
ZnSO4	3.023e-03	3.128e-03	-2.520	-2.505	0.015
Al(SO4)2-	1.713e-03	1.309e-03	-2.766	-2.883	-0.117
CuSO4	1.421e-03	1.471e-03	-2.847	-2.833	0.015
MnSO4	1.109e-03	1.148e-03	-2.955	-2.940	0.015

Distribution of species					
Species	Molality	Activity	Log Molality	Log Activity	Log Gamma
Zn(SO4)2-2	9.919e-04	3.383e-04	-3.004	-3.471	-0.467
HSO4-	8.608e-04	6.578e-04	-3.065	-3.182	-0.117
NaSO4-	1.642e-04	1.255e-04	-3.785	-3.901	-0.117
FeHSO4+	4.259e-05	3.255e-05	-4.371	-4.487	-0.117
KSO4-	3.182e-05	2.431e-05	-4.497	-4.614	-0.117
CdSO4	6.975e-06	7.217e-06	-5.156	-5.142	0.015
FeSO4+	4.608e-06	3.521e-06	-5.336	-5.453	-0.117
Cd(SO4)2-2	3.029e-06	1.033e-06	-5.519	-5.986	-0.467
PbSO4	1.535e-06	1.589e-06	-5.814	-5.799	0.015
Fe(SO4)2-	1.167e-06	8.920e-07	-5.933	-6.050	-0.117
AlHSO4+2	6.109e-07	2.083e-07	-6.214	-6.681	-0.467
Pb(SO4)2-2	2.966e-07	1.011e-07	-6.528	-6.995	-0.467
FeHSO4+2	2.009e-08	6.853e-09	-7.697	-8.164	-0.467
Zn	7.601e-03				
Zn+2	3.584e-03	1.206e-03	-2.446	-2.919	-0.473
ZnSO4	3.023e-03	3.128e-03	-2.520	-2.505	0.015
Zn(SO4)2-2	9.919e-04	3.383e-04	-3.004	-3.471	-0.467
ZnCl+	1.142e-06	8.729e-07	-5.942	-6.059	-0.117
ZnOH+	9.848e-10	7.526e-10	-9.007	-9.123	-0.117
ZnCl2	3.837e-10	3.970e-10	-9.416	-9.401	0.015
ZnCl3-	2.472e-13	1.889e-13	-12.607	-12.724	-0.117
Zn(OH)2	2.917e-14	3.019e-14	-13.535	-13.520	0.015
ZnCl4-2	1.151e-16	3.925e-17	-15.939	-16.406	-0.467
Zn(OH)3-	1.761e-22	1.346e-22	-21.754	-21.871	-0.117
Zn(OH)4-2	8.820e-32	3.008e-32	-31.055	-31.522	-0.467

Saturation indices				
Phase	SI	log	IAP	log KT
Al(OH)3(a)	-6.09	5.49	11.58	Al(OH)3
Alunite	-0.37	-0.30	0.08	KAl3(SO4)2(OH)6
Anglesite	-0.70	-8.55	-7.85	PbSO4
Anhydrite	-0.19	-4.53	-4.33	CaSO4
Cd(OH)2	-13.01	0.64	13.65	Cd(OH)2
CdSO4	-7.90	-7.57	0.33	CdSO4
Fe(OH)3(a)	-2.91	15.29	18.20	Fe(OH)3
FeS(ppt)	-55.99	-95.32	-39.33	FeS
Gibbsite	-3.29	5.49	8.78	Al(OH)3
Goethite	2.99	15.29	12.30	FeOOH
Gypsum	0.06	-4.53	-4.59	CaSO4:2H2O
H2(g)	-22.80	-22.76	0.04	H2
H2S(g)	-55.81	-99.24	-43.43	H2S
Hausmannite	-31.86	32.13	63.99	Mn3O4
Hematite	7.07	30.58	23.51	Fe2O3
Jarosite-K	-2.52	29.11	31.62	KFe3(SO4)2(OH)6
Mackinawite	-55.26	-95.32	-40.07	FeS
Manganite	-10.84	14.50	25.34	MnOOH
Melanterite	-1.95	-4.31	-2.36	FeSO4:7H2O
O2(g)	-41.63	45.51	87.14	O2
Pb(OH)2	-8.90	-0.33	8.56	Pb(OH)2
Pyrite	-82.16	-171.80	-89.65	FeS2
Pyrochroite	-12.08	3.12	15.20	Mn(OH)2
Pyrolusite	-17.42	25.88	43.29	MnO2:H2O
Sphalerite	-48.58	-95.86	-47.28	ZnS
Sulfur	-39.07	-76.48	-37.41	S
Zn(OH)2(e)	-8.12	3.38	11.50	Zn(OH)2

{End of simulation}

{Reading input data for simulation 2}

{No memory leaks}

Water Quality Parameters and their Significance

(Information you may wish to provide homeowners)

Constituent or Physical Property	Source or Cause	Significance
Calcium (Ca) and Magnesium (Mg)	Dissolved from almost all soils and rocks but especially from limestone, dolomite, and gypsiferous sediments. Ca and/or Mg are found in large quantities in some brines; Mg is present in large quantities in sea water.	Cause most of the hardness and scale-forming properties of water; soap consuming. Waters low in calcium and magnesium are desired for electroplating, tanning, dyeing, textile, and electronics manufacturing.
Sodium (Na) and Potassium (K)	Dissolved from almost all rocks and soils. Found in ancient brines, some industrial brines, sea water, and sewage.	Large amounts give a salty taste when combined with chloride. Moderate quantities have little effect upon usefulness of water for most purposes. Sodium may cause foaming in steam boilers, and a high sodium adsorption ratio may limit the water for irrigation. Concentrations greater than 270 mg/L may be harmful to persons on sodium-restricted diets.
Iron (Fe)	Dissolved from almost all rocks and soils. May also be derived from iron pipes, pumps, and other equipment.	On exposure to air, iron in groundwater oxidizes to reddish-brown sediment. More than about 0.3 mg/L stains laundry and fixtures. Objectionable for food processing, beverages, dying, bleaching, ice manufacture, brewing, and other processes. Iron and manganese should not together exceed 0.3 mg/L. Larger quantities cause unpleasant taste and favor growth of iron bacteria, but do not endanger health. Excessive iron may also interfere with the efficient operation of exchange-silicate water softeners. Iron may be removed from water by aeration of the water followed by settling or filtration.
Manganese (Mn)	Dissolved from some rocks and soils. Not as common as iron. Large quantities often associated with high iron content and acid waters.	Same objectionable features as iron. Causes dark-brown or black stain. Iron and manganese should not exceed 0.3 mg/L for taste and aesthetic reasons.

Constituent or Physical Property	Source or Cause	Significance
Silica (SiO_2)	Dissolved from almost all rocks and soils, usually in small amounts (5 to 30 mg/L), but often more from acidic volcanic rocks.	Forms hard scale in pipes and boilers. Carried over in steam of high-pressure boilers to form deposits on blades of steam turbines. Inhibits deterioration of zeolite water softeners.
Bicarbonate (HCO_3^-) and Carbonate (CO_3^{-2})	Action of carbon dioxide in water on carbonate rocks such as limestone and dolomite, and oxidation of organic carbon.	Bicarbonate and carbonate produce alkalinity. Bicarbonates of calcium and magnesium in steam boilers and hot-water facilities form scale and release carbon dioxide gas.
Chloride (Cl^-)	Dissolved from rocks and soils. Present in sewage and found in large amounts in ancient brines, sea water, and industrial brines.	Chloride salts in excess of 100 mg/L give a salty taste to water. When combined with calcium and magnesium may increase the corrosive activity of water. It is recommended that chloride content should not exceed 250 mg/L.
Sulfate (SO_4^{-2})	Dissolved from rocks and soils containing gypsum, iron sulfides, and other sulfur compounds. Usually present in some industrial wastes.	Sulfate in water containing calcium forms hard scale in steam boilers. In large amounts, sulfate in combination with other ions gives a bitter taste to water. Concentrations above 250 mg/L may have a laxative effect, but 500 mg/L is considered safe. Some calcium sulfate is beneficial in the brewing process. Domestic waters in Montana containing as much as 1,000 mg/L sulfate are for drinking in the absence of a less mineralized water supply.
Nitrate (NO_3^-)	Decaying organic matter, sewage, nitrates in soil, and fertilizers.	Concentrations greater than the local average may suggest pollution. High concentrations are generally characteristic of individual wells and not whole aquifers. Nitrate has been shown to be helpful in reducing intercrystalline cracking of boiler steel. Nitrate encourages the growth of algae and other organisms, which produce undesirable tastes and odors. There is evidence that more than about 10 mg/L (as N) may cause a type of methemoglobinemia ("blue babies") in infants, which may be fatal.

Constituent or Physical Property	Source or Cause	Significance
Fluoride (F^-)	Dissolved in small to minute quantities from most rocks and soils. Most hot and warm springs contain more than the recommended concentration of fluoride.	Fluoride in drinking water reduces the incidence of tooth decay in children when the water is consumed during the period of enamel calcification, but it may cause mottling of teeth, depending upon the concentration of fluoride, the age of the child, the amount of drinking water consumed, and the susceptibility of the individual. 0.8 to 1.7 mg/L is optimal, depending upon air temperature.
Hydrogen-ion activity (pH)	Acids, acid-generating salts, and free carbon dioxide lower pH. Carbonates, bicarbonates, hydroxides, silicates, and botates raise the pH.	The pH is a measure of the activity of the hydrogen ions (H^+). A pH of 7.0 indicates neutrality of a solution. Values higher than 7.0 denote increasing alkalinity; values lower than 7.0 indicate increasing acidity. Corrosiveness of water generally increases with decreasing pH, but excessively alkaline waters may also attack metals. Accurate pH can be made only at the well. Laboratory values willvary somewhat from the real value. A pH range between 6.0 and 8.5 is acceptable and is normal for most waters in Montana.
Dissolved Solids	Chiefly mineral constituents dissolved from rocks and soils. Includes all material that is in solution in the water.	Dissolved solids should not exceed 1,000 mg/L, but 1,000 mg/L is acceptable for drinking water if no other supply is available. Amounts exceeding 1,000 mg/L are unacceptable for most uses.
Specific Conductance	Dissolved mineral in the water.	Specific conductance is a measurement of the water's capacity to conduct an electric current. This varies with the temperature and the degree of ionization of the dissolved constituents. When measured in micromhos/cm or microsiemens/cm, it is generally 1.0 to 1.5 times the total dissolved-solids content.

Constituent or Physical Property	Source or Cause	Significance
Hardness as $CaCO_3$	In most water nearly all hardness is due to calcium and magnesium. All of the metallic cations besides the alkaline earths also cause hardness.	Hard water consumes soap before a lather will form, deposits soap curd on bathtubs, and forms scale in boilers, water heaters, and pipes. Hardness equivalent to the bicarbonate and carbonate content is called carbonate hardness. Any hardness in excess of this is called non-carbonate hardness. Waters of hardness as much as 60 mg/L are termed soft; 61 to 120 mg/L moderately hard; 121 to 180 mg/L hard; and more that 180 mg/L very hard.
Alkalinity	Formed in the presence of certain anions in solution. Some organic materials may also produce alkalinity.	Alkalinity is an indicator of the relative amounts of carbonate, bicarbonate, and hydroxide ions and some anions (acid ligands).
Sodium Adsorption Ratio	SAR is defined by the equation $SAR = Na/([Ca+Mg]/2)^{0.5}$ Where the concentrations are expressed in milliequivalents per liter (meq/L).	High sodium concentrationcombined with low alkaline-earth element concentration usually reduces soil tilth and affects plant growth.
Hydrogen Sulfide (H_2S)	Natural decomposition of organic material and from the reduction of sufates.	Causes objectionable odor when in concentration above 1 mg/L and taste when in excess of 0.05 mg/L Presence may limit water usefulness in the food and beverage industry.
Trace metals	Dissolved from rocks and soils. Some metals may be released from plumbing piping, etc. Check the limits for your area.	Limits are usually recommended for health reasons. Limits for drinking water normally are conservative, and higher concentrations may be permitted if the water is the best available supply (e.g., copper).

Appendix D

Values of *W*(*u*) and *u* for the Theis Nonequilibrium Equation

Values of *W(u)* and *u* for the Theis Nonequilibrium Equation

N \ u	N x E-12	N x E-11	N x E-10	N x E-09	N x E-08	N x E-07	N x E-06	N x E-05	N x E-04	N x E-03	N x E-02	N x E-01	N
1.0	27.054	24.751	22.449	20.146	17.844	15.541	13.238	10.936	8.633	6.332	4.038	1.823	0.219
1.5	26.648	24.346	22.043	19.741	17.438	15.135	12.833	10.530	8.228	5.927	3.637	1.465	0.100
2.0	26.361	24.058	21.756	19.453	17.150	14.848	12.545	10.243	7.940	5.639	3.355	1.223	0.049
2.5	26.138	23.835	21.532	19.230	16.927	14.625	12.322	10.019	7.717	5.417	3.137	1.044	0.024
3.0	25.955	23.653	21.350	19.047	16.745	14.442	12.140	9.837	7.535	5.235	2.959	0.906	0.013
3.5	25.801	23.499	21.196	18.893	16.591	14.288	11.986	9.683	7.381	5.081	2.810	0.794	0.007
4.0	25.668	23.365	21.062	18.760	16.457	14.155	11.852	9.550	7.247	4.948	2.681	0.702	0.0038
4.5	25.550	23.247	20.945	18.642	16.339	14.037	11.734	9.432	7.130	4.831	2.568	0.625	0.0021
5.0	25.444	23.142	20.839	18.537	16.234	13.931	11.629	9.356	7.024	4.726	2.468	0.560	0.0011
5.5	25.349	23.047	20.744	18.441	16.139	13.836	11.534	9.231	6.929	4.631	2.378	0.503	0.00064
6.0	25.262	22.960	20.657	18.354	16.052	13.749	11.447	9.144	6.842	4.545	2.295	0.454	0.00036
6.5	25.182	22.879	20.577	18.274	15.972	13.669	11.367	9.064	6.762	4.465	2.22	0.412	0.0002
7.0	25.108	22.805	20.503	18.200	15.898	13.595	11.292	8.990	6.688	4.392	2.151	0.374	0.00012
7.5	25.039	22.736	20.434	18.131	15.829	13.526	11.223	8.921	6.619	4.323	2.087	0.340	0.000066
8.0	24.974	22.672	20.369	18.067	15.764	13.461	11.159	8.856	6.555	4.259	2.027	0.311	0.000037
8.5	24.914	22.611	20.309	18.006	15.703	13.401	11.098	8.796	6.494	4.199	1.971	0.284	0.000022
9.0	24.857	22.554	20.251	17.949	15.646	13.344	11.041	8.739	6.437	4.142	1.919	0.260	0.000012
9.5	24.803	22.500	20.197	17.895	15.592	13.290	10.987	8.685	6.383	4.089	1.870	0.239	0.000007

Appendix E

Using the Guelph Permeameter Model 2800

Setting Up

The permeameter is typically stored in a carrying case that includes the permeameter, augering equipment, a hole brush, and a collapsible water jug. Once you are on site and located, the first step is to open the case and assemble the augering tool. There are two bits; one is for making (augering) the hole, which has 2 tapered "arms" extending from the canlike portion (which collects the cuttings), while the other looks like a can with the bottom intact, but a cut has been made in the bottom and a piece bent out from the bottom, which is used to create a flat bottom in the hole. First assemble the rod and handle. The first bit is attached to the rod with a pin and clip. Then the hole is started and deepened in roughly 6-in. (15-cm) increments to allow for cuttings expansion in the collection portion. Stop a few inches (roughly 6 in. or 15 cm) short of the desired first test depth and finish the hole with the second bit. The normal augering bit creates a tapered hole so initially you are just excavating the outer edges of the hole with this "sizing" bit, but as you go deeper, you will be excavating the entire hole. Remove the bit and the cuttings from the collection can.

Remove the sizing bit and attach the brush to the rod, and brush the bottom of the hole once (twice if really clay-rich and sticky, you can tell this from the cuttings that you removed from the sizing bit). This is to remove smearing on the hole wall that occurs in clay-rich soils. Avoid excessive brushing as it will increase the hole diameter and cause you to have erroneous results (too great a permeability, the calculations are based upon a 6-cm diameter hole). Measure and record the depth of the hole.

Now it is time to assemble the permeameter. Take the legs, chain, and tripod base out of the case. Insert the legs and attach the chain to the legs; the chain prevents the legs from spreading excessively. Put a sheet of clean paper on the ground and center the tripod over the paper (to keep the outlet tip clean). Now locate the support tube. Is there a thin air-tube fitting in it? If there is, it should come to about 6 inches (15 cm) from the top of the support tube. If not, look at the air tubes. The standard permeameter has three air tubes of different lengths. The shortest one goes on top of the unit, the intermediate length one goes in the support tube, and the longest one goes into the reservoir body. Look at the tips. One is different, it is the air inlet tip that goes at the bottom of the support tube assembly, the other two tubes have rubber connectors that joint the tubes together. Put the intermediate length air tube into the support tube with the rubber tip down so that it rests on the water outlet tip (the closed end of the support tube). Lift the tripod up, put the support tube on the paper, and lower tripod over the tube so that the plain "open end" is above the tripod. Locate the tripod bushing and slide it over the support tube and down to the tripod base; the nonflared portion slides into the gap between gap between the support tube and the top of the tripod base.

At this point, you may move the tripod to the hole and add the reservoir assembly there (provided that someone holds the support tube so that it isn't pressed hard into the bottom of the hole) or add the reservoir assembly with the support tube still resting on the paper. The latter approach is preferred, except for assembly under very windy conditions.

Locate the reservoir section. Is the air tube in it? If it is, this long air tube (this one is the most fragile under compression because of its length) should have the rubber fitting projecting below the bottom (the end with a knob on it). If there are two persons assembling the unit, one person holds the reservoir over the support tube while the second person lowers the air tube and connects it to the air tube in the support tube. If you are doing this alone or if the tube was taken out, you must remove the cap from the top of the reservoir and then remove the air tube from the reservoir and connect this air tube to the one in the support tube. You may have to tip the tripod to center the support tube air fitting so that the connector can slip around it. This leaves nearly 3 ft (91 cm) of air tube projecting upward and you have to slip the reservoir assembly over it. Short people may need an ice chest or some other device (i.e., standing on a truck tailgate) so that the reservoir can be carefully lowered over the air tube. The reservoir contains an inner cylinder, and the air tube will be within the inner cylinder after the reservoir is lowered over the air tube. The bottom of the reservoir has a rubber fitting that will slide over the top of the support tube.

Sliding the reservoir section onto the support tube requires downward pressure (this is a watertight friction fit). If the unit was being assembled in the hole, put a piece of paper on the ground next to the hole and lift the assembly out, so that the support tube tip is on the paper and snug reservoir down on the support tube. Be very careful doing this that you don't let the wind blow the reservoir unit over while doing this, as it will bend (and crimp) or break the reservoir air tube at the point where it exits the support tube. Put the unit back over the hole and lower the reservoir and connected support tube so that the support tube base (water outlet tip) is in the bottom of the hole.

Now it is time to put the cap on the reservoir. Check that the neoprene tube is slipped over the vacuum port access tube and is flush with the cap and that the ring clamp is installed to close the "doubled over" neoprene tube. You may need to add vacuum grease to the o-ring seal where the cap slides down over the short length of air tube that extends above the top of the reservoir. Before putting the cap on, raise and lower the air tube 1 in. or so (2 to 3 cm) to be sure that the air tube tip is firmly inserted into the water outlet tip. This seals the system so that water is not released before you intend for it to be released. Now slide the cap down over the air tube and snug it down on the top of the reservoir body.

Next, locate the short air tube, the well height scale (a short, small-diameter, clear plastic cylinder with graduation marks), and the well-height indicator (a small rubber plug that slips over the air tube, you probably had to remove this when you removed the reservoir cap). Slide the well-height indicator down around the air tube to the cap (with the flat surface down and the rounded surface up), add the last air-tube fitting, and lower the well-scale cylinder over the air tube and seat it against the cap.

It's now time to look at the setup. Press down on the air tube to ensure that the air inlet tip is still seated. Next look to see that the well-height indicator (you read the bottom of this rubber cylinder) reads 0.5 cm (5 mm). If it is less than 5 mm, check to see that the well-scale cylinder was fully slid down to the cap. If that isn't the problem, remove the cylinder and raise the indicator until it does read 5 mm when the cylinder is reinstalled. If the reading is greater than 5 mm, and all fittings are flush, just write that value down as the "zero" value and raise the air tube during the tests 4.5 and 9.5 cm above this "new zero" value (the extra 5 mm of head comes from the design of the air inlet tip, which does not quite reach the bottom of the water outlet tip). Next, make sure that the vacuum port's neoprene tube is pushed down to the cap and bent over and sealed with clamping ring. Last, check to see that the fill plug is inserted in the cap.

We're almost ready to fill the reservoir cylinder with water (see Figure 12.5 or the Guelph manual figure for "pumping" the water into the reservoir). It is preferred to fill the unit before putting it in the augered hole, but if you are working alone or in strong wind you may first wish to move the unit to the auger hole. The valve is adjusted by the knob at the bottom of the reservoir. This knob has a small indentation on the outer edge of the knob. When that indentation is down only the inner reservoir contributes water; this is appropriate for slow-draining soils. When the indentation is at the top, both the inner and outer reservoirs will contribute water; this is the filling position and the position used for more rapidly draining soils. With the knob indentation in the up position, fill the water jug with *clean* water, remove the filling plug and use the jugs spout to fill the reservoir cylinder completely, letting air bubbles escape before reinserting the fill plug. If you have failed to seat the air inlet tip into the water outlet tip, the permeameter will start leaking and the water level will not be stable but slowly declines; you will see this leakage on the paper if the unit has not yet been moved to the auger hole. If this occurs, seat it properly to stop the leakage, and recheck the well-height indicator value. Lastly, adjust the reservoir knob to the slow (down) position if you expect "tight" soil or are uncertain what to expect. (Did you look at the last 6 in. worth of auger cuttings? You should have a rough idea of what to expect.)

Running the Tests

Once everything is positioned and you are ready to start the test with the reservoir knob set for the expected soil condition, record the water level and then gently raise the air inlet tube 4.5 cm and start timing the run. Record the water level every 2 minutes until the rate has stabilized (typically on the order of 20 minutes) and you get three consecutive rate readings (change in head in reservoir (cm)/time in seconds) that are the same. Then raise the air tube until the well height indicator is 5 cm higher (to the 10-cm mark if your zero value was 0.5 cm) and restart the timing and recording of water levels.

Once the rate has stabilized again, you have completed the test at this depth. If you desire to test additional depth, repeat the augering process discussed earlier. You will probably be deep enough (15 in. or 38 cm) to dispense with the tripod and just use the tripod bushing in the hole (see manual). Extensions for the unit to permit you to go deeper than the roughly 24-in. (61-cm) depth that can be achieved with the standard unit.

Calculations

You need to have recorded the *Y* and *X* values off of the permeameter valve knob; these are the cross-sectional areas or the inner and the inner plus outer reservoirs, respectively. The values will be approximately 2.2 and 35 cm^2, respectively. The method employed uses the assumption that you doubled the head drive between the first and second test, but more specifically, that you have a 6-cm diameter hole and 5- and 10-cm heads in the hole and are using the Richards equation corrected for the hole diameter and head heights. What you calculate are a "field saturated" hydraulic conductivity, K_{fs}, which is roughly ½ of the saturate hydraulic conductivity because the entrained air from the pore spaces has not all escaped or been dissolved and a matric flux potential. However, this field-saturated hydraulic conductivity value is more realistic of the maximum effective hydraulic conductivity during a typical recharge event.

Below are data from a student exercise conducted near Butte. The cuttings from the bottom of the hole could be rolled into "snakes" about 1.5 to 1.75 inches (3.8 to 4.4 cm) long before they started to disintegrate, indicating a reasonable clay content in the floodplain "soil" (actually trucked in fill material as part of a reclamation project). Additionally, free water was observed on the land surface in a depression near the test hole. Consequently, only the inner reservoir was used by the students, but either they didn't have enough reservoir capacity or patience to run the tests to "stable readings." However, because of soil variability, the numbers might not have stabilized, even if they had used both reservoirs and been capable of doing longer test runs.

25 March 2000 Silver Bow Creek

Reading No.	Time (min)	Height of water in well at 5 cm water level in reservoir (cm)	Rate of water level change at 5 cm	Height of water in well at 10 cm height of water in reservoir (cm)	Rate of water level change at 10 cm
0	0	3.00		5.40	
1	2	11.50	8.50	9.00	3.60
2	4	21.50	10.00	17.70	8.70
3	6	30.00	8.50	31.80	14.10
4	8	37.80	7.80	41.70	9.90
5	10	45.30	7.50	52.70	11.00
6	12	53.60	8.30	63.50	10.80
7	14	61.50	7.90	74.00	10.50
8	16	69.00	7.50		
9	18	76.00	7.00		

Using 7.5 cm per 2 minute increment for the 5-cm height test and 10.7 cm per 2 minute increment for the 10-cm height test yields "average" values of 0.0625 and 0.08917 cm/sec, respectively. These values are then used with the equations for standard Guelph model 2800 Permeameter from Soilmoisture®:

$$K_{fs} = [(.0041)(2.18\ \text{cm}^2)(.08917\ \text{cm/sec})] - [(.0054)(2.18\ \text{cm}^2)(.0625\ \text{cm/sec})]$$

$$K_{fs} = [0.000797\ \text{cm}^3/\text{sec}] - [0.000736\ \text{cm}^3/\text{sec}] = 6.125 \times 10^{-5}\ \text{cm}^3/\text{sec}$$

Note that the results are supposed to come out in cm/sec. Thus, the values 0.0041 and 0.0054 must have units of cm^{-2}, which were not listed in the manual. This can be clarified by reading the Reynolds and Elick (1986) paper on a prototype of this unit.

Second, note that the resulting field-saturated hydraulic conductivity is a pretty small number, being roughly equivalent to a saturated K of 1.2×10^{-4} cm/sec, and is the difference between two numbers an order of magnitude larger. This suggests that the sorptivity should be large.

The matric flux potential, $_m$, calculation is similar to the K_{fs} calculation:

$$_m = [(.0572)(2.18\ \text{cm}^2)(.0625\ \text{cm/sec})] - [(.0237)\ (2.18\ \text{cm}^2)(.08917\ \text{cm/sec})]$$

$$_m = [0.007794\ \text{cm}^3/\text{sec}] - [0.004607\ \text{cm}^3/\text{sec}] = 3.1865 \times 10^{-3}\ \text{cm}^3/\text{sec}$$

This calculation is supposed to come out in cm^2/sec, so the equation constants 0.0572 and 0.0237 must have units of cm^{-1}.

The sorptivity calculation is a bit of a fudge, as you need to know the saturated and field-condition water content for the *undisturbed* soil, and the manual suggests estimating it (and provides a guess of 0.15 as the difference between the two values; the Reynolds and Elrick article suggests time-domain reflectometry or gravimetric determination with soil cores, but uses a Δ of 0.15 in the example). Because of the small K_{fs} value and the standing water nearby, a more appropriate guess might be 0.1 for this site. The equation to be used is:

$$S = [\ 2 \times \Delta \quad \times \quad m]^{1/2}$$

$$S = \text{sqrt}[\ 2(0.10)(3.1865 \times 10^{-3}\ \text{cm}^2/\text{sec})] = (6.373 \times 10^{-4})^{0.5} = 2.52 \times 10^{-2}\ \text{cm/sec}^{1/2}$$

Don't be disturbed by the unit $\text{cm/sec}^{0.5}$, this is correct. The value for the sorptivity is reasonable, based on the field setting and information. Going further into soil characterization with the conductivity-pressure head relationship is beyond the scope of this text and, in fact, the matric flux potential and sorptivity calculations may have been more than the average reader cared for.

Reference

Reynolds, W. D., and Elrick, D. E., 1986. A Method for Simultaneous in Situ Measurement in The Vadose Zone of Field-Saturated Hydraulic Conductivity, Sorptivity, and the Conductivity-Pressure Head Relationship. *Ground Water Monitoring Review*, Vol. 6, No. 1, pp. 84–95.

Glossary

acquisition First of the three divisions of geophysical exploration (data acquisition). The process of designing geophysical surveys recording and recording geophysical measurements.

aggrading A stream reach building up sediments in its channel or floodplain by being supplied more sediment than it can carry. It is a stream reach characterized by active sedimentation.

amplitude Maximum departure of a wave from the average value (Sheriff 1991).

anisotropic *See* isotropy

aplite A dike rock of granitic composition than has very fine-grained crystals. The appearance is usually very felsic or light colored.

approach velocity (V_a) As groundwater moves towards the well screen the velocity increases. If the velocity exceeds a critical limit, finer particles from the aquifer will become entrained and move into the gravel packing. This critical limit is known as the approach velocity (V_a). It is not the same as an entrance velocity because it is specific discharge at the damage zone of the borehole wall.

aquicludes Units of extreme low permeability that form barriers to fluid flow. They are usually six orders of magnitude or less than surrounding units.

aquifer A formation, part of a formation, or group of formations that contain sufficient saturated permeable material to yield significant quantities of water to wells or springs.

aquifer hydraulics Evaluation of aquifer properties by stressing the aquifer and measuring and analyzing the response. Stressing the aquifer may be conducted through pumping tests or slug tests in the field. The drawdown response is plotted and analyzed to estimate the hydraulic properties.

aquitard A saturated unit of low hydraulic conductivity that can store and slowly transmit groundwater either upward or downward depending on the vertical hydraulic gradient.

aquitards Units of low hydraulic conductivity that can store and release groundwater at a slow rate.

arrays Geometrical arrangement of sources or receivers used in a geophysical survey. Certain arrangements can be used enhance particular qualities of the recording.

artesian *See* confined aquifer

average linear velocity (V_{ave}) *See* seepage velocity

backwashing A process of well development. In the development process water in the well casing is lifted by air for several minutes followed by shutting off the air for several minutes. As this is repeated, groundwater comes from the formation through the screen into the well casing and then from the well casing back out into the formation once again.

barrier boundaries Boundaries that inhibit groundwater flow. Examples are faults, bedrock, or thinning of an aquifer unit.

batholiths Igneous rock masses than are greater than 40 square miles (100 km^2).

bed A subunit of a member with distinguishable enough characteristics to be mappable in the field. An example would be a coal bed, within the member of a formation.

bolt method A method of protecting a mini-piezometer from crimping while pounding it to its completion depth. A bolt is taped to the bottom of the mini-piezometer, hence the name.

borehole dilution A type of tracer test where the volume of tracer introduced to a well is diluted by the volume of groundwater within the well bore.

Bouwer and Rice A method used to estimate the hydraulic conductivity in a partially penetrating or fully penetrating well in a unconfined aquifer from slug-test data.

breccia pipe A zone of fractured or angular rock materials that became fragmented during the migration of pressurized gasses towards the land surface along a plane or zone of weakness. The origin of the volatile gasses are associated with a magma source.

bulk density Density of a geologic medium that may be different than the densities of its constituents.

caldera A ring shaped depression structure formed from the collapse of a magma chamber. The diameter of the crater rim usually exceeds 1 mile (1.6 km).

chargeability One of several units of induced polarization in the time domain. The ratio of initial decay voltage (or secondary voltage) to primary voltage (Sheriff 1991).

circulation A term indicating whether drill cutting and fluids are returning to the land surface. Fluids are injected at the bit, which in turn lift the cutting and fluid to the surface.

columnar jointing A phenomenon that occurs as lava contracts to from a solid. During the cooling process, polygonal prismatic shapes form to accommodate the escaping heat. The polygonal prismatic shapes extend down through the thickness of the lava flow to form columns.

compressional Seismic wave motion described by particle motion in the same direction as the wavefront propagation direction. Compressional waves are also called *P*-waves.

conceptual model A simplified representation of the groundwater system that indicates flow directions and boundary conditions affecting fluid flow.

conductance The product of conductivity and thickness. Indication of ease of current flow in a medium.

conductivity The ability of a material to conduct electrical current. In isotropic material, the reciprocal of resistivity (Sheriff 1991). Units are siemens per meter.

cone of depression A depression in the water table or potentiometric surface surrounding a pumping well. The shape of this depression is similar to an inverted cone.

confined aquifer An aquifer that is isolated by have confining layers to maintain the pressure in the system at a pressure greater than atmospheric pressure. This causes the water levels in cased wells to rise above the top of the aquifer. The pressure results form the weight of the water elevation near the recharge area to be propagated throughout the system.

confining layers Layers that have a hydraulic conductivity two to three orders of magnitude less than a layer above or below it.

contact metamorphism A recrystallization of country rock into new compositions as a result of a nearby magma source. In this zone, high temperatures and low pressures are characteristic.

converging test A tracer test where the tracer is introduced under a nonstressed condition, but the sampling well is pumped.

crest-stage gauge A gauging apparatus useful in estimating peak flood events. A tube with grounded up cork or another floating material adheres to the sides of a tube at the place of maximum stream stage.

critical refraction The angle of incidence for which the refracted wave grazes the surface of contact between two media of different velocities.

cross-line Characterizes the direction perpendicular to a measurement direction (see in-line).

current density Current per unit cross-sectional area, determined by the velocity and density of charge carriers (Sheriff 1991).

Darcian velocity *See* specific discharge

Darcy's Law An equation that relates the volume of water per time moving through a given cross-sectional area of aquifer.

dart valve A valve on the bottom of a bailer that opens, when compressed, to allow cuttings and fluids into the bailer. The valve closes when the weight of the bailer is relaxed when lifted off of the bottom of the hole.

degrading The process of lowering a surface by removing sediments via air or water. Degrading streams are actively eroding sediments. This occurs from a stream with too flat a gradient for the volume of discharge.

delayed yield The second or delayed Theisian response after a recharge response indicated on the field plot of time-drawdown data during a pumping test. During the initial part of the time-drawdown plot the early-time data follow the Theis curve. This is followed by a flattening of the drawdown curve below the characteristic Theis curve indicating a recharge response. At late time, the data climb back on a secondary Theis curve once again.

diagenesis The process whereby unconsolidated sediments turn into solid rock. This occurs through burial, compaction, replacement, metamorphism, and cementation.

dikes Tabular igneous features that are discordant or cut across existing rocks.

dip The degree of inclination of a tilted bed from a horizontal plane.

discharge The loss of water from an aquifer system through an upward gradient in hydraulic head.

distance-drawdown plots Plots of drawdown (on arithmetic scale) versus distance (in log scale) indicating the drawdown at a pumping well, observation wells, and the distance of the range of influence. These plots can also be used to estimate the hydraulic properties of aquifers.

diverging test A tracer test where initial injection of a point-source occurs, but pumping of the sampling well is not performed.

drawdown The amount of water level change from the static water level position during a pumping test.

drawworks A heavy cable with the top section laterally connected to a heavy chain along the drill rig mast that is designed to lift the drill string. The driller can raise and lower the drill string by the drawworks while the kelly bar slides past the kelly bushings in the rotary table.

dual-wall reverse circulation A drilling method where drill fluids are injected down the outer passageway of the drill pipe and the cuttings and fluids are drawn up through the center passageway. The inner sleeve is sealed by an O ring. During this method, the only place the drilling fluids contact the formation is near the bit. This allows a continuous sampling of the formation with minimal disturbance and reduces the chances of cross contamination.

dynamic viscosity (μ) It is a value equal to the shear stress divided by the velocity gradient, indicating a fluids resistance to flow.

effective radius (r_e) The distance from the center of a drillhole to the borehole wall where the formation was disturbed by drilling.

effective porosity (η_e) That proportion of porosity available to fluid flow. The effective porosity if often estimated by the value of specific yield.

effective grain size (d_{10}) The diameter of a sieved soil at 10% finer by weight.

Eh or E_H A measure of the oxidation status of the aqueous system, represented as: $E_H = E_H^o + (2.303RT/nF) \log_{10} [P_i\{ox\}^{n_i}/P_j\{red\}^{n_j}]$, where n is the number of electrons involved in the reaction, F is the Faraday's constant, R is the gas constant, T is the temperature in degrees Kelvin, $P_i\{ox\}^{n_i}$ is the product of the activities of the products of the reaction (as written, by convention the oxidized form of the element is on the product side of the reaction and the reduced form is on the reactant side; note that all components *except* the electrons are present in these terms) and $P_j\{red\}^{n_j}$ is the product of the activities of the reactants in the reaction as written. E_H^o is the potential measured (in volts) when all of the species in the reaction are present at unit activities (the reference state for the reaction) and the rest of the terms are the correction for constituents not being present at unit activities. If all products and reactants are present at unit activities ($a_I = 1$), the right-hand term becomes the log of 1, which is zero, and the terms other than E_H^o disappear.

elastic The ability to return to original shape after removal of a distorting stress (Sheriff 1991). Characterizes the ability of most rocks in response to a small disturbance.

electromagnetic surveying A method in which the magnetic and/or electric fields associated with artificially generated subsurface currents are measured. In general, electromagnetic methods are those in which the electric and magnetic fields in the earth satisfy the diffusion equation but not Laplace's equation or the wave equation (Sheriff 1991).

electromagnetic Referring to measurement methods which make use of electric and magnetic properties.

entropy Classically defined by physical chemists for reversible reactions at a uniform temperature as: $dS = dQ/T$, where Q is heat, T is absolute temperature, and S is entropy. In the real world all reactions are irreversible and a second term is added: $dS = dQ/T + dQ'/T$, where $dQ' < 0$ because of the nonideal (irreversibility) nature of real systems. In explaining entropy to geologists, it is often better to start off with stating that it is a measure of the disorder or randomness of the system studied. The use of entropy by Back and Hanshaw (1971) is that it is an indication of the degradation of the system or an index of the exhaustion of the system. They convert head change to potential energy available as heat, and derive the entropy from the head change (always positive values), and convert the water chemistry data into degree of saturation with respect to calcite, dolomite, and gypsum getting entropy values that are always negative.

eolian Pertaining to deposits formed under the influence of the wind.

equipotential lines Lines of equal hydraulic head.

evapotranspiration The amount of moisture loss from the transpiration of plants *and* the evaporation from soils.

exfoliation The process by which concentric layers of rock the fracture and peel off in sheets parallel the rock mass similar to onion layers. This is very characteristic in granites.

extrusive The process of eruption and the rocks that form on the surface.

facies The nature of sedimentary or metamorphic rocks reflective of the condition their environment when they formed. It represents an areally restricted part of a formation with a particular characteristic. In sedimentary rocks, a coarse-grained facies may grade into a medium-grained facies as the energy in the environment decreases. In metamorphic rocks a greenschist facies may grade into an amphibolite facies as the temperature and pressure conditions increase.

falling-head test A slug test where the slug is instantaneously placed below the static water level in the well bore, displacing the water upward. Measurements of the falling water levels over time are recorded. In this case groundwater is entering the formation from the well.

first break The first recognizable part of a seismic trace attributable to seismic wave travel from a known source.

first arrival *See* first break.

float method A crude method of estimating stream velocity by tossing a float into a stream and measuring the time it takes to move a measured distance. This is repeated several times to obtain an average. A factor is multiplied by the result to account for drag and vegetative effects.

flow net A two-dimensional representation of steady-state groundwater flow. The flow net consists of intersecting lines of equal hydraulic head and associated flow lines.

fluvial Pertaining to a river or deposits formed by a river.

fluvial plain A relatively planar feature containing the active stream channel, floodplain and associated fluvially derived sediments.

foliation A planar textural term related to any rock, but is especially characteristic in metamorphic rocks. In this case there is a well defined orientation structure of minerals that formed during recrystallization. The minerals form perpendicular to the applied stresses.

footwall The rock mass below a fault plane.

formation A body of rock that is mappable in the field. Usually a formation consists of a similar lithology or group of lithologies formed in the same depositional environment.

forward modeling The process of calculating predicted data by inputting selected parameters into a desired model (see inverse modeling). For example, calculating seismic traveltimes by performing calculations from earth parameters.

frequency The inverse of period. The units of frequency are inverse seconds, inverse ms, etc. Inverse seconds are given the unit hertz (Hz). One Hz represents one cycle per second.

gaining stream A stream that receives discharging water from a groundwater system. Discharge occurs because the head in the groundwater system is higher than the stage elevation of the stream.

geophones Instruments consisting of a magnet and coil arrangement used to record earth movement for recording by a seismograph. Transforms seismic energy into electrical energy (Sheriff 1991).

graben An extensional valley bounded on both sides by normal faults.

grading An engineering term pertaining to the division of sediments into grain sizes. Well grade indicates a wide range of grains sizes present, where poorly graded sediments indicate a narrow range of grain sizes.

group of formations A body of rocks composed of two or more formations.

hanging wall The rock mass above a fault plane.

head waves The wave created in medium 1 by a critically refracted wave in medium 2.

heaving sands Unconsolidated sands that forge into the borehole under pressure. The pressure deferential often results from the difference in fluid level in the casing compared to sands underlying clay zones.

heterogeneous *See* homogeneity.

hidden layer An earth layer which cannot be detected by seismic refraction methods. A layer of lower velocity lying beneath a layer of higher-velocity high-frequency approximation (Sheriff 1991).

homogeneity Relating to the physical properties of an aquifer from point A to point B, including packing, thickness, cementation. Homogenous units have similar properties from point A to point B and heterogenous units differ in physical properties from point A to point B.

homogeneous *See* homogeneity.

horsts A elongate uplifted crustal unit left from the down-dropping of rock masses on either side from extension.

Hvorslev method A slug test analytical method used to estimate the hydraulic conductivity of an aquifer in the screened portion of a well. In can be used for confined or unconfined aquifers.

hydraulic conductivity (*K*) A value representing the relative ability of water to move through a geologic material of a given permeability. The specific weight and kinematic viscosity are also considered in determining the hydraulic conductivity.

hydraulic gradient The slope of the potentiometric surface. It is the rate of change of head per length of flow in a given direction.

hydraulic head The sum of the elevation head, pressure head, and velocity head at a given point in an aquifer.

hydrostratigraphy The division of stratigraphic units into aquifers and confining layers.

hyporheic zone That portion of the saturated zone in which surface water and groundwater mix. Physical, geochemical, and biological evidence of intermixing are used to define the hyporheic zone. Mixing can occur at many scales, through the stream bed, along the banks, and across meander bends. If seismic waves are characterized as having very high frequencies, the wavefronts will have very high curvature and can be described as arcs of circles with small radii. For high enough frequencies producing small enough radii, the wavefronts approximate the seismic ray.

induced polarization An exploration method involving measurement of the slow decay of voltage in the ground following the cessation of an excitation current pulse (time-domain method) or low-frequency (below 100 Hz) variations of earth impedance (Sheriff 1991).

in-line Characterizes the direction along a measurement direction such as along a line of geophones

image well theory A theory used to simulate the effects of boundaries encountered during a pumping test. To simulate the boundary effect an imaginary well is placed on the other side of the boundary at an equal distance away from the pumping well to the boundary. The sign of the imaginary well (pumping or discharging) depends on the boundary type and whether the well is pumping or injecting. If the boundary is a barrier condition the imaginary well has the same sign as the real well. If the boundary condition is a recharge condition the imaginary well is has the opposite sign of the real well.

infiltration The process of precipitation waters migrating into the soil horizon.

intercept time Time obtained by extrapolating the refraction alignment on a refraction *t-x* plot back to zero offset (Sheriff 1991).

interpretation Third of the three divisions of geophysical exploration (data interpretation). Estimating geological and lithological parameters from processed geophysical measurements.

intrinsic permeability (k_i) The ability of a fluid to move through a porous medium under a given gradient independent of the properties of the fluid. It is generally a function of grain size.

intrusive A process of injection into existing rock and igneous rocks that form from magma under the earth's surface.

inverse modeling The process of estimating model or desired parameters from measured data (see forward modeling). For example, estimating earth parameters from seismic traveltimes.

isotropy The ability of a fluid to move in a given direction at a given point within an aquifer. Fluids that can move equally in all directions are said to be isotropic and those that differ in one direction over another are said to be anisotropic.

joints Undisplaced fractures in rocks that have been subjected to tectonic forces.

kelly A steel bar on the drill rig that connects to the drill string. The power from the rig turns the rotary table which is transmitted to the kelly bar that rotates the drill pipe and bit.

kilobars A unit of pressure denoting 1,000 bars of pressure; one bar is a centimeter-gram-second unit of pressure, equal to 1,000,000 dynes per square centimeter.

knock-out plate or plug A plate placed near the bit of a hollow-stem auger to keep cutting from entering the center cavity of the hollow stem. The plate is knocked out during monitoring well completion.

lag time The time delay it takes for drill cuttings near the bit to appear at the surface.

lava Naturally occurring molten rock materials that flow above the land surface.

law of reflection The angle of incidence equals the angle of reflection. Geometric description of wave reflection from an interface.

leakance The rate of discharge of groundwater through an aquitard.

lineaments Linear features visible on the topographic surface that extend for more than 1 mile (1.6 km). These features are often believed to reflect crustal structure.

longitudinal dispersivity A coefficient representing the mixing that takes place along the direction of the flow path.

losing stream A stream that recharges a groundwater system through the loss of surface water. Surface water is lost because the stage is at a higher elevation than the local groundwater system.

lystric A normal fault that begins at a steep angle near the surface whose angle flattens with depth.

mafic Dark or related to rocks containing ferromagnesium minerals.

magma Naturally occurring molten rock materials occurring under the land surface.

magnetotellurics A method in which orthogonal components of the horizontal electric and magnetic fields induced by natural primary sources are measured simultaneously as a function of frequency (Sheriff 1991).

match point A selected point where values on the Theis curve and the overlaid time-drawdown data coincide. The match point defines where the well function, value of $1/u$, the time, and drawdown are to be used in the Theis equation. The most convenient place to select a match point is where the well function and the value for $1/u$ equals 1.0.

member A part of a formation with distinguishable enough characteristics to be mappable in the field. For example, a member may have unique fossil assemblages.

metasomatism A process where the compositions of rocks change from chemically active pore fluids without changes in the features or structures of the rock. The fluids are active in removing and replacing minerals.

mini-piezometers Small-diameter wells with short slot lengths (less than 6 inches, 15 cm) used to measure hydraulic head. A series of these driven to different depths are useful in evaluating vertical gradients.

noise Any unwanted signal. For example, in geophysical exploration, 60 Hz electrical noise, wind noise, traffic noise, pump noise.

normal type curve A type curve where values of the well function and values for u are plotted forming a characteristic curve in log-log scale.

normal fault A fault forming under extensional conditions where the hanging wall moves down relative to the footwall.

optimum offset The optimum source/receiver separation or offset that results in seismic reflections that have minimal interference with other types of seismic waves.

peak Maximum excursion (positive) of a seismic wavelet or pulse (Sheriff 1991).

perched aquifer An aquifer that occurs above the regional water table with a vadose zone beneath it.

period The time for one cycle of a periodic signal. The time corresponding to one wavelength. Period can be measured in s, ms, micros, etc.

permeability The ability of a porous medium to transmit fluid under a given gradient.

phenocrysts Visible crystals in an igneous rock that are conspicuously larger than the surrounding matrix material.

pitless adapters A fitting from the riser pipe discharge line into a welded port in the casing allowing water to be discharged in a water line to a house. These are placed in areas below the frost line. Wells without pitless adapters have discharge lines that make a 90° turn with an elbow towards the house in a pit dug below the frost line.

porosity (η) The volume of void space within earth materials. Primary porosity occurs with the formation of a rock mass. Secondary porosity represents void spaces that occur after the rock mass formed.

potentiometric surface A surface estimated by the level to which cased wells will rise. This surface represents total head, which includes elevation head and pressure head.

price meter A meter used to perform stream gaging in larger deeper streams. The Price meter is used to determine stream velocities per fraction of cross-sectional area. It has a horizontal wheel approximately 5 inches (12.7 cm) in diameter, with small cups attached. The wheel rotates in the current on a cam attached to the spindle. An electrical contact creates a "click" sound with each rotation. The clicks are transmitted to a headset worn by the field person to obtain a clear signal. The clicks are counted for 30 or 60 seconds or sent to a direct readout meter. The number of clicks are compared with a calibration curve to obtain the velocity.

processing The second of the three divisions of geophysical exploration (data processing). Enhancing geophysical measurements and eliminating unwanted signals to aid geophysical interpretation.

profiling A geophysical survey in which the measuring system is moved about an area with the objective of characterizing lateral variations in the subsurface (Sheriff 1991).

pseudosection A plot of electrical measurements or calculations, often of apparent resistivity or induced polarization, as a function of position and electrode separation (which is indirectly related to the depth of investigation) (Sheriff 1991).

pumice A volcanic froth that has hardened into rock. Pumice forms where volatile gasses are expelled into lava. It is often sufficiently buoyant to float on water.

pygmy meter A small version of the price meter, used in shallower streams. *See* price meter.

range of influence The distance away from the pumping well to where the drawdown is zero.

ray A geometric construction everywhere perpendicular to wavefronts (in isotropic media). Used to characterize seismic wave travel paths.

recharge The addition of water to a groundwater system results in a downward gradient in hydraulic head.

recharge boundaries Boundaries that supply groundwater. Examples include hydraulically connected streams, coarser-grained units, or a thickening of the aquifer.

recirculation test A tracer test where both wells are stressed. The injection well receives "clean" or pumped water from another source.

recovery data The recovery response of drawdown after a pumping has stopped.

reflection Energy or wave from a seismic source which has been reflected (returned) from an acoustic-impedance contrast or series of contrasts within the earth (Sheriff 1991).

refraction The change in direction of a seismic ray upon passing into a medium with a different velocity (Sheriff 1991). Arises from Snell's law.

regional metamorphism A term referring to the metamorphism of an extensive region from tectonic activity. There may be a sequence metamorphic rock types that occur as a result of changes in temperature, pressure, and fluids associated with colliding tectonic plates.

residual drawdown plot A residual drawdown plot is constructed by plotting the ratio of the time since pumping started (t) to time since pumping stopped (t') (t/t') verses residual drawdown (s') to estimate aquifer transmissivity.

residual drawdown (s') The amount of drawdown recovery since pumping stopped.

resistance Opposition to the flow of electric current. Resistance takes into account resistivity and the geometry of the material.

resistivity The property of a material that resists the flow of electrical current. Units are ohmmeters.

resolution The ability to separate two features which are close together. The minimum separation of two bodies before their individual identities are lost on the resultant map or cross section (Sheriff 1991).

reverse A fault forming under compressional conditions where the hanging wall moves up relative to the footwall.

reverse type curve The more commonly used form of the Theis curve. In this case values of the well function and values for $1/u$ are plotted forming a characteristic curve in log-log scale.

rifting A condition where tectonic forces extend or pull apart. This results in cracking, volcanism, and graben valleys.

riparian zone The riparian zone is commonly referred to as the interface between the terrestrial and aquatic interface. It is characterized by trees and other large vegetation and generally does not contain surface water except during episodic floods. In smaller stream systems, the riparian zone may be characterized by willows or other shrubs.

rising-head test A slug test where the slug is extracted form the well bore, displacing the static water level downward. Measurements of the rising recovering water levels over time are recorded. In this test recovering groundwater comes from the formation.

runoff The amount of precipitation that moves down a topographic slope that can not enter the soil horizon.

sand line An auxiliary cable on a drill rig other than the main cable used to retrieve pipe, pick up casing, other lift other heavy tools.

saturated thickness (*b*) The thickness of saturated permeable material in an unconfined aquifer from the water table to lowermost confining layer. In a confined aquifer it is the thickness of saturated permeable material between two confining layers. In a pumping test it represents the contributing saturated thickness to the well screen, which is a function of pumping time and geologic conditions.

scoria A highly vesicular cindery zone at the tops an bottoms of lava flows from escaping gasses and burning vegetation.

seepage velocity The average rate of movement of a fluid over a given distance. It is calculated by dividing the specific discharge by the effective porosity.

seepage meter A seepage meter is an open-bottomed container that is inserted into the stream sediments to measure seepage rates. The time it takes a volume of water to flow into or out of a bag connected to the container is measured to estimate the seepage.

seismic source A device that releases energy, such as an explosion or an air gun release (Sheriff 1991). Any of a variety of devices used to create seismic, electrical or other pulses for propagation through the earth.

seismic velocity Speed of propagation of a seismic disturbance through a medium. *P*-wave velocities are approximately twice as large as *S*-wave velocities.

seismic section A plot of seismic data along a line. The vertical scale is usually arrival time but sometimes depth (Sheriff 1991).

seismic tomography A method for estimating earth parameters from a multitude of observations using combinations of source and receiver locations. For example, CAT scans in medical imaging (see inverse modeling).

seismograph An instrument or system used to record seismic wave disturbances. Used in conjunction with geophones.

shear Seismic wave motion described by particle motion in a plane perpendicular to the wavefront propagation direction. Shear waves are also called *S*-waves.

sills Tabular igneous features that are concordant or are injected parallel to existing rocks.

sinks Extractions of groundwater from a system, such as pumping wells, and surface water discharge.

Snell's Law Describes the change in propagation direction of a seismic wave when it crosses a medium. The direction change depends on the seismic velocities of the media.

sorting The process of dividing sediments according to size during transport. This is done through a variety of processes, such water, wind, or ice. Well sorted indicates the sizes are similar and poorly sorted indicates a wide range of grain sizes represented.

sounding A geophysical survey in which the measuring system is moved about an area with the objective of characterizing vertical variations in the subsurface.

sources Contributions of water to a groundwater system, such as precipitation, injection wells, or imported water.

specific weight (γ) The weight of a given substance per unit volume.

specific yield (S_y) The ratio of volume of water from a saturated rock mass to the total volume that was yielded by gravity drainage.

specific retention (S_r) The ratio of the volume of water retained against the force of gravity to the total volume of rock.

specific storage (S_s) The volume of water that will yield due to compression of the mineral skeleton and decompression of water in a confined aquifer. It is also known as elastic storage.

specific discharge The apparent velocity of a fluid in an aquifer under laminar conditions that would flow if no geologic materials were present.

specific gravity The ratio of the density of a volume material to the density of an equal volume of water at a specified temperature.

specific capacity (S_c) The pumping rate in gallons per minute divided by the drawdown in feet. It is an expression of the productivity of a well and varies with time.

speed of propagation Speed at which a pulse or disturbance is transmitted through earth materials. Typically used to describe seismic or electromagnetic wave propagation.

spontaneous potential Observation of the static natural voltage existing between sets of points on the ground, sometimes caused by the electrochemical effects of ore bodies (Sheriff 1991).

stacking The process of making a composite trace by combining traces from different records. The constituent traces meet the condition of being reflected from the same earth point.

static water level The ambient water level elevation in a cased well where the aquifer has not been stressed.

storage coefficient (*S*) The volume of an aquifer takes into or releases from storage per unit surface area of an aquifer per unit change in head. In an unconfined aquifer it is estimated to be the specific yield and in a confined aquifer it represents the specific storage multiplied by the saturated thickness.

storativity (*S*) *See* storage coefficient.

strain Rock deformation as a result of stresses applied. Examples are folding or faulting or rocks.

strike The directional orientation of a structure resulting with intersection of a horizontal plane with an inclined plane.

support-rod method A method of protecting a mini-piezometer from crimping while pounding it to its completion depth by inserting a metal support rod inside the mini-piezometer. The support rod is retracted after the desired depth is achieved.

synoptic survey A detailed stream gaging survey along a reach of stream, accounting for every gain or loss form diversion ditches or incoming ditches along the path. It is performed as a flux balance approach to stream flow. One can also identify losing or gaining stretches if the values exceed the error limitations of the equipment.

***t-x* plot** A plot of times for a selected seismic event (e.g., first arrival) versus distance (e.g., geophone offset from the shot).

transversal dispersivity A coefficient representing the mixing that takes place normal or transverse to the direction of the flow path.

tectonics The forces within the earth's crust which create structures or crustal movement.

tellurics Of the earth. A natural electrical earth current of low frequency which extends over a large region (Sheriff 1991).

three-point problem A geometric method of determining the strike and dip from three accurately determined positions of a structural surface.

tortuosity The actual sinuous flow path length divided by the straight-line distance.

trace A record of the data from one seismic channel, one electromagnetic channel, etc. (Sheriff 1991).

tracer test A field test where a traceable substance is introduced into the groundwater system at an injection point and tracked and measured to estimate hydraulic or hydrochemical properties.

transmissivity (T) The product of the hydraulic conductivity and the saturated thickness. It represents the ability of a given thickness of aquifer under a given gradient to transmit fluids.

tremie pipe A small-diameter pipe used as an access tube to place gravel packing materials around a well screen.

tripping in A drilling term referring to connecting the drill pipe and running it back into a drill hole after changing the drill bit or some other task requiring the drill pipe to be extracted.

tripping out A drilling term referring to the process of extracting the drill pipe from a drill hole.

trough Minimum excursion (maximum negative) of a seismic wavelet or pulse.

unconfined aquifer An aquifer whose upper surface is at atmospheric pressure.

uniformity coefficient (C_u) The ratio of the diameter of a given sieved soiled at 60% finer by weight to the diameter at 10% finer by weight (effective grain size). Values less than 4 indicate well-sorted soils and values greater than 6 indicate poorly sorted soils.

vesicular An igneous rock texture associated with trapped gas bubbles.

viscosity The physical property of a fluid to resist flow. It is a measure of the shear stress to shear strain.

water table The top of the saturated zone of an unconfined aquifer where the pressure is at atmospheric pressure.

water-table aquifer *See* unconfined aquifer.

wave A disturbance that is propagated through the body or on the surface of a medium without involving net movement of material. Waves are usually characterized by periodicity (Sheriff 1991). For example, seismic waves, electromagnetic waves.

wavefronts The surface over which the phase of a traveling wave disturbance is the same. The wavefront follows the path defined by the seismic ray (see ray).

wavelength The distance between successive similar points on two adjacent cycles of a single frequency wave (Sheriff 1991). Used to quantify resolution in geophysical exploration.

well function ($W_{(u)}$) It is the infinite series term of the Theis equation. The well function includes relationships between the radial distance of an observation well and the hydraulic properties of the aquifer.

zero flux plane Basically, this is a "split the difference" approach to estimating the fraction of the soil moisture that actually recharges the groundwater system. It requires a bell-shaped curve of total hydraulic head (data from the tensiometers) versus depth. This approach presumes that the drainable soil moisture below the peak in the bell curve will continue moving downward to the water table, and that soil moisture above the peak will be drawn upward and be evapotranspired.

zero-offset The situation in which the source is at the same location as the receiver.

Reference

Sheriff, R. E., 1991. *Encyclopedic Dictionary of Exploration Geophysics.* Society of Exploration Geophysicists, Tulsa, OK.

Index

A

B

C

D

E

F

G

H

I

N

O

P

R

S

T

U

V

W

Y

Z

About the Authors

Willis D. Weight is a Professor of Geological Engineering and the head of the hydrogeology program at Montana Tech of the University of Montana. He has taught over 25 professional development courses in hydrogeology, and has held several positions as field hydrogeologist and hydrogeology consultant since 1982. He received as B.S. in Engineering Geology from Brigham Young University in 1980, and a Ph.D. (1989) from the University of Wyoming. He is registered as a Professional Engineer in Montana and Idaho.

John L. Sonderegger is a Professor Emeritus in the Geological Engineering Department at Montana Tech of the University of Montana and an adjunct Associate Professor in the Earth Sciences Department at Montana State University, with interests in rock-water interaction and groundwater chemistry. He has previously worked for the Montana Bureau of Mines and Geology and the Georgia and Alabama Geological Surveys, and received B.S.I. (1966), M.S. (1969), and Ph.D. (1974) degrees from the University of Wisconsin, University of Alabama, and New Mexico Institute of Mining and Technology, respectively.